Thermodynamics

For Engineering Technologists

ENGINEERING TECHNOLOGY

A Series of Textbooks

EDITOR

Gerald A. Rath

Wichita State University
Wichita, Kansas

1. Thermodynamics for Engineering Technologists, *by Robert K. Will*

Other Volumes in Preparation

Thermodynamics

For Engineering Technologists

Robert K. Will, P.E.
College of Engineering
Virginia Polytechnic Institute
and State University
Blacksburg, Virginia

Marcel Dekker, Inc. **New York and Basel**

Library of Congress Cataloging in Publication Data

Will, Robert K [Date]
Thermodynamics for engineering technologists.

(Engineering technology ; 1)
Includes index.
1. Thermodynamics. I. Title. II. Series.
TJ265.W48 621.4'021 79-21021
ISBN 0-8247-6864-7

COPYRIGHT © 1979 by MARCEL DEKKER, INC. ALL RIGHTS RESERVED

Neither this book nor any part may be reproduced or transmitted in any form or by any means, electronic or mechanical, including photocopying, microfilming, and recording, or by any information storage and retrieval system, without permission in writing from the publisher.

MARCEL DEKKER, INC.
270 Madison Avenue, New York, New York 10016

Current printing (last digit):
10 9 8 7 6 5 4 3

PRINTED IN THE UNITED STATES OF AMERICA

PREFACE

This book is the result of notes and experience gained by the author in teaching thermodynamics, over the last six years, in a Baccalaureate Mechanical Engineering Technology program. The students were graduates of an Associate in Applied Science program and their mathematics backgrounds did not include calculus. Since calculus is not required in order to use this textbook, it is recommended for use in either two year Associate degree programs or four year Baccalaureate programs. It should also prove of value in survey courses for nonmechanical Technology students as well as to those majoring in Industrial Arts. For Technology students who have completed calculus it is recommended that the instructor modify the material that treats areas under curves and the use of maximum and minimum concepts. These modifications will be quite simple and involve only a small fraction of the subjects covered.

There is sufficient material for a three quarter or two semester course. For survey or shorter courses most of the important concepts are covered in the first parts of chapters dealing with applications. After covering the fundamentals of thermodynamics it is possible to tailor the material to the needs of any specialized group of students.

The order of presentation of the material has evolved from 25 years of experience and much trial and error. It is felt that the presentation of basic thermodynamic principles should be followed at the earliest possible time by applications to real devices. For this reason the concept of the perfect gas as a working substance is presented first. This is immediately followed by a discussion of the internal combustion engine. New working substances are then developed as needed, progressing from semiperfect gases to vapors, and finally to mixtures. Real devices using these materials are then covered in detail.

The Second Law of Thermodynamics is presented near the conclusion of the book. It seems fitting that this topic be used to tie together previous material and to serve as a comment on the limitations of First Law analysis. The final chapter is concerned with the thermodynamics of fluid flow. Since many technology programs offer this material as a separate course it may be either included or omitted.

Two systems of measurement are used throughout the text. Problems are worked in the English system, but conversions and answers are provided in the metric (SI) system. For a technology student an understanding of the magnitudes of metric units is desirable, but as a practical matter, the English system is the system in which most work is currently being performed in this country. Tabular values, charts, and instruments, as used in existing installations, show little likelihood of a quick conversion to the SI system.

As a final note, the author feels that thermodynamics should be taught as a problem solving course. Much reliance has been placed on the pV, TS, and hs diagrams as teaching aids. Generous assignment of problems and an explanation of the solutions of these problems by the instructor is basic to the understanding of the subject.

I would like to acknowledge both direct and indirect assistance in the preparation of this text. Indirectly I have been aided by Virgil Moring Faires whose excellent Applied Thermodynamics (revised edition, Macmillan, 1949) provided my initial training in this subject. To James Bernard Jones, for many years Head of the Department of Mechanical Engineering at Virginia Polytechnic Institute, I give thanks for his patience and understanding of the foibles of a beginning instructor. To Joseph Herbert Moore, who as Head of the Division of Engineering Technology allowed me the freedom to prepare and search for a technique suitable for the presentation of this material to the Engineering Technologist, thanks are also given. To Victoria P. Trump deep appreciation is expressed for the typing, proofreading, and patience in decoding what is undoubtedly the world's worst penmanship. Finally, to my wife, Sarah S. Will, for her encouragement and understanding during the preparation of this work.

Robert K. Will

CONTENTS

LIST OF SYMBOLS

A - area (ft^2)
A_c - indicator card area ($in.^2$)
AF - air-fuel ratio
a - acceleration (ft/sec^2)
a - acoustic velocity (ft/sec)
BHP - brake horsepower
COD - coefficient of discharge
Cyl - number of cylinders
c - specific heat (B/lbm-R)
c - clearance (%)
c_d - discharge coefficient
D - diameter (ft)
E - total energy (B, B/sec)
E - voltage (volts)
E - error (%)
e - efficiency (%)
e_c - compressor efficiency (%)
e_e - engine efficiency (%)
e_g - regenerator efficiency (%)
e_m - mechanical efficiency (%)
e_p - pump efficiency (%)
e_t - thermal efficiency (%)
e_t - turbine efficiency (%)
e_v - volumetric efficiency (%)
F - force (lbf)
F - velocity of approach factor
FA - fuel-air ratio
FE - flow energy (B, B/sec)
FHP - friction horsepower
f - interpolation factor
g - acceleration due to gravity (ft/sec^2)
g_0 - constant ($lbm\text{-}ft/lbf\text{-}sec^2$)
H - total enthalpy (B, B/sec)
HHV - higher heating value (B/lbm)
HP - horsepower
HV - heating value (B/lbm)
h - specific enthalpy (B/lbm)
h - height (ft)
I - current (A)
IHP - indicated horsepower
J - Joule's constant (ft-lbf/B)
KE - kinetic energy (B, B/sec)
k - c_p/c_v
L - length (ft)
L_c - indicator card length (in.)
LHV - lower heating value (B/lbm)
M - molecular weight
M - mass ($lbf\text{-}sec^2/ft$)
M - Mach number
m - mass (lbm, lbm/sec)
m - moisture content (%)

m' - air supplied (lbm, lbm/sec)
m_{sa} - stoichiometric air (lbm)
N - engine speed (rpm)
N′ - power strokes (rpm)
n - exponent in $pV^n = c$
n - moles of working substance
P - power (ft-lbf/sec, B/sec)
PE - potential energy (B, B/sec)
p - pressure (psi)
p_m - mean effective pressure (psi)
p_r - pressure term, gas tables
Q - heat (B, B/sec)
Q_A - available heat (B)
Q_U - unavailable heat (B)
R - specific gas constant (ft-lbf/lbm-R)
R - radius (ft)
$\bar{R}$ - universal gas constant
r_c - cutoff ratio
r_k - isentropic compression ratio
r_p - pressure ratio
S - total entropy (B/R)
SF - scale factor (lbf/in.^{3})
SG - specific gravity
s - specific entropy (B/lbm-R)
T - temperature (R)
T - torque (ft-lbf)
t - temperature (F)
U - total internal energy (B, B/sec)
u - specific internal energy (B/lbm)
V - total volume (ft^3)
$\bar{V}$ - velocity (ft/sec)
V′ - actual air supplied (ft^3)
V_C - clearance volume (ft^3)
V_D - displacement volume (ft^3)
Vol - volume flow (ft^3/sec)
v - specific volume (ft^3/lbm)
v_r - volume term, gas tables
W - work (B, B/sec)
W_c - compressor work (B/lbm)
W_t - turbine work (B/lbm)
x - quality (%)
x - volumetric analysis (%)
y - gravimetric analysis (%)
Z - elevation (ft)

γ - specific weight (lbf/ft^3)
γ - coefficient of performance
ΔH_f - heat of formation (B/mole)
ΔH_R - enthalpy of reaction (B/lbm)
ΔS_p - entropy production (B/R)
ρ - density (lbm/ft^3)
φ - relative humidity (%)
φ - entropy term, gas tables
ω - absolute humidity
$\oint dQ/T$ - inequality of Clausius

1

INTRODUCTION

SCOPE OF THERMODYNAMICS

Thermodynamics is concerned with the properties of matter, the various forms of energy, and the concepts of heat and work. The interactions among heat, work, and energy may be used to define either a thermodynamic process or a thermodynamic cycle. Processes or cycles are constructed to explain the performance of engineering devices such as internal combustion engines, external combustion engines, power plants, air conditioning processes, combustion processes, and systems in which work is converted to heat or heat to work.

The type of thermodynamics that we shall study is referred to as classical thermodynamics. It is based on the average behavior of a large number of particles or molecules that comprise the system being studied. For gases this behavior was defined before thermodynamics was developed and has been verified by repeated experiments. As a result of these experiments the nature of the internal structure of the atom or molecule is not critical to an understanding of thermodynamics.

Our study of thermodynamics will therefore be based on a small number of different experiments. Since the experiments are reproducible, their results are referred to as thermodynamic laws.

THERMODYNAMIC SYSTEMS, SURROUNDINGS, AND BOUNDARIES

In many engineering problems it is preferable to deal with a component of an engineering device rather than attempting to study the entire device. For example, in the study of an automobile we could concentrate on the transmission, the brakes, the engine, a single cylinder in the engine, or any other part of the car. In the study of a steam power plant we could

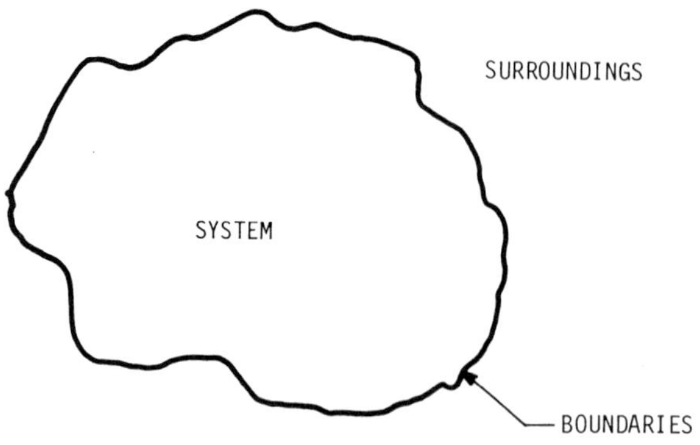

Figure 1.1

study the boiler, the turbine, the condenser, the pumps, or the various heat exchangers within the plant.

The separation of the engineering device into its component parts may be accomplished by means of boundaries that isolate and restrict the area under study. In thermodynamics these boundaries are used to separate the system from the surroundings (Fig. 1.1). The system is that part of the thermodynamic problem that we wish to analyze. Everything outside the system is termed the surroundings, and the totality of the system and the surroundings is known as the universe.

The boundaries of a system may be real or imaginary as well as being fixed in space or movable. In our study of thermodynamics the boundaries will either contain a fixed mass of material or enclose a fixed volume of space.

OPEN, CLOSED, AND ISOLATED SYSTEMS

In an open system (Fig. 1.2) matter is free to enter and to leave the system. While matter enters only at plane 1 and leaves only at plane 2, it is possible to have multiple entrance and exit planes in an open thermodynamic system. The boundaries at planes 1 and 2 are imaginary, while the pipe walls represent real boundaries for the system. Even though the boundaries at the entrance and exit are imaginary, their location is usually dictated by pressure or temperature measurements being located at these planes. At the inlet and exit various forms of energy may be added to or deleted from the system. Along the pipe walls, between the inlet and exit planes, heat may be added to or removed from the system. If a pump is inserted between the inlet and exit planes, work may be added to

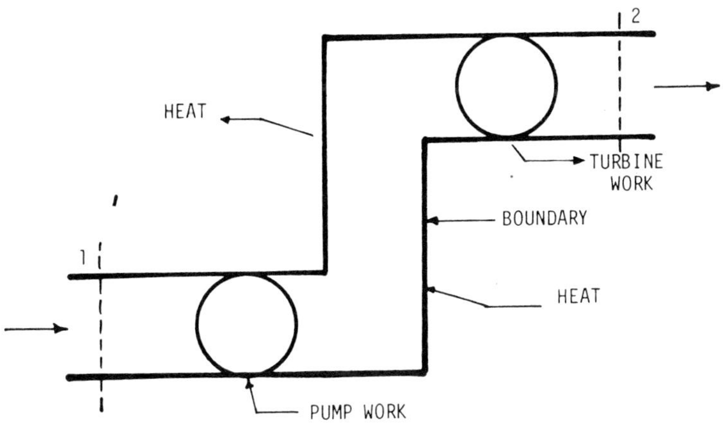

Figure 1.2

the system. If, instead of a pump, a turbine is inserted between these planes, then work may be taken from the system.

A particular type of open system is the steady-flow system. In a steady-flow system the quantity of matter entering the system in a given time period is exactly balanced by the quantity of matter leaving the system. In effect, this means that the quantity of matter within the system remains constant.

A further restriction on the steady-flow open system is implied by the term steady state system. In this type of system, the properties at any given point within the system are constant with respect to time. The properties at the system inlet are therefore constant, as are those at the system exit. However, the properties at the inlet and exit do not have to have the same numerical values.

In our discussion and use of the open system, it will be understood that the system is steady state. While the open system has both real and imaginary boundaries, these boundaries will always contain the same volume of space. The system boundaries are therefore fixed in space instead of being movable.

In a closed system (Fig. 1.3) matter is not permitted to enter or leave the system. The closed system as defined by the cylinder walls and piston face has real boundaries. As in the open system, heat may be added to or removed from the system. As the piston moves between its extreme positions, work may be done on the surroundings or on the system. In an internal combustion engine, the system does work on the surroundings as

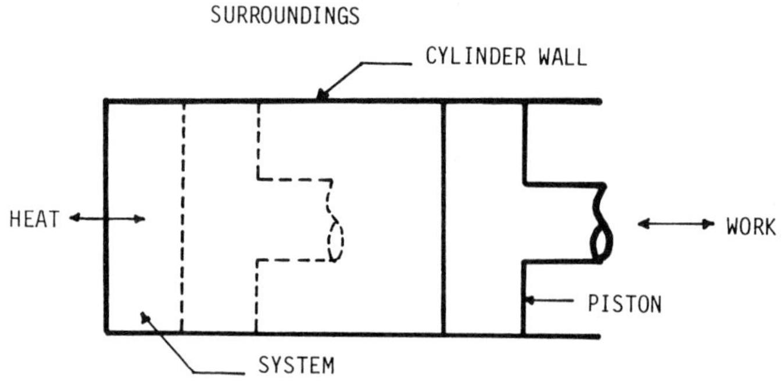

Figure 1.3

the piston moves from top to bottom dead center. When the system moves from bottom to top dead center, as in the compression stroke, work is done on or added to the system. The numerical values of the energy terms may vary as heat or work is transferred into or out of the closed system.

A closed system has real boundaries, and the boundaries may be movable rather than being fixed as in the open system.

The isolated system is not generally required in the study of engineering devices. An isolated system is a closed system with fixed boundaries, which has no transfer of heat or work. In spite of these restrictions, certain properties may change their numerical values depending on the type of isolated system being studied.

WORKING SUBSTANCE

Regardless of the type of system being studied, matter must reside within the system. This matter is defined as the working substance. The function of the working substance is to receive, transport, or disperse heat, work, or energy. The working substance is also necessary to alter energy from one type to another or to convert heat and work. The working substance that we shall employ will be restricted to fluids. A fluid may be a gas, a liquid, or a vapor or any mixture of these substances. It is permissible for the working substance to change phase within a thermodynamic system. In a steam power plant, water, as a liquid, is pumped into the boiler. In the boiler the liquid is converted to a vapor or a mixture of vapor and liquid. After passing through the turbine the vapor or vapor mixture is converted back into a liquid in the condenser.

PROPERTIES

Thermodynamic properties that are often measured or evaluated in engineering problems include pressure, temperature, density, and specific volume. These are basic properties whose level or assigned numerical value can often be determined by instrumentation. From the basic properties it is possible to define other properties that are useful in our thermodynamic studies.

There are two types of properties, intensive and extensive. Intensive properties are independent of the size of the thermodynamic system. If a stable system of a given size is divided into any number of parts, the value of an intensive property is unchanged and is the same in each part.

An extensive property, as mass or total volume, would be affected by any division of the system. An extensive property may be converted to an intensive property by dividing by the mass or volume of the system.

If the total volume of a system is divided by the mass of the system, the result is the specific volume. In equation form

$$v = V/m \tag{1.1}$$

where V is the total volume of the system, m is the mass of the system, and v is the specific volume.

The density is defined as the reciprocal of the specific volume or

$$\rho = 1/v \tag{1.2}$$

where ρ is the density of the system.

Pressure, temperature, density, and specific volume are all intensive properties.

Intensive properties are used to present tabular values of properties since the use of this type of property eliminates the mass or volume as a factor in the tabulation.

The preceding properties are all thermodynamic properties. Two important mechanical properties encountered in the open system are the height and velocity of the working substance.

STATE

The state of a working substance is defined by its properties. In thermodynamics two independent thermodynamic properties are required to define the thermodynamic state of a system. Independent properties are those that are capable of assuming different values when some other property has a constant value. From Eq. (1.2), density and specific volume are dependent properties since a given value for either property fixes the value of the other property.

Once the state of a system is established, all other thermodynamic properties may be calculated from equations or tabular data.

BASIC UNITS

To assign numerical values to properties it is necessary to have a system of units. Since most tabular data and most instruments employed in this country still operate in the English system, problems will be solved in that system. It is also apparent that eventually the SI or international system of units will be adopted in this country; answers to problems will also be expressed in that system.

In thermodynamics the basic units are

Unit	English	SI
Length	Foot (ft)	Meter (m)
Force	Pound force (lbf)	Newton (N)
Mass	Pound mass (lbm)	Kilogram (kg)
Time	Second (sec)	Second (s)

The terms within the parentheses are the preferred notation for each basic unit in either system.

In the English system some difficulty is encountered when the pound is used as a unit. Unfortunately, the pound has two meanings, one related to mass and the other to force. Newton's equation may be used to distinguish between the pound force and the pound mass. In equation form,

$$F = Ma \tag{1.3}$$

where F is the force measured in pounds force (lbf), M is the mass, and a is the acceleration of the dynamic system (ft/sec^2).

For a nonmoving or static system, Eq. (1.3) may be rewritten as

$$F = Mg \tag{1.4}$$

where g is the acceleration due to gravity (ft/sec^2).

The mass of the system is further defined by

$$M = m/g_0 \tag{1.5}$$

where m is the mass measured in pounds mass (lbm) and g_0 is an arbitrary constant.

Substitution of the mass in Eq. (1.5) into Eq. (1.4) yields

$$F = m(g/g_0) \tag{1.6}$$

From Eq. (1.6), the units of the constant g_0 are

$$g_0 = (m/F)g\ [(\text{lbm}/\text{lbf})(\text{ft}/\text{sec}^2)]$$
$$= (\text{lbm-ft})/(\text{lbf-sec}^2)$$

On the surface of the Earth the acceleration due to gravity has the numerical value of 32.2 (ft/sec^2). The acceleration due to gravity varies slightly with either latitude or elevation on the Earth's surface.

If the arbitrary constant g_0 is assigned the value of 32.2, then the magnitude of the pound mass and the pound force will be equal on the Earth's surface. The mass of a system is independent of gravity, but the value of the force will vary with gravity or the acceleration of a dynamic system.

Example 1.1. A man has a mass of 150 lbm on the surface of the Earth. Determine the force he exerts due to gravity.

From Eq. (1.6)

$$F = m(g/g_0) = 150(32.2/32.2) = 150\ (\text{lbf})$$

This is the force that the man exerts due to the Earth's gravitational field in English units.

In the SI system* (see Appendix A for conversion factors),

$$1\ (\text{lbm}) = 4.536\text{E} - 01 = 0.4536\ (\text{kg})$$

E - 01 is scientific notation requiring that the given conversion factor be divided by 10^1 or 10.

For a mass of 150 lbm

$$m = 150(0.4536) = 68.04\ (\text{kg})$$

The length conversion is given by

$$1\ (\text{ft}) = 3.048\text{E} - 01 = 0.3048\ (\text{m})$$

Therefore

$$g = 32.2\ (\text{ft}/\text{sec}^2) = 32.2(0.3048) = 9.81\ (\text{m}/\text{s}^2)$$

---

*Appendix A also includes details of converting from the English to the SI system for all thermodynamic terms.

Since there is no arbitrary constant when Newton's equation is expressed in SI units, from Eq. (1.4)

$$F = Mg = 68.04(9.81) = 667.5 \text{ (kg-m/s}^2\text{)}$$

$$= 667.5 \text{ (N)}$$

This is the force that the man exerts due to the Earth's gravitational field in SI units.

The above answer may be checked since

$$1 \text{ (lbf)} = 4.448\text{E} + 00 = 4.448 \text{ (N)}$$

and

$$F = 150 \text{ (lbf)} = 150(4.448) = 667.2 \text{ (N)}$$

The slight difference between the force expressed in SI units is caused by rounding the value of g in the English system.

Example 1.2. A man exerts a force due to gravity of 180 lbf on the Earth's surface. What is his mass and what would be his force and mass on a planet where $g = 16.1$ ft/sec^2?

From Eq. (1.6)

$$m = F(g_0/g) = 180(32.2/32.2) = 180 \text{ (lbm)}$$

$$= 180(0.4536) = 81.6 \text{ (kg)}$$

The man has a mass of 180 lbm or 81.6 kg on the Earth's surface and everywhere in the universe. For a planet where $g = 16.1$ ft/sec^2,

$$F = m(g/g_0) = 180(16.1/32.2) = 90 \text{ (lbf)}$$

$$= 90(4.448) = 400.3 \text{ (N)}$$

The force exerted by a body is dependent on the gravitational field in which it is located.

COMPOSITE UNITS

The acceleration due to gravity is a composite of the basic units of length and time. Other composite units are required to define heat, work, and power.

Work is defined as the product of the force and the distance through which the force acts or

$$W = \text{force} \times \text{distance [(lbf)(ft) or ft-lbf]}$$

in the English system or

$$W = (\text{Newton})(\text{meter})\ (\text{Nm})$$

$$= \text{joule}\ (\text{J})$$

so that the measure of work in the SI system is the joule.

Heat, in the English system, is measured in British thermal units (B). The British thermal unit is approximately the amount of heat required to raise the temperature of one pound of water one degree Fahrenheit.

In the English system the conversion factor between heat and work units is Joule's constant, where

$$J = 778\ (\text{ft-lbf/B})$$

that is, 1 B is equivalent to 778 ft-lbf. No conversion factor is required in the SI system since heat and work are both measured in joules.

Note that J in the English system is a conversion factor and that in the SI system it is a unit for heat or work.

Power is the time rate of doing work, where

$$P = \text{work/time}\ (\text{ft-lbf/sec})$$

$$= \text{joule/second}\ (\text{J/s})$$

$$= \text{watt}\ (\text{W})$$

The watt and the ft-lbf/sec are both small quantities of power so that the kilowatt and horsepower are commonly used as units, where

$$1\ (\text{kW}) = 1000\ (\text{W})$$

$$1\ (\text{HP}) = 550\ (\text{ft-lbf/sec}) = 33{,}000\ (\text{ft-lbf/min})$$

Other useful conversion factors include

$$1\ (\text{kW}) = 3412\ (\text{B/hr})$$

$$1\ (\text{HP}) = 2545\ (\text{B/hr})$$

$$1\ (\text{HP}) = 0.7457\ (\text{kW})$$

Example 1.3. A force of 200 lbf acts through a distance of 120 ft. Determine the work and the heat equivalent of the work. If the work takes place in 15 sec, determine the power.

$$W = \text{force} \times \text{distance} = 200(120) = 24{,}000\ (\text{ft-lbf})$$

$$= 24{,}000(1.356\text{E} + 00) = 32{,}540\ (\text{J})$$

$$= 3.2540\text{E} + 04\ (\text{J})$$

The heat equivalent of the work is

$$Q = W/J = 24{,}000/778 = 30.84 \text{ (B)}$$

$$= 30.84(1.054E + 03) = 32{,}510 \text{ (J)}$$

Q is the accepted symbol for heat, as is W for work.

The slight difference in heat and work in the SI system is caused by the rounding off of Joule's constant in the English system.

If the work takes place in 15 sec,

$$P = \text{work/time} = 24{,}000/15 = 1600 \text{ (ft-lbf/sec)}$$

$$= 1600/778 = 2.06 \text{ (B/sec)}$$

$$= 1600/550 = 2.91 \text{ (HP)}$$

or in the SI system

$$P = 32{,}540/15 = 2169 \text{ (J/s)}$$

$$= 2169 \text{ (W)} = 2.169 \text{ (kW)}$$

From Appendix A

$$1 \text{ (HP)} = 7.457E + 02 \text{ (W)}$$

so that

$$2.91 \text{ (HP)} = 2.91(745.7) = 2170 \text{ (W)}$$

$$= 2.170 \text{ (kW)}$$

PROPERTY MEASUREMENTS AND NUMERICAL VALUES

Having adopted a set of units, we now find it possible to assign numerical values to the basic thermodynamic properties. For the specific volume, by use of Eq. (1.1)

$$v = V/m$$

and the numerical value may be obtained by measuring the total volume and the mass of the material. If a hydrometer is used to measure the specific gravity of a fluid, the density may be obtained from

$$SG = \rho/\rho_{H_2O} \tag{1.7}$$

where SG is the specific gravity of the fluid, ρ is the density of the fluid, and ρ_{H_2O} is the density of water at standard conditions.

The density of water at standard conditions is

$$\rho_{H_2O} = 62.4\ (\mathrm{lbm/ft^3})$$
$$= 62.4(1.602E + 01) = 62.4(16.02)$$
$$= 1000\ (\mathrm{kg/m^3})$$

The specific weight of a substance is defined by

$$\gamma = F/V \tag{1.8}$$

where γ is the specific weight of the fluid, F is the force exerted by the fluid, and V is the volume occupied by the fluid.

By use of Eq. (1.6)

$$\gamma = F/V = (m/V)(g/g_0) = \rho(g/g_0) \tag{1.9}$$

In locations where $g = 32.2\ \mathrm{ft/sec^2}$ the numerical values of the density and the specific weight are equal.

By use of Eq. (1.9)

$$\rho = \gamma(g_0/g)$$

and by substitution into Eq. (1.7)

$$SG = \rho/\rho_{H_2O} = (\gamma/\gamma_{H_2O})(g_0/g)(g/g_0) = \gamma/\gamma_{H_2O} \tag{1.10}$$

Example 1.4. Mercury has a specific gravity of 13.6. Determine its density, specific volume, and specific weight when $g = 32.2\ \mathrm{ft/sec^2}$.

From Eq. (1.7)

$$\rho = SG\rho_{H_2O} = 13.6(62.4) = 848.6\ (\mathrm{lbm/ft^3})$$
$$= 848.6(16.02) = 13{,}600\ (\mathrm{kg/m^3})$$

From Eq. (1.9)

$$\gamma = \rho(g/g_0) = 848.6(32.2/32.2)$$
$$= 848.6\ (\mathrm{lbf/ft^3})$$

Since

$$1\ (\mathrm{lbf}) = 4.448\ (\mathrm{N})$$

and

$$1\ (\mathrm{ft^3}) = 0.0283\ (\mathrm{m^3})$$

then

$$1\ (\mathrm{lbf/ft^3}) = 4.448/0.0283 = 157.2\ (\mathrm{N/m^3})$$

The SI equivalent of the specific weight is

$$\gamma = 848.6(157.2) = 1.334E + 05\ (N/m^3)$$

Example 1.5. Repeat Example 1.4 if $g = 16.1\ ft/sec^2$.

Since the density and specific volume are a function of the mass, their values are unchanged. The specific weight, based on Eq. (1.9), is

$$\gamma = \rho(g/g_0) = 848.6(16.1/32.2) = 424.3\ (lbf/ft^3)$$
$$= 0.667E + 05\ (N/m^3)$$

The temperature of a substance may be measured by a variety of instruments such as liquid in glass, resistance, and bimetallic thermometers. Temperature may also be determined by means of thermocouples and optical and radiation pyrometers.

Two measurement scales are in current usage; for the English system the Fahrenheit scale is the standard, and in the SI system the Celsius scale is the standard. The old name for the Celsius scale was centigrade. These two scales may be defined by means of the ice and steam points of water at standard pressure, where

Scale	Ice Point	Steam Point
Celsius (C)	0	100
Fahrenheit (F)	32	212

Since 100 degrees (C) between the ice and steam points on the Celsius scale are equivalent to 180 degrees (F) between the same points on the Fahrenheit scale, then 1 degree Celsius is equal to 180/100 or 1.8 degrees Fahrenheit. Because the Fahrenheit scale is offset 32 degrees, the relationship between the two systems is

$$t_C = (1/1.8)(t_F - 32) \qquad (1.11)$$

or

$$t_F = 1.8t_C + 32 \qquad (1.12)$$

When the laws of perfect gases are considered it becomes apparent that neither the Celsius nor Fahrenheit scale is usable for engineering calculations. What is required is an absolute temperature scale, and in the English system of units

$$T_R = t_F + 460 \qquad (1.13)$$

and for the SI system

$$T_K = t_C + 273 \tag{1.14}$$

where t_F is the Fahrenheit temperature (F), t_C is the Celsius temperature (C), T_R is the Rankine temperature (R), and T_K is the Kelvin temperature (K).

The lowercase t is used for the measurement systems, and the uppercase T is used for the absolute temperature scales.

Example 1.6. Express 60 F in terms of C, R, and K.

From Eq. (1.11)

$$t_C = (1/1.8)(t_F - 32) = (1/1.8)(60 - 32)$$
$$= 15.6 \text{ (C)}$$

From Eq. (1.13)

$$T_R = t_F + 460 = 60 + 460 = 520 \text{ (R)}$$

and from Eq. (1.14)

$$T_K = t_C + 273 = 16 + 273 = 289 \text{ (K)}$$

The absolute temperatures are also related by

$$T_R = 1.8T_K \tag{1.15}$$

so that

$$T_R = 1.8(289) = 520 \text{ (R)}$$

Pressure is defined by the equation

$$p = F/A \tag{1.16}$$

where p is the pressure, F is the force, and A is the area on which the force acts.

In the English system pressure is measured in lbf/in.^{2}, and for engineering calculations this is normally converted to lbf/ft^2. These two terms are often abbreviated to psi or psf.

Pressure may be measured by means of manometers, Bourdon tubes, diaphragms, bellows, piezoelectric crystals, and many other instruments. Instruments for measuring pressure usually compare the pressure in the system with the pressure of the air surrounding the instrument. Since the pressure of the air surrounding the instrument, referred to as barometric or atmo-

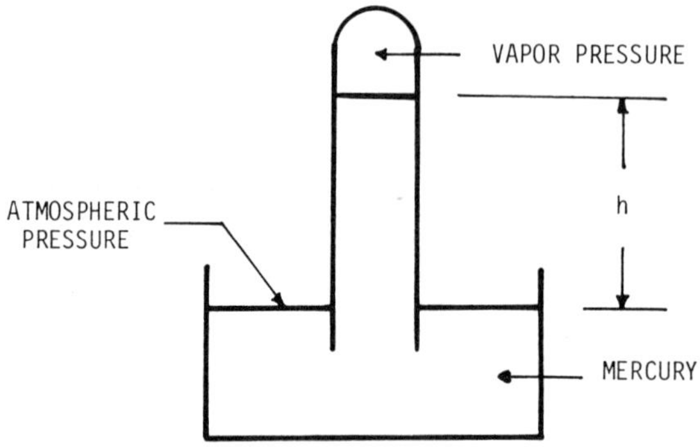

Figure 1.4

spheric pressure, is a variable, it is necessary to define an absolute pressure scale.

A closed manometer (Fig. 1.4) may be used to measure the atmospheric pressure. Mercury is used in this instrument, which is in effect a barometer, because of its high density and low vapor pressure. The height of the mercury in the manometer tube is a measure of the atmospheric pressure. To convert the height of a fluid column to a pressure measurement, consider the system shown in Fig. 1.5.

From Eq. (1.16)

$$p = F/A$$

and by use of Eq. (1.6)

$$p = (m/A)(g/g_0) \tag{1.17}$$

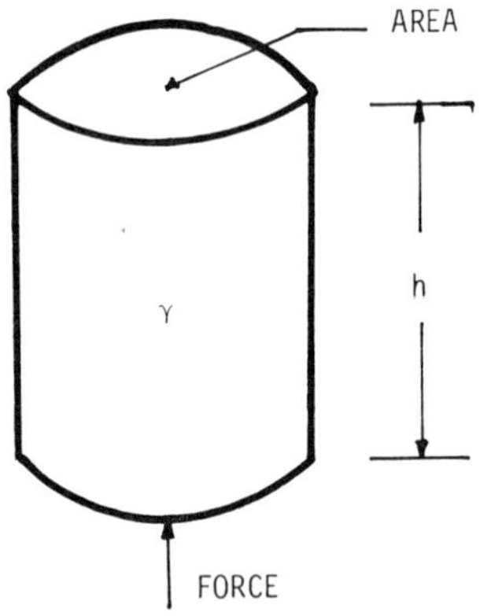

Figure 1.5

By combining Eqs. (1.1) and (1.2)

$$\rho = 1/v = m/V$$

so that the mass within the section is

$$m = \rho V = \rho Ah$$

Substitution of the mass equation into Eq. (1.17) yields

$$p = (\rho Ah/A)(g/g_0) = \rho h(g/g_0)$$

Finally by use of Eq. (1.9)

$$p = \gamma h \tag{1.18}$$

where h is the height of the liquid column.

Since the specific weight γ is a function of the fluid temperature, it is sometimes necessary to measure the temperature to obtain highly accurate pressure measurements.

Example 1.7. Determine the pressure exerted by a mercury column 1 in. in height when $g = 32.2$ ft/sec^2.

From Eq. (1.10)

$$\gamma = SG\gamma_{H_2O} = 13.6(62.4) = 848.6 \text{ (lbf/ft}^3\text{)}$$

and from Eq. (1.18)

$$p = \gamma h = 848.6(1/12) = 70.72 \text{ (lbf/ft}^2\text{)}$$

$$= 70.72(1/12)^2 \text{ [(lbf/ft}^2\text{)(ft/in.)}^2\text{]}$$

$$= 0.491 \text{ (lbf/in.}^2 \text{ or psi)}$$

The unit of pressure in the SI system is the Pascal (Pa), where

$$1 \text{ (psi)} = 6.895\text{E} + 03 \text{ (Pa)}$$

Therefore 1 in. of mercury exerts a pressure of

$$p = 0.491(6895) = 3385 \text{ (Pa)}$$

$$= 3.385 \text{ (kPa)}$$

Example 1.8. Determine the pressure for a mercury column 1 in. in height when $g = 16.1$ ft/sec^2.

From Eq. (1.9)

$$\gamma = \rho(g/g_0) = 13.6(62.4)(16.1/32.2)$$

$$= 424.3 \text{ (lbf/ft}^2\text{)}$$

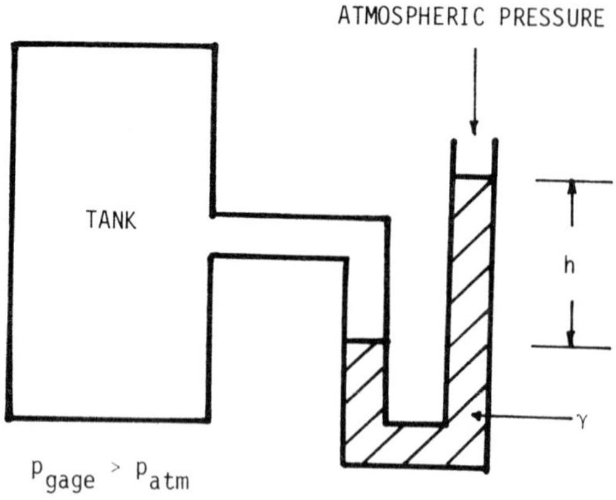

Figure 1.6

and

$$p = \gamma h = 424.3(1/12) = 35.36 \ (\text{lbf/ft}^2)$$

The pressure, which is based on a force measurement, is a function of the acceleration due to gravity.

In all succeeding problems and discussions, we shall assume that the acceleration due to gravity is 32.2 ft/sec^2.

Pressures above atmospheric pressure are referred to as gage pressures. A typical gage pressure measurement, using a manometer, is shown in Fig. 1.6. Gage pressures are normally measured in lbf/in.^{2}, which is written as psig. To convert a gage pressure to an absolute pressure when the atmospheric pressure is known, use

$$p_{abs} = p_{atm} + p_{gage} \tag{1.19}$$

Pressures below atmospheric may be measured by the system shown in Fig. 1.7. These pressures are referred to as vacuum pressures and are normally measured in inches of mercury. To convert a vacuum pressure to an absolute pressure when the atmospheric pressure is known, use

$$p_{abs} = p_{atm} - p_{vac} \tag{1.20}$$

In Eqs. (1.19) and (1.20), p_{abs} is the absolute pressure, p_{atm} is the atmospheric pressure, p_{gage} is the gage pressure, and p_{vac} is the vacuum pressure.

The relationship of the various pressure terms is shown in Fig. 1.8.

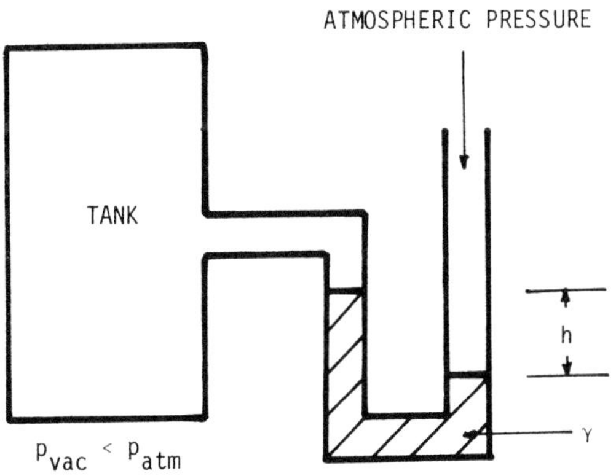

Figure 1.7

For engineering calculations the instrument reading in psig or in.-Hg-vac must be converted to lbf/ft^2-absolute or Pascal absolute.

Example 1.9. Given an atmospheric pressure of 29 in.-Hg and a gage pressure of 20.0 psig, determine the absolute pressure.

The atmospheric pressure is

$$p_{atm} = 0.491(29) = 14.2 \text{ (psi)}$$

where 1 in.-Hg is 0.491 psi.

From Eq. (1.19)

$$p_{abs} = p_{atm} + p_{gage} = 14.2 + 20.0 = 34.2 \text{ (psia)}$$

$$= 34.2(144) = 4925 \text{ (psfa)}$$

$$= 235{,}800 \text{ (Pa-a)} = 235.8 \text{ (kPa-a)}$$

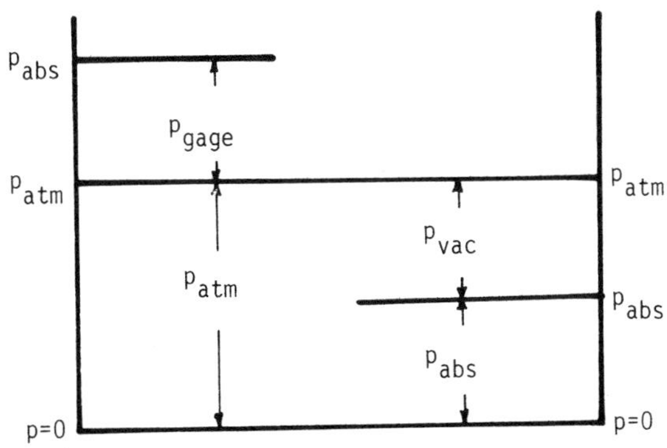

Figure 1.8

The final a in the above quantities designates the pressure as being an absolute value.

Example 1.10. Given an atmospheric pressure of 29 in.-Hg and a vacuum pressure of 20 in.-Hg-vac, determine the absolute pressure.

From Eq. (1.20)

$$p_{abs} = p_{atm} - p_{vac} = 29 - 20 = 9 \text{ (in.-Hg-a)}$$

$$= 0.491(9) = 4.42 \text{ (psia)} = 636 \text{ (psfa)}$$

$$= 30{,}500 \text{ (Pa-a)} = 30.5 \text{ (kPa-a)}$$

Example 1.11. Given an atmospheric pressure of 29 in.-Hg and a vacuum pressure of 30 ft of water, determine the absolute pressure.

From Eq. (1.18)

$$p_{vac} = \gamma h = 62.4(30) = 1872 \text{ (psf-vac)}$$

$$= 13.0 \text{ (psi-vac)}$$

From Eq. (1.20)

$$p_{abs} = p_{atm} - p_{vac} = 14.2 - 13.0 = 1.2 \text{ (psia)}$$

$$= 172.8 \text{ (psfa)} = 8270 \text{ (Pa-a)} = 8.27 \text{ (kPa-a)}$$

Neither absolute temperatures nor absolute pressures can assume negative values.

PROBLEMS

1.1. A body exerts a force of 200 lbf on a planet where $g = 45$ ft/sec^2. The same body exerts a force of 100 lbf on a second planet. Determine the mass of the body and the value of g on the second planet.

1.2. A person riding in a car has a mass of 150 lbm. The car, traveling at 60 miles/hr, strikes a tree and stops in 1.0 sec. Determine the force the person exerts on his seat belt.

1.3. A body on a planet where $g = 64.4$ ft/sec^2 has a mass of 100 lbm. The body is moved vertically upward 5 ft in 2 sec. Determine the work, the heat equivalent of the work, and the horsepower or kilowatts required to raise the body.

1.4. A body on a planet where g = 20 ft/sec^2 has a volume of 0.5 ft^3 and a mass of 200 lbm. Determine the density, specific gravity, and the specific weight.

1.5. Convert 20 C to F, K, and R.

1.6. A Fahrenheit and a Celsius thermometer are inserted into a tank. The readings on the two thermometers are the same. Determine the temperature.

1.7. Given an atmospheric pressure of 29.5 in.-Hg, convert the following instrument readings to absolute pressure: (a) 10 psig and (b) 20 in.-Hg-vac.

1.8. A manometer is connected to a tank on one side and is open to the atmosphere on the opposite side. Atmospheric pressure is 14.70 psi. The pressure in the tank is less than the atmosphere, and the manometer deflection is 54 in. If g = 32.2 ft/sec^2 and the manometer fluid specific gravity is 0.8, determine the absolute pressures.

GENERAL ENERGY EQUATION

In this chapter we shall discover that the analysis of an open thermodynamic system is basically an accounting of the mass flow and the energy, heat, and work terms. For a steady state system the equation of continuity may be used to balance the mass flow. The general energy equation will be developed to balance the energy, heat, and work terms.

EQUATION OF CONTINUITY

Given the open system shown in Fig. 2.1, the mass rate of flow through section 1 is

$$m_1 = \text{density(volume/time)} \tag{2.1}$$

$$= \rho_1 \text{Vol}_1$$

$$= (\text{lbm/ft}^3)(\text{ft}^3/\text{sec}) = \text{lbm/sec}$$

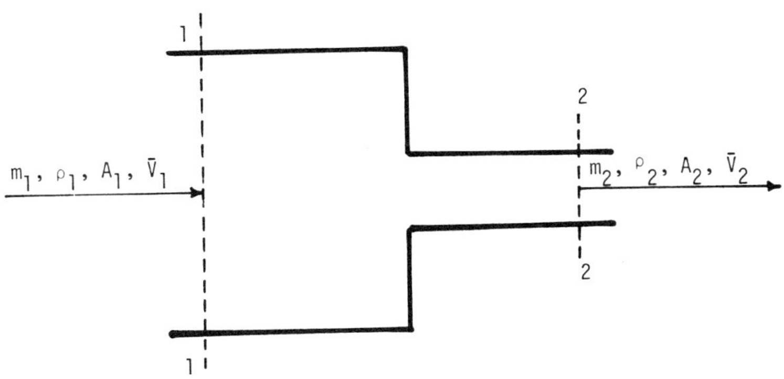

Figure 2.1

The volume rate of flow is

$$\text{Vol}_1 = (\text{area})(\text{velocity}) = A_1\overline{V}_1 \qquad \text{or} \qquad \text{ft}^2(\text{ft/sec}) = \text{ft}^3/\text{sec} \tag{2.2}$$

Substitution of Eq. (2.2) into Eq. (2.1) yields

$$m_1 = \rho_1 A_1 \overline{V}_1 \tag{2.3}$$

where m_1 is the mass rate of flow (lbm/sec), ρ_1 is the fluid density (lbm/ft^3), A_1 is the pipe area (ft^2), and $\overline{V}_1$ is the velocity at plane 1 (ft/sec).

If the velocity is measured in ft/min or ft/hr, then the corresponding mass flow will be expressed in lbm/min or lbm/hr.

The same symbol m will be used for the mass in a closed system or for the mass flow in an open system since it is generally obvious, from the statement of the problem, which term is being evaluated.

The mass rate of fluid leaving the system is

$$m_2 = \rho_2 A_2 \overline{V}_2 \tag{2.4}$$

For a steady-flow system

$$m_1 = m_2$$

or, dropping the subscripts for the mass flow,

$$m = \rho_1 A_1 \overline{V}_1 = \rho_2 A_2 \overline{V}_2 \tag{2.5}$$

Equation (2.5) is the equation of continuity.

Example 2.1. Given the system in Fig. 2.1 where the diameters are 3 in. at plane 1 and 2 in. at plane 2, the working substance is water for which $\rho = 62.4$ lbm/ft^3. The inlet velocity is 10 ft/sec. Determine the mass rate of flow and the exit velocity.

From Eq. (2.3), the mass rate of flow into the system is

$$\begin{aligned} m_1 &= \rho_1 A_1 \overline{V}_1 = \rho_1[(\pi/4)D_1{}^2]\overline{V}_1 \\ &= 62.4(\pi/4)(3/12)^2(10) \\ &= 30.6 \text{ (lbm/sec)} \\ &= 13.9 \text{ (kg/s)} \end{aligned}$$

For steady flow, $m_1 = m_2$, and from Eq. (2.4), the velocity at the exit is

$$\begin{aligned} \overline{V}_2 &= m_2/(\rho_2 A_2) = 30.6/[62.4(\pi/4)(2/12)^2] \\ &= 22.5 \text{ (ft/sec)} \\ &= 6.86 \text{ (m/s)} \end{aligned}$$

The velocity at the system exit may also be calculated directly from the equation of continuity [Eq. (2.5)], so that

$$\overline{V}_2 = (\rho_1 A_1 V_1)/(\rho_2 A_2)$$

Since water is incompressible, $\rho_1 = \rho_2$, and

$$\overline{V}_2 = \overline{V}_1(A_1/A_2)$$

For circular pipes leading into and out of the system,

$$A_1/A_2 = [(\Pi/4)(D_1^2)]/[(\Pi/4)(D_2^2)]$$

$$= (D_1/D_2)^2$$

and the leaving velocity may be calculated from

$$\overline{V}_2 = \overline{V}_1(D_1/D_2)^2 = 10[(3/12)/(2/12)]^2$$

$$= 10(1.5)^2 = 22.5 \text{ (ft/sec)}$$

Example 2.2. Assume that air is the working substance (Fig. 2.1), where

Property	Inlet	Exit	Units
Diameter	3.0	—	in.
Velocity	10.0	20.0	ft/sec
Density	0.073	0.060	lbm/ft^3

Determine the mass rate of flow and the exit diameter.

From Eq. (2.3), the mass rate of flow is

$$m_1 = \rho_1 A_1 \overline{V}_1 = 0.073(\Pi/4)(3/12)^2(10)$$

$$= 0.0358 \text{ (lbm/sec)}$$

$$= 0.0162 \text{ (kg/s)}$$

From Eq. (2.5), the area at the exit is

$$\rho_1 A_1 \overline{V}_1 = \rho_2 A_2 \overline{V}_2$$

or

$$A_2 = (\rho_1 A_1 \overline{V}_1)/(\rho_2 \overline{V}_2)$$

$$= [0.073(\Pi/4)(3/12)^2(10)]/[0.060(20)]$$

$$= 0.02986 \text{ (ft}^2\text{)}$$

For a circular pipe

$$A_2 = (\Pi/4)D_2^2$$

and

$$D_2^2 = (4/\Pi)A_2 = (4/\Pi)0.02986$$

$$= 0.0380 \text{ (ft}^2)$$

so that

$$D_2 = 0.195 \text{ (ft)} = 2.34 \text{ (in.)}$$

$$= 0.0594 \text{ (m)}$$

Example 2.3. Given is the multiple-flow system in Fig. 2.2. Data for the system are

Property	1	2	3	Units
Diameter	3.0	2.0	2.5	in.
Density	0.07	0.08	0.05	lbm/ft^3
Velocity	10.0	40.0	–	ft/sec

Determine the direction, the mass rate of flow, and the velocity at plane 3. The mass flow of fluid into the system at plane 1 is

$$m_1 = \rho_1 A_1 \bar{V}_1 = 0.07(\Pi/4)(3/12)^2(10)$$

$$= 0.0344 \text{ (lbm/sec)}$$

The mass flow of fluid from the system at plane 2 is

$$m_2 = \rho_2 A_2 \bar{V}_2 = 0.08(\Pi/4)(2/12)^2(40)$$

$$= 0.0698 \text{ (lbm/sec)}$$

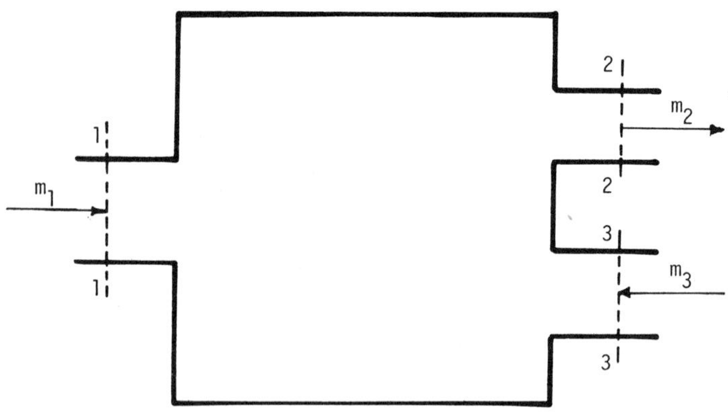

Figure 2.2

For steady flow, the mass flow into the system must be balanced by the flow out of the system. Since the flow out of the system at plane 2 exceeds the flow into the section at plane 1, the flow at plane 3 must be into the system, or

$$m_1 + m_3 = m_2$$

or

$$m_3 = m_2 - m_1 = 0.0698 - 0.0344$$

$$= 0.0354 \text{ (lbm/sec)}$$

$$= 0.0161 \text{ (kg/s)}$$

Since

$$m_3 = \rho_3 A_3 \overline{V}_3$$

then

$$\overline{V}_3 = m_3/(\rho_3 A_3)$$

$$= 0.0354/[0.05(\Pi/4)(2.5/12)^2]$$

$$= 20.8 \text{ (ft/sec)}$$

$$= 6.34 \text{ (m/s)}$$

In our study of open systems the first step in the solution of a problem is to balance the mass flow into and out of the system. The second step is to balance the energy, heat, and work terms. To begin our discussion of the second step, we must first define the types of energy that are normally present in open thermodynamic systems.

ENERGY

Energy is a property of the working substance in a thermodynamic system and may appear in many different forms. Energy is present in a compressed spring, in batteries, in fuels, in electrical circuits, and in magnetic fields. In thermodynamic problems the forms of energy generally encountered include potential, kinetic, internal, and flow energy. In thermodynamic problems, the energy level is measured with reference to an arbitrary datum plane, and the value of the energy at the datum plane is zero. In many problems the numerical value of the energy at a state point is unimportant, but the change in energy between state points is required for the solution of the problem.

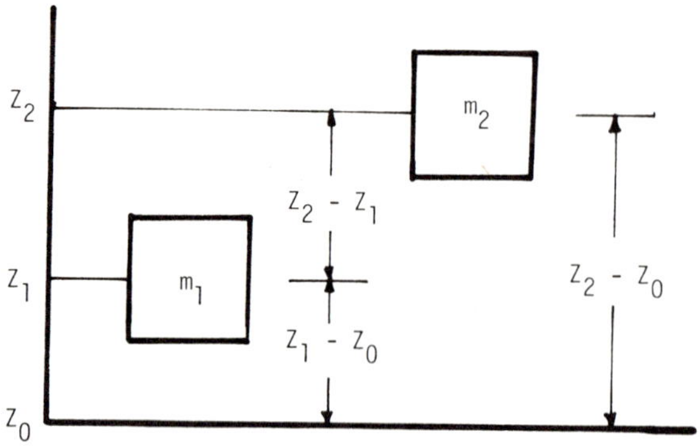

Figure 2.3

POTENTIAL ENERGY

The energy that a body possesses by virtue of its mass and position with reference to a datum plane is referred to as its potential energy. In Fig. 2.3, the potential energy is

$$PE = (\text{force})(\text{height}) = F(Z - Z_0) \quad (\text{ft-lbf})$$

From Eq. (1.6) and by substituting for the force,

$$PE = m(g/g_0)(Z - Z_0)$$

At state 1 in Fig. 2.3

$$PE_1 = m_1(g/g_0)(Z_1 - Z_0)$$

and at state 2

$$PE \; = m_2(g/g_0)(Z_2 - Z_0)$$

In many cases, $m_1 = m_2 = m$, so that

$$PE_2 - PE_1 = m(g/g_0)[(Z_2 - Z_0) - (Z_1 - Z_0)]$$

or

$$\Delta PE = PE_2 - PE_1 = m(g/g_0)(Z_2 - Z_1) \tag{2.6}$$

where ΔPE is the change in potential energy as the body moves from position 1 to position 2. The change in potential energy is independent of the location of the datum plane Z_0. Potential energy is a mechanical rather than a thermodynamic property.

KINETIC ENERGY

The energy that a body possesses by virtue of its mass and its velocity is referred to as kinetic energy. The kinetic energy is

$$KE = [(\text{mass})(\text{velocity})^2]/2g_0$$

$$= \text{lbm}(\text{ft/sec})^2[(\text{lbf-sec}^2)/(\text{lbm-ft})] = \text{ft-lbf}$$

At state 1

$$KE_1 = (m_1/2g_0)(\overline{V}_1^2 - \overline{V}_0^2)$$

and at state 2

$$KE_2 = (m_2/2g_0)(\overline{V}_2^2 - \overline{V}_0^2)$$

When $m_1 = m_2 = m$

$$\Delta KE = KE_2 - KE_1 = (m/2g_0)(\overline{V}_2^2 - \overline{V}_1^2) \tag{2.7}$$

Again the velocity selected for the energy datum plane does not affect the change in kinetic energy. In any case, logic dictates that the kinetic energy datum plane be set at $\overline{V}_0 = 0$. Kinetic energy is a mechanical property.

FLOW ENERGY

Flow energy is the energy associated with the fluid as it crosses the boundaries of the open system (Fig. 2.4). The flow energy at a state

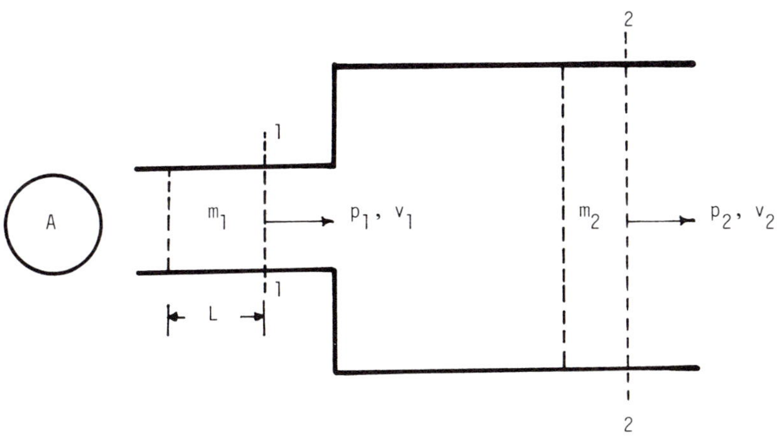

Figure 2.4

point is

$$FE = (\text{force})(\text{distance}) \quad (\text{ft-lbf})$$

From Eq. (1.16), where $F = pA$, then

$$FE = FL = pAL$$

and since

$$V = AL$$

then

$$FE = pV = (\text{pressure})(\text{volume})$$
$$= (\text{lbf/ft}^2)\text{ft}^3 = \text{ft-lbf}$$

Flow energy is the energy required to force the working substance into the system or to remove it from the system.

From Eq. (1.1), $V = mv$, so that

$$FE = mpv$$

and from Eq. (1.2), where $v = 1/\rho$,

$$FE = mp/\rho$$

For state 1

$$FE_1 = m_1(p_1 - p_0)v_1$$

and at state 2

$$FE_2 = m_2(p_2 - p_0)v_2$$

Selecting $p_0 = 0$ as the datum plane requires that p_1 and p_2 be expressed in absolute units.

Again when $m_1 = m_2 = m$

$$\Delta FE = FE_2 - FE_1 = m(p_2v_2 - p_1v_1) \tag{2.8}$$
$$= m[(p_2/\rho_2) - (p_1/\rho_1)]$$

The flow energy is a thermodynamic property.

BERNOULLI'S EQUATION

Given is the system in Fig. 2.5, which includes potential, kinetic, and flow energy at the system entrance and exit. The total energy in the system

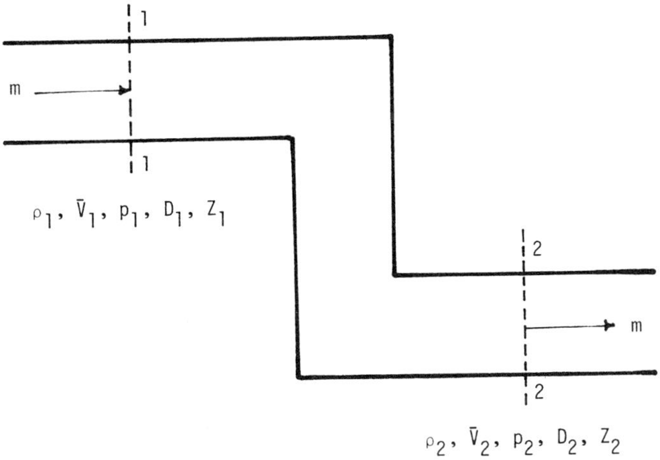

Figure 2.5

must be conserved (does not increase or decrease), so that

$$\Delta PE + \Delta KE + \Delta FE = 0$$

From Eqs. (2.6), (2.7), and (2.8)

$$(mg/g_0)(Z_2 - Z_1) + (m/2g_0)(\bar{V}_2^2 - \bar{V}_1^2) + m[(p_2/\rho_2) - (p_1/\rho_1)] = 0$$

Since m appears in each term, it may be omitted, and by rearranging the order of the terms

$$(g/g_0)Z_1 + (\bar{V}_1^2/2g_0) + (p_1/\rho_1) = (g/g_0)Z_2 + (\bar{V}_2^2/2g_0) + (p_2/\rho_2)$$

Multiplying by g_0/g gives

$$Z_1 + (\bar{V}_1^2/2g) + p_1(g_0/g)(1/\rho_1) = Z_2 + (\bar{V}_2^2/2g) + p_2(g_0/g)(1/\rho_2)$$

From Eq. (1.9), $\gamma = \rho(g/g_0)$, so that

$$Z_1 + (\bar{V}_1^2/2g) + (p_1/\gamma_1) = Z_2 + (\bar{V}_2^2/2g) + (p_2/\gamma_2) \qquad (2.9)$$

This is Bernoulli's equation, where each term in the equation has the units ft.

Example 2.4. Given the system in Fig. 2.5, where

Property	1	2	Units
Diameter	3	1.5	in.
Pressure	20	–	psia
Density	60	80.0	lbm/ft^3
Velocity	20	–	ft/sec
Height	40	10.0	ft

determine the pressure at the system exit.

By use of the equation of continuity [Eq. (2.5)],

$$\rho_1 A_1 \overline{V}_1 = \rho_2 A_2 \overline{V}_2$$

so that

$$\begin{aligned}\overline{V}_2 &= (\rho_1/\rho_2)(A_1/A_2)\overline{V}_1 \\ &= (\rho_1/\rho_2)(D_1/D_2)^2\overline{V}_1 \\ &= (60/80)(3/1.5)^2(20) \\ &= 60 \text{ (ft/sec)} \quad \text{or} \quad 18.3 \text{ (m/s)}\end{aligned}$$

When $g = 32.2$ ft/sec^2 the specific weight and density are numerically equal.

From Eq. (2.9)

$$\begin{aligned}p_2/\gamma_2 &= (Z_1 - Z_2) + [(\overline{V}_1^2 - \overline{V}_2^2)/2g] + (p_1/\gamma_1) \\ &= (40 - 10) + \{(20^2 - 60^2)/[(2)(32.2)]\} + [20(144)/60]\end{aligned}$$

Note that the pressure in psia has been converted to psfa.

$$\begin{aligned}p_2/\gamma_2 &= 30.0 - 49.7 + 48.0 \\ &= 28.3 \text{ (ft)}\end{aligned}$$

Then

$$\begin{aligned}p_2 &= 28.3\gamma_2 = 28.3(80) = 2264 \text{ (psfa)} \\ &= 15.7 \text{ (psia)} \quad \text{or} \quad 108.3 \text{ (kPa-a)}\end{aligned}$$

INTERNAL ENERGY

Internal energy is the energy within the working substance and is dependent on the velocity of the particles that make up the working substance or their positions relative to each other. For a single-phase working substance, such as a solid, liquid, or gas, the internal energy is a function of the temperature of the working substance. As the temperature increases, the particles, generally molecules, that make up the working substance increase in velocity, and the velocity of these particles defines the internal energy level.

For a two-phase working substance in which the material is melting or evaporating, an increase in internal energy produces more of the liquid or vapor phase. Since the molecules are farther apart in these phases, the increase in internal energy has served to move the molecules farther apart. For the freezing and condensation process, the internal energy of the working substance decreases as more of the solid or liquid phases are formed.

The symbol for the total internal energy, which is an extensive property, is U. The specific internal energy, which is an intensive property, is the result of dividing the total internal energy by the mass of the system.

In equation form,

$$u = U/m \tag{2.10}$$

The datum plane for internal energy depends on the temperature scale being employed in the problem. For some working substances, absolute zero is used; in others 32 F becomes the datum plane. Any value of the temperature would be acceptable as long as it is consistently employed throughout the problem.

Example 2.5. Given the system in Fig. 2.5, where the working substance is water and

Property	Inlet	Exit	Units
Diameter	3.0	1.5	in.
Pressure	40.0	20.0	psia
Density	62.4	62.4	lbm/ft^3
Velocity	20.0	—	ft/sec
Height	40.0	10.0	ft

determine the change in internal energy.

The exit velocity may be obtained from the equation of continuity [Eq. (2.5)], where

$$\bar{V}_2 = (\rho_1/\rho_2)(A_1/A_2)\bar{V}_1 = (\rho_1/\rho_2)(D_1/D_2)^2\bar{V}_1$$

$$= (62.4/62.4)(3.0/1.5)^2(20)$$

$$= 80.0 \text{ (ft/sec)} = 24.4 \text{ (m/s)}$$

The mass flow from Eq. (2.3) is

$$m_1 = \rho_1 A_1 \overline{V}_1 = \rho_1 (\Pi/4)(D_1)^2 \overline{V}_1$$

$$= 62.4(\Pi/4)(3.0/12.0)^2(20)$$

$$= 61.3 \text{ (lbm/sec)}$$

$$= 27.8 \text{ (kg/s)}$$

The change in potential energy may be calculated by means of Eq. (2.6), so that

$$\Delta PE = m(g/g_0)(Z_2 - Z_1)$$

$$= 61.3(32.2/32.2)(10 - 40)$$

$$= -1840 \text{ (ft-lbf/sec)} = -2.36 \text{ (B/sec)}$$

$$= -2487 \text{ (W)}$$

The change in kinetic energy may be calculated by means of Eq. (2.7), so that

$$\Delta KE = (m/2g_0)(\overline{V}_2^2 - \overline{V}_1^2)$$

$$= (61.3/64.4)(80.0^2 - 20.0^2)$$

$$= 5711 \text{ (ft-lbf/sec)} = 7.34 \text{ (B/sec)}$$

$$= 7736 \text{ (W)}$$

The change in flow energy may be calculated by means of Eq. (2.8), so that

$$\Delta FE = m[(p_2/\rho_2) - (p_1/\rho_1)]$$

$$= 61.3(144)[(20/62.4) - (40/62.4)]$$

$$= -2830 \text{ (ft-lbf/sec)} = -3.64 \text{ (B/sec)}$$

$$= -3837 \text{ (W)}$$

The energy in this system must be conserved, and the sum of the energy terms will be equal to zero or

$$\Delta PE + \Delta KE + \Delta FE + \Delta U = 0$$

therefore

$$\Delta U = -\Delta PE - \Delta KE - \Delta FE$$
$$= 2.36 - 7.34 + 3.64$$
$$= -1.34 \text{ (B/sec)}$$
$$= -1412 \text{ (W)}$$

GENERAL ENERGY EQUATION

From the discussion of energy terms it should be noted that the energy level of each term is evaluated at the plane where it enters or leaves the system. In Fig. 2.6, four other quantities have been added to the open thermodynamic system. Work could be added to the system by means of a pump or removed from the system by means of a turbine. Heat could also be added or removed by means of heating or cooling coils. If these quantities are expressed in units consistent with the energy terms, then the sum of the heat, energy, and work terms into the system must equal the sum of these terms leaving the system, so that

$$Q_{in} + PE_1 + KE_1 + FE_1 + U_1 + W_{in} = Q_{out} + PE_2 + KE_2 + FE_2 + U_2 + W_{out} \tag{2.11}$$

To reduce the number of terms, let

$$Q = Q_{in} - Q_{out} \tag{2.12}$$

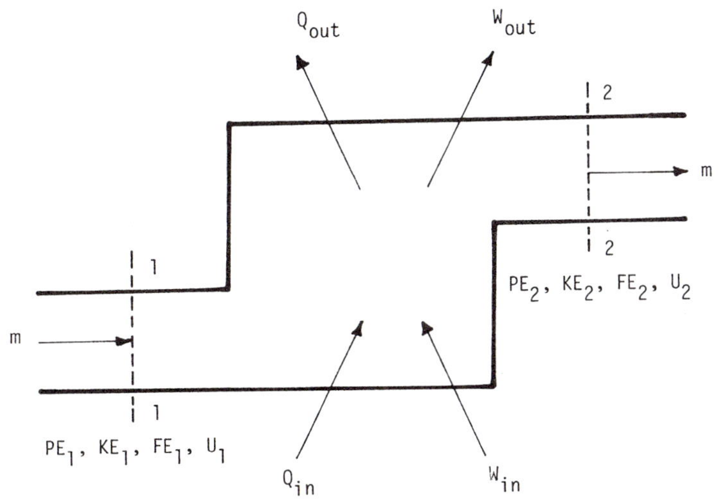

Figure 2.6

so that if Q is positive, the net transfer of heat is into the system. Also let

$$W_{sf} = W_{out} - W_{in} \tag{2.13}$$

and a positive value of the steady-flow work indicates that the net transfer of work is out of the system.

When Eqs. (2.12) and (2.13) are substituted into Eq. (2.11) the result is known as the general energy equation or

$$Q = (PE_2 - PE_1) + (KE_2 - KE_1) + (FE_2 - FE_1) + (U_2 - U_1) + W_{sf}$$

$$= \Delta PE + \Delta KE + \Delta FE + \Delta U + W_{sf} \tag{2.14}$$

The energy terms may be grouped where

$$\Delta E = \Delta PE + \Delta KE + \Delta FE + \Delta U \tag{2.15}$$

so that the general equation reduces to

$$Q = \Delta E + W_{sf} \tag{2.16}$$

ΔE is the total energy change for the system.

Example 2.6. Given the system in Fig. 2.6, where

Property	Inlet	Exit	Units
Diameter	6.0	3.0	in.
Pressure	14.0	30.0	psia
Density	62.4	62.4	lbm/ft^3
Internal energy	60.0	60.05	B/lbm
Height	0.0	20.0	ft
Velocity	10.0	—	ft/sec

$Q_{in} = 6.0$ (B/lbm) $Q_{out} = 6.02$ (B/lbm)

determine the steady-flow work W_{sf}.

From Eq. (2.3), the mass flow is

$$m_1 = \rho_1 A_1 \bar{V}_1 = \rho_1 (\Pi/4)(D_1)^2 \bar{V}_1$$

$$= 62.4(\Pi/4)(6/12)^2(10) = 122.5 \text{ (lbm/sec)}$$

$$= 55.6 \text{ (kg/s)}$$

From Eq. (2.5), the exit velocity is

$$\bar{V}_2 = (\rho_1/\rho_2)(A_1/A_2)\bar{V}_1 = (\rho_1/\rho_2)(D_1/D_2)^2\bar{V}_1$$
$$= (62.4/62.4)(6/3)^2(10) = 40 \text{ (ft/sec)} = 12.2 \text{ (m/s)}$$

From Eq. (2.6), the change in potential energy is

$$\Delta PE = m(g/g_0)(Z_2 - Z_1)$$
$$= 122.5(32.2/32.2)(20 - 0)$$
$$= 2450 \text{ (ft-lbf/sec)} = 3.15 \text{ (B/sec)}$$
$$= 3.32 \text{ (kW)}$$

From Eq. (2.7), the change in kinetic energy is

$$\Delta KE = (m/2g_0)(\bar{V}_2^2 - \bar{V}_1^2)$$
$$= (122.5/64.4)(40^2 - 10^2)$$
$$= 2853 \text{ (ft-lbf/sec)} = 3.67 \text{ (B/sec)}$$
$$= 3.86 \text{ (kW)}$$

From Eq. (2.8), the change in flow energy is

$$\Delta FE = m[(p_2/\rho_2) - (p_1/\rho_1)]$$
$$= 122.5(144/62.4)(30 - 14)$$
$$= 4523 \text{ (ft-lbf/sec)} = 5.81 \text{ (B/sec)}$$
$$= 6.13 \text{ (kW)}$$

From Eq. (2.10), the internal energy is

$$U = mu$$

so that

$$\Delta U = m(\Delta u) = m(u_2 - u_1)$$
$$= 122.5(60.05 - 60.0) = 6.12 \text{ (B/sec)}$$
$$= 6.45 \text{ (kW)}$$

From Eq. (2.15), the total energy change is

$$\Delta E = \Delta PE + \Delta KE + \Delta FE + \Delta U$$
$$= 3.15 + 3.67 + 5.81 + 6.12$$
$$= 18.75 \text{ (B/sec)}$$
$$= 19.76 \text{ (kW)}$$

From Eq. (2.12), the net heat transfer is

$$Q = Q_{in} - Q_{out} = 6.0 - 6.02 = -0.02 \text{ (B/lbm)}$$

$$= 122.5(-0.02) \text{ [(lbm/sec)(B/lbm)]}$$

$$= -2.45 \text{ (B/sec)}$$

From Eq. (2.16), the steady-flow work is

$$W_{sf} = Q - \Delta E = -2.45 - 18.75 = -21.20 \text{ (B/sec)}$$

$$= -22.34 \text{ (kW)}$$

$$= -29.99 \text{ (HP)}$$

Since W_{sf} is negative, the power must be added to the system.

In open system problems, the steady-flow work is normally solved for in terms of power units.

MULTIPLE-FLOW SYSTEMS

When there are two or more inlets or exits for the open system, it is necessary to evaluate the energy terms for each inlet or exit. The total amount of energy entering or leaving the system may then be calculated.

Example 2.7. Given the system in Fig. 2.2 and Example 2.3, where

Property	1	2	3	Units
Direction of flow	In	Out	In	—
Mass flow	0.0344	0.0698	0.0354	lbm/sec
Density	0.07	0.08	0.05	lbm/ft^3
Velocity	10.0	40.0	20.8	ft/sec
Internal energy	5.0	10.0	5.0	B/lbm
Height	0.0	40.0	-50.0	ft
Pressure	20.0	15.0	20.0	psia

Net heat transfer = 120 (B/min)

determine the steady-flow work W_{sf}.

The potential energy for each inlet and exit plane is

$$PE_1 = m_1(g/g_0)(Z_1/J)$$

$$= 0.0344(32.2/32.2)(0/778) = 0.000 \text{ (B/sec)}$$

$$PE_2 = m_2(g/g_0)(Z_2/J)$$

$$= 0.0698(32.2/32.2)(40/778) = 0.004 \text{ (B/sec)}$$

$$PE_3 = m_3(g/g_0)(Z_3/J)$$

$$= 0.0354(32.2/32.2)(-50/778) = -0.002 \text{ (B/sec)}$$

The kinetic energy for each inlet and exit plane is

$$KE_1 = (m_1/2g_0)(\bar{V}_1^2/J)$$

$$= (0.0344/64.4)(10^2/778) = 0.000 \text{ (B/sec)}$$

$$KE_2 = (m_2/2g_0)(\bar{V}_2^2/J)$$

$$= (0.0698/64.4)(40^2/778) = 0.002 \text{ (B/sec)}$$

$$KE_3 = (m_3/2g_0)(\bar{V}_3^2/J)$$

$$= (0.0354/64.4)(20.8^2/778) = 0.000 \text{ (B/sec)}$$

The flow energy for each inlet and exit plane is

$$FE_1 = (m_1/J)(p_1/\rho_1)$$

$$= (0.0344/778)144(20/0.07) = 1.819 \text{ (B/sec)}$$

$$FE_2 = (m_2/J)(p_2/\rho_2)$$

$$= (0.0698/778)144(15/0.08) = 2.422 \text{ (B/sec)}$$

$$FE_3 = (m_3/J)(p_3/\rho_3)$$

$$= (0.0354/778)144(20/0.05) = 2.621 \text{ (B/sec)}$$

The internal energy at each inlet and exit plane is

$$U_1 = m_1u_1 = 0.0344(5) = 0.172 \text{ (B/sec)}$$

$$U_2 = m_2u_2 = 0.0698(10) = 0.698 \text{ (B/sec)}$$

$$U_3 = m_3u_3 = 0.0354(5) = 0.177 \text{ (B/sec)}$$

The total energy at each inlet and exit plane is

$$E_1 = PE_1 + KE_1 + FE_1 + U_1 = 0.000 + 0.000 + 1.819 + 0.172$$

$$= 1.991 \text{ (B/sec)}$$

$$E_2 = PE_2 + KE_2 + FE_2 + U_2 = 0.004 + 0.002 + 2.422 + 0.698$$

$$= 3.126 \text{ (B/sec)}$$

$$E_3 = PE_3 + KE_3 + FE_3 + U_3 = -0.002 + 0.000 + 2.621 + 0.177$$

$$= 2.796 \text{ (B/sec)}$$

The change in total energy is

$$\Delta E = E_{out} - E_{in} = E_2 - (E_1 + E_3)$$

$$= 3.126 - (1.991 + 2.796)$$

$$= -1.661 \text{ (B/sec)}$$

The net heat transfer is

$$Q = 120 \text{ (B/min)} = 2.000 \text{ (B/sec)}$$

and the steady-flow work is

$$W_{sf} = Q - \Delta E = 2.000 + 1.661 = 3.661 \text{ (B/sec)}$$

$$= 5.18 \text{ (HP)}$$

$$= 3.86 \text{ (kW)}$$

Since W_{sf} is positive, the system will produce the above amount of power.

GENERAL ENERGY EQUATION FOR THE CLOSED SYSTEM

Consider Eq. (2.14) for the open system where

$$Q = \Delta PE + \Delta KE + \Delta FE + \Delta U + W_{sf}$$

For a nonflow system, the change in potential energy is negligible. Since matter does not cross the system boundaries, there is no change in kinetic or flow energy. Equation (2.14) reduces to

$$Q = \Delta U + W_{nf} \tag{2.17}$$

where W_{nf} is the nonflow work for the closed system.

LOCATION OF HEAT, WORK, AND ENERGY

For an open system energy terms are evaluated at inlet and exit planes of the system. Heat and work are transported into or out of the system between the inlet and exit planes.

For a closed system energy terms are evaluated at the initial and final states of the system. Heat and work are transported into or out of the system between these states.

PROBLEMS

2.1. See Fig. 2.1. The diameter at the exit is 4 in. and the exit velocity and density are 30 mph and 0.1 lbm/ft^3. The inlet velocity and density are 10 mph and 0.08 lbm/ft^3. Determine the mass flow and the inlet diameter.

2.2. Given a system with two inlets and two exits where

Plane	1	2	3	4	Units
Direction of flow	In	In	Out	Out	–
Diameter	6.0	3.0	5.0	–	in.
Mass flow	–	–	–	0.4	lbm/sec
Density	0.1	0.1	0.15	0.15	lbm/ft^3
Velocity	–	40.0	5.0	15.0	ft/sec

supply the missing quantities in the table of data.

2.3. A car traveling at 50 mph has a mass of 3000 lbm. The car strikes an obstacle that directs it vertically upwards. Determine the height to which the car rises.

2.4. If the inlet and exit pressures in Prob. 2.1 are 100 and 80 psia, determine the flow energy at the inlet and exit planes.

2.5. Given the system in Fig. 2.5 where

Property	1	2	Units
Diameter	4.0	2.0	in.
Pressure	20.0	15.0	psia
Density	62.4	62.4	lbm/ft^3
Velocity	–	–	ft/sec
Height	60.0	-20.0	ft

determine the entering and leaving velocity.

2.6. If the leaving velocity in Prob. 2.5 is 30 ft/sec, determine the mass flow and the new exit diameter. Determine ΔPE, ΔKE, ΔFE, and ΔU.

2.7. Given the system in Fig. 2.6, where

Property	Inlet	Exit	Units
Diameter	4.0	4.0	in.
Pressure	50.0	15.0	psia
Density	62.4	62.4	lbm/ft^3
Internal energy	40.0	40.1	B/lbm
Height	60.0	0.0	ft
Velocity	30.0	–	ft/sec

determine m, $\overline{V}_2$, ΔPE, ΔKE, ΔFE, ΔU, ΔE, and Q. W_{sf} = 30 HP.

2.8. Given the multiple-flow system in Fig. 2.2, where

Property	1	2	3	Units
Diameter	3.0	3.0	3.0	in.
Density	0.1	0.08	0.09	lbm/ft^3
Velocity	20.0	10.0	–	ft/sec
Internal energy	50.0	60.0	40.0	B/lbm
Height	20.0	20.0	20.0	ft
Pressure	20.0	15.0	15.0	psia

Q = -720 B/hr

determine the direction, the mass rate of flow, and the velocity at plane 3. Also calculate ΔE and W_{sf} for the open system.

3

FIRST LAW OF THERMODYNAMICS AND WORK

The topics to be discussed in this chapter are processes, cycles, the First Law of Thermodynamics, point functions, work, path functions, and the pV plane.

PROCESSES

A process is defined by the values that the thermodynamic properties assume between the initial and final states in a system. In thermodynamics a particular type of process is often assumed and is designated as a reversible process. Several criteria must be met for a process to be reversible. The most important requirements are that the process is frictionless and that heat is not transferred across a finite temperature difference. Another restriction is that the volume of a closed system be changed slowly to avoid fluid turbulence. No real process is reversible, but the reversible process does serve as a measure of the changes in properties—energy, heat, or work—that would occur under ideal or theoretical conditions.

Graphical techniques are often used in thermodynamics to illustrate reversible processes. Several different sets of coordinates may be employed; among these sets, the use of the pressure and volume is of major importance. Figure 3.1 illustrates three reversible processes, as plotted on the pV plane, where the initial state is 1 and the final state is 2. Process 1-2a is a constant pressure or isobaric process. Process 1-2b is a constant volume or isometric process. For process 1-2c, both the pressure and the volume vary between the initial and final states.

Unless otherwise noted, processes will be assumed to be reversible for all problems and discussions.

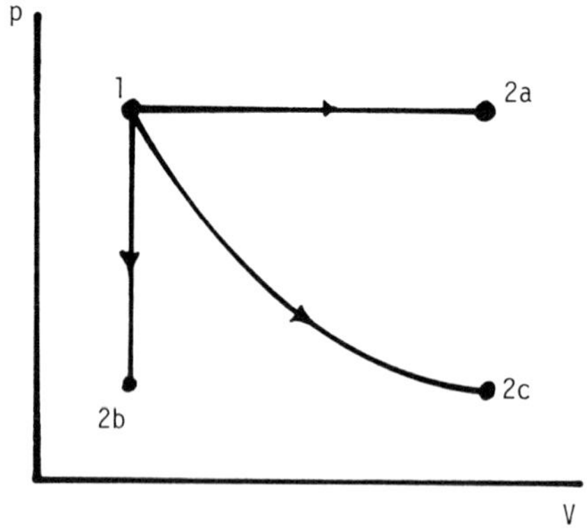

Figure 3.1

CYCLES

If a series of processes is connected so that the final state of the last process is identical to the initial state of the first process, the series of processes constitutes a cycle (Fig. 3.2). Since the working substance that undergoes a cycle is returned to its initial state, the change in property values for a complete cycle is zero. Therefore, for a cycle, $\Delta PE = 0$, $\Delta KE = 0$, $\Delta FE = 0$, $\Delta U = 0$, and $\Delta E = 0$.

By applying the general energy equation [Eq. (2.16)] to the processes in Fig. 3.2,

$$Q_{12} = \Delta E_{21} + W_{12}$$

where Q_{12} is the heat transfer, W_{12} is the work, and ΔE_{21} is the change in energy terms as the process proceeds from state 1 to state 2. Likewise

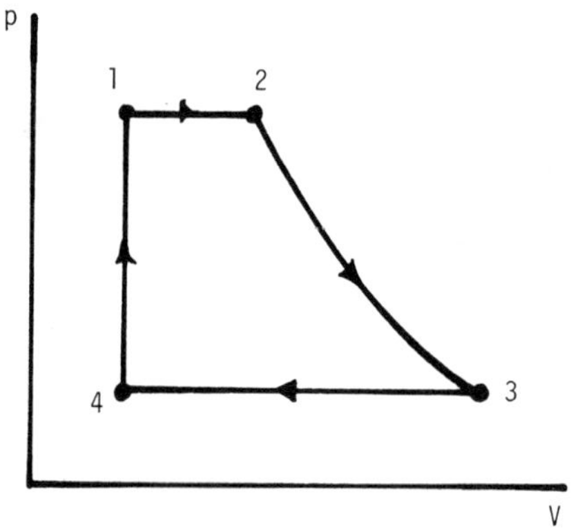

Figure 3.2

$$Q_{23} = \Delta E_{32} + W_{23}$$

$$Q_{34} = \Delta E_{43} + W_{34}$$

and

$$Q_{41} = \Delta E_{14} + W_{41}$$

Adding these four equations yields

$$Q_{12} + Q_{23} + Q_{34} + Q_{41} = \Delta E_{21} + \Delta E_{32} + \Delta E_{43} + \Delta E_{14} + W_{12} + W_{23} + W_{34} + W_{41}$$

For a cycle each energy term must return to its original value so that

$$\Delta E_{21} + \Delta E_{32} + \Delta E_{43} + \Delta E_{14} = 0$$

and

$$Q_{12} + Q_{23} + Q_{34} + Q_{41} = W_{12} + W_{23} + W_{34} + W_{41}$$

Let

$$\Sigma Q = Q_{12} + Q_{23} + Q_{34} + Q_{41}$$

which is the heat transfer for the cycle, and

$$\Sigma W = W_{12} + W_{23} + W_{34} + W_{41}$$

which is the work for a cycle; therefore

$$\Sigma Q = \Sigma W \tag{3.1}$$

The above relationship is valid for either the open or closed thermodynamic system and represents a mathematical statement of the First Law of Thermodynamics.

FIRST LAW OF THERMODYNAMICS

The sum of the net heat transferred in a cycle is equivalent to the sum of the net work transferred in a cycle. The First Law of Thermodynamics does not depend on the processes being reversible.

POINT FUNCTIONS

In Fig. 3.3 a cycle, plotted with respect to any two properties, is proposed that proceeds from state 1 to state 2 along path A. The cycle is completed by the process from 2 to 1 along path B.

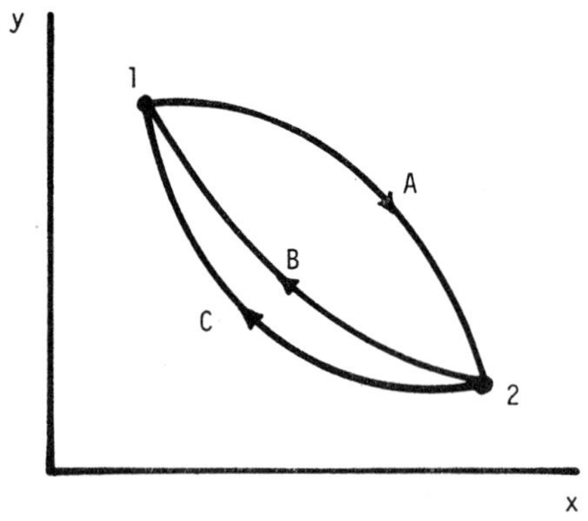

Figure 3.3

From Eq. (2.16), for an open or closed system, from state 1 to state 2, along path A

$$Q_{1A2} = \Delta E_{2A1} + W_{1A2} \tag{3.2}$$

where the letter A is used to define the path. ΔE_{2A1} represents the change of any single energy term or any combination of energy terms. Likewise, for the process from 2 to 1 along path B

$$Q_{2B1} = \Delta E_{1B2} + W_{2B1} \tag{3.3}$$

Adding Eqs. (3.2) and (3.3) yields

$$Q_{1A2} + Q_{2B1} = \Delta E_{2A1} + \Delta E_{1B2} + W_{1A2} + W_{2B1} \tag{3.4}$$

and for the cycle

$$\Sigma Q_{AB} = Q_{1A2} + Q_{2B1}$$

and

$$\Sigma W_{AB} = W_{1A2} + W_{2B1}$$

Substituting these equations into Eq. (3.4) gives

$$\Sigma Q_{AB} = \Delta E_{2A1} + \Delta E_{1B2} + \Sigma W_{AB} \tag{3.5}$$

For the cycle, by use of Eq. (3.1), where

$$\Sigma Q_{AB} = \Sigma W_{AB}$$

then

$$\Delta E_{2A1} + \Delta E_{1B2} = 0$$

so that

$$\Delta E_{2A1} = -\Delta E_{1B2} \tag{3.6}$$

Now consider a second cycle with the same process from state 1 to state 2 by path A but where the process from state 2 to state 1 proceeds along path C. Therefore

$$Q_{1A2} = \Delta E_{2A1} + W_{1A2}$$

and

$$Q_{2C1} = \Delta E_{1C2} + W_{2C1}$$

As before

$$\Sigma Q_{AC} = Q_{1A2} + Q_{2C1}$$

and

$$\Sigma W_{AC} = W_{1A2} + W_{2C1}$$

and by adding the process equations

$$\Sigma Q_{AC} = \Delta E_{2A1} + \Delta E_{1C2} + \Sigma W_{AC}$$

Again for the cycle

$$\Sigma Q_{AC} = \Sigma W_{AC}$$

and

$$\Delta E_{2A1} = -\Delta E_{1C2} \tag{3.7}$$

Combining Eqs. (3.6) and (3.7) gives

$$\Delta E_{2A1} = -\Delta E_{1B2} = -\Delta E_{1C2}$$

or

$$\Delta E_{1B2} = \Delta E_{1C2}$$

The change in energy from state 2 to state 1 is the same for both processes B and C. Therefore, the change in the value of an energy term is independent of the path between the two states. A function that is dependent on the initial and final points and that is independent of the path connecting these points is a point function.

All energy terms are point functions, and all point functions are properties. Therefore, energy terms are properties.

NONFLOW WORK

Consider the closed system in Fig. 3.4. Work is the product of the force and the distance in the direction of the force. By assuming that the pressure within the cylinder is constant,

$$\Delta W_{nf} = F\ \Delta L$$

and since

$$F = pA$$

then

$$\Delta W_{nf} = pA\ \Delta L$$

but the change in volume due to the movement of the piston is

$$\Delta V = A\ \Delta L$$

so that

$$\Delta W_{nf} = p\ \Delta V$$

Any initial and final state, as plotted on the pV plane, may be connected by use of a constant pressure and constant volume process with an intermediate state A (Fig. 3.5). For process 1A where the pressure is constant

$$W_{1A\text{-}nf} = p_1(V_A - V_1)$$

For process A2 the volume is constant and the piston does not move, so that $\Delta V = 0$ and

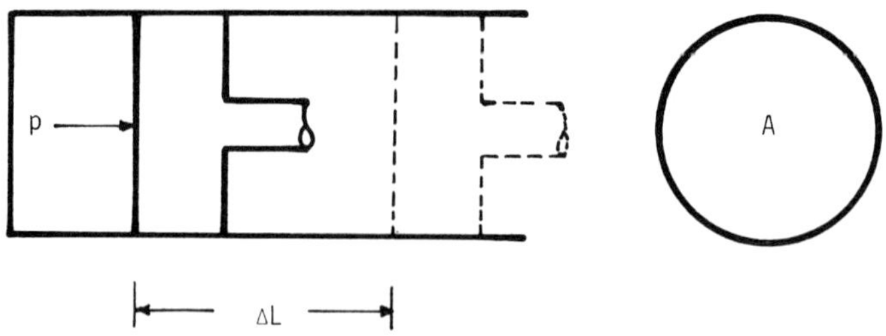

Figure 3.4

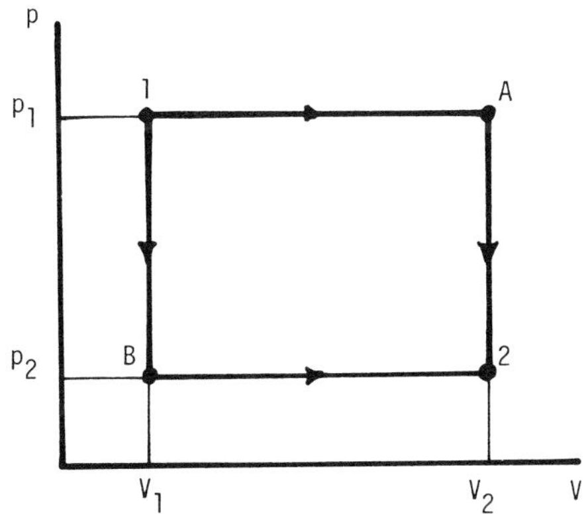

Figure 3.5

$$W_{A2\text{-}nf} = 0$$

The work of the two processes from state 1 to state 2 is

$$(W_{1A} + W_{A2})_{nf} = p_1(V_A - V_1)$$

but

$$V_A = V_2$$

so that

$$W_{1A2\text{-}nf} = (W_{1A} + W_{A2})_{nf} = p_1(V_2 - V_1) \tag{3.8}$$

Now assume two processes from state 1 to state 2 using the intermediate state B, so that

$$W_{1B\text{-}nf} = 0$$

and

$$W_{B2\text{-}nf} = p_2(V_2 - V_B)$$

or

$$W_{1B2\text{-}nf} = (W_{1B} + W_{B2})_{nf} = p_2(V_2 - V_B)$$

From Fig. 3.5,

$$V_B = V_1$$

or

$$W_{1B2\text{-}nf} = p_2(V_2 - V_1) \tag{3.9}$$

If the nonflow work is a point function, it would be independent of the path between the initial and final states, and

$$W_{1A2\text{-}nf} = W_{1B2\text{-}nf}$$

or

$$p_1(V_2 - V_1) = p_2(V_2 - V_1)$$

so that

$$p_1 = p_2$$

But from Fig. 3.5, it is apparent that p_1 is not equal to p_2; therefore the nonflow work is not a point function. The nonflow work is a path function and is dependent on the process or processes used to connect the initial and final state points.

Example 3.1. Given Fig. 3.5, where p_1 = 200 psia, p_2 = 20 psia, V_1 = 1 ft³, and V_2 = 5 ft³, determine $W_{1A2\text{-}nf}$ and $W_{1B2\text{-}nf}$.

$$W_{1A2\text{-}nf} = p_1(V_2 - V_1) = 200(144)(5 - 1) = 115{,}200 \text{ (ft-lbf)}$$

$$= 148.1 \text{ (B)} = 1.561\text{E} + 05 \text{ (J)}$$

Also

$$W_{1B2\text{-}nf} = p_2(V_2 - V_1) = 20(144)(5 - 1) = 11{,}520 \text{ (ft-lbf)}$$

$$= 14.8 \text{ (B)} = 1.561\text{E} + 04 \text{ (J)}$$

To determine a general relationship for nonflow work as a path function between an initial state 1 and a final state 2, consider the process in Fig. 3.6. The work for a small increment of the process is given by

$$\Delta W_{nf} = p\,\Delta V$$

where p is the constant average pressure for the volume increment ΔV. The total work is the sum of all the volume increments between the initial and final states, so that

$$W_{nf} = \Sigma \Delta W_{nf} = \Sigma p\,\Delta V$$

This also represents the area below the process curve on the pV plane. As ΔV is decreased, the value of the work approaches

$$W_{nf} = \int_{V_1}^{V_2} p\,dV \tag{3.10}$$

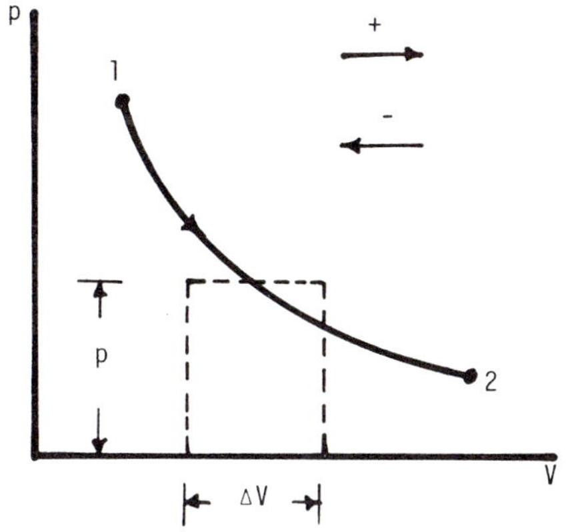

Figure 3.6

When the process takes place in the direction of increasing system volume, the work is positive and represents work done by the system. Negative work, when the final volume is smaller than the initial volume, represents work done on the system. Since the pressure is variable for the process in Fig. 3.6, the relationship between the pressure and volume must be known before Eq. (3.10) can be solved. The most useful thermodynamic relationship is

$$pV^n = C \tag{3.11}$$

where n is a constant for the process and may be any real, positive, or negative number. C is a positive constant.

Solving Eq. (3.11) for the pressure gives

$$p = C/V^n$$

and by substitution in Eq. (3.10)

$$W_{nf} = \int_{V_1}^{V_2} (C/V_n)\, dV$$

From calculus

$$W_{nf} = C\left(\frac{V^{1-n}}{1-n}\right)_{V_1}^{V_2} = C\left(\frac{V_2^{\,1-n} - V_1^{\,1-n}}{1-n}\right)$$

$$= (CV_2^{\,1-n} - CV_1^{\,1-n})/(1-n)$$

By use of Eq. (3.11)

$$p_1V_1^{\,n} = p_2V_2^{\,n} = C$$

or

$$W_{nf} = (p_2V_2^{\,n}V_2^{\,1-n} - p_1V_1^{\,n}V_1^{\,1-n})/(1 - n)$$

$$= (p_2V_2 - p_1V_1)/(1 - n), \qquad n \neq 1 \qquad (3.12)$$

If $n = 1$, then $p_1V_1 = p_2V_2$, and Eq. (3.12) reduces to

$$W_{nf} = 0/0$$

which is intermediate. Equation (3.12) may not be used when $n = 1$.

For $n = 1$, $pV = C$, or $p = V/C$, by use of Eq. (3.10),

$$W_{nf} = \int_{V_1}^{V_2} (V/C)\, dV$$

or

$$W_{nf} = C(\ln V)\Big|_{V_1}^{V_2} = C(\ln V_2 - \ln V_1)$$

$$= C \ln(V_2/V_1)$$

Again for $n = 1$

$$p_1V_1 = p_2V_2 = C$$

and

$$W_{nf} = p_1V_1 \ln(V_2/V_1) = p_2V_2 \ln(V_2/V_1)$$

$$= p_1V_1 \ln(p_1/p_2) = p_2V_2 \ln(p_1/p_2), \qquad n = 1 \qquad (3.13)$$

ln is the shortened notation for $\log_e$, which is the natural logarithm or logarithm to the base e, where $e = 2.71828$.

For students who have not studied calculus, it is sufficient to use the two nonflow work equations where Eq. (3.12) requires that $n \neq 1$ and Eq. (3.13) may be used only when $n = 1$.

Example 3.2. Given $n = 0.5$, $p_1 = 50$ psia, $V_1 = 8$, and $V_2 = 2$ ft^3, determine p_2 and W_{nf}. From Eq. (3.11)

$$pV^n = C$$

or

$$p_1V_1^n = p_2V_2^n$$

and

$$p_2 = p_1(V_1/V_2)^n = 144(50)(8/2)^{0.5}$$

$$= 14{,}400 \text{ (psfa)}$$

$$= 100 \text{ (psia)}$$

$$= 689.5 \text{ (kPa-a)}$$

From Eq. (3.12), since $n \neq 1$,

$$W_{nf} = (p_2V_2 - p_1V_1)/(1 - n)$$

$$= 144\{[100(2) - 50(8)]/(1 - 0.5)\}$$

$$= -57{,}600 \text{ (ft-lbf)}$$

$$= -74.0 \text{ (B)} = -78.0 \text{ (kJ)}$$

Since the nonflow work is negative, the work is done on the closed system and the volume must decrease for the given process.

Example 3.3. Given $p_1 = 40$ and $p_2 = 20$ psia and $V_1 = 8$ and $V_2 = 10$ ft^3, determine n and W_{nf}, using

$$p_1/p_2 = (V_2/V_1)^n$$

or

$$40/20 = (10/8)^n$$

and

$$2 = 1.25^n$$

Taking the natural logarithm of both sides of the equation gives

$$\ln(2) = n \ln(1.25)$$

or

$$n = \ln(2)/\ln(1.25) = 0.6931/0.2231$$

$$= 3.106$$

Again by use of Eq. (3.12), since $n \neq 1$,

$$W_{nf} = (p_2V_2 - p_1V_1)/(1 - n)$$

$$= 144\{[20(10) - 40(8)]/(1 - 3.106)\}$$

$$= 8200 \text{ (ft-lbf)}$$

$$= 10.5 \text{ (B)} = 11.1 \text{ (kJ)}$$

Since the final volume is greater than the initial volume, the nonflow work is positive, and the work is done by the system.

<u>Example 3.4</u>. Given $p_1 = 20$ and $p_2 = 40$ psia and $V_1 = 10$ and $V_2 = 5$ ft^3, determine n and W_{nf}. From Eq. (3.11)

$$p_2/p_1 = (V_1/V_2)^n$$

or

$$40/20 = (10/5)^n$$

and

$$2 = 2^n$$

so that

$$n = 1$$

By use of Eq. 3.13, since n = 1,

$$W_{nf} = p_1V_1 \ln(V_2/V_1)$$

$$= 144(20)(10) \ln(5/10)$$

$$= 28,800 \ln(0.5)$$

$$= -28,800 \ln(2)$$

$$= (-28,800)(0.6931)$$

$$= -19,960 \text{ (ft-lbf)}$$

$$= -25.7 \text{ (B)} = -27.1 \text{ (kJ)}$$

Note that $V_1 > V_2$, so that $W_{nf} < 0$, and the work is done on the closed system.

NONFLOW WORK ON THE pV PLANE

For a constant pressure process (Fig. 3.7) the work is the area below the process path from the initial state 1 to the final state 2, or

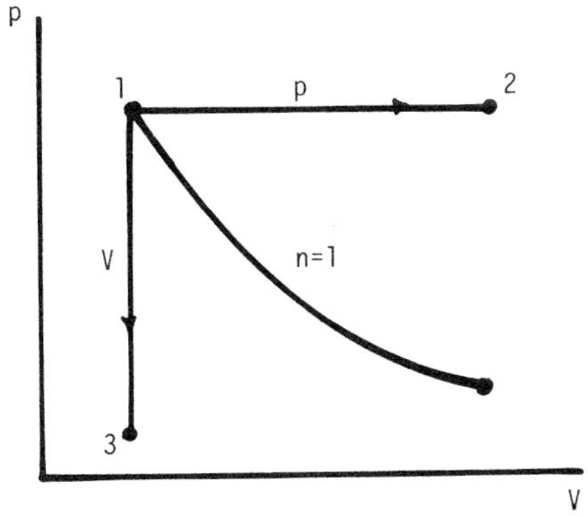

Figure 3.7

$$W_{nf} = p(V_2 - V_1)$$

which can be compared to Eq. (3.12):

$$W_{nf} = (p_2V_2 - p_1V_1)/(1 - n) = [p(V_2 - V_1)]/(1 - n)$$

By equating the two nonflow work equations

$$p(V_2 - V_1) = [p(V_2 - V_1)]/(1 - n)$$

or

$$1 = 1/(1 - n)$$

and

$$1 - n = 1$$

so that

$$n = 0$$

For a constant volume process (Fig. 3.7)

$$W_{nf} = 0$$

and

$$W_{nf} = (p_3V_3 - p_1V_1)/(1 - n) = [V(p_3 - p_1)]/(1 - n)$$

but $V \neq 0$ and $p_3 - p_1 \neq 0$; therefore $1 - n$ must be an extremely large number, or mathematically $n \to \infty$ (infinity). For $n = 1$, $pV = C$, which is the equation of a hyperbola. These three processes, starting at a common initial state 1, are sketched in Fig. 3.7.

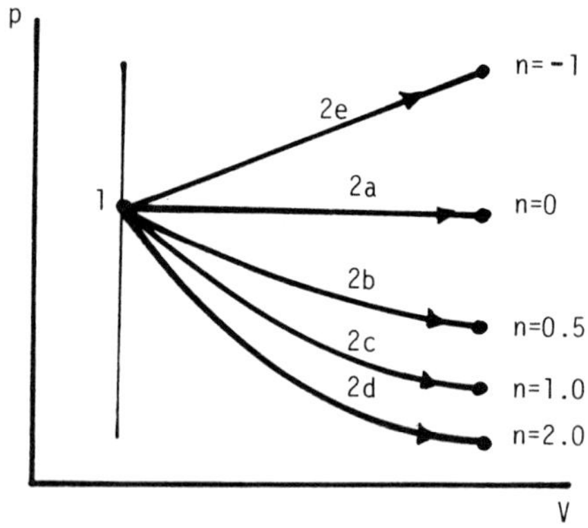

Figure 3.8

Example 3.5. Given p_1 = 10 psia and V_1 = 10 and V_2 = 40 ft^3, determine the final pressure when (a) n = 0, (b) n = 0.5, (c) n = 1.0, (d) n = 2.0, and (e) n = -1.0. From Eq. (3.11)

$$pV^n = C$$

or

$$p_2 = p_1(V_1/V_2)^n = 10(10/40)^n$$

(a) $p_2 = 10(10/40)^0 = 10$ (psia)
(b) $p_2 = 10(10/40)^{0.5} = 5$ (psia)
(c) $p_2 = 10(10/40)^1 = 2.5$ (psia)
(d) $p_2 = 10(10/40)^2 = 0.625$ (psia)
(e) $p_2 = 10(10/40)^{-1} = 40$ (psia)

These processes are illustrated in Fig. 3.8. The paths may be extended for processes where V_2 is less than V_1.

NONFLOW CYCLIC WORK

Given the cycle in Fig. 3.9, the process work is the area below the curve on the pV plane so that

$$W_{12\text{-nf}} = p_1(V_2 - V_1) = \text{area}(12ab1)$$

$$W_{23\text{-nf}} = 0$$

$$W_{34\text{-nf}} = p_3(V_4 - V_3) = \text{area}(34ba3)$$

$$W_{14\text{-nf}} = 0$$

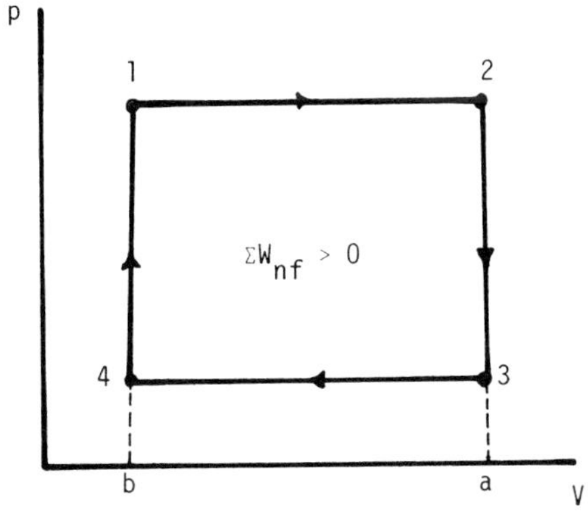

Figure 3.9

Adding the work terms yields the work of the cycle or

$$\Sigma W_{nf} = (W_{12} + W_{23} + W_{34} + W_{41})_{nf}$$

$$= p_1(V_2 - V_1) + p_3(V_4 - V_3)$$

Since $V_4 = V_1$ and $V_3 = V_2$,

$$\Sigma W_{nf} = p_1(V_2 - V_1) + p_3(V_1 - V_2) = (p_1 - p_3)(V_2 - V_1) = \text{area}(12341)$$

Since $p_1 > p_3$ and $V_2 > V_1$, $\Sigma W_{nf} > 0$.

The cyclic nonflow work is positive, and the net work is done by the system. The area within the figure is equivalent to the work of the cycle. Cycles that produce work are referred to as engines, and on the pV plane the rotation is clockwise.

If the cycle is reversed, as in Fig. 3.10, then

$$W_{12\text{-}nf} = p_1(V_2 - V_1)$$

$$W_{23\text{-}nf} = 0$$

$$W_{34\text{-}nf} = p_3(V_4 - V_3) = p_3(V_1 - V_2)$$

$$W_{41\text{-}nf} = 0$$

Adding the work terms gives

$$\Sigma W_{nf} = p_1(V_2 - V_1) + 0 + p_3(V_1 - V_2) + 0$$

$$= (p_1 - p_3)(V_2 - V_1)$$

Since $p_1 < p_3$ and $V_2 > V_1$, $\Sigma W_{nf} < 0$.

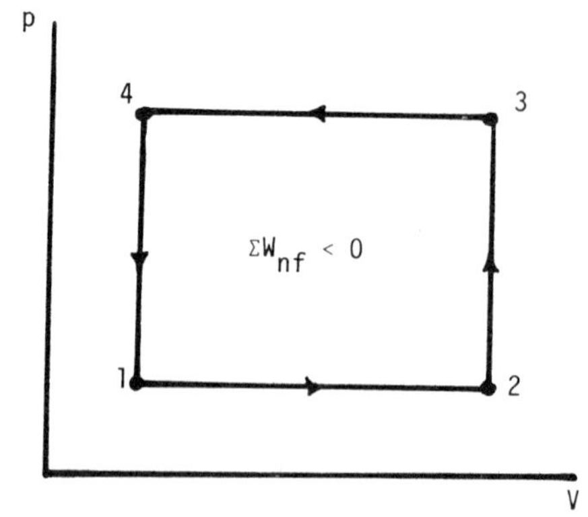

Figure 3.10

Since the cyclic nonflow work is negative, the net work is done on the system. The area within the figure is again equivalent to the cyclic work. Cycles that absorb work are referred to as pumps, and on the pV plane the rotation is counterclockwise.

Example 3.6. Given is the cycle in Fig. 3.11, where $p_1 = p_2 = 100$ psia, $V_1 = V_4 = 1\ \text{ft}^3$, $p_3 = p_4 = 10$ psia, $V_2 = 5\ \text{ft}^3$, and n for process 23 = 1. Determine the cyclic work.

For the constant pressure process from state 1 to state 2

$$W_{12\text{-nf}} = p_1(V_2 - V_1) = 144(100)(5 - 1) = 57{,}600\ (\text{ft-lbf})$$

$$= 74.0\ (\text{B})$$

For the process from state 2 to state 3 where n = 1

$$p_2V_2 = p_3V_3$$

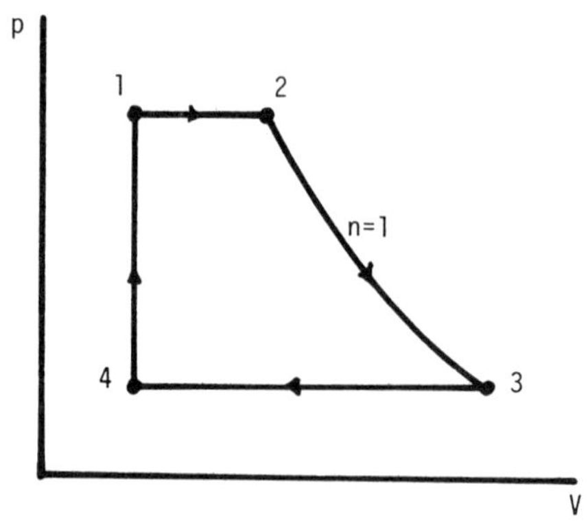

Figure 3.11

or

$$V_3 = V_2(p_2/p_3) = 5(100/10) = 50 \text{ (ft}^3\text{)}$$

and

$$W_{23\text{-nf}} = p_2V_2 \ln(V_3/V_2) = 144(100)(5) \ln(50/5)$$

$$= 72{,}000 \ln(10) = 72{,}000(2.303) = 166{,}000 \text{ (ft-lbf)}$$

$$= 213.1 \text{ (B)}$$

For the constant pressure process from state 3 to state 4

$$W_{34\text{-nf}} = p_3(V_4 - V_3) = 144(10)(1 - 50) = -70{,}600 \text{ (ft-lbf)}$$

$$= -90.7 \text{ (B)}$$

For the constant volume process from state 4 to state 1

$$W_{41\text{-nf}} = 0$$

The cyclic work, which must be positive since the rotation on the pV plane is clockwise, is

$$\Sigma W_{nf} = (W_{12} + W_{23} + W_{34} + W_{41})_{nf} = 74.0 + 213.1 - 90.7 - 0.0$$

$$= 196.4 \text{ (B)}$$

STEADY-FLOW PROCESS WORK

From the general energy equation for a steady-flow process [Eq. (2.14)],

$$W_{sf} = Q - \Delta U - \Delta FE - \Delta KE - \Delta PE$$

and for a nonflow process [Eq. (2.17)],

$$W_{nf} = Q - \Delta U$$

If the two processes have the same initial and final states and follow the same path between these states,

$$W_{sf} = W_{nf} - \Delta FE - \Delta KE - \Delta PE$$

When the mechanical properties ΔKE and ΔPE are negligible

$$W_{sf} = W_{nf} - \Delta FE$$

From Eq. (2.8), the flow energy is

$$\Delta FE = m(p_2v_2 - p_1v_1) = p_2V_2 - p_1V_1$$

so that

$$W_{sf} - W_{nf} = -(p_2V_2 - p_1V_1) \tag{3.14}$$

For $n \neq 1$, from Eq. (3.12),

$$W_{sf} = W_{nf} - (p_2V_2 - p_1V_1)$$

$$= [(p_2V_2 - p_1V_1)/(1 - n)] - (p_2V_2 - p_1V_1)$$

$$= [n(p_2V_2 - p_1V_1)]/(1 - n)$$

or

$$W_{sf} = nW_{nf} \tag{3.15}$$

For $n = 1$,

$$p_2V_2 = p_1V_1$$

and from Eq. (3.14)

$$W_{sf} = W_{nf}$$

so that Eq. (3.15) may be applied for all real values of n when ΔKE, $\Delta PE = 0$.

For a constant volume process, $n \rightarrow \infty$,

$$W_{nf} = 0$$

and Eq. (3.14) becomes

$$W_{sf} = -V(p_2 - p_1)$$

For a constant pressure process

$$W_{nf} = p(V_2 - V_1)$$

and Eq. (3.14) becomes

$$W_{sf} - p(V_2 - V_1) = -p(V_2 - V_1)$$

or

$$W_{sf} = 0$$

For the constant volume process the steady-flow work is the negative of the area in back of the curve on the pV plane when ΔKE, $\Delta PE = 0$ (Fig. 3.12).

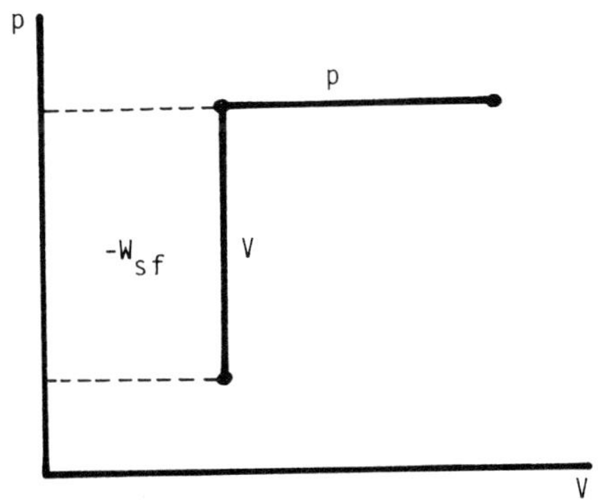

Figure 3.12

For the constant pressure process the area in back of the curve is zero when ΔKE, $\Delta PE = 0$. A proof, similar to that used for the nonflow work, could be employed to prove that steady-flow work is a path function and not a point function. Steady-flow work is not a thermodynamic property.

STEADY-FLOW CYCLIC WORK

Since the steady-flow and nonflow process work are related by

$$W_{sf} = W_{nf} - \Delta FE - \Delta KE - \Delta PE$$

for a cycle

$$\Sigma W_{sf} = \Sigma W_{nf} - \Sigma(\Delta FE + \Delta KE + \Delta PE)$$

and the sum of the property terms for a cycle must be

$$\Sigma(\Delta FE + \Delta KE + \Delta PE) = 0$$

Therefore, for any cycle

$$\Sigma W_{sf} = \Sigma W_{nf}$$

and the area enclosed within the figure on the pV plane is both the steady-flow and the nonflow work.

PROBLEMS

3.1 Determine the final volume and the nonflow work for a process where $n = 1.2$ and $p_1 = 50$ and $p_2 = 25$ psia. The initial volume is 10 ft^3.

3.2 Determine the value of n and the nonflow work for a process where $p_1 = 50$ and $p_2 = 100$ psia. The volume decreases from 25 to 5 ft^3.

3.3 Determine the final pressure and the nonflow work for a process where $n = 1$, $p_1 = 20$ psia, and $V_1 = 10$ and $V_2 = 5$ ft^3.

3.4 Sketch the processes in Probs. 3.1, 3.2, and 3.3 on the pV plane.

3.5 Determine the steady-flow work for Probs. 3.1, 3.2, and 3.3 when $\Delta KE, \Delta PE = 0$.

3.6 Given is a three-process cycle where $V_1 = 10$ and $V_2 = 5$ ft^3. The value of n for the process from state 1 to state 2 is 1, and $p_1 = 20$ psia. The process from state 2 to state 3 is a constant pressure process, and the cycle is closed by a constant volume process. Determine the nonflow and steady-flow work for each process (ΔKE, $\Delta PE = 0$). Determine the nonflow and steady-flow cyclic work. Sketch the cycle on the pV plane.

3.7 Given a four-process cycle where

Process	Data
1-2	$V = C$, $p_1 = 20$ psia, $p_2 = 40$ psia, $V_1 = 5$ ft^3
2-3	$p = C$, $V_3 = 1$ ft^3, $\Delta KE = -10$ B
3-4	$n = 1.5$
4-1	$p = C$, $\Delta KE = 10$ B

determine the nonflow and steady-flow work (ΔKE, $\Delta PE = 0$ except where otherwise noted for each process). Determine the nonflow and steady-flow cyclic work. Sketch the cycle on the pV plane.

3.8 A cycle is represented by a circle, with counterclockwise rotation, on the pV plane. The scale of the pressure and volume axis is as follows: For pressure, 1 in. = 100 psia, and for volume, 1 in. = 10 ft^3. If the radius of the circle is 1 in., determine the nonflow and steady-flow work for the cycle.

4

HEAT

In Chap. 3 the concept of nonflow and steady-flow work as path functions was studied. In this chapter we shall establish that the heat transfer into or out of a thermodynamic system is also a path function. The specific heat as a physical property of the working substance will be defined. Entropy, a most important thermodynamic property, will also be defined. Finally the enthalpy, a thermodynamic property, will be expressed in terms of other thermodynamic properties.

HEAT

The heat, which may be transferred into or out of an open or closed thermodynamic system, is one component of either the steady-flow or nonflow process equation. For the open or steady-flow system, the heat term appeared in the general energy equation [Eq. (2.16)],

$$Q = \Delta E + W_{sf}$$

where $\Delta E = \Delta PE + \Delta KE + \Delta FE + \Delta U$.

For the closed or nonflow system, from Eq. (2.17), the heat term is

$$Q = \Delta U + W_{nf}$$

To show that heat is a path function a proof similar to that used to prove that the nonflow work is a path function may be employed. Given the initial and final states 1 and 2 in Fig. 4.1 and by assuming that a composite path from state 1 to state 2, by use of the intermediate state A, is utilized, then for a closed or nonflow system

$$Q_{1A} = \Delta U_{A1} + W_{1A\text{-}nf} \tag{4.1}$$

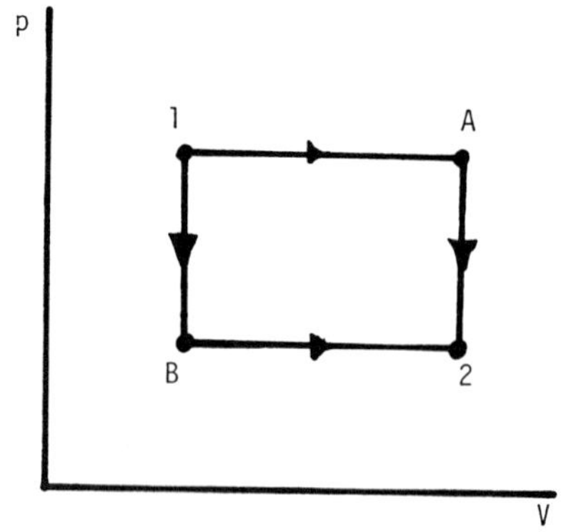

Figure 4.1

For the constant pressure process from state 1 to state A, the nonflow work is

$$W_{1A\text{-}nf} = p_1(V_A - V_1) = p_1(V_2 - V_1)$$

where $V_A = V_2$.

For the constant volume process from state A to state 2

$$Q_{A2} = \Delta U_{2A} + W_{A2\text{-}nf} = \Delta U_{2A} \tag{4.2}$$

since $W_{A2\text{-}nf} = 0$ for the constant volume process.

As a result of adding Eqs. (4.1) and (4.2)

$$Q_{1A2} = \Delta U_{A1} + \Delta U_{2A} + p_1(V_2 - V_1)$$

The internal energy, as a property, is independent of the path connecting the initial and final state points or

$$\Delta U_{21} = \Delta U_{A1} + \Delta U_{2A}$$

and the heat transfer from state 1 to state 2, by use of the intermediate state A, is

$$Q_{1A2} = \Delta U_{21} + p_1(V_2 - V_1) \tag{4.3}$$

Now assume that the initial and final state points are connected by a constant volume process from state 1 to state B and a constant pressure process from state B to state 2, so that

$$Q_{1B} = \Delta U_{B1} + W_{1B\text{-}nf} = \Delta U_{B1}$$

and

$$Q_{B2} = \Delta U_{2B} + W_{B2\text{-}nf} = \Delta U_{2B} + p_2(V_2 - V_B)$$

$$= \Delta U_{2B} + p_2(V_2 - V_1)$$

since $V_B = V_1$.

Adding the previous two heat equations results in

$$Q_{1B2} = \Delta U_{B1} + \Delta U_{2B} + p_2(V_2 - V_1) \tag{4.4}$$

and since the internal energy is a property

$$\Delta U_{21} = \Delta U_{B1} + \Delta U_{2B}$$

then

$$Q_{1B2} = \Delta U_{21} + p_2(V_2 - V_1) \tag{4.5}$$

By subtracting Eq. (4.5) from Eq. (4.3)

$$Q_{1A2} - Q_{1B2} = \Delta U_{21} + p_1(V_2 - V_1) - \Delta U_{21} - p_2(V_2 - V_1)$$

$$= (p_1 - p_2)(V_2 - V_1)$$

From Fig. 4.1 it is apparent that $p_1 - p_2 \neq 0$ and $V_2 - V_1 \neq 0$ so that $Q_{1A2} \neq Q_{1B2}$. This proves that the quantity of heat transferred is dependent on the intermediate states A or B and is therefore dependent on the path selected to connect the initial and final state points.

Heat is a path function. In thermodynamics, the only path functions are heat, nonflow work, and steady-flow work.

SPECIFIC HEAT

To determine the quantity of heat added to an open or closed thermodynamic system in which there is no change in phase, note that

$$Q = mc(T_2 - T_1) \tag{4.6}$$

where Q is the heat transfer (B), m is the mass of the working substance in the system (lbm), T_1 and T_2 are the initial and final state temperatures (R), and c is the specific heat.

The units of the specific heat are

$$c = Q/[m(T_2 - T_1)] \text{ (B/lbm-R)} \tag{4.7}$$

If a steady-flow system is under study, the units of the mass m will be lbm/sec so that the units of the heat transfer Q are B/sec.

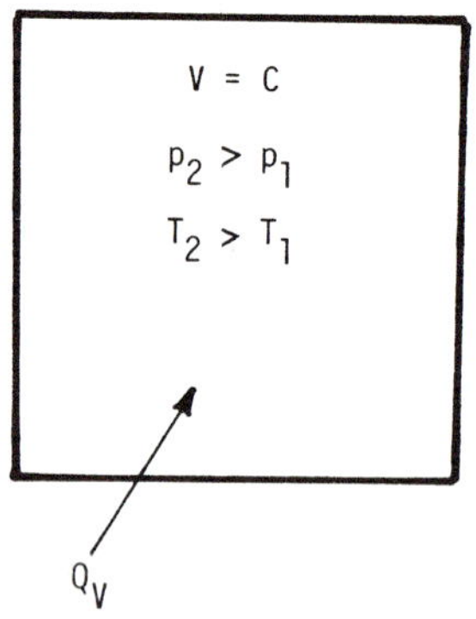

Figure 4.2

Since Eq. (4.6) may be applied to either the open or closed system, it is not necessary to consider nonflow heat as a distinct quantity which differs from steady-flow heat. The equations and concepts that will be established for nonflow heat may be applied to steady-flow heat.

For solids and liquids, Eq. 4.6 or 4.7 may be used because the specific heats for these working substances are essentially independent of the manner in which the heat is transferred.

For gases, a different situation exists, and the value of the specific heat is dependent on the process in which the heat is added or removed from the system.

Consider the closed or nonflow constant volume system shown in Fig. 4.2. The working substance is a gas, and Eq. (4.6) must be modified, so that

$$Q_v = mc_v(T_2 - T_1) \tag{4.8}$$

where Q_v is the heat transferred at constant volume and c_v is the specific heat at constant volume. The definition of the terms m, T_1, and T_2 are unchanged, and the units of c_v are B/lbm-R.

For a constant pressure system (Fig. 4.3) the heat transfer is given by

$$Q_p = mc_p(T_2 - T_1) \tag{4.9}$$

where Q_p is the heat transferred at constant pressure and c_p is the specific heat at constant pressure.

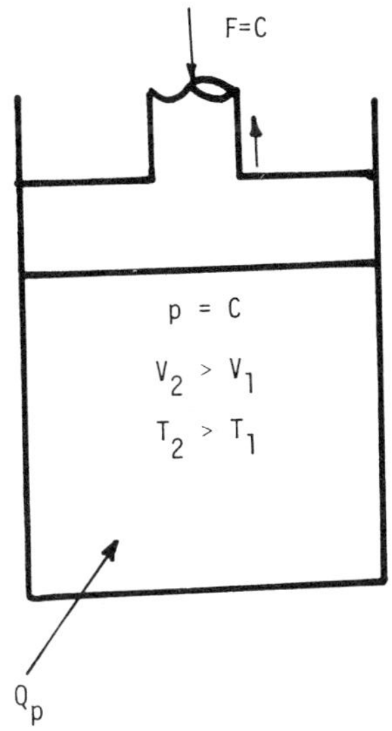

Figure 4.3

JOULE'S LAW

To establish that a difference must exist between the numerical values of the specific heat at constant pressure and the specific heat at constant volume, consider the system, as proposed by Joule, shown in Fig. 4.4. The initial state of the system is defined by the properties of the gaseous working substance in container 1. These properties include the pressure, volume, and temperature. The valve is opened, and the gas occupies both containers within the system. Note that the working substance does not

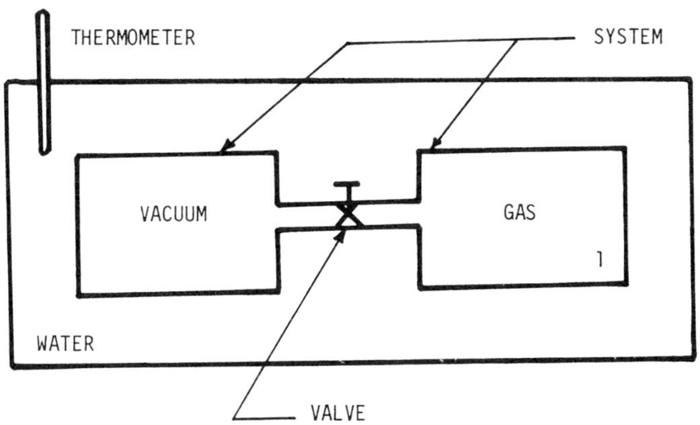

Figure 4.4

cross the system boundaries. After the gas occupies both containers and is no longer in motion, the state is defined by the final pressure, temperature, and volume in the system. For this nonflow system

$$Q = \Delta U + W_{nf}$$

Since there was no capability for nonflow work in the system,

$$W_{nf} = 0$$

Temperature measurements of the surrounding water established that the temperature remained constant; therefore there was no heat transfer with respect to the water, or $Q = 0$, which also implies that the initial and final temperatures of the gas were unchanged so that

$$T_1 = T_2$$

From the nonflow process equation,

$$\Delta U = 0$$

since Q and W_{nf} are zero.

But the pressure and volume of the initial and final states were different; therefore, since

$$\Delta U = 0$$

and $p_1 \neq p_2$ and $V_1 \neq V_2$, Joule's law may be stated.

Joule deduced that the change in internal energy of a gas is independent of the changes in the pressure and volume and is only a function of the temperature change. Stated in equation form,

$$\Delta U = U_2 - U_1 = mc_v(T_2 - T_1) \tag{4.10}$$

for all processes using a perfect gas as the working substance.

Joule's experiment was confined to gaseous materials and does not apply to a process in which there is a complete or partial change of phase of the working substance.

COMPARISON OF c_p AND c_v

By using Joule's experiment, the nonflow process equation for a gas may be rewritten, so that

$$Q = \Delta U + W_{nf} = mc_v(T_2 - T_1) + W_{nf} \tag{4.11}$$

Assume that the systems in Fig. 4.2 and 4.3 contain the same mass of identical working substance and that sufficient heat is added to cause the same temperature rise in both systems.

For the constant volume system (Fig. 4.2)

$$W_{nf} = 0$$

and from Eq. (4.11)

$$Q_v = mc_v(T_2 - T_1) \tag{4.12}$$

Equation (4.12) for the nonflow process is identical to Eq. (4.8), which was used to define c_v for an open or closed system.

For the constant pressure system (Fig. 4.3)

$$W_{nf} = p(V_2 - V_1) \neq 0$$

since the addition of heat causes the piston to rise where $V_2 > V_1$.

Therefore, from Eq. (4.11)

$$Q_p = mc_v(T_2 - T_1) + p(V_2 - V_1) \tag{4.13}$$

Subtracting Eq. (4.12) from Eq. (4.13) gives

$$Q_p - Q_v = p(V_2 - V_1)$$

and since

$$V_2 - V_1 > 0$$

then

$$Q_p > Q_v$$

By use of Eqs. (4.8) and (4.9)

$$mc_p(T_2 - T_1) > mc_v(T_2 - T_1)$$

and from the restrictions placed on the two systems

$$c_p > c_v$$

For a gas the numerical value of the specific heat at constant pressure must exceed the numerical value of the specific heat at constant volume. Since the addition of heat to the system, which is positive, causes a temperature rise in the constant pressure or constant volume process, the values of c_p and c_v are positive physical properties.

The ratio of the specific heat at constant pressure to the specific heat at constant volume is defined as a physical property of the working substance so that

$$k = c_p/c_v \tag{4.14}$$

where $k > 1$.

Solids and liquids do possess different values for c_p and c_v, but their differences are so small that for engineering calculations they may be considered to be equal.

Values of c_p and c_v for selected gases are given in Appendix B.

Example 4.1. Given 5 lbm of air that receives a quantity of heat sufficient to raise the temperature from 60 to 240 F, determine Q_p, Q_v, ΔU, and W_{nf}.

From Appendix B the values of the specific heat for air are $c_p = 0.24$ and $c_v = 0.1714$ B/lbm-R.

From Eq. (4.10), for all processes,

$$\Delta U = U_2 - U_1 = mc_v(T_2 - T_1) = 5(0.1714)(700 - 520)$$

$$= 154.3 \text{ (B)} \quad \text{or} \quad 162.6 \text{ (kJ)}$$

For the constant pressure process, from Eq. (4.9)

$$Q_p = mc_p(T_2 - T_1) = 5(0.24)(700 - 520)$$

$$= 216.0 \text{ (B)} \quad \text{or} \quad 227.7 \text{ (kJ)}$$

The nonflow work from Eq. (2.17) is

$$W_{nf} = Q_p - \Delta U = 216.0 - 154.3$$

$$= 61.7 \text{ (B)} \quad \text{or} \quad 65.0 \text{ (kJ)}$$

For the constant volume process, from Eq. (4.8) or (4.12)

$$Q_v = mc_v(T_2 - T_1) = 5(0.1714)(700 - 520)$$

$$= 154.3 \text{ (B)} \quad \text{or} \quad 162.6 \text{ (kJ)}$$

As before, for the constant volume process

$$W_{nf} = Q_v - \Delta U = 154.3 - 154.3 = 0$$

ENTROPY

Nonflow work, a path function, was found to be

$$W = \int_{V_1}^{V_2} p\, dV$$

which is the area below the process curve on the pV plane for a reversible process. Heat, which also has been proved to be a path function, will be plotted on the temperature-entropy (TS) plane such that

$$Q = \int_{S_1}^{S_2} T\, dS \tag{4.15}$$

and the area below the process curve on the TS plane is the heat transfer for the open or closed system (Fig. 4.5) for a reversible process.

Since the heat transfer has been defined by means of Eq. (4.6) or

$$Q = mc(T_2 - T_1)$$

it is more useful to write Eq. (4.15) as

$$\Delta S = S_2 - S_1 = \int_{T_1}^{T_2} (dQ/T) \tag{4.16}$$

Entropy is a thermodynamic property, and the change in entropy between an initial and final state is defined by Eq. (4.16).

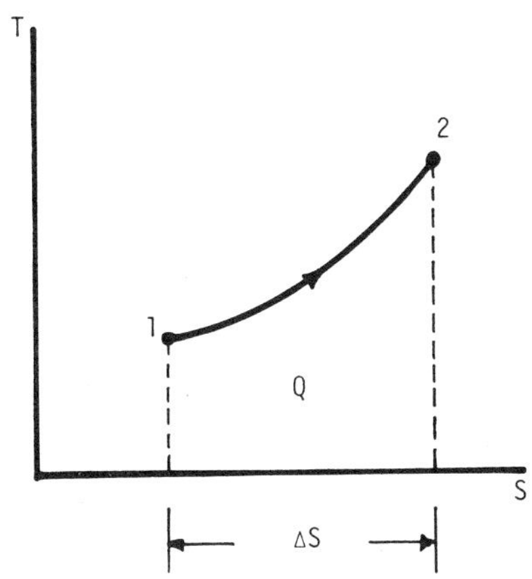

Figure 4.5

From Eq. (4.16), the units of entropy are

$$\Delta S = S_2 - S_1 = \int_{T_1}^{T_2} (dQ/T) \quad \text{or} \quad (B/R)$$

and the specific entropy is defined by

$$S = ms$$

so that the units of the specific entropy are

$$s = S/m = B/\text{lbm-R}$$

If Eq. (4.6) is written in differential form,

$$dQ = mc\ dT$$

then Eq. (4.16) becomes

$$\Delta S = S_2 - S_1 = \int_{T_1}^{T_2} (dQ/T) = \int_{T_1}^{T_2} (mc\ dT/T)$$

$$= mc \ln(T_2/T_1) \tag{4.17}$$

Equation (4.17) may be applied to determine the entropy change for a liquid or solid working substance.

For a gaseous working substance and a constant pressure process, the entropy change is

$$\Delta S = S_2 - S_1 = mc_p \ln(T_2/T_1) \tag{4.18}$$

and for a constant volume process

$$\Delta S = S_2 - S_1 = mc_v \ln(T_2/T_1) \tag{4.19}$$

where c_p and c_v replace c in Eq. (4.17).

For a constant temperature, isothermal process, Eq. (4.16) becomes

$$\Delta S = S_2 - S_1 = Q/T \tag{4.20}$$

Example 4.2. Given is 5 lbm of air as a working substance. Heat is added such that the temperature increases from 60 to 240 F. Determine ΔS and the heat transfer for a constant pressure process and for a constant volume process. Plot the two process paths on the TS plane.

For the constant pressure process, by use of Eq. (4.18),

$$\Delta S = S_2 - S_1 = mc_p \ln(T_2 - T_1) = 5(0.24) \ln(700/520)$$

$= 0.357\ (B/R)$

Note that the absolute temperature must be used in Eq. (4.18).

The heat transfer for the constant pressure process [Eq. (4.9)] is

$$Q_p = mc_p(T_2 - T_1) = 5(0.24)(700 - 520)$$

$$= 216\ (B)$$

For the constant volume process, by use of Eq. (4.19),

$$\Delta S = S_2 - S_1 = mc_v \ln(T_2/T_1) = 5(0.1714) \ln(700/520)$$

$$= 0.255\ (B/R)$$

and the heat transfer from Eq. (4.8) is

$$Q_v = mc_v(T_2 - T_1) = 5(0.1714)(700 - 520)$$

$$= 154.3\ (B)$$

The two processes are plotted on the TS plane in Fig. 4.6. By assuming that both processes start at the same initial state, point 1, and since the entropy change for the constant pressure exceeds that of the constant volume process, the slope of the process curve on the TS plane must be greater for the constant volume process than it is for the constant pressure process. The area below the process curve must be larger for the constant pressure process than it is for the constant volume process, since $Q_p > Q_v$. For both processes the slope must be positive since

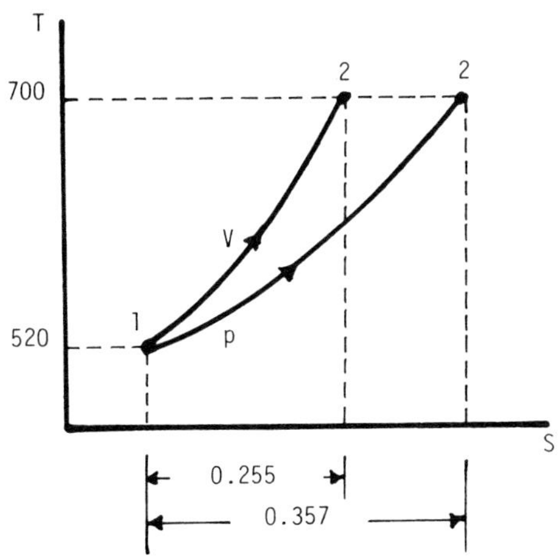

Figure 4.6

an increase in the value of the absolute temperature results in a positive value for the change in entropy.

Example 4.3. Given is 5 lbm of air which undergoes a process for which $\Delta S = 0.3$ B/R. The initial temperature is 60 F. Determine ΔS in SI units, T_2, and Q for a constant pressure, constant volume, and isothermal process. Plot the three processes on the TS plane.

Since

$$1 \text{ (B)} = 1.054E + 03 \text{ (J)}$$

and

$$1 \text{ (R)} = 1/1.8 \text{ (K)}$$

then

$$\text{B/R} = 1.897E + 03 \text{ (J/K)}$$

or

$$\Delta S = 0.3 \text{ (B/R)} = 569.1 \text{ (J/K)}$$

For the constant pressure process

$$\Delta S = mc_p \ln(T_2/T_1)$$

$$0.3 = 5(0.24) \ln(T_2/520)$$

$$\ln(T_2/520) = 0.250$$

$$T_2/520 = e^{0.250} = 1.284$$

or

$$T_2 = 668 \text{ (R)}$$

Then

$$Q_p = mc_p(T_2 - T_1) = 5(0.24)(668 - 520)$$

$$= 177.6 \text{ (B)} \quad \text{or} \quad 187.2 \text{ (kJ)}$$

For the constant volume process

$$\Delta S = mc_v \ln(T_2/T_1)$$

$$0.3 = 5(0.1714) \ln(T_2/520)$$

$$\ln(T_2/520) = 0.350$$

$$T_2/520 = e^{0.350} = 1.419$$

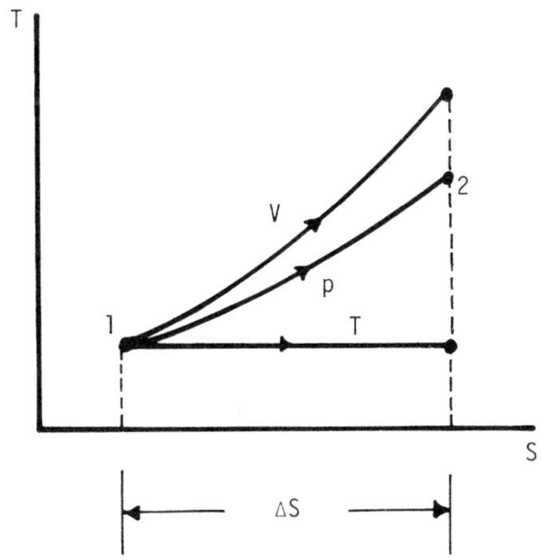

Figure 4.7

or

$$T_2 = 738 \text{ (R)}$$

Then

$$Q_v = mc_v(T_2 - T_1) = 5(0.1714)(738 - 520)$$

$$= 186.8 \text{ (B)} \quad \text{or} \quad 196.9 \text{ (kJ)}$$

For the constant temperature or isothermal process

$$T_1 = T_2 = 520 \text{ (R)}$$

and from Eq. (4.20)

$$Q = T\,\Delta S = 520(0.3)$$

$$= 156.0 \text{ (B)} \quad \text{or} \quad 164.4 \text{ (kJ)}$$

The three processes are plotted in Fig. 4.7. Note that the area below the process curve on the TS plane is greatest for the constant volume process and is the least for the isothermal process when ΔS is constant.

Example 4.4. Given is a three-process cycle, using air, consisting of a constant entropy, isentropic process where the temperature increases from 40 to 540 F. The second process is constant pressure, and the cycle closes with a constant volume process. Sketch the process on the TS plane, and determine the heat transfer for each process. Also determine the cyclic heat and cyclic work for the cycle.

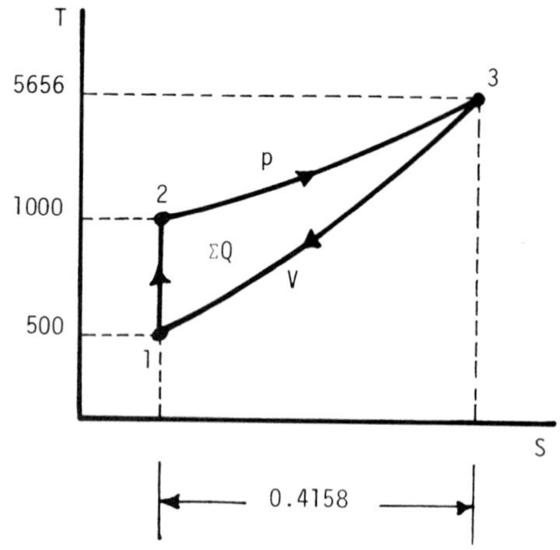

Figure 4.8

The cycle is sketched in Fig. 4.8. Since the slope of the constant pressure process is less than that of the constant volume process, state point 3 must have a higher entropy value than state point 1 or 2. The entropy change for a cycle is zero, so that

$$\Delta S_{21} + \Delta S_{32} + \Delta S_{13} = 0$$

But for the constant entropy process

$$\Delta S_{21} = 0$$

so that

$$\Delta S_{32} = -\Delta S_{13}$$

or

$$mc_p \ln(T_3/T_2) = -mc_v \ln(T_1/T_3)$$

$$= mc_v \ln(T_3/T_1)$$

$$0.24 \ln(T_3/1000) = 0.1714 \ln(T_3/500)$$

$$1.4 \ln(T_3/1000) = \ln(T_3/500)$$

$$\ln(T_3/1000)^{1.4} = \ln(T_3/500)$$

$$(T_3/1000)^{1.4} = T_3/500$$

$$T_3^{0.4} = 31.70$$

$$T_3 = 5656 \text{ (R)}$$

Then

$$\Delta s_{32} = s_3 - s_2 = c_p \ln(T_3/T_2) = 0.24 \ln(5656/1000)$$

$$= 0.4158 \text{ (B/lbm-R)}$$

and

$$\Delta s_{13} = s_1 - s_3 = c_v \ln(T_1/T_3) = 0.1714 \ln(500/5656)$$

$$= -0.4158 \text{ (B/lbm-R)}$$

The heat transfer for each process is, for the isentropic process,

$$Q_{12} = 0$$

since the area below the isentropic process path is zero.

$$Q_{23} = c_p(T_3 - T_2) = 0.24(5656 - 1000) \qquad p = c$$

$$= 1117 \text{ (B/lbm)}$$

and

$$Q_{31} = c_v(T_1 - T_3) = 0.1714(500 - 5656) \qquad v = c$$

$$= -884 \text{ (B/lbm)}$$

For the cycle,

$$\Sigma Q = Q_{12} + Q_{23} + Q_{31} = 0 + 1117 - 884$$

$$= 233 \text{ (B/lbm)}$$

By the First Law of Thermodynamics

$$\Sigma W = \Sigma Q = 233 \text{ (B/lbm)}$$

Clockwise rotation for a cycle on the TS plane produces a positive value for the cyclic work so that the device is classified as an engine.

The efficiency of a heat engine is defined by the equation

$$e = \Sigma W/Q_{in} \tag{4.21}$$

where e is the cycle efficiency, ΣW is the work of the cycle, and Q_{in} is the positive heat terms in the cycle.

For Example 4.4

$$e = \Sigma W/Q_{in} = 233/1117$$

$$= 0.209 \quad \text{or} \quad 20.9\%$$

The area inside the closed figure on the TS plane is the heat transfer for the cycle.

ENTHALPY

The enthalpy of a working substance is defined by the equation

$$H = (U + pV)/J \tag{4.22}$$

and the change in enthalpy is given by

$$\Delta H = H_2 - H_1 = [(U_2 + p_2V_2)/J] - [(U_1 + p_1V_1)/J]$$

$$= (U_2 - U_1) + [(p_2V_2 - p_1V_1)/J] \text{ (B)}$$

The specific enthalpy is

$$h = H/m \text{ (B/lbm)}$$

For a steady-flow or open system, the change in flow energy is

$$\Delta FE = (p_2V_2 - p_1V_1)/J$$

so that

$$\Delta H = \Delta U + \Delta FE$$

The steady-flow process equation

$$Q = \Delta PE + \Delta KE + \Delta FE + \Delta U + W_{sf}$$

reduces to

$$Q = \Delta H + \Delta PE + \Delta KE + W_{sf} \tag{4.23}$$

Consider the steady-flow device in Fig. 4.9, where the working substance is a gas. The potential energies entering and leaving the system are equal since both the inlet and exit pipes are at the same elevation. The inlet and exit diameters may be selected so that the inlet and exit velocities are equal. Therefore, for this system $\Delta PE = 0$ and $\Delta KE = 0$, and the steady-flow process equation becomes

$$Q = \Delta H + W_{sf}$$

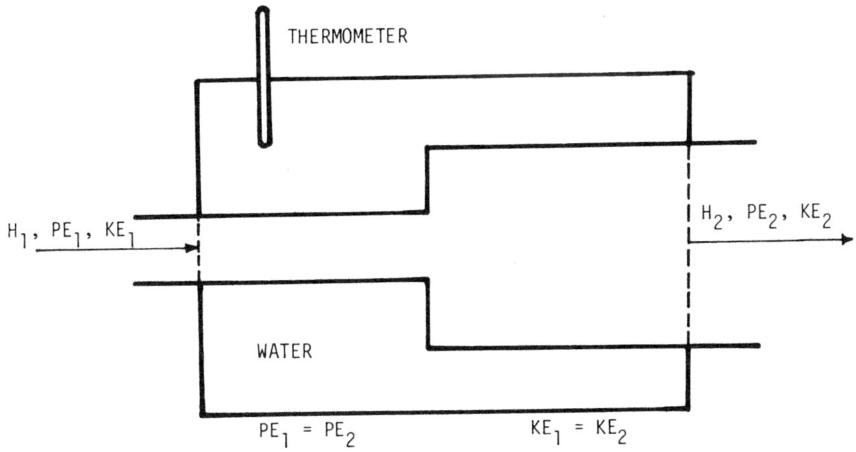

Figure 4.9

Measurements of the surrounding water temperature, as in Joule's experiment, established that there was no heat transfer, and since there is no capability, such as a turbine or a pump, for steady-flow work,

$$\Delta H = 0$$

Since both the pressure and density or specific volume of the gaseous working substance vary, the enthalpy is not a function of these properties. Therefore, the enthalpy is a function of the temperature, and in mathematical terms

$$\Delta H = H_2 - H_1 = mc_p(T_2 - T_1) \tag{4.24}$$

for all perfect gas processes. Equation (4.24) may be applied to all fluid systems but does not apply for processes in which there is a total or partial change of phase.

Example 4.5. Determine the change in enthalpy when 5 lbm of air receives a quantity of heat sufficient to raise the temperature from 60 to 240 F. From Example 4.1, with the same temperature rise, $\Delta U = 154.3$ B and $Q_p = 216.0$ and $Q_v = 154.3$ B. The work for the constant volume process was

$$W_{nf} = 0$$

and for the constant pressure process

$$W_{nf} = 61.7 \text{ B}$$

For both the constant volume and constant pressure processes

$$\Delta H = mc_p(T_2 - T_1) = 5(0.24)(700 - 520)$$

$$= 216.0 \text{ (B)} = Q_p$$

CONSTANT SPECIFIC HEATS

In the previous discussions, the values of c_p and c_v, as physical constants, have been assumed to have a fixed numerical value. The following equations are dependent on this assumption:

$$Q_v = mc_v(T_2 - T_1) \qquad (4.8), (4.12)$$

$$Q_p = mc_p(T_2 - T_1) \qquad (4.9)$$

$$\Delta U = mc_v(T_2 - T_1) \qquad (4.10)$$

$$\Delta S = mc_p \ln(T_2/T_1), \qquad p = c \qquad (4.18)$$

$$\Delta S = mc_v \ln(T_2/T_1), \qquad v = c \qquad (4.19)$$

$$\Delta H = mc_p(T_2 - T_1) \qquad (4.24)$$

PROBLEMS

4.1 Given 3 lbm of argon receiving a quantity of heat sufficient to raise the temperature from 40 to 80 F, determine ΔU, ΔH, ΔS, Q, and W_{nf} for a constant pressure and constant volume process. Sketch the processes on the pV and TS planes.

4.2 A quantity of air is reduced from 150 to 80 F by removal of 100 B of heat from the system during a constant pressure process. Determine the mass of air in the system and ΔU, ΔH, ΔS, and W_{nf}. Sketch the process on the pV and TS planes.

4.3 Ten lbm of water undergoes a process for which $\Delta S = -2.38$ B/R. The initial water temperature is 200 F, and the specific heat of water is 1.0 B/lbm-R. Determine the final temperature.

4.4 Air undergoes a cycle consisting of a constant pressure process in which the temperature is raised from 40 to 540 F. The second process is isothermal, and the final process is isentropic. Determine the heat transfer for each process in the cycle and the cyclic heat and cyclic work. Sketch the cycle on the TS plane. Is this device a heat engine or a heat pump?

4.5 Air undergoes a cycle which consists of an isentropic process from 40 to 240 F. The second process is isothermal, and the cycle closes with a constant volume process. Sketch the cycle on the TS plane. Determine the heat transfer for each process and the cyclic heat and work. Determine the cycle efficiency.

5

PERFECT GASES

In this chapter we shall define the perfect gas, which is a most useful working substance in many engineering applications. The relationship between various physical constants for a perfect gas will be derived. Reversible processes, for a perfect gas between an initial and final state, will be discussed and equations derived for the calculation of the energy changes, heat, and nonflow and steady-flow work. Three general entropy equations, relating the entropy change between the initial and final states, will be established.

PERFECT GAS

A perfect gas is defined as a working substance that obeys the perfect gas law and whose specific heats have fixed numerical values.

GAS RELATIONSHIPS

Experiments by Boyle and Charles established the relationship between the absolute pressure, the absolute temperature, and the total volume for certain types of processes.

Boyle established that, for a constant temperature or isothermal process,

$$p_1V_1 = p_2V_2 \tag{5.1}$$

When the thermodynamic process path is defined by

$$pV^n = c$$

then

$$p_1V_1^n = p_2V_2^n = c$$

and a direct comparison with Eq. (5.1) establishes that for an isothermal process $n = 1$.

Charles provided the following equations: For a constant pressure process

$$V_2/V_1 = T_2/T_1 \tag{5.2}$$

and for a constant volume process

$$p_2/p_1 = T_2/T_1 \tag{5.3}$$

It has been previously shown that for a constant pressure process $n = 0$ and that for a constant volume process $n \to \infty$.

The need for an absolute temperature scale may be illustrated by an experiment performed on a constant volume system. When heat is added to such a system, the temperature increase was from 100 to 400 F. Readings of the pressure indicated an increase from 20 to 30.71 psia. By use of Eq. (5.3)

$$p_2/p_1 = T_2/T_1$$

$$30.71/20.0 \neq 400/100$$

where the temperatures are inserted into Eq. (5.3) in degrees Fahrenheit. The ratios of the absolute pressures and the Fahrenheit temperatures are

$$1.536 \neq 4$$

Since the pressures are related to a zero datum plane, the error must be in the selection of the temperature scale. This is readily apparent since the value of p_2 would approach infinity as t_1 approaches 0 F.

Let

$$p_2/p_1 = (t_2 + c)/(t_1 + c)$$

so that

$$1.536 = (400 + c)/(100 + c)$$

and

$$c = 459.7 \text{ (F)}$$

so that

$$T(R) = t(F) + 460$$

A similar experiment where the temperatures are measured in degrees Celsius yields

$$T(K) = t(C) + 273$$

PERFECT GAS LAW

Given are initial state 1 and a final state 2 so that a composite process path from state 1 to state 2 by use of the intermediate state X may be constructed (Fig.5.1). Since the pressure, volume, and temperature are properties or point functions, intermediate state X or Y will result in the same relationship between the initial and final properties. The process from state 1 to state X is a constant pressure process, and the process from state X to state 2 is a constant volume process.

For the constant pressure process, from Eq. (5.2),

$$V_1/V_X = T_1/T_X, \qquad p = c$$

and for the constant volume process, from Eq. (5.3),

$$p_X/p_2 = T_X/T_2, \qquad V = c$$

Solving these equations for T_X results in

$$T_X = T_1V_X/V_1 = p_XT_2/p_2$$

but $p_1 = p_X$ and $V_X = V_2$, so that

$$T_X = T_1V_2/V_1 = p_1T_2/p_2$$

or

$$p_1V_1/T_1 = p_2V_2/T_2$$

By dividing by the mass of the gas in the closed system

$$p_1V_1/mT_1 = p_2V_2/mT_2$$

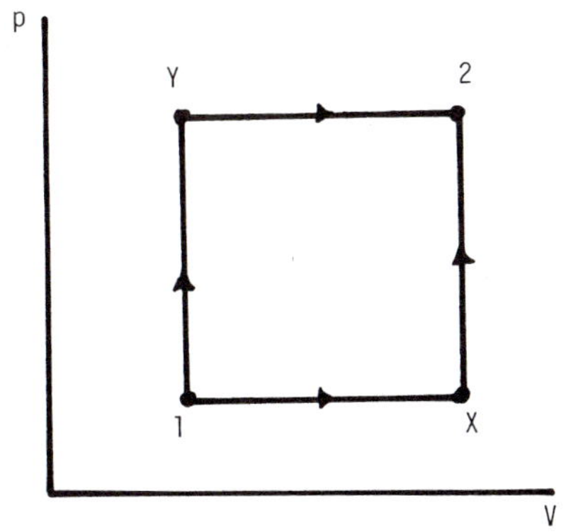

Figure 5.1

since $V_1/m = v_1$ and $V_2/m = v_2$, so that

$$p_1v_1/T_1 = p_2v_2/T_2 = R \tag{5.4}$$

where R is a physical property having a fixed numerical value for each gas or each mixture of gases. Values of R are listed in Appendix B for various gases.

The units of R are

$$R = pv/T = [(lbf/ft^2)(ft^3/lbm)]/R$$

$$= ft\text{-}lbf/lbm\text{-}R$$

The value of R for a gas or mixture of gases may be calculated from

$$R = \bar{R}/M$$

where $\bar{R}$ is the universal gas constant or

$$\bar{R} = 1545 \text{ (ft-lbf/lbmole-R)}$$

and M is the molecular weight of the gas or gas mixture.

Multiplying Eq. (5.4) by the mass of the working substance in the system gives

$$pvm/T = mR$$

or

$$pV = mRT \tag{5.5}$$

Both Eqs. (5.4) and (5.5) represent a mathematical statement of the perfect gas law. The perfect gas law is a powerful tool for the solution of a wide variety of engineering problems.

Example 5.1. Determine the volume occupied by 5 lbm of air at a pressure of 20 psia and a temperature of 60 F. From Eq. (5.5)

$$V = mRT/p = [5(53.3)(520)]/[20(144)] = 48.1 \text{ (ft}^3\text{)}$$

Note that the absolute temperature is used and that the pressure is converted from psia to psfa in the perfect gas equation.

Example 5.2. Assume that the gas in Example 5.1 has a molecular weight of 20. Determine the volume. By solving for R and from Eq. (5.5)

$$R = \bar{R}/M = 1545/20 = 77.2 \text{ (ft-lbf/lbm-R)}$$

and

$$V = mRT/p = [5(77.2)(520)]/[20(144)]$$

$$= 69.7\ (\text{ft}^3) \quad \text{or} \quad 1.97\ (\text{m}^3)$$

Example 5.3. Determine the number of small containers that can be filled from the large air tank (Fig. 5.2). Given: $V_{tank} = 120\ \text{ft}^3$, $t_{tank} = 80$ F, and $p_{tank} = 100$ psia and $V_{container} = 1\ \text{ft}^3$, $t_{container} = 70$ F, and $p_{container} = 20$ psia. The mass of air initially in the large tank may be determined by use of the perfect gas law or

$$m_{t1} = (p_1V_1/RT_1)_t = [100(144)(120)]/[53.3(540)]$$

$$= 60.0\ (\text{lbm})$$

When the pressure in the tank is reduced to 20 psia, no more small containers may be filled from the large tank. The mass of air remaining in the large tank for a tank pressure of 20 psia is

$$m_{t2} = (p_2V_2/RT_2)_t = [20(144)(120)]/[53.3(540)]$$

$$= 12.0\ (\text{lbm})$$

It has been assumed that the air has been removed from the tank slowly so that the temperature in the tank remains constant.

The mass of air that can be used to fill the small containers is

$$\Delta m = m_{t1} - m_{t2} = 60.0 - 12.0 = 48.0\ (\text{lbm})$$

The mass of air required to fill each small container is

$$m_c = (pV/RT)_c = [20(144)(1)]/[53.3(530)]$$

$$= 0.102\ (\text{lbm})$$

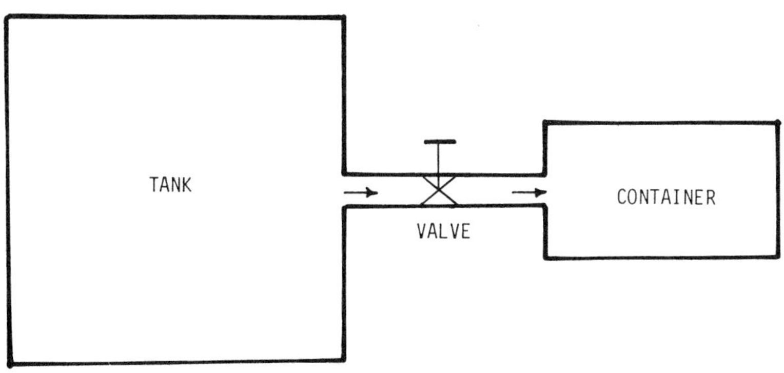

Figure 5.2

The number of containers that can be filled is

$$N = \Delta m/m_c = 48/0.102$$

$$= 470$$

Example 5.4. Determine the mass that can be lifted by a spherical balloon having a diameter of 30 ft^3. The balloon is filled with hydrogen at 20 psia and 80 F. Surrounding air has properties of 14.7 psia and 60 F, and the balloon weighs 200 lbm.

The volume of the spherical balloon is

$$V = (\Pi/6)D^3 = (\Pi/6)(30)^3 = 14,140 \text{ (ft}^3\text{)}$$

The weight of the hydrogen in the balloon is

$$m_{H_2} = pV/RT = [20(144)(14,140)]/[767(540)]$$

$$= 98.3 \text{ (lbm)}$$

The weight of the air displaced by the balloon is

$$m_{air} = pV/RT = [14.7(144)(14,140)]/[53.3(520)]$$

$$= 1080 \text{ (lbm)}$$

The gross lift of the balloon is

$$m_g = m_{air} - m_{H_2} = 1080 - 98 = 982 \text{ (lbm)}$$

and the net lift of the balloon is

$$m_{net} = 982 - 200 = 782 \text{ (lbm)} \quad \text{or} \quad 355 \text{ (kg)}$$

Example 5.5. Determine the net lift if the balloon in Example 5.4 is filled with helium.

The weight of the helium in the balloon is

$$m_{He} = pV/RT = [20(144)(14,140)]/[386(540)]$$

$$= 195.4 \text{ (lbm)}$$

The gross lift of the balloon is

$$m_g = m_{air} - m_{He} = 1080 - 195 = 885 \text{ (lbm)}$$

and the net lift of the balloon is

$$m_{net} = 885 - 200 = 685 \text{ (lbm)} \quad \text{or} \quad 311 \text{ (kg)}$$

The helium balloon lifts 87.5% as much mass as the hydrogen balloon, and the loss of lifting effort is 97 lbm or 44 kg. The advantage in using helium is that helium is inert and therefore nonexplosive.

Example 5.6. Given is the system consisting of two air containers shown in Fig. 5.3. Both containers hold air with $p_A = 40$ psia, $t_A = 60$ F, $V_A = 30$ ft^3, $p_B = 100$ psia, $t_B = 140$ F, and $V_B = 80$ ft^3. In the same manner as Joule's experiment, $Q = 0$ and $W_{nf} = 0$. Determine the pressure and temperature after the valve is opened.

From the nonflow process equation

$$Q = \Delta U + W_{nf}$$

and

$$\Delta U = 0$$

since Q, $W_{nf} = 0$. The internal energy change is

$$m_A c_{vA}(T - T_A) + m_B c_{vB}(T - T_B) = 0$$

where

$$m_A = p_A V_A/RT_A = [40(144)(30)]/[53.3(520)]$$

$$= 6.23 \text{ (lbm)}$$

and

$$m_B = p_B V_B/RT_B = [100(144)(80)]/[53.3(600)]$$

$$= 36.02 \text{ (lbm)}$$

or

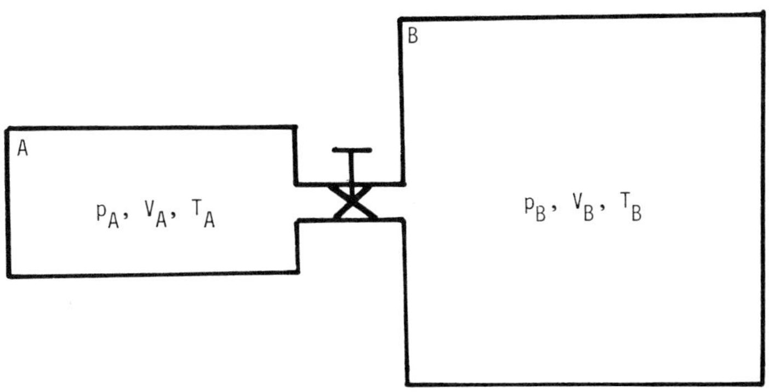

Figure 5.3

$$6.23(0.1714)(T - 520) + 36.02(0.1714)(T - 600) = 0$$

resulting in T = 588 R or 327 K.

By use of the perfect gas law

$$p = mRT/V = [(6.23 + 36.02)53.3(588)]/[(30 + 80)144]$$

$$= 83.6 \text{ (psia)} \quad \text{or} \quad 576.4 \text{ (kPa-a)}$$

This problem is an example of an irreversible process since the air in passing from container B to container A could have done useful work. Even though the heat transfer is zero, there is an increase in the entropy of the system. Such a process is referred to as adiabatic, while a reversible adiabatic process is a constant entropy or isentropic process. This is an example of the use of an isolated system.

PROPERTY RELATIONSHIPS

The nonflow process equation is

$$Q = \Delta U + W_{nf}$$

This equation is valid for any working substance and any type, reversible or irreversible, of nonflow process.

By use of Eq. (4.10), where c_v has a fixed numerical value,

$$\Delta U = U_2 - U_1 = mc_v(T_2 - T_1)$$

assuming that the working substance does not undergo a change of phase.

For a constant pressure nonflow reversible process

$$Q = mc_p(T_2 - T_1)$$

and

$$W_{nf} = p(V_2 - V_1)$$

and by substitution in the process equation

$$mc_p(T_2 - T_1) = mc_v(T_2 - T_1) + p(V_2 - V_1)/J$$

$$= mc_v(T_2 - T_1) + p_2V_2 - p_1V_1)/J$$

For a perfect gas

$$mc_p(T_2 - T_1) = mc_v(T_2 - T_1) + [mR(T_2 - T_1)]/J$$

or

$$c_p = c_v + (R/J) \tag{5.6}$$

From the definition of the physical property k,

$$k = c_p/c_v$$

or

$$c_p = kc_v$$

Substitution in Eq. (5.6) gives

$$kc_v = c_v + (R/J)$$

and

$$c_v = R/[J(k - 1)] \tag{5.7}$$

and

$$c_p = kR/[J(k - 1)] \tag{5.8}$$

For a monoatomic gas, one atom per molecule, such as argon and helium,

$$k = 5/3 = 1.667$$

For a diatomic gas, two atoms per molecule, such as air, oxygen, and nitrogen,

$$k = 7/5 = 1.400$$

Example 5.7. A diatomic gas has a molecular weight of 29. Determine c_p and c_v for the gas. The value of the specific gas constant is

$$R = \bar{R}/M = 1545/29 = 53.3 \text{ (ft-lbf/lbm-R)}$$

and from Eq. (5.7)

$$c_v = 53.3/[778(1.4 - 1)] = 0.1714 \text{ (B/lbm-R)}$$

$$= 0.717 \text{ (kJ/kg-K)}$$

The value of c_p is

$$c_p = kc_v = 1.4(0.1714) = 0.24 \text{ (B/lbm-R)}$$

$$= 1.004 \text{ (kJ/kg-K)}$$

The value of R in SI units is

$$R = c_p - c_v = 1.004 - 0.717 = 0.287 \text{ (kJ/kg-K)}$$

The values of c_p, c_v, and R are those of air. However, any mixture of diatomic gases which has an effective molecular weight of 29 would have the same physical property values.

CONSTANT PRESSURE PROCESS

We shall now consider four reversible processes, using the perfect gas as the working substance, which correspond to the four coordinates used to plot the pV and TS diagrams. These processes may then be connected to form cycles which are the theoretical models for real engineering devices. For the constant pressure process

$$n = 0$$

$$V_2/V_1 = T_2/T_1$$

$$U_2 - U_1 = mc_v(T_2 - T_1)$$

$$H_2 - H_1 = mc_p(T_2 - T_1) = k(U_2 - U_1)$$

$$S_2 - S_1 = mc_p \ln(T_2/T_1)$$

$$Q = mc_p(T_2 - T_1) = \Delta H$$

$$W_{nf} = Q - \Delta U = p(V_2 - V_1)/J$$

$$W_{sf} = Q - \Delta H - \Delta KE - \Delta PE$$

or

$$W_{sf} = Q - \Delta H = 0$$

if $\Delta KE, \Delta PE = 0$.

Example 5.8. Nitrogen at a pressure, temperature, and volume of 50 psia, 340 F, and 20 ft^3 undergoes a constant pressure process to a final temperature of 60 F. Determine V_2, ΔU, ΔH, ΔS, Q, W_{nf}, W_{sf} ($\Delta KE = 0$, $\Delta PE = 0$), and W_{sf} ($\Delta KE = 30$ B, $\Delta PE = 0$). Sketch the process on the pV and TS diagrams.

From the perfect gas equation

$$m = p_1V_1/RT_1 = [50(144)(20)]/[55.2(800)]$$

$$= 3.26\ (\text{lbm}) \quad \text{or} \quad 1.48\ (\text{kg})$$

$$V_2 = V_1(T_2/T_1) = 20(520/800)$$

$$= 13.0\ (\text{ft}^3) \quad \text{or} \quad 0.368\ (\text{m}^3)$$

For all processes

$$U_2 - U_1 = mc_v(T_2 - T_1) = 3.26(0.178)(520 - 800)$$

$$= -162.5\ (\text{B}) \quad \text{or} \quad -171.3\ (\text{kJ})$$

and

$$H_2 - H_1 = mc_p(T_2 - T_1) = 3.26(0.249)(520 - 800)$$

$$= -227.3\ (\text{B}) \quad \text{or} \quad -239.6\ (\text{kJ})$$

Also

$$H_2 - H_1 = k\ \Delta U = 1.399(-162.5) = -227.3\ (\text{B})$$

For the constant pressure process, $p_1 = p_2$,

$$S_2 - S_1 = mc_p \ln(T_2/T_1) = 3.26(0.249) \ln(520/800)$$

$$= -0.350\ (\text{B/R}) \quad \text{or} \quad -663\ (\text{J/K})$$

and

$$Q_p = mc_p(T_2 - T_1) = \Delta H = -227.3\ (\text{B})$$

The work terms are

$$W_{nf} = Q - \Delta U = -227.3 + 162.5 = -64.8\ (\text{B}) \quad \text{or} \quad -68.3\ (\text{kJ})$$

or

$$W_{nf} = p(V_2 - V_1)/J = [144(50)(13.0 - 20)]/778 = -64.8\ (\text{B})$$

and

$$W_{sf} = Q - \Delta H = 0 \qquad (\Delta KE,\ \Delta PE = 0)$$

or

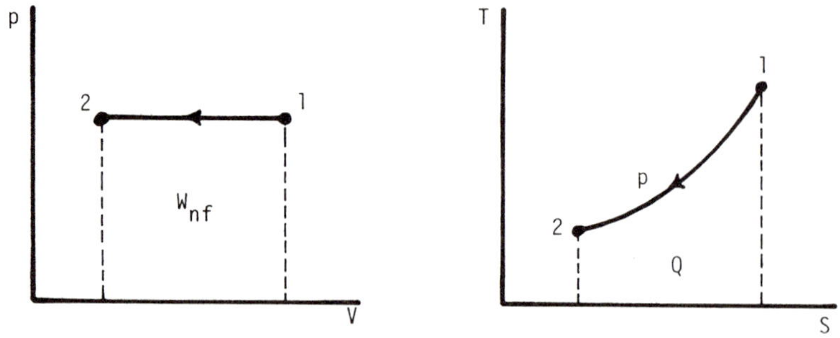

Figure 5.4

$$W_{sf} = nW_{nf} = 0$$

since n = 0.

For steady-flow problems where $\Delta KE \neq 0$ or $\Delta PE \neq 0$

$$W_{sf} = Q - \Delta H - \Delta KE - \Delta PE = -227.3 + 227.3 - 30 - 0 = -30 \text{ (B)}$$

when $\Delta KE = 30$ B and $\Delta PE = 0$.

The pV and TS diagrams are plotted in Fig. 5.4. Since $V_2 < V_1$, the process must proceed as a horizontal line, p = c, from right to left on the pV plane. Therefore, W_{nf} must be negative, and the nonflow work is added to the system.

The negative of the area in back of the curve on the pV plane represents the steady-flow work when ΔKE, $\Delta PE = 0$ and the area in back of a horizontal line is zero.

From the problem data $T_2 < T_1$, which for a constant pressure process results in $\Delta S < 0$. Therefore, the process curve on the TS plane must slope downward from right to left, and the value of the heat transfer must be negative. The heat is transferred from the system to the surroundings.

CONSTANT VOLUME PROCESS

For a constant volume process, $V_1 = V_2$ and $n \to \infty$,

$$p_2/p_1 = T_2/T_1$$

$$U_2 - U_1 = mc_v(T_2 - T_1)$$

$$H_2 - H_1 = mc_p(T_2 - T_1) = k\ \Delta U$$

$$S_2 - S_1 = mc_v \ln(T_2/T_1)$$

$$Q = mc_v(T_2 - T_1) = \Delta U$$

$$W_{nf} = Q - \Delta U = 0$$

$$W_{sf} = Q - \Delta H - \Delta KE - \Delta PE$$

or

$$W_{sf} = Q - \Delta H = -V(p_2 - p_1)/J \qquad (\Delta KE,\ \Delta PE = 0)$$

Example 5.9. Five lbm of air at a pressure and temperature of 15 psia and 40 F undergoes a constant volume process to a final pressure of 30 psia. Determine V, T_2, ΔU, ΔH, ΔS, Q, W_{nf}, and W_{sf} (ΔKE, $\Delta PE = 0$). Sketch the process on the pV and TS planes.

From the perfect gas law [Eq. (5.5)]

$$V = mRT_1/p_1 = [5(53.3)(500)]/[15(144)]$$

$$= 61.7\ (\text{ft}^3) \quad \text{or} \quad 1.75\ (\text{m}^3)$$

$$T_2 = T_1(p_2/p_1) = 500(30/15)$$

$$= 1000\ (\text{R}) \quad \text{or} \quad 556\ (\text{K})$$

For all processes

$$U_2 - U_1 = mc_v(T_2 - T_1) = 5(0.1714)(1000 - 500)$$

$$= 428.5\ (\text{B}) \quad \text{or} \quad 451.6\ (\text{kJ})$$

and

$$H_2 - H_1 = mc_p(T_2 - T_1) = 5(0.24)(1000 - 500)$$

$$= 600\ (\text{B}) \quad \text{or} \quad 632.4\ (\text{kJ})$$

Also

$$H_2 - H_1 = k\ \Delta U = 1.4(428.5) = 600\ (\text{B})$$

For the constant volume process

$$S_2 - S_1 = mc_v \ln(T_2/T_1) = 5(0.1714) \ln(1000/500)$$

$$= 0.594\ (\text{B/R}) \quad \text{or} \quad 1127\ (\text{J/K})$$

and

$$Q_v = mc_v(T_2 - T_1) = \Delta U = 428.5\ (\text{B})$$

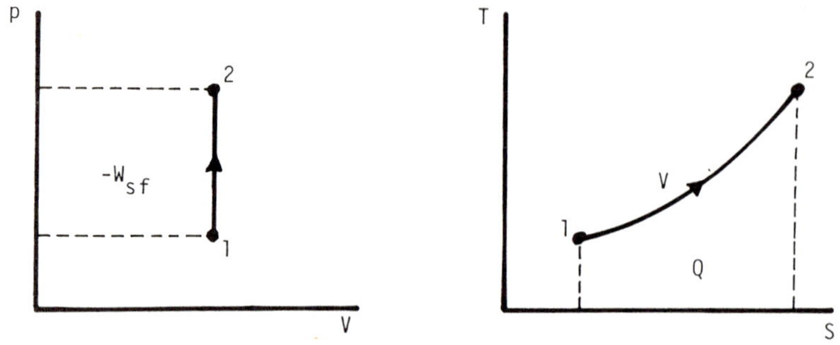

Figure 5.5

The work terms are $W_{nf} = 0$ and for ΔKE, $\Delta PE = 0$,

$$W_{sf} = Q - \Delta H = 428.5 - 600$$

$$= -171.5 \text{ (B)} \quad \text{or} \quad -180.8 \text{ (kJ)}$$

Also

$$W_{sf} = -V(p_2 - p_1)/J = -[61.7(144)(30 - 15)]/778$$

$$= -171.3 \text{ (B)}$$

The pV and TS diagrams are sketched in Fig. 5.5. For the constant volume process on the pV plane, the curve is vertically upward since $p_2 > p_1$. The area below the curve, W_{nf}, is zero. The steady-flow work is negative since $p_2 > p_1$ and is represented by the negative of the area in back of the process curve on the pV plane when ΔKE, $\Delta PE = 0$.

On the TS diagram, $T_2 > T_1$ and $\Delta S > 0$; therefore the curve must proceed upwards from left to right. Since $\Delta S > 0$, the heat transfer is positive, and heat is added to the system.

CONSTANT TEMPERATURE PROCESS

For the constant temperature or isothermal process, $T_1 = T_2$,

$$n = 1$$

$$p_1V_1 = p_2V_2$$

$$U_2 - U_1 = mc_v(T_2 - T_1) = 0$$

$$H_2 - H_1 = mc_p(T_2 - T_1) = 0$$

$$W_{nf} = (p_1V_1/J)\ \ln(V_2/V_1), \qquad n = 1$$

$$Q = \Delta U + W_{nf} = W_{nf}$$

$$S_2 - S_1 = Q/T$$

$$W_{sf} = Q - \Delta H - \Delta KE - \Delta PE$$

or

$$W_{sf} = Q - \Delta H = Q = W_{nf} \qquad (\Delta KE,\ \Delta PE = 0)$$

Example 5.10. Air at a pressure and temperature of 40 psia and 140 F undergoes an isothermal process to a pressure of 20 psia. Determine v_1, v_2, Δu, Δh, W_{nf}, Q, Δs, and W_{sf} (ΔKE, $\Delta PE = 0$). Sketch the process on the pV and TS planes.

This problem must be worked on a specific basis since there are insufficient data available for the determination of the mass of the working substance.

From the perfect gas law [Eq. (5.4)]

$$v_1 = RT_1/p_1 = [53.3(600)]/[40(144)]$$

$$= 5.55\ (ft^3/lbm) \quad \text{or} \quad 0.347\ (m^3/kg)$$

$$v_2 = v_1(p_1/p_2) = 5.55(40/20)$$

$$= 11.10\ (ft^3/lbm) \quad \text{or} \quad 0.694\ (m^3/kg)$$

For the isothermal process $\Delta u = 0$ and $\Delta h = 0$.

The heat and nonflow work terms are

$$Q = W_{nf} = (p_1v_1/J)\ \ln(v_2/v_1) = (RT_1/J)\ \ln(p_1/p_2)$$

$$= [53.3(600)/778]\ \ln(40/20)$$

$$= 28.49\ (B/lbm)$$

$$= 66.3\ (kJ/kg)$$

The steady-flow work when ΔKE, $\Delta PE = 0$ is

$$W_{sf} = Q - \Delta H = 28.49\ (B/lbm)$$

which verifies $W_{sf} = nW_{nf}$ where n = 1 for the isothermal process.

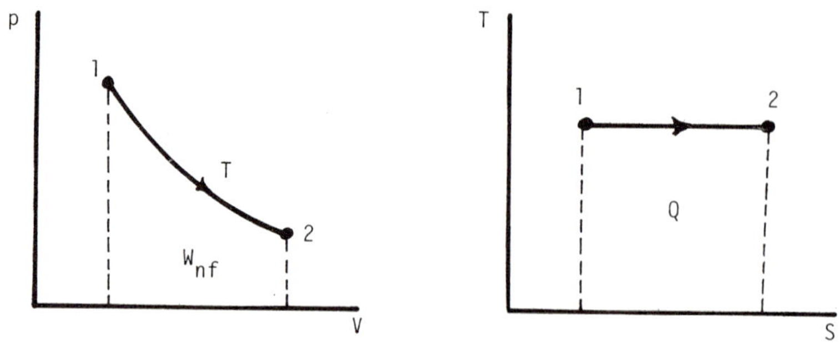

Figure 5.6

The entropy change for the isothermal process is

$$S_2 - S_1 = Q/T = 28.49/600$$
$$= 0.0475 \text{ (B/lbm-R)} \quad \text{or} \quad 198.7 \text{ (J/kg-K)}$$

The pV and TS diagrams are sketched in Fig. 5.6. For n = 1, the process curve on the pV plane is a hyperbole, and since $v_2 > v_1$, the non-flow work is positive. Since $p_1 > p_2$, the steady-flow work is also positive. On the TS diagram, the process curve is horizontal. Since ΔS is positive, the heat transfer is also positive.

CONSTANT ENTROPY PROCESS

For the constant entropy or isentropic process, $S_2 = S_1$,

$$p_1V_1/T_1 = p_2V_2/T_2$$

$$U_2 - U_1 = mc_v(T_2 - T_1)$$

$$H_2 - H_1 = mc_p(T_2 - T_1) = k(U_2 - U_1)$$

$$S_2 - S_1 = 0$$

$$Q = 0$$

$$W_{nf} = Q - \Delta U = -\Delta U$$

$$W_{sf} = Q - \Delta H - \Delta KE - \Delta PE$$

or

$$W_{sf} = Q - \Delta H = -\Delta H = -k\,\Delta U \quad (\Delta KE, \Delta PE = 0)$$

To determine the value of the constant n in

$$pV^n = c$$

remember that the nonflow work can be expressed by

$$W_{nf} = (p_2V_2 - p_1V_1)/[J(1 - n)], \qquad n \neq 1$$

The value of n cannot be 1 since this is the value of n for the isothermal process. By using the perfect gas law, the nonflow work is

$$W_{nf} = [mR(T_2 - T_1)]/[J(1 - n)]$$

but the nonflow work for the isentropic process is

$$W_{nf} = -\Delta U = -mc_v(T_2 - T_1)$$

so that

$$-mc_v(T_2 - T_1) = [mR(T_2 - T_1)]/[J(1 - n)]$$

or

$$c_v = R/[J(n - 1)]$$

But from Eq. (5.7)

$$c_v = R/[J(k - 1)]$$

Therefore, for the isentropic process

$$n = k$$

and

$$p_1V_1^{\ k} = p_2V_2^{\ k}$$

or

$$p_2/p_1 = (V_1/V_2)^k \tag{5.9}$$

From the perfect gas law

$$p_2/p_1 = V_1T_2/V_2T_1$$

and by substitution into Eq. (5.9)

$$(V_1/V_2)(T_2/T_1) = (V_1/V_2)^k$$

or

$$T_2/T_1 = (V_1/V_2)^k(V_1/V_2)^{-1} = (V_1/V_2)^{k-1} \tag{5.10}$$

In like manner, it may be shown that

$$T_2/T_1 = (p_2/p_1)^{k-1/k} \tag{5.11}$$

Equations (5.9), (5.10), and (5.11) are the relationship between the pressure, volume, and temperature for an isentropic process.

Example 5.11. A diatomic gas has a pressure and volume of 20 psia and 10 ft^3. The final volume is 1 ft^3. Determine the final pressure for an isothermal and an isentropic process. Sketch the two processes on the pV plane.

For the isothermal process

$$p_2 = p_1(V_1/V_2) = 20(10/1) = 200 \text{ (psia)}$$

and for the isentropic process, from Eq. (5.9),

$$p_2 = p_1(V_1/V_2)^k = 20(10)^{1.4} = 502 \text{ (psia)}$$

where k = 1.4 for a diatomic gas.

The process is sketched on the pV plane in Fig. 5.7. It is apparent that the isentropic process curve has a greater slope than the isothermal process curve on the pV diagram.

Example 5.12. Five lbm of air at 20 psia and 40 F undergoes an isentropic process to 240 F. Determine V_1, V_2, p_2, ΔU, ΔH, ΔS, Q, W_{nf}, and W_{sf} (ΔKE, ΔPE = 0). Sketch the isentropic process on the pV and TS planes.

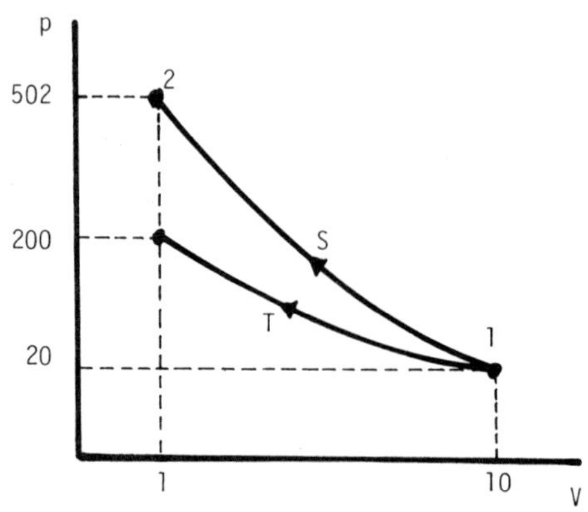

Figure 5.7

From the perfect gas law

$$V_1 = mRT_1/p_1 = [5(53.3)(500)]/[20(144)]$$

$$= 46.27\ (ft^3) \quad or \quad 1.31\ (m^3)$$

From Eq. (5.10)

$$V_2 = V_1(T_1/T_2)^{1/(k-1)} = 46.27(500/700)^{1/0.4}$$

$$= 19.95\ (ft^3) \quad or \quad 0.565\ (m^3)$$

The final pressure may be obtained from Eq. (5.11), where

$$p_2 = p_1(T_2/T_1)^{k/(k-1)} = 20(700/500)^{1.4/0.4}$$

$$= 64.9\ (psia) \quad or \quad 447.5\ (kPa\text{-}a)$$

or by use of the perfect gas law,

$$p_2 = mRT_2/V_2 = [5(53.3)(700)]/[144(19.95)]$$

$$= 64.9\ (psia)$$

For any process

$$U_2 - U_1 = mc_v(T_2 - T_1) = 5(0.1714)(700 - 500)$$

$$= 171.4\ (B) \quad or \quad 180.7\ (kJ)$$

and

$$H_2 - H_1 = mc_p(T_2 - T_1) = 5(0.24)(700 - 500)$$

$$= 240\ (B) \quad or \quad 253\ (kJ)$$

or

$$H_2 - H_1 = k(U_2 - U_1) = 1.4(171.4) = 240\ (B)$$

For the isentropic process, $Q = 0$ and $S_2 - S_1 = 0$. The nonflow work may be calculated from

$$W_{nf} = Q - \Delta U = -\Delta U = -171.4\ (B)$$

or

$$W_{nf} = (p_2V_2 - p_1V_1)/[J(1 - k)] = [mR(T_2 - T_1)]/[J(1 - k)]$$

$$= [5(53.3)(700 - 500)]/[778(1 - 1.4)]$$

$$= -171.3\ (B)$$

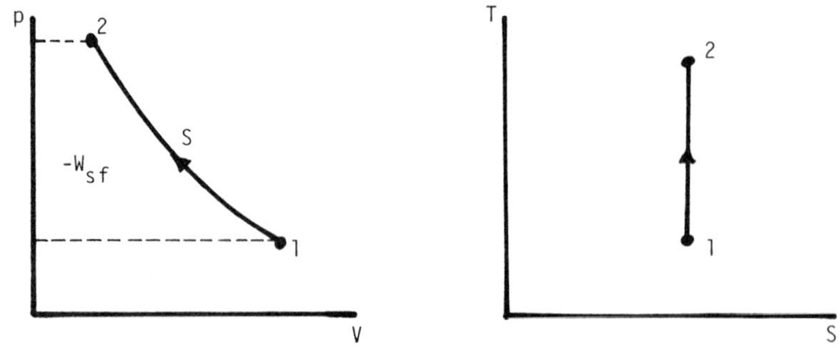

Figure 5.8

The steady-flow work may be calculated from (ΔKE, $\Delta PE = 0$)

$$W_{nf} = Q - \Delta H = -\Delta H = -240 \text{ (B)}$$

$$= kW_{nf} = 1.4(-171.3) = -240 \text{ (B)}$$

The isentropic process is sketched in Fig. 5.8. Since $V_2 < V_1$, the nonflow work must be negative. Since $p_2 > p_1$, the steady-flow work is also negative. To satisfy these property requirements, the process curve on the pV diagram must be upward from right to left. On the TS diagram, the isentropic process is represented by a vertical line, and the heat transfer is zero. Since $T_2 > T_1$, the process curve must be vertically upward.

PROCESSES ON THE pV AND TS PLANES

Given an initial state point ①, the possible final state points for a constant pressure, constant volume, isothermal, and isentropic process are illustrated in Fig. 5.9. The values of n for each process are located

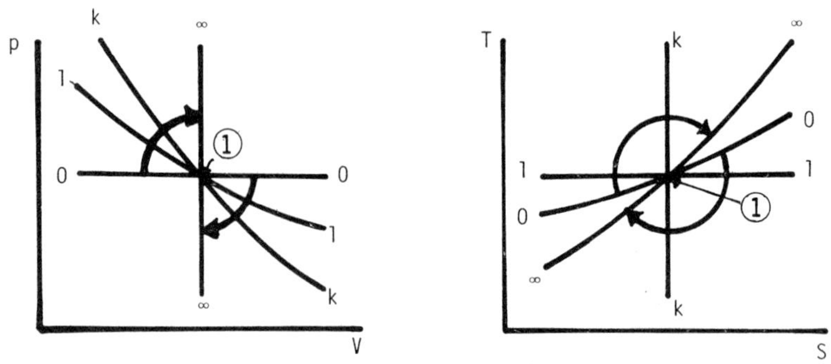

Figure 5.9

along the process path. Notice that the value n increases in the clockwise direction and that these processes are located in the second and fourth quadrants of the diagram. The first and third quadrants are reserved for negative values of n on the pV diagram.

On the TS diagram, increasing values of n also proceed in the clockwise direction, and the remaining portion of the diagram is again reserved for the negative values of n.

POLYTROPIC PROCESS

While the four constant parameter processes are certainly the most important processes for engineering applications, there remains the possibility that the value of n for a process is not 0, 1, or k or does not approach infinity. Such processes occur when the system is heated or cooled in such a manner that n differs from these values. A good example of such a process is the cooling of a reciprocating air compressor where the values of n may be of the order of 1.2 to 1.3.

A process that is not constant pressure, constant volume, isothermal, or isentropic is a polytropic process. The relationships among the pressure, volume, and temperature may be obtained from the isentropic process, where k is replaced by n.

From Eq. (5.9), for the polytropic process,

$$p_1V_1^{\,n} = p_2V_2^{\,n} \tag{5.12}$$

and from Eq. (5.10)

$$T_2/T_1 = (V_1/V_2)^{n-1} \tag{5.13}$$

Also, from Eq. (5.11)

$$T_2/T_1 = (p_2/p_1)^{n-1/n} \tag{5.14}$$

Other polytropic relationships include

$$U_2 - U_1 = mc_v(T_2 - T_1)$$

$$H_2 - H_1 = mc_p(T_2 - T_1)$$

$$W_{nf} = (p_2V_2 - p_1V_1)/[J(1 - n)], \qquad n \neq 1$$

and

$$W_{sf} = nW_{nf}$$

if ΔKE, $\Delta PE = 0$.

For the nonflow process and for a perfect gas

$$Q = \Delta U + W_{nf}$$

$$= mc_v(T_2 - T_1) + [p_2V_2 - p_1V_1]/[J(1 - n)]$$

$$= mc_v(T_2 - T_1) + [mR(T_2 - T_1)]/[J(1 - n)]$$

$$= [m(T_2 - T_1)](c_v + \{R/[J(1 - n)]\})$$

From Eq. (5.6)

$$c_p = c_v + (R/J)$$

or

$$R/J = c_p - c_v$$

and since $c_p = kc_v$,

$$R/J = kc_v - c_v = (k - 1)c_v$$

The equation for the heat transfer is

$$Q = [m(T_2 - T_1)]\{c_v + [(k - 1)c_v/(1 - n)]\}$$

$$= [m(T_2 - T_1)][(c_v - nc_v + kc_v - c_v)/(1 - n)]$$

$$= m\left[c_v\left(\frac{k - n}{1 - n}\right)\right]\left(T_2 - T_1\right)$$

If the polytropic specific heat c_n is defined by

$$c_n = c_v[(k - n)/(1 - n)] \tag{5.15}$$

the heat transfer becomes

$$Q = mc_n(T_2 - T_1) \tag{5.16}$$

The change in entropy may be calculated from

$$\Delta S = \int_{T_1}^{T_2} dQ/T = \int_{T_1}^{T_2} (mc_n\ dT)/T$$

$$= mc_n \ln(T_2/T_1) \tag{5.17}$$

If the heat transfer can be evaluated, the nonflow and steady-flow work can be calculated from

$$W_{nf} = Q - \Delta U$$

and

$$W_{sf} = Q - \Delta H - \Delta KE - \Delta PE$$

Equation (5.15) may be used to evaluate the specific heat for the four constant parameter processes.

For n = 0, the pressure is constant, and from Eq. (5.15)

$$c_n = c_v[(k - n)/(1 - n)] = c_v k = c_p$$

For n = 1, the temperature is constant, so that

$$c_n = c_v[(k - n)/(1 - n)] = c_v(k/0)$$

or $c_n \to$ infinity. This is necessary since from Eq. (5.16)

$$Q = mc_n(T_2 - T_1)$$

and the heat transfer is the area below the horizontal line on the TS plane, but $T_2 - T_1 = 0$. Therefore, Eq. (5.16) cannot be applied because $Q = (\infty)(0)$, which is indeterminate.

For n = k, the entropy is constant, and

$$c_n = c_v[(k - n)/(1 - n)] = 0$$

and since $T_1 \neq T_2$,

$$Q = mc_n(T_2 - T_1) = 0$$

As $n \to \infty$ the volume is constant, and

$$c_n = c_v[(k - n)/(1 - n)]$$

$$= c_v\left[\left(\frac{k}{n} - 1\right)/\left(\frac{1}{n} - 1\right)\right]$$

$$= c_v$$

To compare W_{sf} and W_{nf} when ΔKE, $\Delta PE = 0$, use

$$\frac{W_{sf}}{W_{nf}} = \frac{Q - \Delta H}{Q - \Delta U} = \frac{mc_n(T_2 - T_1) - mc_p(T_2 - T_1)}{mc_n(T_2 - T_1) - mc_v(T_2 - T_1)}$$

$$= \frac{c_n - c_p}{c_n - c_v} = \frac{c_n - kc_v}{c_n - c_v} = \frac{c_v[(k - n)/(1 - n)] - kc_v}{c_v[(k - n)/(1 - n)] - c_v}$$

$$= \frac{[(k - n)/(1 - n)] - k}{[(k - n)/(1 - n)] - 1} = \frac{k - n - k + nk}{k - n - 1 + n} = \frac{nk - n}{k - 1} = \frac{n(k - 1)}{k - 1} = n$$

which verifies Eq. (3.15).

Example 5.13. 0.5 lbm of hydrogen at a pressure and temperature of 15 psia and 800 F undergoes a polytropic process for which $n = 0.5$ to a pressure of 20 psia. Determine V_1, V_2, T_2, ΔU, ΔH, c_n, ΔS, Q, W_{nf}, and W_{sf} (ΔKE, $\Delta PE = 0$). Sketch the process on the pV and TS planes.

From the perfect gas law

$$V_1 = mRT_1/p_1 = [0.5(767)(1260)]/[15(144)]$$

$$= 223.7 \text{ (ft}^3\text{)} \quad \text{or} \quad 6.34 \text{ (m}^3\text{)}$$

From Eq. (5.12)

$$V_2 = V_1(p_1/p_2)^{1/n} = 223.7(15/20)^2$$

$$= 125.8 \text{ (ft}^3\text{)} \quad \text{or} \quad 3.56 \text{ (m}^3\text{)}$$

From the perfect gas law

$$T_2 = p_2V_2/mR = [20(144)(125.8)/0.5(767)]$$

$$= 945 \text{ (R)} \quad \text{or} \quad 525 \text{ (K)}$$

The temperature of the final state may also be determined by means of Eq. (5.14) when

$$T_2 = T_1(p_2/p_1)^{n-1/n} = 1260(20/15)^{-1}$$

$$= 945 \text{ (R)}$$

For any process

$$U_2 - U_1 = mc_v(T_2 - T_1) = 0.5(2.465)(945 - 1260)$$

$$= -388.2 \text{ (B)} \quad \text{or} \quad -409.2 \text{ (kJ)}$$

and

$$H_2 - H_1 = k(U_2 - U_1) = 1.4(-388.2)$$

$$= -543.5 \text{ (B)} \quad \text{or} \quad -572.8 \text{ (kJ)}$$

From Eq. (5.15)

$$c_n = c_v[(k - n)/(1 - n)] = 2.465[1.4 - 0.5)/(1 - 0.5)]$$

$$= 4.437 \text{ (B/lbm-R)} \quad \text{or} \quad 18.56 \text{ (kJ/kg-K)}$$

and from Eq. (5.16)

$$Q = mc_n(T_2 - T_1) = 0.5(4.437)(945 - 1260)$$

$$= -698.8 \text{ (B)} \quad \text{or} \quad -736.5 \text{ (kJ)}$$

From Eq. (5.17)

$$S_2 - S_1 = mc_n \ln(T_2/T_1) = 0.5(4.437) \ln(945/1260)$$

$$= -0.6382 \text{ (B/R)} \quad \text{or} \quad 1211 \text{ (J/K)}$$

The work terms are easily obtained from

$$W_{nf} = Q - \Delta U = -698.8 + 388.2$$

$$= -310.6 \text{ (B)} \quad \text{or} \quad -327.4 \text{ (kJ)}$$

or with somewhat more difficulty from

$$W_{nf} = (p_2V_2 - p_1V_1)/[J(1 - n)]$$

$$= [(144/778)(1/0.5)][20(125.8) - 15(223.7)]$$

$$= -310.8 \text{ (B)}$$

The steady-flow work, when ΔKE, $\Delta PE = 0$, is

$$W_{sf} = Q - \Delta H = -698.8 + 543.5$$

$$= -155.3 \text{ (B)} \quad \text{or} \quad -163.7 \text{ (kJ)}$$

or

$$W_{sf} = nW_{nf} = 0.5(-310.6) = -155.3 \text{ (B)}$$

The polytropic process, for $n = 0.5$, is sketched in Fig. 5.10. For $n = 0.5$ the process path must be between the constant pressure and isothermal processes. Since $p_2 > p_1$, the path must be upward from right to left on the pV plane. Since $V_1 > V_2$, the nonflow work is negative, and

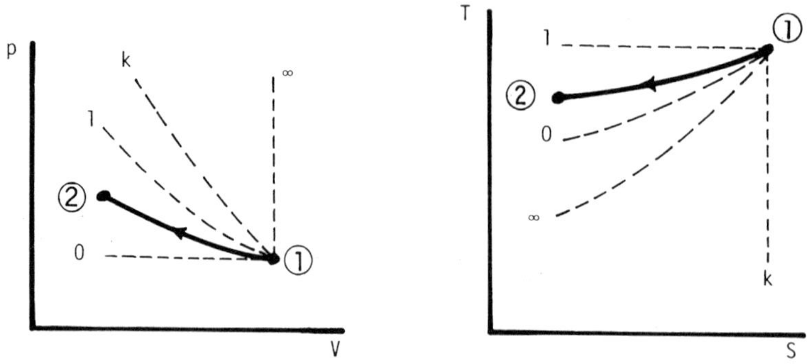

Figure 5.10

since $p_2 > p_1$, the steady-flow work is also negative. Since $T_2 < T_1$, the curve must be downward from right to left on the TS plane, and for $\Delta S < 0$ the value of the heat transfer is negative.

Example 5.14. Two lbm of air undergoes a polytropic process from $p_1 = 40$ psia and $V_1 = 10\ ft^3$ to $p_2 = 18$ psia and $V_2 = 20\ ft^3$. Determine T_1, T_2, ΔU, ΔH, c_n, Q, ΔS, W_{nf}, and W_{sf} ($\Delta KE = 10$ B). Sketch the process on the pV and TS planes.

From the perfect gas law

$$T_1 = p_1V_1/mR = [144(40)(10)]/[2(53.3)] = 540\ (R)$$

and

$$T_2 = p_2V_2/mR = [144(18)(20)]/[2(53.3)] = 486\ (R)$$

For all processes

$$\Delta U = mc_v(T_2 - T_1) = 2(0.1714)(486 - 540)$$

$$= -18.5\ (B) \quad \text{or} \quad -19.5\ (kJ)$$

$$\Delta H = mc_p(T_2 - T_1) = 2(0.24)(486 - 540)$$

$$= -25.9\ (B) \quad \text{or} \quad -27.3\ (kJ)$$

For the polytropic process

$$p_1/p_2 = (V_2/V_1)^n$$

$$40/18 = (20/10)^n$$

so that

$$n = \ln(40/18)/\ln(2) = 1.152$$

The polytropic specific heat is

$$c_n = c_v[(k - n)/(1 - n)] = 0.1714[(1.4 - 1.152)/(1 - 1.152)]$$

$$= -0.2796 \text{ (B/lbm-R)} \quad \text{or} \quad -1.170 \text{ (kJ/kg-K)}$$

Negative values of c_n are valid and result from $k > n > 1$. The heat transfer is

$$Q = mc_n(T_2 - T_1) = 2(-0.2796)(486 - 540)$$

$$= 30.2 \text{ (B)} \quad \text{or} \quad 31.8 \text{ (kJ)}$$

and the entropy change is

$$\Delta S = mc_n \ln(T_2/T_1) = 2(-0.2796) \ln(486/540)$$

$$= 0.0589 \text{ (B/R)} \quad \text{or} \quad -111.7 \text{ (J/K)}$$

The work terms are

$$W_{nf} = Q - \Delta U = 30.2 + 18.5 = 48.7 \text{ (B)} \quad \text{or} \quad 51.3 \text{ (kJ)}$$

and

$$W_{sf} = Q - \Delta H - \Delta KE = 30.2 + 25.9 - 10 = 46.1 \text{ (B)} \quad \text{or} \quad 48.6 \text{ (kJ)}$$

This process is plotted on the pV and TS planes in Fig. 5.11. For $n = 1.152$, the process lies between the isothermal and isentropic process curves. Since $V_2 > V_1$, the nonflow work is positive. Since $p_2 < p_1$, the steady-flow work would be 56.1 B neglecting ΔKE or 46.1 B including ΔKE. Since the entropy change is positive, the heat transfer is also positive.

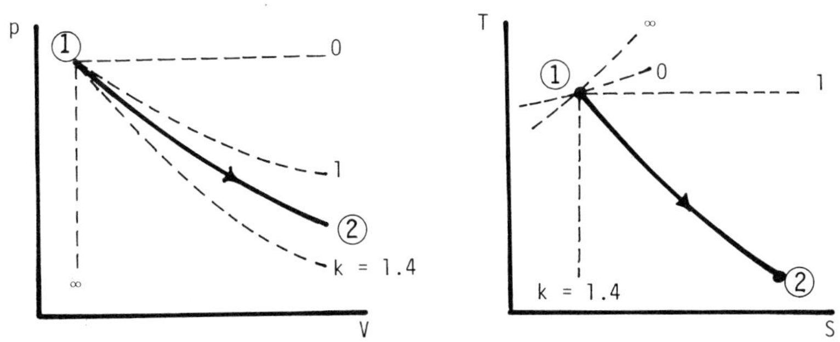

Figure 5.11

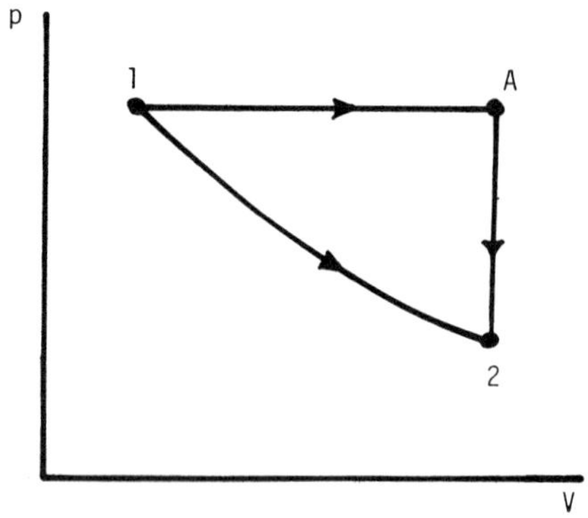

Figure 5.12

GENERAL ENTROPY EQUATIONS

Changes in entropy have previously been calculated by use of special equations for each process, including the use of the polytropic specific heat for the polytropic process. Given the polytropic process between the initial and final states as sketched in Fig. 5.12, this process may be replaced by a composite constant pressure and constant volume process. Since entropy is a property, the changes in entropy for the polytropic process and for the composite process will be identical. For the constant pressure process 1A,

$$S_A - S_1 = mc_p \ln(T_A/T_1)$$

and for the constant volume process A2,

$$S_2 - S_A = mc_v \ln(T_2/T_A)$$

Adding the above equations yields

$$S_2 - S_1 = mc_p \ln(T_A/T_1) + mc_v \ln(T_2/T_A)$$

but

$$V_A/V_1 = V_2/V_1 = T_A/T_1, \qquad p = C$$

since $V_2 = V_A$ and

$$p_2/p_A = p_2/p_1 = T_2/T_A, \qquad V = C$$

where $p_1 = p_A$.

The entropy equation may now be written

$$S_2 - S_1 = mc_p \ln(V_2/V_1) + mc_v \ln(p_2/p_1) \tag{5.18}$$

This is a general entropy equation that relates the absolute pressure and volumes; c_p and c_v must be constant, fixed value, physical properties.

From the perfect gas equation

$$p_1V_1/T_1 = p_2V_2/T_2 \quad \text{or} \quad V_2/V_1 = p_1T_2/p_2T_1$$

Substitution into Eq. (5.18) yields

$$\begin{aligned} S_2 - S_1 &= mc_p \ln(p_1T_2/p_2T_1) + mc_v \ln(p_2/p_1) \\ &= mc_p \ln(T_2/T_1) + mc_p \ln(p_1/p_2) + mc_v \ln(p_2/p_1) \\ &= mc_p \ln(T_2/T_1) - mc_p \ln(p_2/p_1) + mc_v \ln(p_2/p_1) \\ &= mc_p \ln(T_2/T_1) - m(c_p - c_v) \ln(p_2/p_1) \end{aligned}$$

but $c_p - c_v = R/J$ so that

$$S_2 - S_1 = mc_p \ln(T_2/T_1) - (mR/J) \ln(p_2/p_1) \tag{5.19}$$

This is a general entropy equation that relates the absolute temperature and pressures. If $p_2 = p_1 = C$, the second term in Eq. (5.19) is zero, and

$$S_2 - S_1 = mc_p \ln(T_2/T_1), \qquad p = C$$

In like manner a third general entropy equation may be derived which gives

$$S_2 - S_1 = mc_v \ln(T_2/T_1) + (mR/J) \ln(V_2/V_1) \tag{5.20}$$

For a constant volume process the second term in Eq. (5.20) is zero, and

$$S_2 - S_1 = mc_v \ln(T_2/T_1), \qquad V = C$$

Since entropy is a property, these general equations may be used to calculate the entropy change of a perfect gas for open or closed systems and for any process, reversible or irreversible.

Example 5.15. Determine the entropy change, by use of the general equations, for Example 5.14 where $p_1 = 40$ psia, $V_1 = 10$ ft^3, $T_1 = 540$ R, $p_2 = 18$ psia, $V_2 = 20$ ft^3, and $T_2 = 486$ R.

In Example 5.14 the entropy change was calculated by means of

$$\Delta S = mc_n \ln(T_2/T_1) = 0.0589 \text{ (B/R)}$$

for m = 2 lbm of air and $c_n = -0.2796$ B/R.

By using Eq. (5.18)

$$\Delta S = mc_p \ln(V_2/V_1) + mc_v \ln(p_2/p_1)$$

$$= 2(0.24) \ln(20/10) + 2(0.1714) \ln(18/40)$$

$$= 0.3327 - 0.2737 = 0.0590 \text{ (B/R)}$$

or by the use of Eq. (5.19)

$$\Delta S = mc_p \ln(T_2/T_1) - (mR/J) \ln(p_2/p_1)$$

$$= 2(0.24) \ln(486/540) - [2(53.3)/778] \ln(18/40)$$

$$= -0.0506 + 0.1094 = 0.0588 \text{ (B/R)}$$

and finally by use of Eq. (5.20)

$$\Delta S = mc_v \ln(T_2/T_1) + (mR/J) \ln(V_2/V_1)$$

$$= 2(0.1714) \ln(486/540) + [2(53.3)/778] \ln(20/10)$$

$$= -0.0361 + 0.0950 = 0.0589 \text{ (B/R)}$$

IRREVERSIBLE ADIABATIC PROCESS

For an isentropic or reversible adiabatic process, $Q = 0$ and $\Delta S = 0$. For an irreversible adiabatic process, $Q = 0$, but $\Delta S \neq 0$.

Example 5.16. Consider the apparatus sketched in Fig. 5.13. The pressure, volume, and temperature of the air in container A are $p_A = 100$ psia, $t_A = 40$ F, and $V_A = 20$ ft^3, and container B has a volume of 30 ft^3.

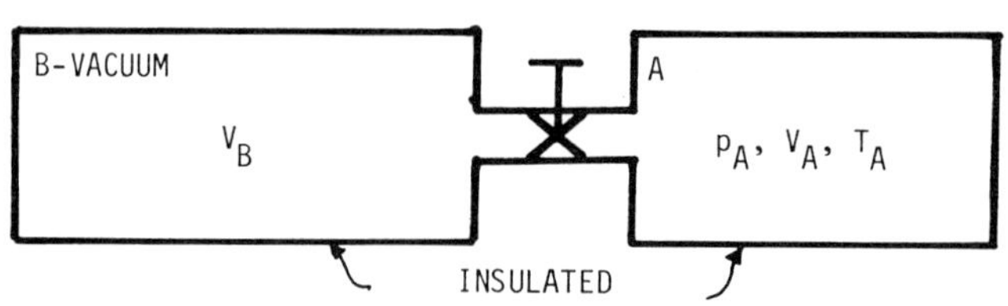

Figure 5.13

The mass of the air in the system is

$$m_a = p_A V_A / RT_A = [144(100)(20)/53.3(500)]$$

$$= 10.81 \text{ (lbm)}$$

When the valve is open the air fills both containers, and $Q = 0$ for an insulated container and $W_{nf} = 0$, since no work device is included in the apparatus.

From the nonflow process equation

$$Q = \Delta U + W_{nf}$$

and

$$\Delta U = 0$$

or

$$mc_v(T - T_A) = 0$$

and the temperature remains constant such that

$$T = T_A = T_2$$

In essence this is Joule's experiment.

The final pressure is

$$p_2 = mRT/V = [10.81(53.3)(500)/(20 + 30)144]$$

$$= 40.0 \text{ (psia)}$$

From Eq. (5.18)

$$\Delta S = mc_p \ln(V_2/V_1) + mc_v \ln(p_2/p_1)$$

$$= 10.81(0.24) \ln(50/20) + 10.81(0.1714) \ln(40/100)$$

$$= 0.6795 \text{ (B/R)}$$

From Eq. (5.19)

$$\Delta S = mc_p \ln(T_2/T_1) - (mR/J) \ln(p_2/p_1)$$

$$= 10.81(0.24) \ln(500/500) - [10.81(53.3)/778] \ln(40/100)$$

$$= 0.6786 \text{ (B/R)}$$

From Eq. (5.20)

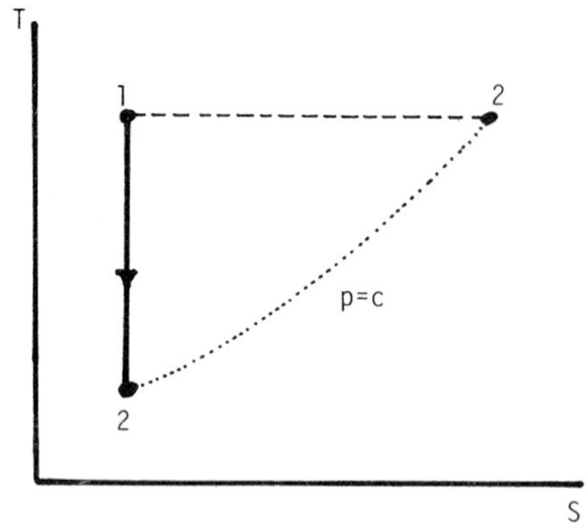

Figure 5.14

$$\Delta S = mc_v \ln(T_2/T_1) + (mR/J) \ln(V_2/V_1)$$
$$= 10.81(0.1714) \ln(500/500) + [10.81(53.3)/778] \ln(50/20)$$
$$= 0.6786 \text{ (B/R)}$$

so that all three values of ΔS are essentially the same.

The isentropic and irreversible adiabatic processes between the same pressure limits are plotted in Fig. 5.14. The irreversible adiabatic process is also an isothermal process and is illustrated by use of a dashed line instead of a solid line. Dashed lines are used to indicate that the process is irreversible, and the areas below or behind the irreversible curves on the pV and TS planes do not represent the heat or work terms.

EQUATIONS FOR IRREVERSIBLE AND REVERSIBLE PROCESSES

Properties are point functions; if the initial and final states are known, any equation that represents the property changes between these states is valid. For properties, the evaluation of the property change by use of a reversible process is valid for an irreversible process.

Heat and work are path functions; equations such as

$$W_{nf} = (p_2V_2 - p_1V_1)/[J(1 - n)], \qquad n \neq 1$$
$$= (p_1V_1/J) \ln(V_2/V_1), \qquad n = 1$$
$$W_{sf} = nW_{nf} \quad \text{and} \quad Q = mc_n(T_2 - T_1)$$

may not be used for irreversible processes

Process equations for the open or closed system such as

$$Q = \Delta U + W_{nf}$$

or

$$Q = \Delta H + \Delta KE + \Delta PE + W_{sf}$$

are valid for all processes. Again, unless specifically stated to be an irreversible process, all processes are reversible.

PROBLEMS

5.1 Determine the volume occupied by 5 lbm of carbon dioxide, CO_2, at a pressure and temperature of 50 psia and 80 F.

5.2 A rigid tank has a volume of 60 ft^3 and contains air at a pressure and temperature of 20 psia and 80 F. After 200 B of heat is added to the tank, determine the temperature and the pressure in the tank. Air is slowly bled from the tank, so that the temperature remains constant, until the pressure returns to its original value. Determine the amount of air removed. The sealed tank is then cooled to its original temperature. Determine the final pressure in the tank.

5.3 A monoatomic gas has a molecular weight of 20. Determine the value of the physical constants R, c_p, and c_v.

5.4 Five lbm of air at a pressure and temperature of 100 psia and 40 F undergoes a constant pressure process to a temperature of 240 F. Determine V_1, V_2, ΔU, ΔH, ΔS, Q, W_{nf}, and W_{sf} (ΔKE = -40 B, ΔPE = 0). Sketch the process on the pV and TS planes.

5.5 Oxygen at a pressure, temperature, and volume of 40 psia, 80 F, and 20 ft^3 undergoes a constant volume process to a pressure of 60 psia. Determine T_2, ΔU, ΔH, ΔS, Q, W_{nf}, and W_{sf} (ΔKE, ΔPE = 0). Sketch the processes on the pV and TS planes.

5.6 Two lbm of helium at a pressure and volume of 100 psia and 30 ft^3 undergoes an isothermal process to a volume of 10 ft^3. Determine T, p_2, ΔU, ΔH, Q, ΔS, W_{nf}, and W_{sf} (ΔKE, ΔPE = 0). Sketch the processes on the pV and TS planes.

5.7 Air at a pressure and temperature of 20 psia and 40 F undergoes an isentropic process to a pressure of 60 psia. Determine v_1, v_2, T_2, Δu, Δh, Δs, Q, W_{nf}, and W_{sf} (ΔKE, ΔPE = 0). Sketch the processes on the pV and TS planes.

5.8 0.5 lbm of air undergoes a polytropic process for which n = 1.2. The temperature increases from 40 F to 540 F. Determine V_2/V_1, p_2/p_1, ΔU, ΔH, c_n, ΔS, Q, W_{nf}, and W_{sf}. Sketch the processes on the pV and TS planes.

5.9 Hydrogen undergoes a process in which the absolute pressure and volume are doubled. The initial temperature is 40 F. Determine n, T_2, Δu, Δh, c_n, Δs, Q, W_{nf}, and W_{sf} (ΔKE, ΔPE = 0). Sketch the processes on the pV and TS planes.

5.10 Determine ΔS for the process in Prob. 5.8 by means of Eqs. (5.18), (5.19), and (5.20).

5.11 Determine Δs for the process in Prob. 5.9 by means of Eqs. (5.18), (5.19), and (5.20).

6

INTERNAL COMBUSTION ENGINES

In this chapter we shall study the theoretical cycles that are the models for the spark-ignited and compression-ignited internal combustion engines. Three different cycles will be presented; the Otto cycle is the model for the spark-ignited engine, the Diesel cycle is the model for the low-speed compression-ignition engine, and the dual cycle is the model for the high-speed compression-ignition engine. The effects of the volumetric efficiency and a supercharger on the Otto cycle engine will be examined. At the close of this chapter we shall consider the testing techniques employed to determine internal combustion engine performance.

The closed or nonflow system is the basis of the three theoretical cycles used to simulate the spark-ignited or compression-ignited engines. Using the closed cycle as a model implies several important differences between the theoretical model and the real engine. For a closed system, mass does not cross the system boundaries; therefore the theoretical model does not require intake or exhaust valves. Heat is transferred across the system boundaries to approximate the combustion of fuel or the exhaust stroke in the real engine. There is no counterpart for the intake stroke in the theoretical cycle.

OTTO CYCLE

The Otto cycle is the theoretical model for the spark-ignited internal combustion engine, which is the engine used in most automobiles. The system is sketched in Fig. 6.1; when the piston is at state 2 or 3 it is at the top dead-center position. At the thermodynamic state 1 or 4 the piston is at the bottom dead-center position. The working substance in the closed system is assumed to be air acting as a perfect gas. The

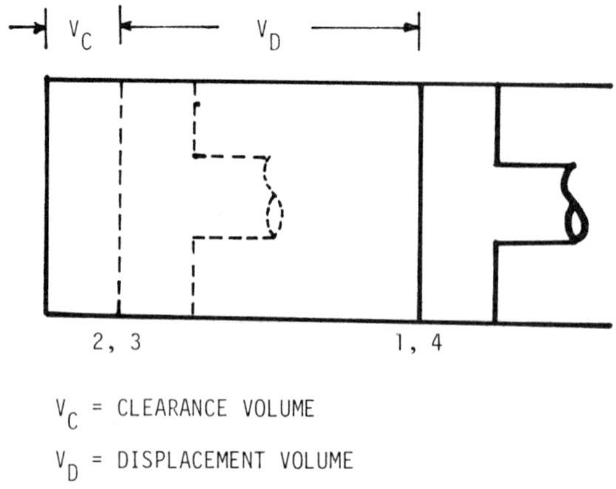

Figure 6.1

theoretical Otto cycle is sketched in Fig. 6.2. The cycle consists of an isentropic compression process from state 1 to state 2. Heat is transferred into the engine during the constant volume process from state 2 to state 3. An isentropic expansion, which represents the power stroke in the real engine, connects states 3 and 4. Heat rejection, which is necessary to close the cycle, occurs in the constant volume process from state 4 to state 1.

An analysis of the cycle can be made with respect to either the pV or TS diagram. For the TS diagram, assuming that all processes are reversible,

$$Q_{12} = 0 \qquad \text{since } S = C$$

$$Q_{23} = mc_v(T_3 - T_2) \qquad \text{since } V = C$$

$$Q_{34} = 0 \qquad \text{since } S = c$$

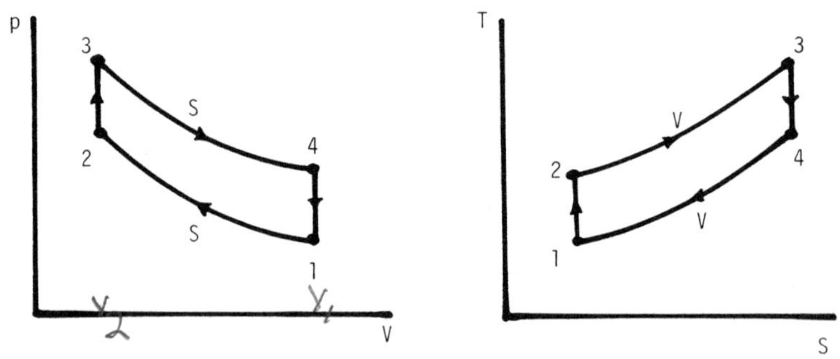

Figure 6.2

and

$$Q_{41} = mc_v(T_1 - T_4) \qquad \text{since } V = C$$

Q_{23} represents the heat added to the engine since $T_3 > T_2$. Q_{41} represents the heat rejected by the engine since $T_1 < T_4$.

The net heat or work for the cycle is

$$\Sigma Q = \Sigma W = Q_{12} + Q_{23} + Q_{34} + Q_{41}$$

$$= mc_v(T_3 - T_2) + mc_v(T_1 - T_4) = mc_v(T_3 - T_2) - mc_v(T_4 - T_1) \quad (6.1)$$

The efficiency of any heat engine cycle is defined by Eq. (4.21),

$$e = \Sigma W/Q_{in}$$

or

$$e = [mc_v(T_3 - T_2) - mc_v(T_4 - T_1)]/[mc_v(T_3 - T_2)]$$

where the Q_{in} term is represented by the left to right process, $\Delta S > 0$, on the TS diagram. For this cycle the heat added is represented by the process from state 2 to state 3.

The cycle efficiency equation may be reduced to

$$e = 1 - [(T_4 - T_1)/(T_3 - T_2)] \quad (6.2)$$

When the pV diagram is used to analyze the Otto cycle

$$W_{12} = (p_2V_2 - p_1V_1)/[J(1 - k)] \qquad \text{since } S = C$$

$$W_{23} = 0 \qquad \text{since } V = C$$

$$W_{34} = (p_4V_4 - p_3V_3)/[J(1 - k)] \qquad \text{since } S = C$$

and

$$W_{41} = 0 \qquad \text{since } V = C$$

The net work or net heat of the cycle is

$$\Sigma W = \Sigma Q = \{(p_2V_2 - p_1V_1)/[J(1 - k)]\} + \{(p_4V_4 - p_3V_3)/[J(1 - k)]\}$$

$$= (p_2V_2 - p_1V_1 + p_4V_4 - p_3V_3)/[J(1 - k)]$$

$$= mR(T_2 - T_1 + T_4 - T_3)/[J(1 - k)]$$

Since, from Eq. (5.7),

$$c_v = R/[J(k - 1)]$$

then

$$\Sigma W = \Sigma Q = -mc_v(T_2 - T_1 + T_4 - T_3) = mc_v(T_3 - T_2) - mc_v(T_1 - T_4)$$

which is identical to the results expressed in Eq. (6.1) based on a TS cycle analysis.

Another parameter of importance in the study of the internal combustion engine is the mean effective pressure. This parameter is often abbreviated to mep or p_m. In Fig. 6.3 the net work of the Otto cycle has been reduced to a rectangle having a height p_m and a width $V_1 - V_2$ such that the area, which represents the work on the pV plane, is the same for the Otto cycle and the rectangle. In equation form

$$p_m = \Sigma W/(V_1 - V_2)$$

or in general

$$p_m = \Sigma W/(V_{max} - V_{min}) \tag{6.3}$$

where V_{max} is the maximum volume in the engine, bottom dead center, and V_{min} is the minimum volume in the engine, top dead center. From Fig. 6.3

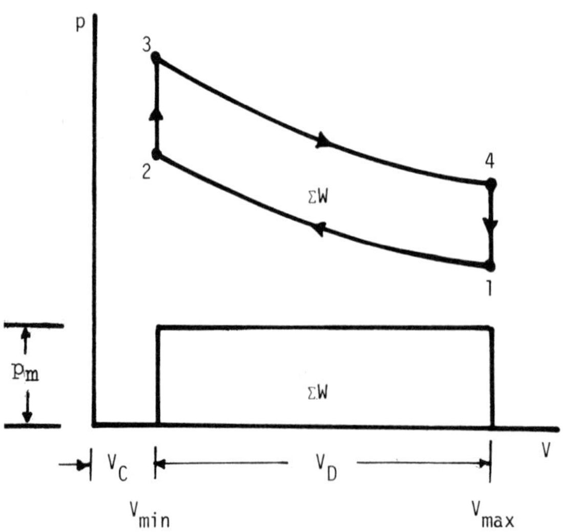

Figure 6.3

$$V_{max} = V_1 = V_C + V_D \tag{6.4}$$

and

$$V_{min} = V_2 = V_C \tag{6.5}$$

where V_D is the volumetric displacement of the engine between the top and bottom dead-center positions of the piston. V_C is the clearance volume or the volume remaining in the engine when the piston is at top dead center.

While Eq. (6.2) is a valid expression for the efficiency of the Otto cycle, it does have one important drawback. In a real engine the temperature at state 1 is directly related to the temperature of the air surrounding the engine. Since the surrounding air temperature is a variable, the determination of the cycle efficiency would require the evaluation of the other three state temperatures. We would prefer to express the cycle efficiency in terms of an engine parameter that is either constant or related to the mode of engine operation.

The compression ratio is a parameter that is fixed for a given engine where

$$r_k = V_1/V_2 \tag{6.6}$$

and r_k is the isentropic compression ratio for the process between system states 1 and 2. By use of Eqs. (6.4) and (6.5)

$$r_k = (V_C + V_D)/V_C \tag{6.7}$$

For the isentropic process between states 1 and 2

$$T_2/T_1 = (V_1/V_2)^{k-1} = r_k^{k-1}$$

or

$$T_2 = T_1 r_k^{k-1}$$

For the isentropic process between states 3 and 4

$$T_3/T_4 = (V_4/V_3)^{k-1} = (V_1/V_2)^{k-1} = r_k^{k-1}$$

since, from the pV diagram, $V_1 = V_4$ and $V_2 = V_3$. Therefore

$$T_3 = T_4 r_k^{k-1}$$

and by substitution into Eq. (6.2)

$$e = 1 - [(T_4 - T_1)/(T_3 - T_2)]$$

$$= 1 - [(T_4 - T_1)/(T_4 r_k^{k-1} - T_1 r_k^{k-1})]$$

$$= 1 - [(T_4 - T_1)/r_k^{k-1}(T_4 - T_1)]$$

$$= 1 - (1/r_k^{k-1}) \tag{6.8}$$

As the compression ratio of the Otto cycle or spark-ignited internal combustion increases, the theoretical cycle efficiency must increase.

Example 6.1. Determine the Otto cycle efficiency and the temperature and pressure at the end of the compression process for an isentropic compression ratio r_k of 4, 5, 6, 7, 8, 9, and 10. Assume the working substance is air, as a perfect gas, so that $k = 1.4$. The temperature and pressures at state 1 are 40 F and 15 psia.

For $r_k = 4$ from Eq. (6.8)

$$e = 1 - (1/r_k^{k-1}) = 1 - [1/(4)^{0.4}] = 1 - 0.574$$

$$= 0.426 \quad \text{or} \quad 42.6\%$$

For the isentropic process between states 1 and 2

$$T_2/T_1 = (V_1/V_2)^{k-1}$$

or

$$T_2 = T_1 r_k^{k-1} = 500(4)^{0.4} = 871 \text{ (R)}$$

Also

$$p_2/p_1 = (V_1/V_2)^K$$

or

$$p_2 = p_1 r_k^k = 15(4)^{1.4} = 104.5 \text{ (psia)}$$

The results of the other compression ratios are listed in the following table:

r_k	e	T_2 (R)	p_2 (psia)
4	42.6	871	104.5
5	47.5	952	142.8
6	51.2	1024	184.3
7	54.1	1089	228.7
8	56.5	1149	275.7
9	58.5	1204	325.1
10	60.2	1256	376.8

A curve of theoretical cycle efficiency as a function of the isentropic compression ratio is shown in Fig. 6.4. Since there is a direct correlation between the efficiency of an actual engine and the theoretical cycle efficiency, the question could be asked, why not increase the compression ratio without limit? In any internal combustion engine an increase in the compression ratio increases the maximum pressure generated in the engine. As the pressure increases the engine must be sturdier and of heavier construction, thereby increasing both weight and expense. For the spark-ignited engine the fuel is normally compressed with the incoming air. As the compression ratio increases a temperature will be reached where the air-gas mixture fires before the spark plug is activated. This results in an expansion of the burning gases against the still upward motion of the piston. Such preignition or detonation increases the stresses on the engine and shortens engine life as well as reducing the power produced by the engine.

Example 6.2. An engine has a cylinder diameter, bore of 3.438 in., and a stroke of 4.25 in. The clearance volume is 6.60 in.^{3} The engine,

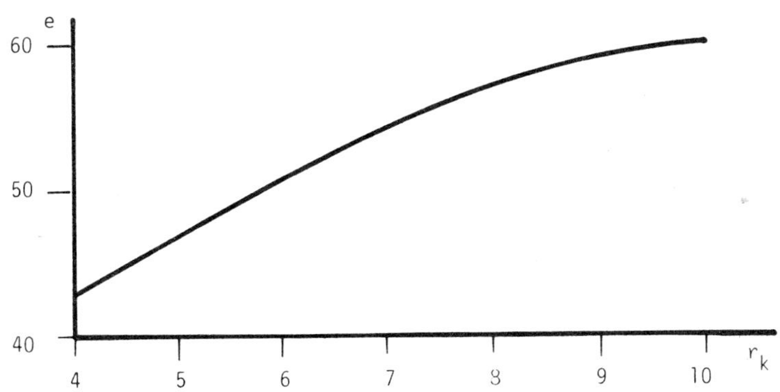

Figure 6.4

operating on the Otto cycle, has a pressure and temperature, at the start of the compression process, of 14.7 psia and 40 F. The temperature after the heat addition process is 2540 F. Determine the cycle efficiency, the net work, and the mean effective pressure.

The engine displacement is

$$V_D = (\Pi/4)(\text{bore})^2(\text{stroke}) = (\Pi/4)D^2L$$

$$= (\Pi/4)(3.438)^2(4.25)$$

$$= 39.45 \text{ (in.}^3)$$

From Eq. (6.7)

$$r_k = (V_C + V_D)/V_C = (6.60 + 39.45)/6.60 = 7.0$$

The efficiency of the cycle, from Eq. (6.8), is

$$e = 1 - (1/r_k^{k-1}) = 1 - [1/(7)^{0.4}] = 0.541$$

For the isentropic compression process

$$T_2 = T_1 r_k^{k-1} = 500(7)^{0.4} = 1089 \text{ (R)}$$

The heat added, during the constant volume process, is

$$Q_{in} = Q_{23} = c_v(T_3 - T_2)$$

$$= 0.1714(3000 - 1089) = 327.5 \text{ (B/lbm)}$$

since the temperature at the end of the heat addition process is 3000 R.

The net heat or net work for the cycle is [Eq. (4.21)]

$$\Sigma W = \Sigma Q = eQ_{in} = 0.541(327.5)$$

$$= 177.2 \text{ (B/lbm)}$$

From the perfect gas equation, the specific volume at the start of the compression process is [Eq. (5.4)]

$$v_1 = RT_1/p_1 = [53.3(500)]/[14.7(144)]$$

$$= 12.59 \text{ (ft}^3\text{/lbm)}$$

Since the compression ratio is defined by Eq. (6.6),

$$r_k = V_1/V_2 = v_1/v_2$$

or

$$v_2 = v_1/r_k = 12.59/7 = 1.80 \ (\text{ft}^3/\text{lbm})$$

From Eq. (6.3), the mean effective pressure is

$$p_m = \Sigma W/(V_{max} - V_{min})$$

$$= [177.2(778)]/(12.59 - 1.80) = 12,780 \ (\text{psf})$$

$$= 88.7 \ (\text{psi})$$

When the work is calculated in B/lbm it must be converted to ft-lbf/lbm. This requires that $V_{max} - V_{min}$ be expressed in ft^3/lbm.

CYCLE COMPARISON

The real four-stroke engine cycle consists of an intake stroke, a compression stroke, a power stroke, and an exhaust stroke. The processes are sketched in Fig. 6.5. It requires two revolutions of the engine to produce a single pV diagram. Two-stroke cycle engines are limited to smaller engines since efficient intake of fuel and exhaust of combustion products as well as lubrication problems exist in this type of engine. The following table compares the theoretical and the four-stroke cycle engines:

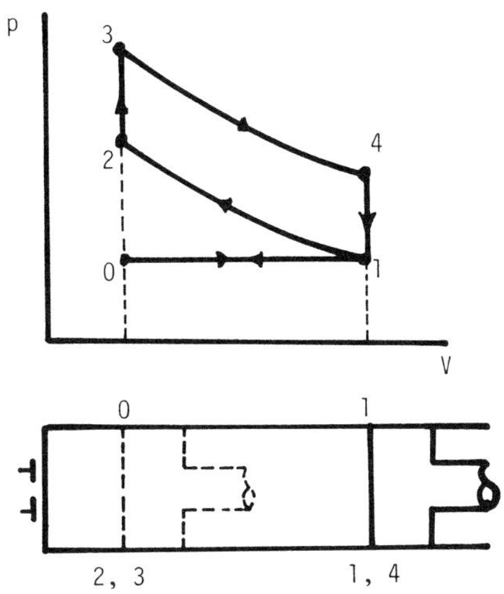

Figure 6.5

Process	Real Engine	Theoretical	Valves	
			Intake	Exhaust
01	Intake	None	Open	Closed
12	Compression	Isentropic compression	Closed	Closed
23	None	Heat addition		
34	Power	Isentropic expansion	Closed	Closed
41	None	Heat rejection		
10	Exhaust	None	Closed	Open

Example 6.3. Assume the engine in Example 6.2 has an engine speed of 3600 rpm when operated on the four-stroke cycle. The engine has six cylinders. Determine the theoretical horsepower of the engine.

From Eq. (6.4)

$$V_1 = V_C + V_D = 6.60 + 39.45 = 46.05 \text{ (in.}^3\text{)}$$

$$= 0.0266 \text{ (ft}^3\text{)}$$

Since $v_1 = 12.59$ ft^3/lbm, the mass of air in the cylinder at the start of the compression process is

$$m = V_1/v_1 = 0.0266/12.59 = 0.00212 \text{ (lbm)}$$

The net work of the cycle is

$$\Sigma W = 0.00212(177.2) = 0.375 \text{ (B)}$$

For an engine speed of 3600 rpm there are 1800 power strokes per minute or

$$\Sigma W = 0.375(1800) = 675 \text{ (B/min)}$$

or

$$= [675(778)]/33{,}000$$

$$= 15.9 \text{ (HP)}$$

and for six cylinders

$$\Sigma W = 6(15.9) = 95.4 \text{ (HP)} \quad \text{or} \quad 71.1 \text{ (kW)}$$

This is the power that the engine would develop if it could operate using reversible processes without friction.

DIESEL CYCLE

In the spark-ignited engine both the fuel and air undergo the compression process, and fuel combustion is initiated by the firing of the spark plug slightly before top dead center. In the Diesel cycle only the air is

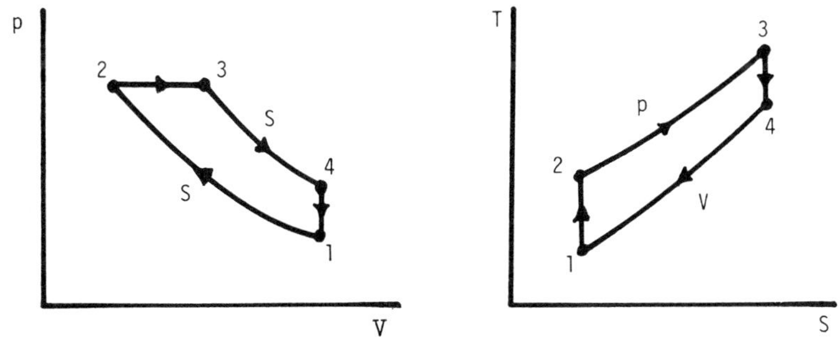

Figure 6.6

compressed, and the fuel is injected at or near to dead center. Compression ratios in the Diesel engine must be larger than those in the Otto cycle since the initial combustion of the fuel is dependent on the temperature at the end of the isentropic compression process. The fuel burns, theoretically at constant pressure, as it is introduced into the engine. The original Diesel engines were slow-speed engines and had relatively large piston bores and strokes. The cycle used as the model for these low-speed engines is shown in Fig. 6.6. The theoretical cycle consists of an isentropic compression, a constant pressure heat addition, an isentropic expansion, and a constant volume rejection of heat process. While the Otto cycle was analyzed from both the pV and TS diagrams, it is sufficient, and generally simpler, to analyze only the heat transfer terms as obtained from the TS diagram. For the isentropic processes the heat transfer is zero so that

$$Q_{in} = Q_{23} = mc_p(T_3 - T_2) \quad \text{where } P = C$$

and

$$Q_{out} = Q_{41} = mc_v(T_1 - T_4) \quad \text{where } V = C$$

The net work or net heat of the cycle is given by

$$\Sigma Q = \Sigma W = Q_{in} + Q_{out} = Q_{12} + Q_{41} = mc_p(T_3 - T_2) - mc_v(T_4 - T_1)$$

The efficiency of the cycle is [Eq. (4.21)]

$$e = \Sigma W/Q_{in} = [mc_p(T_3 - T_2) - mc_v(T_4 - T_1)]/[mc_p(T_3 - T_2)]$$

$$= 1 - [c_v(T_4 - T_1)]/[c_p(T_3 - T_2)]$$

or

$$= 1 - [(T_4 - T_1)/k(T_3 - T_2)] \tag{6.9}$$

As in the Otto cycle, we would prefer to express the cycle efficiency in terms of engine parameters. Also as in the Otto cycle, the isentropic compression ratio is [Eq. (6.6)]

$$r_k = V_1/V_2$$

and

$$T_2 = T_1(V_1/V_2)^{k-1} = T_1 r_k^{k-1}$$

For the process from state 2 to state 3 it is necessary to define a new engine parameter r_c, known as the cutoff ratio. The pressure may be theoretically maintained at the value of state 2 as long as fuel is injected into the engine. When fuel injection stops or is cut off by the throttle setting, the pressure drops inside the cylinder, and the isentropic expansion process is initiated. Therefore

$$r_c = V_3/V_2 = T_3/T_2 \qquad \text{since } p_2 = p_3 = C$$

Since

$$T_3 = r_c T_2$$

then

$$T_3 = T_1 r_c r_k^{k-1}$$

For the isentropic process from state 3 to state 4

$$T_4/T_3 = (V_3/V_4)^{k-1}$$

but

$$V_4 = V_1 \quad \text{and} \quad V_3 = r_c V_2$$

so that

$$T_4/T_3 = (r_c V_2/V_1)^{k-1}$$

and since

$$V_2/V_1 = 1/r_k$$

then

$$T_4 = T_3(r_c/r_k)^{k-1}$$

Using the previously determined value of T_3 yields

$$T_4 = T_1 r_c r_k^{k-1}(r_c/r_k)^{k-1}$$

$$= T_1 r_c^k$$

By substitution of the values of T_2, T_3, and T_4, expressed in terms of r_c and r_k, into Eq. (6.9),

$$e = 1 - [(T_4 - T_1)/k(T_3 - T_2)$$

$$= 1 - [(T_1 r_c^k - T_1)/k(T_1 r_c r_k^{k-1} - T_1 r_k^{k-1})]$$

$$= 1 - \{(r_c^k - 1)/[k r_k^{k-1}(r_c - 1)]\} \quad (6.10)$$

In the Diesel cycles two engine parameters are included in the efficiency equation, the compression ratio and the cutoff ratio. In the Otto cycle only the compression ratio was required to establish the cycle efficiency. The mean effective pressure of the Diesel cycle is defined by Eq. (6.3), which is the same equation as used for the Otto cycle.

Example 6.4. A Diesel cycle, operating with air, has t_1 = 40 F, p_1 = 14 psia, r_k = 15, and r_c = 1.5. Determine the cycle efficiency Q_{in}, net work, and the mean effective pressure.

From Eq. (6.10)

$$e = 1 - \{(r_c^k - 1)/[k r_k^{k-1}(r_c - 1)]\}$$

$$= 1 - \{(1.5^{1.4} - 1)/[1.4(15)^{0.4}(1.5 - 1)]\} = 1 - (0.764/2.068)$$

$$= 0.631 \quad \text{or} \quad 63.1\%$$

From the temperature equations

$$T_1 = 500 \text{ (R)}$$

$$T_2 = T_1 r_k^{k-1} = 500(15)^{0.4} = 1477 \text{ (R)}$$

$$T_3 = T_2 r_c = 1477(1.5) = 2216$$

and

$$T_4 = T_1 r_c^{\,k} = 500(1.5)^{1.4} = 882 \text{ (R)}$$

By use of Eq. (6.9)

$$e = 1 - [(T_4 - T_1)/k(T_3 - T_2)] = 1 - [(882 - 500)/1.4(2216 - 1477)]$$

$$= 0.631 \quad \text{or} \quad 63.1\%$$

The heat added to the cycle is

$$Q_{in} = Q_{23} = c_p(T_3 - T_2) = 0.24(2216 - 1477)$$

$$= 177.4 \text{ (B/lbm)} \quad \text{or} \quad 412.6 \text{ (kJ/kg)}$$

and the net work or net heat of the cycle is

$$\Sigma W = \Sigma Q = eQ_{in} = 0.631(177.4)$$

$$= 111.9 \text{ (B/lbm)} \quad \text{or} \quad 260.3 \text{ (kJ/kg)}$$

The specific volumes are

$$v_1 = RT_1/p_1 = [53.3(500)]/[14(144)]$$

$$= 13.22 \text{ (ft}^3\text{/lbm)}$$

and

$$v_2 = v_1/r_k = 13.22/15 = 0.88 \text{ (ft}^3\text{/lbm)}$$

From Eq. (6.3)

$$p_m = \Sigma W/(V_{max} - V_{min}) = \Sigma W/(v_1 - v_2)$$

$$= [111.9(778)]/[(13.22 - 0.88)144]$$

$$= 49.0 \text{ (psi)} \quad \text{or} \quad 337.9 \text{ (kPa)}$$

This value of p_m is relatively low and indicates that the engine is being operated at about 50% of the full throttle setting.

<u>Example 6.5</u>. Given data for a Diesel cycle, using air, are $r_c = 3.0$, $p_1 = 14$ psia, and $t_1 = 40$ F. Determine and plot engine data as a result of varying the isentropic compression ratio r_k.

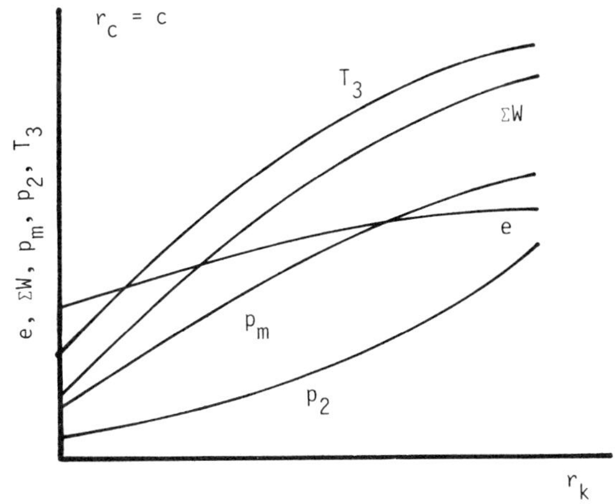

Figure 6.7

r_k	e (%)	Q_{in} (B/lbm)	ΣW (B/lbm)	p_m (psi)	p_2 (psia)	T_3 (R)
8	43.2	551.5	238.2	111.2	257.3	3447
10	48.1	602.9	290.0	131.7	351.6	3768
12	51.7	648.5	335.3	155.4	453.9	4053
14	54.5	698.8	380.8	167.6	563.3	4311
16	56.9	727.7	414.1	180.5	679.0	4548
18	58.9	762.7	449.2	194.4	800.8	4767
20	60.6	795.4	482.0	207.4	928.0	4971

The results of varying the compression ratio are sketched in Fig. 6.7. As shown in Fig. 6.7, for a fixed cutoff ratio, the theoretical cycle efficiency, net work, and mean effective pressure increase as the compression ratio is increased. The maximum pressure and temperature, p_2 and T_3, also increase with increasing compression ratios.

The strength of the engine structure must be increased as the maximum pressure generated in the cycle increases. Even though the maximum temperature, in the cycle, exists for only a short period of time, lubrication problems increase with higher engine temperatures. Each design for the Diesel cycle has a trade-off point where increases in the cycle efficiency and power are balanced by the greater cost of the engine.

Example 6.6. Given data for a Diesel cycle, using air, are r_k = 16, p_1 = 14 psia, and t_1 = 40 F. Determine and plot engine data as a result of varying the cutoff ratio r_c.

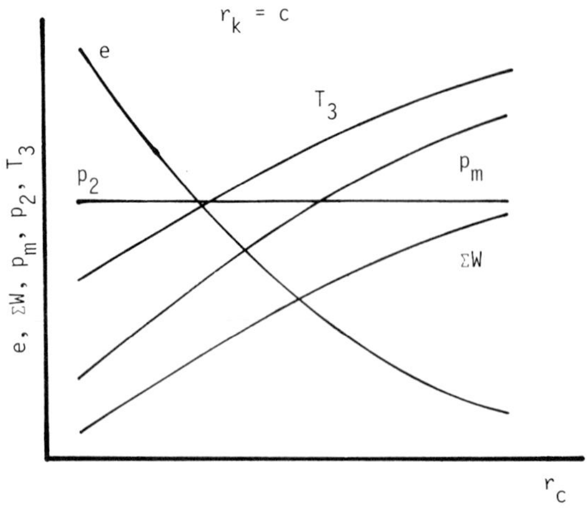

Figure 6.8

r_c	e (%)	Q_{in} (B/lbm)	ΣW (B/lbm)	p_m (psi)	p_2 (psia)	T_3 (R)
1.1	0.664	36.5	24.2	10.6	679.0	1668
1.25	0.654	91.0	59.5	25.9	679.0	1895
1.50	0.640	181.9	116.4	50.7	679.0	2274
1.75	0.626	272.9	170.8	74.5	679.0	2653
2.00	0.614	363.8	223.4	97.4	679.0	3032
2.50	0.591	545.8	322.6	140.6	679.0	3790
3.00	0.569	727.7	414.0	180.5	679.0	4548
3.50	0.550	909.6	500.3	218.1	679.0	5306
4.00	0.532	1092.0	580.9	253.2	679.0	6064

The results of varying the cutoff ratio are shown in Fig. 6.8.

Given a fixed compression ratio, which is the case for all actual engines, the head added to the cycle, the net work of the cycle, the mean effective pressure, and the maximum cycle temperature increase with an increase of the fuel cutoff ratio. The cutoff ratio is controlled by the throttle setting in an automobile or truck.

As the power produced by the engine increases the theoretical cycle efficiency decreases. Again this implies a trade-off between the engine power requirements and the economy of engine operation.

DUAL CYCLE

In the high-speed Diesel engine fuel may be injected slightly before the top dead-center position of the piston. This is simulated in the theoretical cycle by assuming that a portion of the fuel is burned at constant

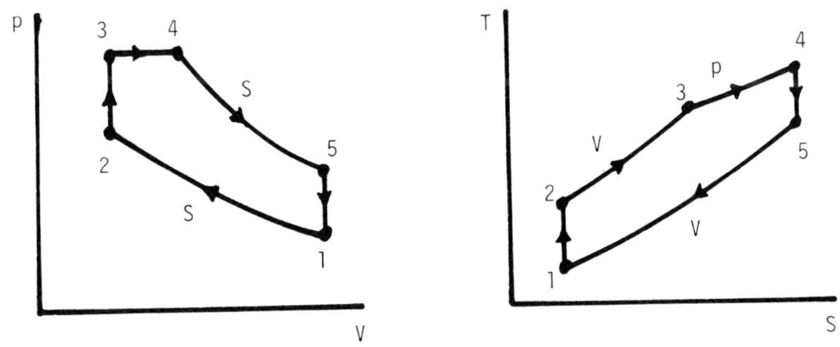

Figure 6.9

volume and the remainder is burned at constant pressure (Fig. 6.9). The fraction of the fuel burned at constant volume or constant pressure is a function of the design of the injection apparatus. The heat transfer into the system is

$$Q_{in} = Q_{23} + Q_{34} = mc_v(T_3 - T_2) + mc_p(T_4 - T_3)$$

and the heat transfer from the system is

$$Q_{out} = Q_{51} = mc_v(T_1 - T_5)$$

The efficiency of the cycle is

$$e = \Sigma Q/Q_{in} = 1 - \{[c_v(T_5 - T_1)]/[c_v(T_3 - T_2) + c_p(T_4 - T_3)]\}$$

$$= 1 - \{(T_5 - T_1)/[(T_3 - T_2) + k(T_4 - T_3)]\} \qquad (6.11)$$

To reduce Eq. (6.11) from temperatures to engine parameters, it is necessary to express each state point temperature in terms of T_1, the state point temperature at the beginning of the isentropic compression process.

For the process 12

$$T_2 = T_1(V_1/V_2)^{k-1} = T_1 r_k^{k-1}$$

For the process 23 it is necessary to define a new engine parameter r_p, the pressure ratio, where

$$r_p = p_3/p_2 = T_3/T_2, \qquad V = C$$

so that

$$T_3 = T_2 r_p = T_1 r_p r_k^{k-1}$$

For the process 34, where the cutoff ratio is defined by

$$r_c = V_4/V_3 = T_4/T_3, \qquad p = C$$

the temperature at state 4 is

$$T_4 = T_3 r_c = T_1 r_p r_k^{k-1} r_c$$

For the process 45, isentropic expansion,

$$T_5 = T_4 (V_4/V_5)^{k-1} = T_4 (r_c V_3/V_1)^{k-1}$$

since

$$V_5 = V_1 \quad \text{and} \quad V_4 = r_c V_3 = r_c V_2$$

$$T_5 = T_4 (r_c V_2/V_1)^{k-1} = T_4 (r_c/r_k)^{k-1}$$

or

$$T_5 = T_1 r_p r_k^{k-1} r_c (r_c/r_k)^{k-1} = T_1 r_c^k r_p$$

Substitution of the temperatures into Eq. (6.11) gives

$$e = 1 - (r_c^k r_p - 1)/\{r_k^{k-1}[(r_p - 1) + kr_p(r_c - 1)]\} \tag{6.12}$$

Example 6.7. A high-speed Diesel engine operating on the dual cycle has the following properties: $p_1 = 14.696$ psia, $t_1 = 40$ F, and $r_k = 16$; also $p_{max} = 1000$ psia and $t_{max} = 2540$ F. Determine the cycle efficiency and the mean effective pressure.

For the isentropic compression process, 12,

$$T_2 = T_1 r_k^{k-1} = 500(16)^{0.4} = 1516 \text{ (R)}$$

assuming that air is the working substance. Since

$$p_2 = p_1 (V_1/V_2)^k = 14.696(16)^{1.4} = 712.8 \text{ (psia)}$$

for the constant volume heat addition process, 23,

$$r_p = p_3/p_2 = p_{max}/p_2 = T_3/T_2$$

and

$$r_p = 1000/712.8 = 1.403$$

or $T_3 = T_2 r_p = 1516(1.403) = 2127$ (R)

For the constant pressure heat addition process, 34,

$$r_c = V_4/V_3 = T_4/T_3 = T_{max}/T_3$$

or

$$r_c = 3000/2127 = 1.410 \quad \text{and} \quad T_4 = 3000 \text{ (R)}$$

For the isentropic expansion process

$$T_5 = T_1 r_c^{\ k} r_p = 500(1.410)^{1.4}(1.403) = 1135 \text{ (R)}$$

The heat terms are

$$Q_{12} = 0 \qquad s = c$$

$$Q_{23} = c_v(T_3 - T_2) = 0.1714(2127 - 1516) = 104.7 \text{ (B/lbm)} \qquad v = c$$

$$Q_{34} = c_p(T_4 - T_3) = 0.24(3000 - 2127) = 209.5 \text{ (B/lbm)} \qquad p = c$$

$$Q_{45} = 0 \qquad s = c$$

$$Q_{51} = c_v(T_1 - T_5) = 0.1714(500 - 1135) = -108.8 \text{ (B/lbm)} \qquad v = c$$

The net heat or work for the cycle is

$$\Sigma Q = \Sigma W = Q_{12} + Q_{23} + Q_{34} + Q_{45} + Q_{51} = 0 + 104.7 + 209.5 + 0 - 108.8$$

$$= 205.4 \text{ (B/lbm)}$$

and

$$Q_{in} = Q_{23} + Q_{34} = 104.7 + 209.5 = 314.2 \text{ (B/lbm)}$$

The cycle efficiency is

$$e = \Sigma W/Q_{in} = 205.4/314.2 = 0.654 \quad \text{or} \quad 65.4\%$$

By means of Eq. (6.12)

$$e = 1 - (r_c^{\ k} r_p - 1)/\{r_k^{\ k-1}[(r_p - 1) + kr_p(r_c - 1)]\}$$

$$= 1 - [(1.410)^{1.4}(1.403) - 1]/\{(16)^{0.4}[(1.403 - 1)] + 1.4(1.403)(1.410 - 1)]\}$$

$$= 1 - (1.270/3663) = 0.653 \quad \text{or} \quad 65.3\%$$

To calculate the mean effective pressure

$$v_1 = RT_1/p_1 = [53.3(500)]/[14.696(144)]$$

$$= 12.593\ (ft^3/lbm)$$

and

$$v_2 = v_1/r_k = 12.593/16 = 0.787$$

so that

$$p_m = \Sigma W/(v_{max} - v_{min}) = \Sigma W/(v_1 - v_2)$$

$$= [205.4(778)]/[(12.593 - 0.787)144]$$

$$= 94.0\ (psi)$$

Example 6.8. The engine in Example 6.7 has a bore of 4.5 in. and a stroke of 5.0 in. and operates at a speed of 3000 rpm on the four-stroke cycle. Determine the theoretical horsepower for a six-cylinder engine.

$$V_D = V_1 - V_2 = (\Pi/4)D^2L = (\Pi/4)(4.5)^2(5)$$

$$= 79.52\ (in.^3) = 0.0460\ (ft^3)$$

Since

$$r_k = V_1/V_2 = (V_D + V_2)/V_2 = (0.0460 + V_2)/V_2 = 16$$

then

$$V_2 = 0.0031\ (ft^3)$$

and

$$V_1 = V_D + V_2 = 0.0460 + 0.0031 = 0.0491\ (ft^3)$$

The mass at the start of the compression process is

$$m = V_1/v_1 = 0.0491/12.593 = 0.00390\ (lbm)$$

The work per cycle is

$$\Sigma W = m\Sigma W = 0.00390(205.4) = 0.800\ (B)$$

The power for one cylinder is

$$\Sigma W = 0.800(3000/2) = 1200\ (B/min) = 28.3\ (HP)$$

and for six cylinders

$$\Sigma W = 6(28.6) = 169.8 \text{ (HP)}$$

Example 6.9. Determine the theoretical cycle efficiency for a slow-speed Diesel cycle with $p_1 = 14.696$ psia, $t_1 = 40$ F, $p_{max} = 1000$ psia, and $t_{max} = 2540$ F.

After the isentropic compression process, 12,

$$p_2/p_1 = (V_1/V_2)^k = r_k^{\ k}$$

so that

$$r_k^{\ k} = 1000/14.696 = 68.04$$

and

$$r_k = 20.4$$

or

$$T_2 = T_1 r_k^{\ k-1} = 500(20.4)^{0.4} = 1670 \text{ (R)}$$

The cutoff ratio is

$$r_c = T_3/T_2 = 3000/1670 = 1.796$$

From Eq. (6.10)

$$e = 1 - \{(r_c^{\ k} - 1)/[kr_k^{\ k-1}(r_c - 1)]\}$$

$$= 1 - \{(1.796^{1.4} - 1)/[1.4(20.4)^{0.4}(1.796 - 1)]\}$$

$$= 0.659 \quad \text{or} \quad 65.9\%$$

Operating between the same initial state and the same maximum pressure and maximum temperature, the dual cycle is 0.5% less efficient than the Diesel cycle.

VOLUMETRIC EFFICIENCY

The Otto, Diesel, and dual cycles are theoretical cycles based on the concept of the closed or nonflow system. Comparisons between real engine operating conditions and the theoretical cycle may be made by evaluating the mass of air in the real engine cylinder at the start of the compression process.

In real engines, as the engine speed increases, the capability of the engine to draw in either air, in the Diesel engine, or an air-fuel mixture, in the Otto cycle, decreases. The decrease is evaluated by means of the volumetric efficiency. The volumetric efficiency is the ratio of the volume of air actually admitted to the engine, measured at atmospheric pressure and temperature, to the displacement volume V_D of the engine or

$$e_V = V'/V_D \tag{6.13}$$

where V' is the volume of air drawn into the engine and V_D is the displacement volume of the engine. The volumetric efficiency may vary from 60% at high engine speeds to 95% at low engine speeds.

Example 6.10. Determine the theoretical horsepower for an Otto cycle engine with a compression ratio of 8.5 and a maximum cycle temperature of 2540 F. The engine bore is 3.438 in. and has a stroke of 4.25 in. The clearance volume is 6.60 in.^{3} The engine operates on the four-stroke cycle at an engine speed of 3600 rpm. The pressure and temperature of the surrounding air are 14.7 psia and 40 F. The engine has six cylinders. For a volumetric efficiency of 100%, the volume of air drawn into the engine is

$$V' = e_V V_D = e_V(\Pi/4)D^2L$$

$$= 1.0(\Pi/4)(3.438)^2(4.25) = 39.45 \text{ (in.}^3)$$

$$= 0.0228 \text{ (ft}^3)$$

The specific volume of the air drawn into the engine is

$$v' = RT'/p'$$

where T' is the temperature of the surrounding air and p' is the pressure of the surrounding air, so that

$$v' = [53.3(500)]/[14.7(144)] = 12.59 \text{ (ft}^3\text{/lbm)}$$

The mass of air drawn into the engine is

$$m' = V'/v' = 0.0228/12.59 = 0.00181 \text{ (lbm)}$$

This is the quantity of the air drawn into the engine during the intake stroke 01 (Fig. 6.10). An amount of air was present in the engine at the start of the intake stroke as given by

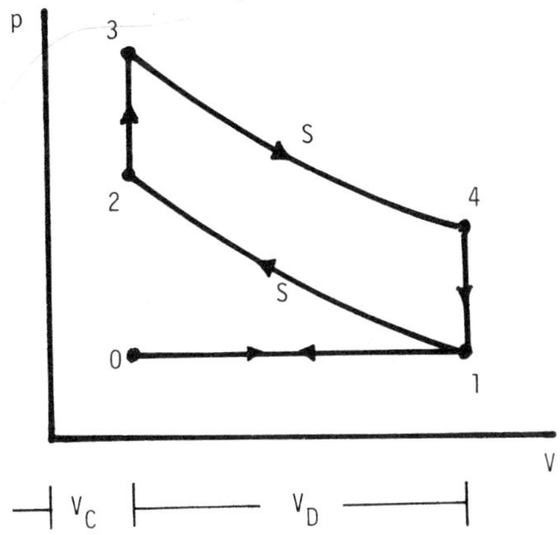

Figure 6.10

$$m_c' = V_2/v' = V_c/v'$$

assuming that v' represents the specific volume of the air in the clearance volume at the start of the intake stroke, so that

$$m_c' = (6.60/1728)/12.59 = 0.00030 \text{ (lbm)}$$

The mass of air in the cylinder at the start of the compression process is

$$m_1' = m' + m_c' = 0.00181 + 0.0030 = 0.00211 \text{ (lbm)}$$

Since

$$r_k = V_1/V_2 = (V_C + V_D)/V_C = (6.60 + 39.45)/6.60$$

$$= 7.0$$

the Otto cycle efficiency is

$$e = 1 - (1/r_k^{k-1}) = 1 - [1/(7.0)^{0.4}] = 0.541$$

and the temperature after the isentropic compression process is

$$T_2 = T_1 r_k^{k-1} = 500(7.0)^{0.4} = 1089 \text{ (R)}$$

The heat added is

$$Q_{in} = Q_{23} = c_v(T_3 - T_2) = 0.1714(3000 - 1089)$$

$$= 327.5 \text{ (B/lbm)}$$

and the net work of the cycle is

$$\Sigma W = eQ_{in} = 0.541(327.5) = 177.2 \text{ (B/lbm)}$$

or

$$\Sigma W = m_1' \, \Sigma W = 0.00211(177.2) = 0.3738 \text{ (B)}$$

The theoretical horsepower of the engine for each cylinder is

$$HP = [0.3738(778)(3600/2)]/33{,}000$$
$$= 15.9$$

and for six cylinders

$$HP = 6(15.9) = 95.4$$

This is the same result as obtained in Example 6.4.

Example 6.11. Repeat Example 6.10 for a volumetric efficiency of 80%. As before

$$V' = e_V V_D = 0.8(\pi/4)(3.438/12)^2(4.25/12)$$
$$= 0.01827 \text{ (ft}^3\text{)}$$

and

$$m' = V'/v' = 0.01827/12.59 = 0.00145 \text{ (lbm)}$$

Also

$$m_1' = m' + m_c' = 0.00145 + 0.00030 = 0.00175 \text{ (lbm)}$$

The net work of the cycle is

$$\Sigma W = m_1' eQ_{in} = 0.00175(0.541)(327.5)$$
$$= 0.3100 \text{ (B)}$$

and the theoretical horsepower is

$$HP = [0.3100(778)(3600/2)(6)]/33{,}000$$
$$= 78.9 \text{ (HP)}$$

Reducing the volumetric efficiency from 100% to 80% reduces the theoretical horsepower from 95.4 to 78.9, which is a 17.3% reduction in the power output of the engine.

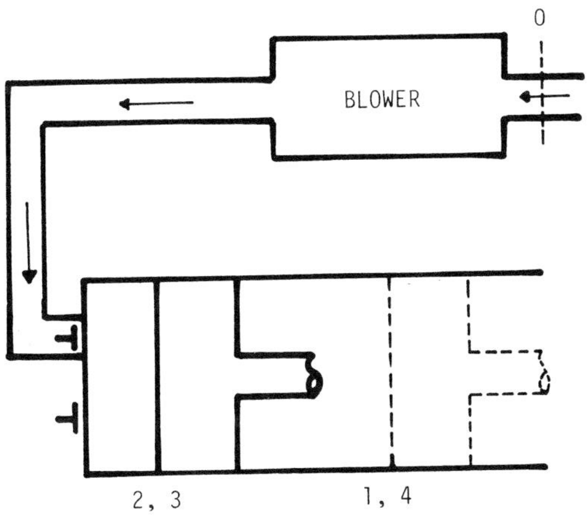

Figure 6.11

SUPERCHARGER

To increase the mass of air intake to the engine in order to combat the effects of volumetric efficiency or, in the case of aircraft, to overcome the decreasing air density with altitude, a supercharger (Fig. 6.11) may be installed.

The supercharger compresses the atmospheric air before it enters the engine, thereby increasing the mass of air in the engine and the theoretical horsepower of the cycle.

Example 6.12. Assume that the Otto cycle engine in Example 6.11 is equipped with a supercharger that raises the inlet pressure to 48 in.-Hg-abs. Determine the theoretical horsepower of the cycle and the effective cycle efficiency.

Assuming that the compression process in the supercharger is isentropic,

$$T_1 = T_0(p_1/p_0)^{(k-1)/k} = 500\{[0.491(48)]/14.7\}^{0.286}$$

$$= 572 \text{ (R)}$$

where the surrounding air properties are 14.7 psia and 40 F.

The specific volume at state point 1 is

$$v_1 = RT_1/p_1 = [53.3(572)]/[0.491(48)(144)]$$

$$= 8.98 \text{ (ft}^3\text{/lbm)}$$

The total volume at state 1 is

$$V_1 = V_C + V_D = 6.60 + (\Pi/4)(3.438)^2(4.25) = 46.05 \text{ (in.}^3)$$

$$= 0.02665 \text{ (ft}^3)$$

and the mass of air at state 1 is

$$m = V_1/v_1 = 0.02665/8.98 = 0.00297 \text{ (lbm)}$$

The temperature after the isentropic compression process is

$$T_2 = T_1 r_k^{k-1} = 572(7)^{0.4} = 1246 \text{ (R)}$$

and if the maximum cycle temperature is maintained at 2540 F,

$$Q_{in} = mc_v(T_3 - T_2) = 0.00297(0.1714)(3000 - 1246)$$

$$= 0.893 \text{ (B)}$$

Since r_k remains constant at 7.0, the cycle efficiency is 54.1%, and the work of the cycle is

$$\Sigma W = eQ_{in} = 0.541(0.893) = 0.483 \text{ (B)}$$

By assuming that the supercharger acts as a steady-flow device,

$$W_{sf} = Q - \Delta H - \Delta KE - \Delta PE$$

and by assuming ΔKE, $\Delta PE = 0$ together with $Q = 0$ for the isentropic process,

$$W_{sf} = -\Delta H = -m'c_p(T_1 - T_0) = -0.00254(0.24)(572 - 500)$$

$$= -0.044 \text{ (B)}$$

where

$$m' = V_D/v_1 = (39.45/12^3)/8.98 = 0.00254 \text{ (lbm)}$$

The net work for the engine is

$$W_{net} = \Sigma W + W_{sf} = 0.483 - 0.044 = 0.439 \text{ (B)}$$

The effective efficiency of the cycle is

$$e = W_{net}/Q_{in} = 0.439/0.893 = 0.492 \quad \text{or} \quad 49.2\%$$

which is 4.9% less than the simple Otto cycle due to the power required to drive the supercharger.

The theoretical horsepower of the engine is

$$HP = [0.439(778)(3600/2)(6)]/33{,}000$$

$$= 111.8 \text{ (HP)}$$

compared to 95.4 HP for the engine with a volumetric efficiency of 100% or 78.9 HP for the engine with a volumetric efficiency of 80%.

TESTING OF INTERNAL COMBUSTION ENGINES

Practical tests for internal combustion engines require the determination of the power developed within the cylinder, the power output of the engine, and the quantity of fuel supplied to the engine. For low-speed engines an indicator may be used to simultaneously record the pressure and volume in the engine cylinder. For high-speed engines a pressure sensor, such as a piezoelectric crystal, may be used to detect the variation in pressure with time. If the engine speed is known, the pressure-time data may be converted to a pressure volume diagram.

The pressure volume diagram obtained from actual engines differs from the Otto cycle due to several factors (Fig. 6.12). Since the cylinder walls are hot, the isentropic compression process in the Otto cycle is replaced by a process in which heat is transferred to the compressed air. The air-fuel mixture in an Otto cycle is fired before top dead center and requires a finite period of time to burn. In the Diesel engine fuel is injected before top dead center, and the combustion process does not occur at constant pressure. The power stroke is not isentropic, again due to heat transfer within the cylinder walls. The pressure drops quickly due to the

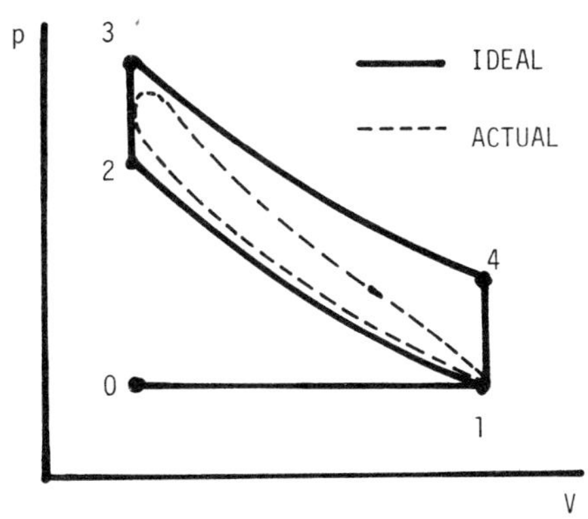

Figure 6.12

opening of the exhaust valve, and the heat is not transferred at constant volume since the exhaust gases are ejected from the engine by the exhaust stroke.

The mean effective pressure for the actual cycle may be obtained from measurements of the area and length of the pressure-volume diagram, commonly called an indicator card.

The mean effective pressure is

$$p_m = (A_c/L_c)SF \tag{6.14}$$

where p_m is the mean effective pressure (psi), A_c is the area of the card (in.), L_c is the length of the card (in.), and SF is the scale factor (lbf/in.^{3}).

The scale factor is necessary since both the pressure and volume scales are a function of the instrumentation used to generate the indicator card.

The power developed depends on the number of power strokes per minute N'. For a four-stroke cycle, N' is one-half the engine speed, and for a two-stroke cycle it is equal to the engine speed:

$$\text{Power} = p_m ALN'$$

$$= (\text{lbf/in.}^2)(\text{in.}^2)(\text{ft})(1/\text{min}) = \text{ft-lbf/min}$$

where p_m is the mean effective pressure (psi), A is the cross-sectional area of the cylinder (in.^{2}), L is the engine stroke (ft), and N' is the power strokes per minute.

The indicated horsepower IHP for each cylinder is

$$\text{IHP} = p_m LAN'/33{,}000$$

and for a multicylinder engine

$$\text{IHP} = (p_m LAN'/33{,}000)(\text{Cyl}) \tag{6.15}$$

where Cyl is the number of engine cylinders.

Example 6.13. Given are the following data from an internal combustion engine test on a four-stroke cycle engine:

Engine speed = 3600 rpm

Bore = 3.438 in.

Stroke = 4.25 in.

Area of card = 1.52 in.^{2}

Length of card = 2.16 in.

Scale factor = 100 lbf/in.^{3}

Number of cylinders = 6

From Eq. (6.14), the mean effective pressure is

$$p_m = (A_c/L_c)SF = (1.52/2.16)100 = 70.4 \text{ (psi)}$$

The cross-sectional area of the cylinder is

$$A = (\pi/4)D^2 = (\pi/4)(3.438)^2 = 9.283 \text{ (in.}^2\text{)}$$

and

$$N' = N/2 = 3600/2 = 1800 \text{ (power strokes/min)}$$

From Eq. (6.15), the IHP of the engine is

$$\text{IHP} = (p_m LAN'/33{,}000)\ (\text{Cyl})$$

$$= \{[70.4(4.25/12)(9.283)(1800)]/33{,}000\}(6)$$

$$= 75.75 \quad \text{or} \quad 56.5 \text{ (kW)}$$

BRAKE HORSEPOWER

The horsepower developed at the crankshaft of an internal combustion engine, or the shaft of any rotative device, may be measured by means of a dynamometer (Fig. 6.13), where the dynamometer acts as an electric generator.

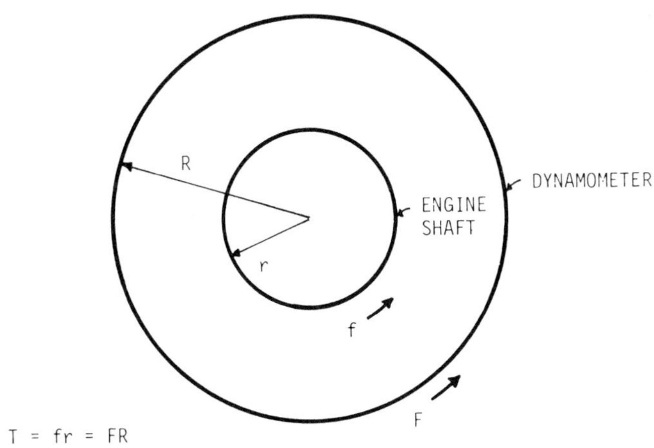

Figure 6.13

The work for one revolution of the dynamometer is

$$W = (\text{force})(\text{distance}) = f(2\Pi r)$$

It would be difficult to measure the force along the engine shaft, but since the torque along the shaft is equal to the torque at any point on the dynamometer,

$$T = FR = fr$$

where F equals the force on the dynamometer (lbf) and R equals the radius of the dynamometer.

The power output of the engine is

$$\text{Power} = 2\Pi rfN = 2\Pi RFN$$

where N is the speed of the dynamometer (rpm).

If the engine is connected directly to the dynamometer, the engine speed and the dynamometer speed are identical. If the transmission is located between the engine and dynamometer, the relationship between the engine speed and the dynamometer speed will be a function of the gear ratio of the transmission. The brake horsepower is defined by

$$BHP = 2\Pi RFN/33{,}000 \tag{6.16}$$

One disadvantage of the internal combustion engine is that it develops its maximum torque at high engine speeds. In practical devices we would normally prefer to have the maximum torque at low speeds. For this reason internal combustion engines require a transmission so that high engine speeds may be converted to low shaft speeds with a high value of the engine torque. Turbines and steam engines develop maximum torque at low speeds.

Example 6.14. Determine the BHP and engine torque for the engine in Example 6.13. Additional data include

Force on dynamometer = 55 (lbf)
Dynamometer radius = 18 (in.)
Dynamometer speed = 3600 (rpm)

From Eq. (6.16), the brake horsepower is

$$\begin{aligned} BHP &= 2\Pi RFN/33{,}000 \\ &= [2\Pi(18/12)(55)(3600)]/33{,}000 \\ &= 56.55\ (\text{HP}) \quad \text{or} \quad 42.2\ (\text{kW}) \end{aligned}$$

The engine torque is

$$T = FR = 55(1.5) = 82.5 \text{ (ft-lbf)}$$

FRICTION HORSEPOWER

The difference between the indicated and brake horsepower represents the loss of power due to friction between the oil and compression rings and the cylinder walls as well as friction in the various bearings. It may also include the power required to drive accessories such as the water pump, the oil pump, air conditioners, etc. The friction horsepower FHP is given by

$$\text{FHP} = \text{IHP} - \text{BHP} \tag{6.17}$$

The mechanical efficiency e_m of an internal combustion engine is given by

$$e_m = \text{BHP}/\text{IHP} \tag{6.18}$$

Example 6.15. Using the results of Examples 6.13 and 6.14, determine the friction horsepower and the mechanical efficiency.

From the previous examples

$$\text{IHP} = 75.75 \quad \text{and} \quad \text{BHP} = 56.55$$

From Eq. (6.17), the friction horsepower is

$$\text{FHP} = \text{IHP} - \text{BHP} = 75.75 - 56.55 \text{ (HP)}$$

$$= 19.20 \text{ (HP)} \quad \text{or} \quad 14.32 \text{ (kW)}$$

and from Eq. (6.18), the mechanical efficiency is

$$e_m = \text{BHP}/\text{IHP} = 56.55/75.75$$

$$= 0.747 \quad \text{or} \quad 74.7\%$$

The instrumentation required to generate the indicator diagram, from which the mean effective pressure may be calculated, is expensive, especially for high-speed engines. An alternative procedure is to use the dynamometer as a motor. At the conclusion of the BHP test, while the engine is still warm, the fuel supply is disconnected from the engine, and the engine is motored at the same speed as in the BHP test. The power required to drive the motor is a good approximation of the friction horsepower.

Example 6.16. If the dynamometer, acting as a motor, has a voltage level of 220 V-DC, determine the current required to drive the motor at a speed of 3600 rpm with a friction horsepower of 19.20. The electric motor efficiency is 90%.

The metric equivalent of the FHP, from Example 6.15, is

$$FHP = 14.32\ (kW) = 14{,}320\ (W) = EI$$

where E is the DC voltage (V) and I is the DC current (A). The current equivalent of the FHP is

$$I = 14{,}320/220 = 65.1\ (A)$$

Since the motor efficiency is 90%, the current supplied to the motor is

$$I = 65.1/0.90 = 72.3\ (A)$$

Example 6.17. Determine the force, acting on the dynamometer as a motor, that is equivalent to 19.20 HP.

$$F = 33{,}000\ FHP/2\pi RN$$

where, from Example 6.14, R = 18 in. and N = 3600 rpm, so that

$$F = [33{,}000(19.20)]/[2\pi(18/12)(3600)]$$
$$= 18.7\ (lbf)$$

This technique eliminates the need to know the value of the motor efficiency since only the force on the dynamometer need be measured.

THERMAL EFFICIENCY

To determine the thermal efficiency of an engine the amount of fuel consumed during an engine test must be evaluated. Either the volume or mass of fuel may be measured. If a volumetric measurement is used, it is also necessary to know the specific gravity of the fuel. It is also necessary to know the heating value HV of the fuel. Most hydrocarbon fuels have approximately 20,000 B of chemical energy in each pound of fuel. The thermal efficiency is defined by

$$e_t = [2545(HP)]/m_f HV \tag{6.19}$$

where HP is the brake or indicated horsepower, m_f is the mass of fuel burned (lbm/hr), and HV is the heating value of the fuel (B/lbm). Since

$$1\ \text{HP} = 33{,}000\ \text{ft-lb/min}$$
$$= 33{,}000/778 = 42.42\ (\text{B/min})$$
$$= 42.42(600) = 2545\ (\text{B/hr})$$

and the units of the numerator of Eq. (6.19) are B/hr. The units of the denominator are

$$m_f\text{HV} = (\text{lbm/hr})(\text{B/lbm}) = \text{B/hr}$$

Example 6.18. From Examples 6.13 and 6.14, BHP = 56.55 and IHP = 75.75. Given are the following fuel consumption data: HV = 20,000 B/lbm, SG = 0.72, and 1.2 gal of fuel is used in a 15-min test. Determine the brake and indicated thermal efficiency.

The volume of fuel used is

$$V_f = 1.2(60/15) = 4.8\ (\text{gal/hr})$$
$$= 4.8/7.481\ [(\text{gal/hr})(\text{ft}^3/\text{gal})]$$
$$= 0.642\ (\text{ft}^3/\text{hr})$$

The mass of fuel is

$$m_f = \rho_f V_f = \text{SG}\rho_{H_2O} V_f$$
$$= 0.72(62.4)(0.642) = 28.83\ (\text{lbm/hr})$$

The brake thermal efficiency may be obtained by using the BHP in Eq. (6.19) so that

$$e_{tb} = [2545(\text{BHP})]/m_f\text{HV}$$
$$= [2545(56.55)]/[28.83(20{,}000)]$$
$$= 0.249 \quad \text{or} \quad 24.9\%$$

In the same manner, the indicator thermal efficiency is

$$e_{ti} = [2545(\text{IHP})]/m_f\text{HV}$$
$$= [2545(75.75)]/[28.83(20{,}000)]$$
$$= 0.334 \quad \text{or} \quad 33.4\%$$

The brake and indicated thermal efficiencies are related by

$$e_m = \text{BHP/IHP} = e_{tb}/e_{ti}$$

The brake thermal efficiency represents the fraction of energy in the fuel that is converted to useful shaft work.

ENGINE EFFICIENCY

The engine efficiency is the ratio of the power produced by the actual engine to the power that would be produced by the same type of engine operating on its theoretical cycle. In equation form,

$$e_e = (\text{power/heat added})_{actual}/(\text{power/heat added})_{theoretical} = e_t/e \quad (6.20)$$

where e_t is the brake or indicated thermal efficiency, e is the theoretical cycle efficiency, and e_e is the engine efficiency.

Example 6.19. The engine used in Example 6.18 had a compression ratio of 7.0 and a theoretical cycle efficiency of 54.1% operating in the Otto cycle. The brake and indicated thermal efficiencies were 24.9% and 33.4%.

From Eq. (6.20), the brake engine efficiency is

$$e_{eb} = e_{tb} = 0.249/0.541 = 0.460 \quad \text{or} \quad 46.0\%$$

and the indicated engine efficiency is

$$e_{ei} = e_{ti} = 0.334/0.541 = 0.617 \quad \text{or} \quad 61.7\%$$

From the brake engine efficiency it is apparent that the actual engine produces 46.0% of the power that would be produced by the theoretical engine.

The engine efficiency is a useful criterion for comparing engines of different sizes or engines operating on different cycles.

PROBLEMS

Air is the working substance in all theoretical cycles.

6.1. An Otto cycle engine has a compression ratio of 9.0. The heat added in the cycle is 300 B/lbm, and the pressure and temperature at the start of the compression process are 14.5 psia and 60 F. Determine the cycle efficiency, the net work of the cycle, the mean effective pressure, p_{max}, and T_{max} for the cycle.

6.2. The engine in Prob. 6.1 has a bore of 3.5 in. and a stroke of 3.75 in. and operates in the four-stroke cycle at a speed of 4000 rpm. For an eight-cylinder engine, determine the theoretical horsepower.

6.3. Determine the temperature at states 2, 3, and 4 for the Otto cycle in Prob. 6.1. Determine the efficiency by means of Eq. (6.2).

6.4. A Diesel cycle engine has a compression ratio of 15 and a maximum cycle temperature of 2540 F. p_1 = 15 psia, and t_1 = 40 F. Determine p_2, r_c, Q_{in}, e, ΣW, and p_m.

6.5. A dual cycle engine has a compression ratio of 15, a maximum pressure of 800 psia, and a maximum temperature of 2540 F. p_1 = 15 psia, and t_1 = 40 F. Determine r_p, r_c, Q_{in}, e, ΣW, and p_m.

6.6. Determine the theoretical horsepower for the Otto cycle engine in Probs. 6.1 and 6.2. e_V = 75%, and the surrounding air pressure and temperature are 14.5 psia and 60 F.

6.7. The engine in Prob. 6.2 is equipped with a supercharger that raises the pressure from 14.5 psia to 30 psia. The compression ratio is 9.0, the air temperature entering the supercharger is 60 F, and the heat added in the cycle is 300 B/lbm. Determine the effective theoretical horsepower of the cycle, the mean effective pressure, the effective cycle efficiency, p_{max}, and T_{max} for the cycle. Compare the answers with Prob. 6.2.

6.8. Given are the following data from an eight-cylinder, four-stroke cycle, spark-ignited engine with a compression ratio of 9.0:

Bore = 3.5 in.
Stroke = 3.75 in.
Engine speed = 4000 rpm
Dynamometer speed = 4000 rpm
Dynamometer radius = 24 in.
Force on dynamometer as a generator = 40 lbf
Force on dynamometer as a motor = 10 lbf
SG of fuel = 0.73
HV of fuel = 20,000 B/lbm
Volume of fuel = 4.25 gal/hr

Determine BHP, FHP, IHP, e_m, e_{tb}, e_{ti}, e_{eb}, and e_{ei}.

7

SEMIPERFECT GAS

The relationship between physical properties and the formulas needed to evaluate thermodynamic properties, heat, and work values for reversible processes were derived in Chap. 5 for the perfect gas. In Chap. 6, we made use of these equations to develop the Otto, Diesel, and dual cycles. For moderately high temperatures the perfect gas requirement that the specific heat has a fixed numerical value is no longer satisfied. For these temperatures the perfect gas equation is still an accurate representation of the relationship among the pressure, volume, and temperature. A gas for which the perfect gas equation is valid but has specific heats that are a function of temperature is described as a semiperfect gas.

For high temperatures and high pressure, both criteria used to define a perfect gas fail. The gas is then described as a real gas, and tabular values that cross-index the pressure and temperature are required to obtain thermodynamic properties. In this book we shall not use the real gas as a working substance.

SEMIPERFECT GASES

For pressures below approximately 1000 psia the properties of gas may be determined if the temperature is known. No equation derived for perfect gases may be used if the equation contains a c_p, c_v, or k term. To evaluate properties it is necessary to determine their values from tabular data as presented in Gas Tables by J. H. Keenan and J. Kaye, John Wiley & Sons, Inc., New York, 1948. An abbreviated list of tabular values for properties is presented in Appendix C.

For temperatures of 500 and 1000 R the tables list, for 1 lb of air,

T (T)	h (B/lbm)	u (B/lbm)	p_r	v_r	φ (B/lbm-R)
500	119.48	85.20	1.0590	174.90	0.58233
1000	240.98	172.43	12.298	30.12	0.75042

By using these tabular values the changes in properties for the constant volume, constant pressure, and isentropic and polytropic processes may be calculated. It is not necessary to use the semiperfect gas tables for the isothermal process since all of the equations for this process may be written independently of c_p, c_v, and k.

In the following examples it will be assumed that the initial state of the air is defined by p_1 = 15.0 psia, T_1 = 500 R, and V_1 = 10.0 ft^3. Corresponding values for the air, assuming a perfect gas relationship, will be included within square brackets.

Example 7.1. For a constant volume process where T_2 = 1000 R, determine p_2, V_2, ΔU, ΔH, ΔS, Q, W_{nf}, and W_{sf} (ΔKE, ΔPE = 0).

For a semiperfect gas, the perfect gas equation may be used, so that the mass of air is

$$m = p_1V_1/RT_1 = [15(144)(10)]/[53.3(500)]$$

$$= 0.811\ [0.811]\ (\text{lbm})$$

and

$$p_2 = p_1(V_1/V_2)(T_2/T_1) = p_1(T_2/T_1)$$

so that

$$p_2 = 15.0(1000/500) = 30.0\ [30.0]\ (\text{ft}^3)$$

Also

$$V_2 = V_1 = 10\ [10]\ (\text{ft}^3)$$

The change in internal energy and the enthalpy are given by

$$U_2 - U_1 = m(u_2 - u_1) = 0.811(172.43 - 85.20)$$

$$= 70.7\ [69.5]\ (\text{B})$$

and

$$H_2 - H_1 = m(h_2 - h_1) = 0.811(240.98 - 119.48)$$

$$= 98.5\ [97.3]\ (B)$$

To determine the change in entropy, consider the general entropy equation [Eq. (5.19)] that relates pressures and temperatures, or

$$S_2 - S_1 = mc_p \ln(T_2/T_1) - (mR/J) \ln(p_2/p_1)$$

and for 1 lb of air

$$s_2 - s_1 = c_p \ln(T_2/T_1) - (R/J) \ln(p_2/p_1)$$

The term $c_p \ln(T_2/T_1)$ cannot be evaluated for a semiperfect gas due to the variation in c_p with temperature. The values of φ in the tables have been calculated so that

$$\varphi_2 - \varphi_1 = c_p \ln(T_2/T_1)$$

and for the semiperfect gas

$$s_2 - s_1 = \varphi_2 - \varphi_1 - (R/J) \ln(p_2/p_1) \tag{7.1}$$

For the present example

$$s_2 - s_1 = (0.75042 - 0.58233) - (53.3/778) \ln(30/15)$$

$$= 0.16809 - 0.04749 = 0.1206\ (B/lbm\text{-}R)$$

and

$$S_2 - S_1 = m(s_2 - s_1) = 0.811(0.1206)$$

$$= 0.0978\ [0.0963]\ (B/R)$$

For a constant volume process

$$W_{nf} = 0\ [0]$$

and the heat transfer is

$$Q = \Delta U + W_{nf} = \Delta U$$

$$= 70.7\ [69.5]\ (B)$$

The steady-flow work may be calculated from

$$W_{sf} = Q - \Delta H - \Delta KE - \Delta PE$$

and for ΔKE, $\Delta PE = 0$

$$W_{sf} = Q - \Delta H = 70.7 - 98.5 = -27.8 \; [-27.8] \; (B)$$

Since the steady-flow work, in the absence of kinetic and potential terms, could have been calculated from

$$W_{sf} = [-V(p_2 - p_1)]/J$$

which does not contain a c_p, c_v, or k term, the steady-flow work is the same for the perfect and semiperfect gases, or

$$W_{sf} = -[10(30 - 15)(144)]/778$$

$$= -27.8 \; (B) \quad \text{or} \quad -29.3 \; (kJ)$$

Example 7.2. Use the same initial data and m = 0.811 lbm; also T_2 = 1000 R. For a constant pressure process, determine p_2, V_2, ΔU, ΔH, ΔS, Q, W_{nf}, and W_{sf} (ΔKE, $\Delta PE = 0$).

For a constant pressure process

$$p_2 = p_1 = 15.0 \; [15.0] \; (psia)$$

and from Charles' law

$$V_2 = V_1(T_2/T_1) = 10(1000/500) = 20.0 \; [20.0] \; (ft^3)$$

Since T_1 and T_2 are identical to those for the constant volume process,

$$U_2 - U_1 = 70.7 \; [69.5] \; (B)$$

and

$$H_2 - H_1 = 98.5 \; [97.3] \; (B)$$

From Eq. (7.1)

$$s_2 - s_1 = \varphi_2 - \varphi_1 - (R/J) \ln(p_2/p_1)$$

where $p_2/p_1 = 1$ so that $\ln(p_2/p_1) = 0$ and

$$s_2 - s_1 = \varphi_2 - \varphi_1 = 0.75042 - 0.58233$$

$$= 0.16809 \; (B/lbm\text{-}R)$$

and

$$S_2 - S_1 = m(s_2 - s_1) = 0.811(0.16809) = 0.1363 \; [0.1349] \; (B/R)$$

The nonflow work may be determined from

$$W_{nf} = p(V_2 - V_1)/J = [15(144)(20 - 10)]/778$$

$$= 27.8 \ [27.8] \ (B)$$

Again in the absence of c_p, c_v, and k property terms the perfect and semiperfect gas nonflow work solutions are identical.

The heat transfer is

$$Q = \Delta U + W_{nf} = 70.7 + 27.8$$

$$= 98.5 \ [97.3] \ (B)$$

or

$$Q = \Delta H = 98.5 \ (B)$$

For ΔKE, $\Delta PE = 0$, the steady-flow work is

$$W_{sf} = Q - \Delta H = 0 \ [0] \ (B)$$

and the solution for the perfect gas is valid for the semiperfect gas.

Example 7.3. Use the same initial data and m = 0.811 lbm. For an isentropic process, with T_2 = 1000 R, determine Q, ΔS, ΔU, ΔH, W_{nf}, W_{sf} (ΔKE, $\Delta PE = 0$), p_2 and V_2.

For an isentropic process, $\Delta S = 0$ and $Q = 0$. Since T_1 and T_2 remain the same as for the constant volume and constant pressure process,

$$\Delta U = 70.7 \ [69.5] \ (B)$$

and

$$\Delta H = 98.5 \ [97.3] \ (B)$$

The nonflow and steady-flow work values are

$$W_{nf} = Q - \Delta U = -70.7 \ [-69.5] \ (B)$$

and

$$W_{sf} = Q - \Delta H - \Delta KE - \Delta PE$$

$$= Q - \Delta H \qquad (\Delta KE, \Delta PE = 0)$$

$$= -98.5 \ [-97.3] \ (B)$$

Since the perfect gas solutions for the work terms are

$$W_{nf} = (p_2V_2 - p_1V_1)/[J(1 - k)]$$

and

$$W_{sf} = kW_{nf}$$

the variation in k with temperature produces a difference between the semi-perfect and perfect gas solutions. In all cases the semiperfect gas solution is more accurate than the perfect gas solution.

To determine the final pressure and volume for the constant entropy process the remaining data in the gas tables, p_r and v_r, are utilized. These two columns of data may be used only for the isentropic process. The appropriate relationships are

$$V_2/V_1 = v_2/v_1 = v_{r2}/v_{r1} \tag{7.2}$$

and

$$p_2/p_1 = p_{r2}/p_{r1} \tag{7.3}$$

For $T_1 = 500$ R, $p_{r1} = 1.0590$ and $v_{r1} = 174.90$, and for $T_2 = 1000$ R, $p_{r2} = 12.298$ and $v_{r2} = 30.12$. Therefore

$$p_2 = p_1(p_{r2}/p_{r1}) = 15(12.298/1.0590)$$

$$= 174.2\ [169.7]\ \text{(psia)}$$

and

$$V_2 = V_1(v_{r2}/v_{r1}) = 10(30.12/174.90)$$

$$= 1.72\ [1.77]\ (\text{ft}^3)$$

<u>Example 7.4</u>. Use the same initial data and m = 0.811 lbm. For a polytropic process with n = 1.2 and $T_2 = 1000$ R, determine p_2, V_2, ΔU, ΔH, W_{nf}, W_{sf} (ΔKE, $\Delta PE = 0$), and Q.

For a semiperfect gas undergoing a polytropic process the relationship

$$p_1V_1^{\,n} = p_2V_2^{\,n} = C$$

is valid. Therefore

$$p_2 = p_1(T_2/T_1)^{n/n-1} = 15(1000/500)^{1.2/0.2}$$

$$= 960\ [960]\ \text{(psia)}$$

and

$$V_2 = V_1(T_1/T_2)^{1/(n-1)} = 10(500/1000)^{1/0.2}$$

$$= 0.312\ [0.312]\ (\text{ft}^3)$$

As in the previous examples with $T_1 = 500$ and $T_2 = 1000$ R,

$$\Delta U = 70.7\ [69.5]\ (\text{B})$$

and

$$\Delta H = 98.5\ [97.3]\ (\text{B})$$

The nonflow work is given by

$$W_{nf} = (p_2V_2 - p_1V_1)/[J(1 - n)]$$

$$= [mR(T_2 - T_1)]/[J(1 - n)]$$

$$= [0.811(53.3)(1000 - 500)]/[778(1 - 1.2)]$$

$$= -138.9\ [-138.9]\ (\text{B})$$

From the nonflow process equation the heat transfer is

$$Q = \Delta U + W_{nf} = 70.7 - 138.9 = -68.2\ [-69.4]\ (\text{B})$$

and the steady-flow work, ΔKE, $\Delta PE = 0$, is

$$W_{sf} = Q - \Delta H = -68.2 - 98.5 = -166.7\ [-166.7]\ (\text{B})$$

The steady-flow work in the absence of kinetic and potential terms may also be calculated from

$$W_{sf} = nW_{nf} = 1.2(-138.9) = -166.7\ (\text{B})$$

Note that the equation

$$Q = mc_n(T_2 - T_1)$$

may not be used for the semiperfect gas since

$$c_n = c_v[(k - n)/(1 - n)]$$

which includes both c_v and k in the polytropic specific heat equation.

CYCLE ANALYSIS USING A SEMIPERFECT GAS

An Otto cycle will be analyzed using the semiperfect gas as a working substance to determine the differences between this solution and the perfect gas solution.

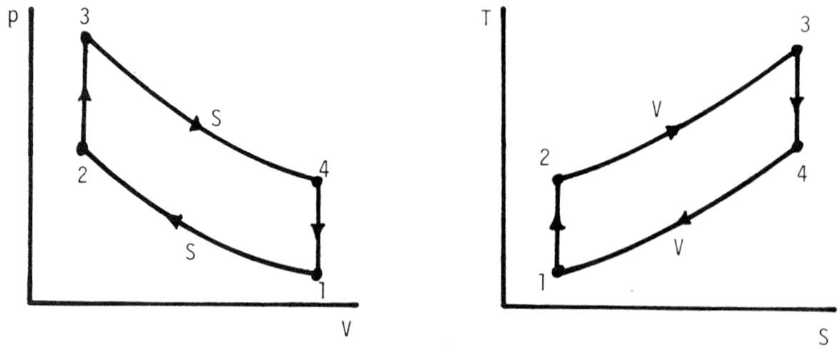

Figure 7.1

Example 7.5. Given is an Otto cycle with $r_k = 7.0$, a bore of 3.438 in., and a stroke of 4.25 in. The clearance volume is 6.60 in.^{3}, $p_1 = 14.7$ psia, $t_1 = 40$ F, and $t_3 = 2540$ F (Fig. 7.1). The engine operates on the four-stroke cycle with a speed of 3600 rpm and has six cylinders. Determine T_2, Q_{in}, W_{net}, e, p_m, and the theoretical horsepower of the cycle. For $T_1 = 500$ R, $v_{r1} = 174.90$ and $u_1 = 85.20$ B/lbm.

The value of v_{r2} may be calculated by means of Eq. (7.2) where for the isentropic process

$$v_{r2} = v_{r1}(V_2/V_1) = v_{r1}/r_k$$

so that

$$v_{r2} = 174.90/7.0 = 24.99$$

To determine the conditions at state 2, bracket the value of v_{r2} with values from the gas tables, Appendix C. For $T_2 = 1060$ R, $u_2 = 183.29$ B/lbm and $v_{r2} = 25.82$, and for $T_2 = 1080$ R, $u_2 = 186.93$ B/lbm and $v_{r2} = 24.58$.

The linear fraction f of the distance between the two tabular values of v_{r2} is

$$f = (25.82 - 24.99)/(25.82 - 24.58)$$
$$= 0.669$$

and the temperature at state 2 is

$$T_2 = 1060 + 0.669(1080 - 1060) = 1073\ [1089]\ (\text{R})$$

and

$$u_2 = 183.29 + 0.669(186.93 - 183.29) = 185.73\ (\text{B/lbm})$$

Values in square brackets are from the perfect gas solution, Example 6.2.

For a temperature $T_3 = 3000$ R, from the tables

$$T_3 = 3000 \text{ R} \qquad u_3 = 585.04 \qquad v_{r3} = 1.1803$$

The heat added to the cycle using a semiperfect gas for the constant volume process is

$$Q_{in} = Q_{23} = u_3 - u_2 = 585.04 - 185.73$$

$$= 399.31 \; [327.5] \; (\text{B/lbm})$$

For the isentropic process from state 3 to state 4

$$v_{r4} = v_{r3}(V_4/V_3) = v_{r3}(V_1/V_2) = v_{r3}r_k$$

$$= 1.1803(7) = 8.262$$

At $T_4 = 1600$ R, $v_{r4} = 8.263$ and $u_4 = 286.06$ B/lbm, and at $T_4 = 1620$ R, $v_{r4} = 7.791$ and $u_4 = 290.04$ B/lbm. The linear factor is

$$f = (8.263 - 8.262)/(8.263 - 7.791) = 0.002$$

and

$$u_4 = 286.06 + 0.002(290.94 - 286.06) = 286.07 \; (\text{B/lbm})$$

The heat rejected during the constant volume process is

$$Q_{out} = Q_{14} = u_1 - u_4 = 85.20 - 286.07$$

$$= -200.87 \; [-150.4] \; (\text{B/lbm})$$

The net work and net heat of the cycle are

$$\Sigma W = \Sigma Q = Q_{in} + Q_{out} = 399.31 - 200.87$$

$$= 198.44 \; [177.2] \; (\text{B/lbm})$$

and the cycle efficiency is

$$e = \Sigma W/Q_{in} = 198.44/399.31$$

$$= 0.497 \; [0.541]$$

Since, by use of the perfect gas equation,

$$v_1 = RT_1/p_1 = [53.3(500)]/[14.7(144)] = 12.59 \; (\text{ft}^3/\text{lbm})$$

and from the definition of the isentropic compression ratio

$$v_2 = v_1/r_k = 12.59/7.0 = 1.80 \; (\text{ft}^3/\text{lbm})$$

the mean effective pressure is

$$p_m = \Sigma W/(V_{max} - V_{min}) = [198.44(778)]/[(12.59 - 1.80)144]$$
$$= 99.4 \; [88.7] \text{ (psia)}$$

The displacement volume is

$$V_D = (\Pi/4)D^2L = (\Pi/4)(3.438)^2(4.25) = 39.45 \text{ (in.}^3\text{)}$$

and

$$V_1 = V_C + V_D = 6.60 + 39.45 = 46.05 \text{ (in.}^3\text{)}$$

so that the mass at the start of the compression process is

$$m = V_1/v_1 = (46.05/1728)/12.59 = 0.00212 \text{ (lbm)}$$

and the theoretical horsepower is

$$HP = [198.44(0.00212)(778)(3600/2)(6)]/33{,}000$$
$$= 107.1 \; [95.4]$$

Comparison of the Otto cycle using a perfect gas and a semiperfect gas as the working substance shows that the cycle efficiency is 4.4% lower for the semiperfect gas analysis, the mean effective pressure is 10.7 psia higher, and the theoretical horsepower increases by 11.7 HP.

PROBLEMS

7.1 Air, as a semiperfect gas, has an initial temperature, pressure, and volume of 100 F, 20 psia, and 20 ft^3. The final temperature is 1540 F. Determine p_2, V_2, ΔU, ΔH, ΔS, Q, W_{nf}, and W_{sf} (ΔKE, $\Delta PE = 0$) for (a) a constant volume process; (b) a constant pressure process; (c) an isentropic process; (d) a polytropic process, $n = 0.5$; (e) a polytropic process, $n = 1.3$; and (f) a polytropic process, $n = 1.5$.

7.2 Rework Prob. 6.4 for air as a semiperfect gas.

8

AIR COMPRESSORS AND AIR ENGINES

Compressed air is required for many industrial operations such as the operation of air engines, drills, and jackhammers. The use of air equipment is recommended for operations that must take place in atmospheres that are capable of supporting combustion, such as many mining operations. These devices can be built so that the tendency for the equipment to spark is eliminated.

RECIPROCATING COMPRESSORS

A piston-cylinder reciprocating air compressor is illustrated in Fig. 8.1. The inlet and exit valves are spring loaded and will open when the designed value for the suction or discharge pressure is reached in the compressor. Unlike the internal combustion engine, a linkage to open and close the

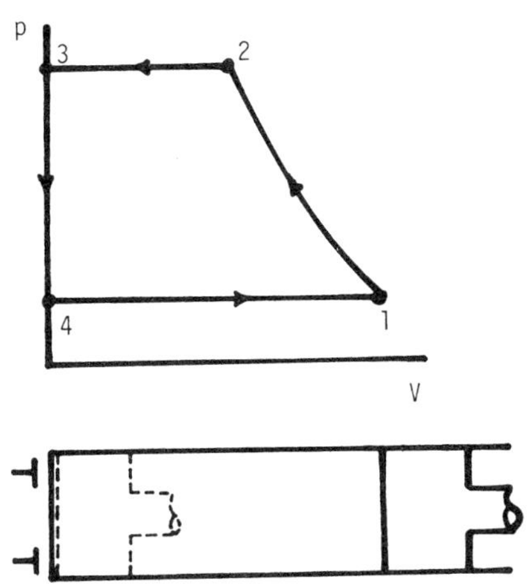

Figure 8.1

valves is not required. As shown on the pV diagram in Fig. 8.1, the top dead-center position of the piston forces all of the air out of the cylinder, and the compressor is described as a reciprocating compressor without clearance.

The nonflow work for the compression process, except for isothermal compression, is given by Eq. (3.12),

$$W_{12} = \frac{p_2V_2 - p_1V_1}{1 - n}, \qquad n \neq 1$$

The work required to discharge the air from the cylinder at the constant discharge pressure is

$$W_{23} = p_2(V_3 - V_2) = -p_2V_2$$

since

$$V_3 = 0$$

For the constant volume process from state 3 to state 4 the work is

$$W_{34} = 0$$

The work done by the air entering the cylinder during the suction stroke is

$$W_{41} = p_1(V_1 - V_4) = p_1V_1$$

since

$$V_4 = 0$$

The cylinder work is

$$\begin{aligned}\Sigma W &= W_{12} + W_{23} + W_{34} + W_{41} \\ &= [(p_2V_2 - p_1V_1)/(1 - n)] - p_2V_2 + p_1V_1 \\ &= (p_2V_2 - p_1V_1)[1/(1 - n) - 1] \\ &= [n/(1 - n)](p_2V_2 - p_1V_1) \qquad (8.1)\end{aligned}$$

Assuming that the working substance is a perfect gas, Eq. (8.1) may be rewritten as

$$\begin{aligned}\Sigma W &= [n/(1 - n)](mRT_2 - mRT_1) \\ &= [mRn/(1 - n)](T_2 - T_1) \\ &= [mRnT_1/(1 - n)][(T_2/T_1) - 1] \qquad (8.2)\end{aligned}$$

For a nonisothermal process

$$T_2/T_1 = (p_2/p_1)^{(n-1)/n}$$

and Eq. (8.2) may be expressed as

$$\Sigma W = [mRnT_1/(1 - n)][(p_2/p_1)^{(n-1)/n} - 1] \quad (8.3)$$

Example 8.1. Determine the cyclic work required to compress air from 15 to 60 psia for an entering air temperature of 40 F and for n = 1.2, 1.4, and 1.6.

The work, per pound mass of working substance, from Eq. (8.3) is

$$\Sigma W/m = [RnT_1/(1 - n)][(p_2/p_1)^{(n-1)/n} - 1]$$

and for n = 1.2

$$\Sigma W/m = [53.3(1.2)(500)/-0.2][(60/15)^{(1.2-1)/1.2} - 1]$$

$$= -160{,}000(0.2599) = -41{,}560 \text{ (ft-lbf/lbm)}$$

$$= -53.4 \text{ (B/lbm)} \quad \text{or} \quad -124.2 \text{ (kJ/kg)}$$

In the same manner, for the isentropic process, where n = k = 1.4,

$$\Sigma W/m = -58.2 \text{ (B/lbm)} \quad \text{or} \quad -135.4 \text{ (kJ/kg)}$$

and for the polytropic process, where n = 1.6, the work is

$$\Sigma W/m = -62.3 \text{ (B/lbm)} \quad \text{or} \quad -144.9 \text{ (kJ/kg)}$$

From the above data the lower values of the constant n require the least amount of work to drive the compressor.

If the compression process from state 1 to state 2 is isothermal, the work for the compression process would be [Eq. (3.13)]

$$W_{12} = p_1V_1 \ln(V_2/V_1)$$

and the remaining three work terms would be unchanged. The cyclic work becomes

$$\Sigma W = W_{12} + W_{23} + W_{34} + W_{41}$$

$$= p_1V_1 \ln(V_2/V_1) - p_2V_2 + p_1V_1$$

but for the isothermal process, where n = 1,

$$p_1V_1 = p_2V_2$$

so that

$$\Sigma W = p_1V_1 \ln(V_2/V_1)$$

$$= mRT_1 \ln(p_1/p_2) \quad (8.4)$$

Example 8.2. Determine the cyclic work in Example 8.1 for n = 1 by use of Eq. (8.4) for the isothermal compressor.

$$\Sigma W/m = RT_1 \ln(p_1/p_2)$$

$$= 53.3(500) \ln(15/60) = -36{,}940 \text{ (ft-lbf/lbm)}$$

$$= -47.5 \text{ (B/lbm)} \quad \text{or} \quad -110.4 \text{ (kJ/kg)}$$

If an isothermal compressor is used, the required compressor cyclic work is less than either of the polytropic processes or the isentropic process. The four cycles are sketched in Fig. 8.2, and the work savings as n is reduced are represented by the reduced work area enclosed within the cycle.

Example 8.3. Determine the temperature at the end of the compression process, the change in internal energy, and the work and heat transfer for the compression process where n = 1.0, 1.2, 1.4, and 1.6. Pressure and temperatures are p_1 = 15 psia, p_2 = 60 psia, and t_1 = 40 F. The working substance is air.

For n = 1.0

$$T_2 = T_1 = 500 \text{ (R)}$$

$$\Delta u = 0$$

and

$$Q_{12} = W_{12} = RT_1 \ln(p_1/p_2) = -47.5 \text{ (B/lbm)}$$

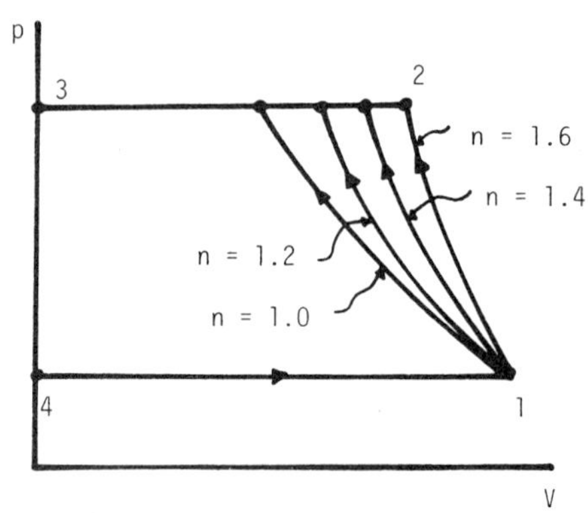

Figure 8.2

For $n = 1.2$

$$T_2 = T_1(p_2/p_1)^{(n-1)/n} = 500(4)^{0.167} = 630 \text{ (R)}$$

$$u_2 - u_1 = c_v(T_2 - T_1) = 0.1714(630 - 500) = 22.3 \text{ (B/lbm)}$$

$$W_{12} = [R(T_2 - T_1)]/[J(1 - n)] = -44.5 \text{ (B/lbm)}$$

$$Q_{12} = (U_2 - U_1) + W_{12} = 22.3 - 44.5 = -22.2 \text{ (B/lbm)}$$

In like manner, for $n = 1.4$, $T_2 = 742$ R, $\Delta u = 41.5$ B/lbm, $W_{12} = -41.5$ B/lbm, and $Q = 0$. For $n = 1.6$, $T_2 = 841$ R, $\Delta u = 58.4$ B/lbm, $W_{12} = -38.9$ B/lbm, and $Q = 19.5$ B/lbm.

Comparative data are listed for the various values of n in the following chart:

n	T_2 (R)	Δu (B/lbm)	W_{12} (B/lbm)	ΣW (B/lbm)
1.0	500	0.0	-47.5	-47.5
1.2	630	22.3	-44.5	-53.4
1.4	742	41.5	-41.5	-58.2
1.6	841	58.4	-38.9	-62.3

As n increases, the temperature of the air leaving the compressor is increased, which is also indicated by the increasing value of Δu for the compression process.

The work for the compression process decreases as the value of n increases, as shown in the chart and by the area below the process curve in Fig. 8.2.

The cyclic work, which is the power required to actually operate the system, decreases with a reduction in the value of n.

In normal operation the compressed air is stored in a tank until it is required for use in an air engine or other air-driven equipment. The storage tank also serves to dampen the pulsations of the air delivered by the reciprocating compressor. In the storage tank the elevated temperature T_2 of the air leaving the compressor and the energy it represents are wasted since the air will assume the ambient temperature in the storage tank.

If this temperature increase can be used or conserved in an insulated tank, the savings in cyclic work obtained by cooling the cylinder walls (Fig. 8.3) and thereby lowering the value of n are not realized, and cylinder wall cooling is not necessary. Since the air is only in the compressor for a short period of time, practical values of n produced by cylinder wall cooling are from $n = 1.2$ to $n = 1.3$.

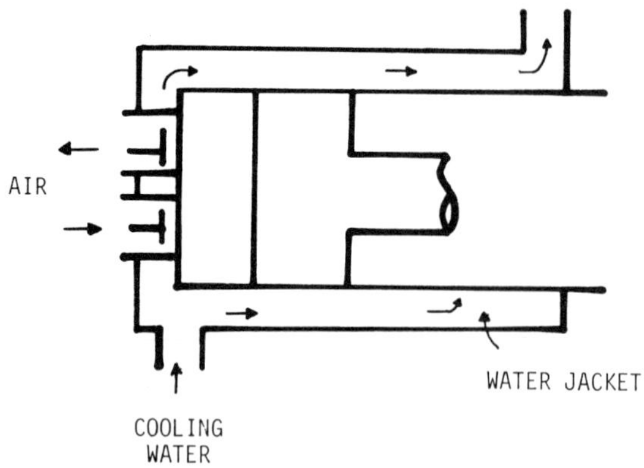

Figure 8.3

EFFECT OF CLEARANCE

From a realistic design standpoint it is not practical to discharge all of the air from the cylinder in a reciprocating compressor. As in the internal combustion engine there is a volume of air remaining in the cylinder when the piston is at the top dead-center position (Fig. 8.4).

The clearance is defined by the equation

$$V_C = cV_D \tag{8.5}$$

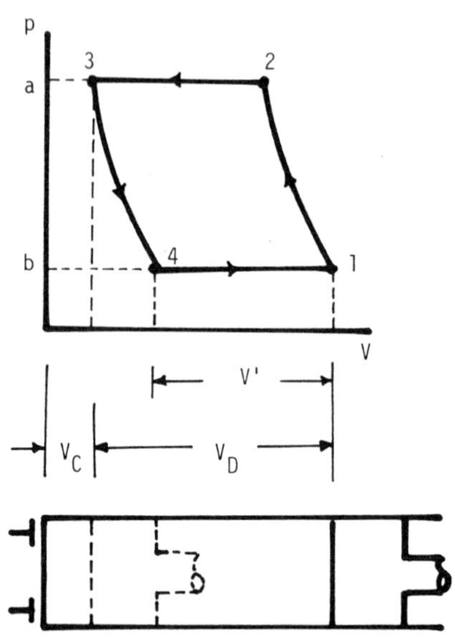

Figure 8.4

or in terms of the state points in Fig. 8.4

$$c = V_3/(V_1 - V_3)$$

The volume of air entering the air compressor is (Fig. 8.4)

$$V' = V_1 - V_4 \tag{8.6}$$

The work for the cycle without clearance, 12ab1, is for the polytropic process [Eq. (8.3)]

$$\Sigma W_t = [mRnT_1/(1 - n)][(p_2/p_1)^{(n-1)/n} - 1]$$

$$= [np_1V_1/(1 - n)][(p_2/p_1)^{(n-1)/n} - 1]$$

The work for a cycle without clearance, 43ab4, is

$$\Sigma W_c = [n'p_4V_4/(1 - n')][(p_3/p_4)^{(n'-1)/n'} - 1]$$

The work for the cycle 1234 is the cyclic work of 12ab1, ΣW_t, less the work of the cycle 43ab4, ΣW_c. Therefore the work of a cycle with clearance is $\Sigma W = \Sigma W_t - \Sigma W_c$,

$$\Sigma W = [np_1V_1/(1 - n)][(p_2/p_1)^{(n-1)/n} - 1]$$

$$- [n'p_4V_4/(1 - n')][(p_3/p_4)^{(n'-1)/n} - 1] \tag{8.7}$$

As shown in Fig. 8.3, $p_2 = p_3$ and $p_1 = p_4$, and, assuming $n = n'$, Eq. (8.7) reduces to

$$\Sigma W = [n/(1 - n)][p_1(V_1 - V_4)][(p_2/p_1)^{(n-1)/n} - 1]$$

or by use of Eq. (8.6)

$$\Sigma W = [n/(1 - n)]p_1V'[(p_2/p_1)^{(n-1)/n} - 1] \tag{8.8}$$

Example 8.4. Determine the cyclic work required to compress air from 15 to 60 psia for an entering air temperature of 40 F when n = 1.2 and the percentage clearance is 5%.

The mass of air entering or leaving the compressor is

$$m' = p_1V'/RT_1$$

so that

$$p_1V' = m'RT_1$$

and by rewriting Eq. (8.8)

$$\Sigma W = [n/(1 - n)]m'RT_1[(p_2/p_1)^{(n-1)/n} - 1]$$

and for each pound of air delivered

$$\Sigma W/m' = [n/(1 - n)]RT_1[(p_2/p_1)^{(n-1)/n} - 1]$$

so that the cyclic work per pound mass of air is

$$\Sigma W/m' = (1.2/-0.2)[53.3(500)][(60/15)^{0.2/1.2} - 1]$$

$$= -41{,}560 \text{ (ft-lbf/lbm)} = -53.4 \text{ (B/lbm)}$$

This is the same quantity of work required for the cycle without clearance. Theoretically, the amount of clearance has no effect on the work required to compress the air. As a practical matter, increasing the percentage clearance increases the displacement of the compressor, and for a given cylinder bore the stroke must increase. As the stroke increases, the losses due to friction between the compression rings used to seal the space between the piston and the cylinder walls must increase.

VOLUMETRIC EFFICIENCY

For the internal combustion engine the volumetric efficiency was defined by Eq. (6.13), where

$$e_V = V'/V_D$$

where V' is the volume of air drawn into the cylinder and V_D is the displacement volume of the cylinder. The same equation may be used to define the volumetric efficiency of a reciprocating compressor where V' is measured at the intake pressure and temperature.

From Eq. (8.6)

$$e_V = V'/V_D = (V_1 - V_4)/V_D$$

But, from Fig. 8.4,

$$V_1 = V_3 + V_D = V_C + V_D$$

and from Eq. (8.5)

$$V_1 = cV_D + V_D$$

so that

$$e_V = (cV_D + V_D - V_4)/V_D = c + 1 - (V_4/V_D)$$

For the process from state 3 to state 4

$$V_4 = V_3(p_3/p_4)^{1/n} = cV_D(p_2/p_1)^{1/n}$$

and the volumetric efficiency is

$$e_V = 1 + c - c(p_2/p_1)^{1/n} \tag{8.9}$$

This is the volumetric efficiency based on the theoretical cycle. The actual volumetric efficiency is less than this value due to fluid turbulence as well as the heating of the incoming air due to contact with the hot cylinder walls.

Example 8.5. Determine the volumetric efficiency of a reciprocating air compressor operating between pressures of 15 and 60 psia where n = 1.2 and the percentage clearance is 5%.

From Eq. (8.9)

$$e_V = 1 + c - c(p_2/p_1)^{1/n} = 1 + 0.05 - 0.05(60/15)^{1/1.2}$$

$$= 0.891 \quad \text{or} \quad 89.1\%$$

Example 8.6. Determine the percentage clearance such that the above compressor does not deliver any air. If the compressor does not deliver air, the discharge of air and the intake of air into the compressor are zero or

$$e_V = V'/V_D = 0.0$$

From Eq. (8.9)

$$e_V = 1 + c - c(p_2/p_1)^{1/n} = 1 + c - c(60/15)^{1/1.2} = 0$$

$$3.174c = 1 + c$$

$$2.174c = 1$$

$$c = 0.460 \quad \text{or} \quad 46.0\%$$

If the percentage clearance is 46%, the compressor will not deliver any air.

TWO-STAGE COMPRESSION

Instead of compressing the air from its initial intake pressure to its discharge pressure in a single cylinder, assume that the compression takes place in two stages (Fig. 8.5). An intercooler is also included so that

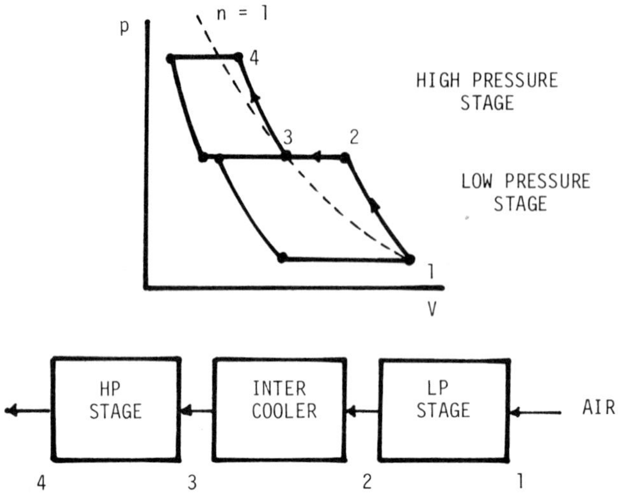

Figure 8.5

the air leaving the low-pressure stage may be cooled before it enters the high-pressure stage.

Assume that the value of n for both compression stages is the same and that the air leaving the intercooler has the temperature of the entering air at state 1.

The cyclic work for the low-pressure stage, with clearance, by means of Eq. (8.8) is

$$\Sigma W_L = [n/(1 - n)]p_1V'[(p_2/p_1)^{(n-1)/n} - 1]$$

$$= [n/(1 - n)]m'RT_1[(p_2/p_1)^{(n-1)/n} - 1]$$

where m′ is the mass of air leaving the compressor.

In like manner for the high-pressure stage

$$\Sigma W_H = [n/(1 - n)]m'RT_3[(p_4/p_3)^{(n-1)/n} - 1]$$

and since we have assumed that $T_1 = T_3$, the total cyclic compression work is

$$\Sigma W_T = \Sigma W_L + \Sigma W_H = [n/(1 - n)]m'RT_1[(p_2/p_1)^{(n-1)/n} + (p_4/p_2)^{(n-1)/n} - 2] \tag{8.10}$$

where $p_2 = p_3$.

Example 8.7. Determine the cyclic work for the two-stage compressor where $p_1 = 15$ and $p_4 = 60$ psia. $n = 1.2$, and the entering air temperature is 40 F. p_2 will assume the values of 20, 25, 30, 35, and 40 psia.

From Eq. (8.10)

$$\Sigma W_T/m' = [n/(1 - n)](RT_1/J)[(p_2/p_1)^{(n-1)/n} + (p_4/p_2)^{(n-1)/n} - 2]$$

and for $p_2 = 20$ psia,

$$\Sigma W_T/m' = (1.2/-0.2)[53.3(500)/778][(20/15)^{0.2/1.2} + (60/20)^{0.2/1.2} - 2]$$

$$= -6.0(34.2)(1.049 + 1.201 - 2)$$

$$= -51.3 \text{ (B/lbm)}$$

By using the same technique for the other values of p_2 the results are

p_2:	20	25	30	35	40 (psia)
$\Sigma W_T/m'$:	-51.3	-50.5	-50.3	-50.5	-50.8 (B/lbm)

For the conditions given in this problem the minimum cyclic work occurs when $p_2 = 30$ psia.

Example 8.8. Determine the work of a single-stage compressor in Example 8.7. When $p_2 = 30$ psia, determine the work of each stage of compression.

For a single stage of compression, $p_2 = p_4 = 60$ psia, and

$$\Sigma W_T/m' = [n/(1 - n)](RT_1/J)[(p_2/p_1)^{(n-1)/n} + (p_4/p_2)^{(n-1)/n} - 2]$$

$$= (1.2/-0.2)[53.3(500)/778][(60/15)^{0.2/1.2} + (60/60)^{0.2/1.2} - 2]$$

$$= -6.0(34.2)(1.260 + 1 - 2) = -53.4 \text{ (B/lbm)}$$

The savings in cyclic work by using two-stage compression is 2.1 B/lbm or 4.1%.

For $p_2 = 30$ psia

$$p_2/p_1 = 30/15 = 2 \quad \text{and} \quad p_4/p_2 = 60/30 = 2$$

so that at the minimum work point

$$\Sigma W_L = \Sigma W_H = -50.3/2 = -25.15 \text{ (B/lbm)}$$

and each stage of the compressor does the same quantity of work.

While the intermediate pressure for minimum work can be established by trial and error, it is also possible to determine this pressure by the use of calculus.

From Eq. (8.10)

$$\Sigma W_T = C[(p_2/p_1)^m + (p_4/p_2)^m - 2]$$

where

$$C = [n/(1 - n)]m'RT_1 \quad \text{and} \quad m = (n - 1)/n$$

which have constant values for a given compressor design. Since p_1 and p_4 have fixed values, taking the derivative of ΣW_T with respect to p_2 yields

$$d(\Sigma W_T)/dp_2 = C[(1/p_1^m)mp_2^{m-1} - p_4^m mp_2^{-m-1}]$$

$$= C[(1/p_1^m)m(p_2^m/p_2) - p_4 m(1/p_2^m p_2)]$$

Setting the derivate equation to zero yields

$$(1/p_1^m)p_2^m = p_4^m(1/p_2^m)$$

or

$$p_2^{2m} = (p_1 p_4)^m$$

and

$$p_2 = (p_1 p_4)^{0.5} \qquad (8.11)$$

This will be the value of the intermediate pressure that requires the least amount of cyclic work, when $T_1 = T_3$ as in Fig. 8.5.

Example 8.9. Determine the optimum value of p_2 when $p_1 = 15$ and $p_4 =$ 60 psia.

From Eq. (8.11)

$$p_2 = (p_1 p_4)^{0.5} = [15(60)]^{0.5} = (900)^{0.5}$$

$$= 30 \text{ (psia)}$$

This is the same value as the trial and error solution.

An increase in the number of stages beyond two will further reduce the cyclic work, but the savings is progressively less for each additional stage. Economically, due to the cost of each extra piston and cylinder, two-stage compression is about the practical limit for air compression in a reciprocating compressor.

Example 8.10. Given is a two-stage air compressor, acting under optimum conditions, that compresses air from 15 psia to 200 psia. The initial air temperature is 60 F, and the percentage clearance, both stages, is 7%. The low-pressure cylinder bore is 10 in., and the stroke is 12 in. The compressor is single acting and has a speed of 400 rpm. Determine the mass of air compressed, the diameter of the high-pressure cylinder, the heat removed by the cylinder jackets for n = 1.25, the heat removed in the intercooler for an intercooler exit temperature of 60 F, and the horsepower required to drive the unit.

The displacement volume of the low-pressure stage is

$$V_{DL} = (\pi/4)D_L^2LN$$

$$= (\pi/4)(10/12)^2(12/12)(400)$$

$$= 218.2 \text{ (ft}^3\text{/min or cfm)}$$

when N is the compressor speed.

From Eq. (8.11) the optimum discharge pressure leaving the low-pressure stage is

$$p_2 = (p_1p_4)^{0.5} = [15(200)]^{0.5} = 54.8 \text{ (psia)}$$

From Eq. (8.9) the volumetric efficiency is

$$e_v = 1 + c - c(p_2/p_1)^{1/n}$$

$$= 1 + 0.07 - 0.07(54.8/15)^{1/1.25}$$

$$= 0.873 \quad \text{or} \quad 87.3\%$$

The volume of air admitted to the low-pressure cylinder is

$$V_L' = e_vV_D = 0.873(218.2)$$

$$= 190.5 \text{ (cfm)}$$

and the mass of air taken in or delivered from the compressor is

$$m' = p_1V'/RT_1$$

$$= [144(15)(190.5)]/[53.3(520)]$$

$$= 14.85 \text{ (lbm/min)}$$

For the high-pressure stage, the volumetric efficiency is

$$e_v = 1 + c - c(p_4/p_2)^{1/n}$$

$$= 1 + 0.07 - 0.07(200/54.8)^{1/1.25}$$

$$= 0.873 \quad \text{or} \quad 87.3\%$$

which is the same as the low-pressure stage since the pressure ratios and the clearances are identical.

The volume of air admitted to the high-pressure cylinder is

$$V_H' = m'RT_3/p_3$$

$$= [14.85(53.3)(520)]/[54.8(144)]$$

$$= 52.2 \text{ (cfm)}$$

where $T_3 = T_1 = 520$ R and $p_3 = p_2 = 54.8$ psia.

The displacement volume of the high-pressure stage must be

$$V_{DH} = V_H'/e_v = 52.2/0.873$$

$$= 59.8 \text{ (cfm)}$$

The cylinder bore for the high-pressure stage is, L = 12 in.,

$$V_{DH} = (\Pi/4)D_H^2LN$$

or

$$59.8 = (\Pi/4)D_H^2(12/12)(400)$$

and

$$D_H = 0.436 \text{ (ft)} \quad \text{or} \quad 5.24 \text{ (in.)}$$

From Eq. (8.5) the clearance volume of the low-pressure stage is

$$V_{CL} = cV_{DL} = 0.07(218.2) = 15.3 \text{ (cfm)}$$

and the volume at state 1 (Fig. 8.4) is

$$V_1 = V_{CL} + V_{DL} = 15.3 + 218.2 = 233.5 \text{ (cfm)}$$

The mass of air compressed in the low-pressure stage is

$$m_1 = p_1V_1/RT_1 = [15(144)(233.5)]/[53.3(520)]$$

$$= 18.2 \text{ (lbm/min)}$$

For the polytropic process from state 1 to state 2, the temperature at state 2 is

$$T_2 = T_1(p_2/p_1)^{(n-1)/n}$$

$$= 520(54.8/15)^{0.25/1.25} = 674 \text{ (R)}$$

and the polytropic specific heat is

$$c_n = c_v[(k - n)/(1 - n)]$$

$$= 0.1714[(1.4 - 1.25)/(1 - 1.25)]$$

$$= -0.1028 \text{ (B/lbm-R)}$$

The heat removed from the air in the low-pressure stage is

$$Q_L = m_1 c_n(T_2 - T_1) = 18.2(-0.1028)(674 - 520)$$

$$= -288.1 \text{ (B/min)}$$

The clearance volume for the high-pressure stage is

$$V_{CH} = cV_{DH} = 0.07(59.8) = 4.2 \text{ (cfm)}$$

and the volume at state 3 is

$$V_3 = V_{CH} + V_{DH} = 4.2 + 59.8 = 64.0 \text{ (cfm)}$$

The mass of air compressed in the high-pressure stage is

$$m_3 = p_3V_3/RT_3 = [144(54.8)(64.0)]/[53.3(520)]$$

$$= 18.2 \text{ (lbm/min)}$$

The temperature of the air at state 4 is

$$T_4 = T_3(p_4/p_3)^{(n-1)/n}$$

$$= 520(200/54.8)^{0.25/1.25} = 674 \text{ (R)}$$

and the heat removed from the air in the high-pressure stage is

$$Q_H = m_3 c_n(T_4 - T_3) = 18.2(-0.1028)(674 - 520)$$

$$= -288.1 \text{ (B/min)}$$

which is the same as the heat removed from the low-pressure cylinder since c_n is the same for both stages of the compressor when n for the low-pressure stage is identical to n for the high-pressure stage.

The heat removed in the constant pressure intercooler is

$$Q_i = m'c_p(T_3 - T_2) = 14.85(0.24)(520 - 674)$$

$$= -548.9 \text{ (B/lbm)}$$

since only the air delivered from the low pressure compressor passes through the intercooler.

From Eq. (8.10)

$$\Sigma W_T = [n/(1 - n)]m'RT_1[(p_2/p_1)^{(n-1)/n} + (p_4/p_2)^{(n-1)/n} - 2]$$

$$= (1.25/-0.25)[14.85(53.3)(520)][(54.8/15)^{0.2} + (200/54.8)^{0.2} - 2]$$

$$= -1.217E06 \text{ (ft-lbf/min)} = -36.9 \text{ (HP)}$$

For single stage compression the required HP is -42.3.

DOUBLE-ACTING COMPRESSORS

Reciprocating compressors are often constructed so that the gas may be compressed on both sides of the piston. When this system is used it is necessary to use a crosshead so that the piston rod travels in a straight line to permit sealing of the shaft (Fig. 8.6). The displacement volume of the cylinder must be decreased by the volume of the piston rod. The purpose of using a double-acting system is to reduce the size of the compressor.

Example 8.11. Determine the diameter of the low-pressure cylinder in Example 8.10 for a double-acting compressor with a 1-in.-diameter piston rod.

The displacement volume for the head end of the low-pressure cylinder is

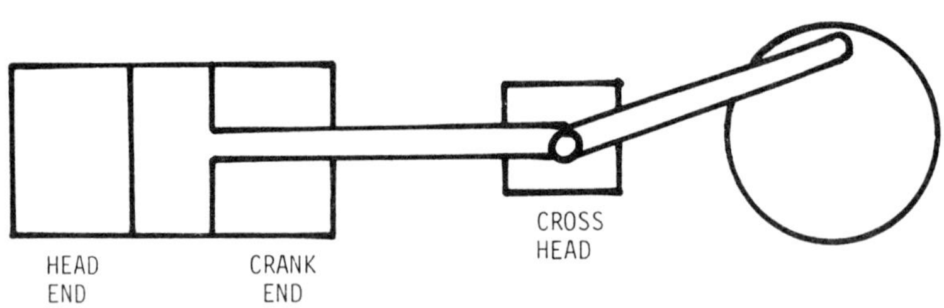

Figure 8.6

$$V_{DLH} = (\pi/4)D^2LN$$

and for the crank end

$$V_{DLC} = (\Pi/4)(D^2 - D_R^2)LN$$

where D_R is the diameter of the piston rod.

The total displacement volume of the low-pressure cylinder is

$$V_{DL} = V_{DLH} + V_{DLC} = [(\Pi/4)LN](D^2 + D^2 - D_R^2)$$

$$= 218.2 \text{ (cfm)}$$

where 218.2 cfm is the required displacement from Example 8.10; therefore

$$[(\Pi/4)(12/12)400][2D^2 - (1/12)^2] = 218.2$$

and

$$D = 0.592 \text{ (ft)} \quad \text{or} \quad 7.10 \text{ (in.)}$$

The cylinder bore is reduced from 10 in. to 7.1 in. by use of a double-acting compressor.

STEADY-FLOW AIR COMPRESSION

The steady-flow process equation for the steady-flow compression of air in an open system (Fig. 8.7) is

$$W_{sf} = Q - \Delta H - \Delta KE - \Delta PE$$

By using a gas as the working substance, the change in potential energy is negligible. Generally the velocities of the air entering and leaving the compressor are of such a magnitude, as well as being nearly equal, that the kinetic energy terms may also be neglected. In a steady-flow compressor, either centrifugal or axial flow, the opportunity to transfer heat to or from the moving airstream does not exist. Since Q is zero and if the process is assumed to be reversible, it must be a reversible adiabatic or isentropic process. The steady-flow work reduces to

$$W_{sf} = -\Delta H$$

which for a perfect gas becomes

$$W_{sf} = -mc_p(T_2 - T_1)$$

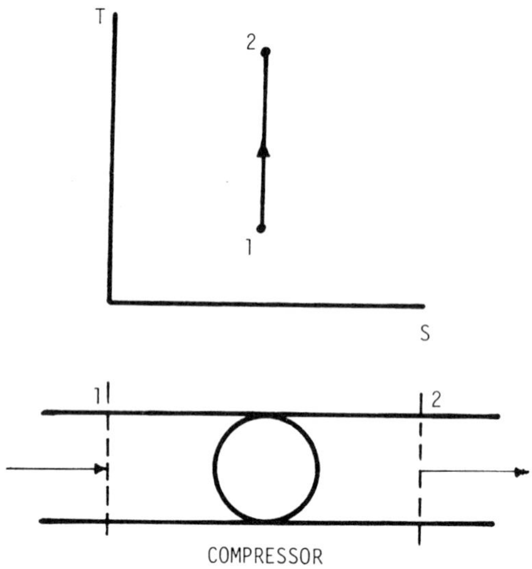

Figure 8.7

Example 8.12. Given the state points of the two-stage compressor in Example 8.10 where $T_1 = 520$ R, $p_1 = 15$ psia, $p_2 = 200$ psia, and $m' = 14.85$ lbm/min, determine the horsepower required to drive the steady-flow compressor.

For an isentropic process

$$T_2 = T_1(p_2/p_1)^{(k-1)/k}$$

$$= 520(200/15)^{0.4/1.4} = 1090 \text{ (R)}$$

The temperature leaving either stage of the reciprocating compressor for the polytropic process, n = 1.25, was 674 R.

The power required to compress the air is

$$W_{sf} = -mc_p(T_2 - T_1) = -(14.85)(0.24)(1090 - 520)$$

$$= -2031 \text{ (B/min)} = -47.9 \text{ (HP)}$$

The horsepower required for the reciprocating compressor was -36.9, which is -11.0 HP or 30.0% less than the power required for the steady-flow compressor. The increase in power required to drive the steady-flow system is due to the higher temperature of the air leaving the steady-flow device. As mentioned previously, if this temperature can be utilized in the device using the compressed air, no loss of power actually occurs. The gas turbine cycle, considered in Chap. 9, is an example of a device where the high compression temperature is fully utilized.

AIR ENGINES: NONEXPANSION

Many air-driven tools, such as jackhammers, operate by admitting a quantity of air which drives the piston from top dead center to bottom dead center in a constant pressure process (Fig. 8.8). At bottom dead center, the intake valve is closed, the exhaust valve is opened, and the pressure in the cylinder drops to atmospheric levels. The piston may be returned to top dead center by a spring mechanism at which point the exhaust valve is closed and the intake valve is opened to complete the cycle. The advantage of this system lies in the ability to use a small cylinder for the production of a relatively large amount of work. The disadvantage is the loss of work that could have been realized if a full or partial expansion process of the high-pressure air had been utilized.

The work of the cycle in Fig. 8.8 is

$$\Sigma W = W_{12} + W_{23} + W_{34} + W_{41}$$

$$= p_1(V_2 - V_1) + 0 + p_3(V_4 - V_3) + 0 = p_1(V_2 - V_1) - p_3(V_2 - V_1)$$

since $V_1 = V_4$ and $V_2 = V_3$, or

$$\Sigma W = (p_1 - p_3)(V_2 - V_1) \tag{8.12}$$

Example 8.13. A jackhammer has a stroke of 2 in. and a cylinder bore of 6 in. Air enters the hammer at a pressure of 200 psia and a temperature

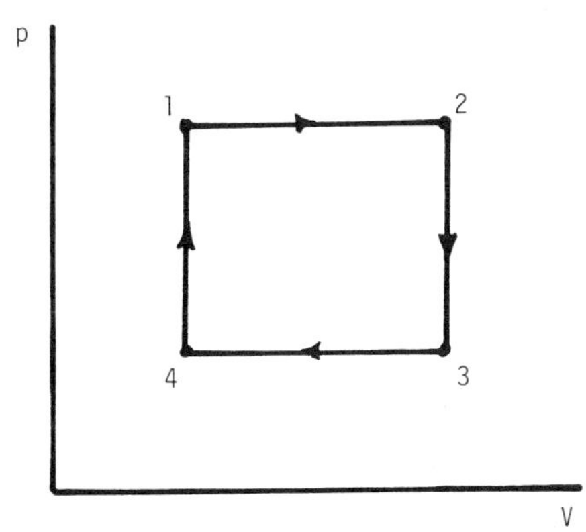

Figure 8.8

of 60 F. The exhaust pressure is 14.7 psia. Determine the force exerted by the hammer, the work per cycle of the hammer, and the mass of air used per cycle.

The force exerted on the piston face is

$$F_p = pA = p(\pi/4)D^2 = 200(144)(\pi/4)(6/12)^2$$

$$= 5654 \text{ (lbf)}$$

The force exerted by the atmosphere, $p_{atm} = 14.7$ psi, is

$$F_{atm} = 144(14.7)(\pi/4)(6/12)^2 = 416 \text{ (lbf)}$$

so that the force exerted by the hammer is

$$F = F_p - F_{atm} = 5654 - 416 = 5238 \text{ (lbf)}$$

The work per cycle, from Eq. (8.12), is

$$\Sigma W = (p_1 - p_3)(V_2 - V_1) = (p_1 - p_3)V_D$$

where V_D is the displacement volume of the hammer, or

$$V_D = (\pi/4)D^2L = (\pi/4)(6/12)^2(2/12)$$

$$= 0.0327 \text{ (ft}^3\text{)}$$

and

$$\Sigma W = 144(200 - 14.7)(0.0327) = 872.5 \text{ (ft-lbf)}$$

$$= 1.12 \text{ (B)}$$

The specific volume of the entering air is

$$v_1 = R_1T_1/p_1 = [53.3(520)/144(200)] = 0.962 \text{ (ft}^3\text{/lbm)}$$

and the mass of air required for each cycle is

$$m = V_D/v_1 = 0.0327/0.962 = 0.0340 \text{ (lbm)}$$

AIR ENGINES: EXPANSION

Consider the air engine sketched in Fig. 8.9. The air is admitted to the engine by means of a constant pressure process from state 4 to state 1. The air expands to exhaust pressure in the isentropic process from state 1 to state 2. The air is discharged from the cylinder by a constant pressure

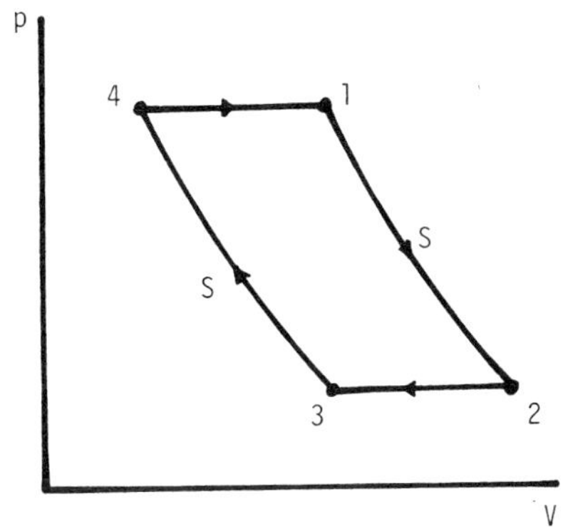

Figure 8.9

process from state 2 to state 3. The air remaining in the cylinder is isentropically compressed to the state of the entering air, process 34. By comparing Figs. 8.4 and 8.9, it is apparent that we have simply reversed the reciprocating air compression cycle to produce an expansion air engine. Equation (8.3) may be used for this device so that

$$\Sigma W/m' = [kRT_1/(1 - k)][(p_2/p_1)^{(k-1)/k} - 1] \tag{8.13}$$

where m′ is the mass of air entering the cylinder.

Example 8.14. An expansion air engine uses 0.0340 lbm of air per cycle which enters at a temperature and pressure of 60 F and 200 psia. The exhaust pressure is 14.7 psia. The cylinder bore is 6 in., and the clearance volume is 18 in.^{3}. Determine the volumes at each point in the cycle, the displacement volume of the engine, the clearance of the engine, the stroke of the engine, and the net work per cycle.

The specific volume of the entering air is

$$v_1 = RT_1/p_1 = [53.3(520)/144(200)] = 0.962 \text{ (ft}^3\text{/lbm)}$$

Since the air is admitted from state 4 to state 1,

$$V_1 - V_4 = m'v_1 = 0.0340(0.962) = 0.0327 \text{ (ft}^3\text{)}$$

Note that the data in this example have been selected so as to duplicate the intake conditions of the jackhammer in Example 8.13.

The volume at state 4 is

$$V_4 = V_c = 18/1728 = 0.0104 \text{ (ft}^3\text{)}$$

so that

$$V_1 = V_4 + 0.0327 = 0.0104 + 0.0327 = 0.0431 \text{ (ft}^3\text{)}$$

For the isentropic process from state 1 to state 2

$$V_2 = V_1(p_1/p_2)^{1/k} = 0.0431(200/14.7)^{1/1.4}$$

$$= 0.2781 \text{ (ft}^3\text{)}$$

The temperature of the air at state 2 is

$$T_2 = T_1(p_2/p_1)^{(k-1)/k} = 520(14.7/200)^{0.4/1.4}$$

$$= 247 \text{ (R)}$$

and the specific volume of the air at state 2 is

$$v_2 = RT_2/p_2 = [53.3(247)]/[144(14.7)]$$

$$= 6.219 \text{ (ft}^3\text{/lbm)}$$

The change in volume necessary to discharge the given mass of air having the specific volume of state 2 is

$$V_2 - V_3 = m'v_2 = 0.0340(6.219)$$

$$= 0.2114 \text{ (ft}^3\text{)}$$

so that

$$V_3 = V_2 - 0.2114 = 0.2781 - 0.2114 = 0.0667 \text{ (ft}^3\text{)}$$

To check the cycle for closure

$$V_4 = V_3(p_3/p_4)^{1/k} = 0.0667(14.7/200)^{1/1.4} = 0.0103 \text{ (ft}^3\text{)}$$

and

$$T_4 = T_3(p_4/p_3)^{1/k} = 247(200/14.7)^{0.4/1.4} = 521 \text{ (R)}$$

for the isentropic compression process from state 3 to state 4.

The displacement of the air engine is

$$V_D = V_2 - V_4 = 0.2781 - 0.0104$$

$$= 0.2677 \text{ (ft}^3\text{)}$$

and the clearance of the engine is

$$V_C = V_4 = cV_D$$

so that

$$c = V_4/V_D = 0.0104/0.2677$$

$$= 0.039 \quad \text{or} \quad 3.9\%$$

The length of the piston stroke may be obtained from

$$V_D = (\Pi/4)D^2L = (\Pi/4)(6/12)^2L = 0.2677$$

or

$$L = 1.364 \text{ (ft)} = 16.4 \text{ (in.)}$$

compared to 2 in. for the jackhammer.

From Eq. (8.13), the work per cycle is

$$\Sigma W/m' = [kRT_1/(1 - k)][(p_2/p_1)^{(k-1)/k} - 1]$$

$$= [1.4(53.3)(520)/(1 - 1.4)][(14.7/200)^{0.4/1.4} - 1]$$

$$= 50{,}990 \text{ (ft-lbf/lbm)} = 65.5 \text{ (B/lbm)}$$

or

$$\Sigma W = 0.0340(65.5) = 2.23 \text{ (B)}$$

compared to 1.12 B for the jackhammer. Expanding the air increases the cyclic work by 49.8%.

AIR ENGINES: PARTIAL EXPANSION

To obtain full expansion of the air in the air engine requires a large displacement volume and therefore a long piston stroke. In high-speed engines, long piston strokes increase the forces acting on the connecting rods and bearings. A reduction in these forces may be obtained by partial expansion of the air in the cylinder. A partial expansion air engine is sketched in Fig. 8.10. When the piston reaches the bottom dead-center position at state 2, the exhaust valve is opened, and the pressure drops to atmospheric pressure.

Example 8.15. Given the data for the air engine in Example 8.14 with a piston stroke of 12 in., determine the net work per cycle of the engine. From Example 8.14, $V_5 = 0.0104$ ft^3, $V_1 = 0.0431$ ft^3, and $p_5 = p_1 = 200$ psia, where the values at state 5 are the values at state 4 in Example 8.14.

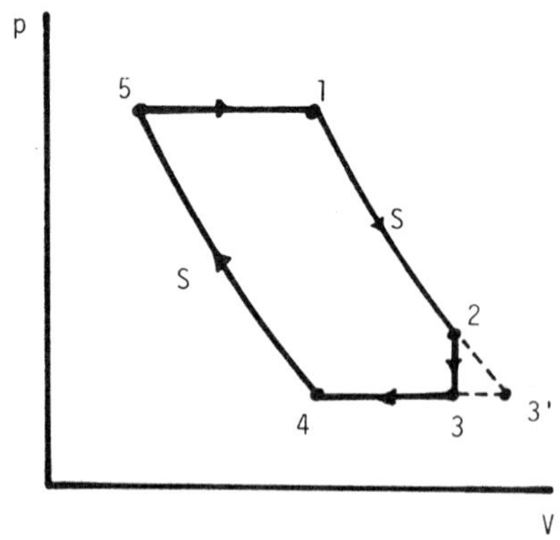

Figure 8.10

The nonflow work of the constant pressure process from state 5 to state 1 is

$$W_{51} = [p_5(V_1 - V_5)]/J = (144/778)200(0.0431 - 0.0104)$$

$$= 1.21 \text{ (B)}$$

The displacement volume of the air engine is

$$V_D = (\Pi/4)D^2L = (\Pi/4)(6/12)^2(12/12) = 0.1963 \text{ (ft}^3\text{)}$$

The volume at state 2 or state 3 is

$$V_3 = V_2 = V_D + V_5 = 0.1963 + 0.0104 = 0.2067 \text{ (ft}^3\text{)}$$

compared to a maximum volume of 0.2781 ft^3 for a full expansion engine.

The pressure at state 2 for an isentropic expansion process is

$$p_2 = p_1(V_1/V_2)^k = 200(0.0431/0.2067)^{1.4}$$

$$= 22.3 \text{ (psia)}$$

Instead of expanding the air to 14.7 psia, the final expansion pressure is 22.3 psia.

The nonflow work for the isentropic process from state 1 to state 2 is given by

$$W_{12} = (p_2V_2 - p_1V_1)/[J(1 - k)]$$

$$= (144/778)[22.3(0.2067) - 200(0.0431)]/(1 - 1.4)$$

$$= 1.86 \text{ (B)}$$

Since the isentropic compression process is unchanged from the full expansion air engine, the volumes at states 3 and 4 are $V_3 = 0.2067\ ft^3$, $V_4 = 0.0667\ ft^3$, and $p_3 = 14.7$ psia, where V_4 is the value of V_3 in Example 8.14.

For the constant pressure process from state 3 to state 4

$$W_{34} = [p_3(V_4 - V_3)]/J = (144/778)14.7(0.0667 - 0.2067)$$

$$= -0.38\ (B)$$

For the isentropic compression process from state 4 to state 5, $p_5 = 200$ psia, $p_4 = 14.7$ psia, $V_5 = 0.0104\ ft^3$, and $V_4 = 0.0667\ ft^3$, so that

$$W_{45} = (p_5V_5 - p_4V_4)/[J(1 - k)]$$

$$= (144/778)[200(0.0104) - 14.7(0.0667)]/(1 - 1.4)$$

$$= -0.51\ (B)$$

The net work of the cycle is

$$\Sigma W = W_{12} + W_{23} + W_{34} + W_{45} + W_{51}$$

$$= 1.86 + 0 - 0.38 - 0.51 + 1.21$$

$$= 2.18\ (B)$$

Therefore the loss due to partial expansion is

$$\Delta W_{loss} = 2.23 - 2.18 = 0.05\ (B)$$

This may be verified since (Fig. 8.10) $W_{23'} = 0.24$ B for the isentropic process and $W_{3'3} = -0.19$ (B) for the constant pressure process, and the loss in work is

$$\Delta W_{loss} = W_{23'} - W_{3'3} = 0.24 - 0.19 = 0.05\ (B)$$

The work per pound mass of air is

$$\Sigma W = 2.18/m' = 2.18/0.0340 = 64.1\ (B/lbm)$$

AIR ENGINES: PARTIAL EXPANSION AND PARTIAL COMPRESSION

Instead of closing the exhaust valve at state 4 in Fig. 8.10, it may be allowed to remain open to a point 4 in Fig. 8.11. The isentropic process will not raise the air pressure to the initial air pressure, so that when the intake air valve is opened an extra amount of air must be admitted to raise the pressure by means of a constant volume process. The use of

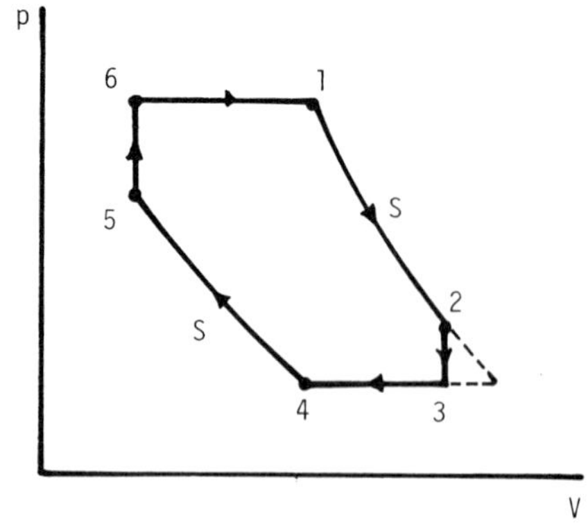

Figure 8.11

partial compression will increase the net work area on the pV diagram, but the work performed per pound of air will be reduced.

Example 8.16. Given are the air engine in Example 8.15 and a pressure at state 5 (Fig. 8.10) of 50 psia. From the previous examples,

	1	2	3	4	5	6
p (psia)	200.0	22.3	14.7	14.7	50.0	200.0
V (ft^3)	0.0431	0.2067	0.2067	—	0.0104	0.0104

The mass of air admitted between states 6 and 1 is 0.0340 lbm.

The volume at state 4 may be obtained from

$$V_4 = V_5(p_5/p_4)^{1/k} = 0.0104(50/14.7)^{1/1.4}$$

$$= 0.0249 \ (ft^3)$$

The nonflow work for each process is

$$W_{12} = (p_2V_2 - p_1V_1)/[J(1 - k) = 1.86 \ (B), \qquad S = C$$

$$W_{23} = 0 \ (B), \qquad V = C$$

$$W_{34} = [p_3(V_4 - V_3)]/J = -0.49 \ (B), \qquad p = C$$

$$W_{45} = (p_5V_5 - p_4V_4)/[J(1 - k)] = -0.07 \ (B), \qquad S = C$$

$$W_{56} = 0 \ (B) \qquad V = C$$

$$W_{61} = [p_6(V_1 - V_6)]/J = 1.21 \ (B), \qquad p = C$$

and

$$\Sigma W = W_{12} + W_{23} + W_{34} + W_{45} + W_{56} + W_{61} = 2.51 \text{ (B)}$$

for the cycle.

For the full expansion engine, $\Sigma W = 2.23$, and for the partial expansion engine, $\Sigma W = 2.18$.

The temperature at state 2 after the isentropic expansion process is

$$T_2 = T_1(p_2/p_1)^{(k-1)/k}$$

$$= 520(22.3/200)^{0.4/1.4} = 278 \text{ (R)}$$

compared to 247 R for the full expansion engine.

The mass of air in the cylinder at state 4 is

$$m_4 = p_4V_4/RT_4 = [144(14.7)(0.0249)/53.3(278)]$$

$$= 0.00355 \text{ (lbm)}$$

where $T_2 = T_3 = T_4$. This is also the mass of air in the cylinder at state 5.

The mass of air in the cylinder at state 6 is

$$m_6 = p_6V_6/RT_6 = [144(200)(0.0104)/53.3(520)]$$

$$= 0.01081 \text{ (lbm)}$$

The mass of air added to raise the pressure from 50 psia at state 5 to 200 psia at state 6 is

$$\Delta m = m_6 - m_4 = 0.01081 - 0.00355$$

$$= 0.0073$$

and the total mass of air required to operate the engine is

$$m_t = m' + \Delta m = 0.0340 + 0.0073 = 0.0413$$

The results for the three cycles are

	Full Expansion	Partial Expansion	Partial Expansion-Partial Compression
ΣW (B)	2.23	2.18	2.51
m (lbm)	0.0340	0.0340	0.0413
ΣW (B/lbm)	65.6	64.1	60.7

While the partial expansion-partial compression engine produces the most work per cycle, it also produces the least amount of work per pound mass of air used to operate the engine.

MINIMUM CYCLE TEMPERATURE

The entering air properties for the air engines in the examples were 60 F and 200 psia. After the expansion process the air temperature was 247 R or -213 F for the full expansion engine and 278 R or -182 F for the partial expansion engine. Since the air is exhausted from the engine at these low temperatures, the air engine may be used to supply cooling air. In general, expansion processes are capable of producing large amounts of cold air, but the technique is not economically competitive with normal cooling devices except in special applications.

Low exhaust air temperatures do present one serious problem. Unless the air is absolutely dry, the moisture will freeze inside the engine and clog the exhaust system. To avoid this problem the air entering the engine must either be preheated or be the result of a compression process which has a high discharge pressure.

The exterior surfaces of the exhaust system will also be subject to ice or frost deposit which may cause operating difficulties.

Example 8.17. Given is the partial expansion-partial compression air engine used in Example 8.16. Determine the temperature at the inlet so that the temperature at state 2 is 40 F. What is the mass of air supplied to the engine?

The required inlet air temperature for the isentropic expansion process from state 1 to state 2 is

$$T_1 = T_2(p_1/p_2)^{(k-1)/k} = 500(200/22.3)^{0.4/1.4}$$

$$= 936 \text{ (R)} \quad \text{or} \quad 476 \text{ (F)}$$

The specific volume of the air entering the engine is

$$v_6 = v_1 = RT_1/p_1 = [53.3(936)]/[200(144)]$$

$$= 1.732 \text{ (ft}^3\text{/lbm)}$$

The mass of air supplied to the engine from states 1 to 6 is

$$m' = m_1 - m_6 = (V_1/v_1) - (V_6/v_6) = (1/v_1)(V_1 - V_6)$$

$$= (1/1.732)(0.0431 - 0.0104) = 0.0189 \text{ (lbm)}$$

compared to 0.0340 lbm of air admitted at 60 F.

Since the value of the pressure and volume at each state point is unchanged, the net work of the cycle is 2.51 B, but the work of the cycle per pound of air admitted becomes

$$\Sigma W = 2.51/(0.0189 + 0.0040) = 109.6 \text{ (B/lbm)}$$

where

$$\Delta m = m_6 - m_4 = 0.0040 \text{ (lbm)}$$

compared to 60.7 B/lbm when the colder air is used.

CLOSURE

While our discussion of compressors and engines has been confined to the use of air as a working substance, all of the material, as presented, is valid for any perfect gas or mixture of perfect gases. By using the appropriate values of the physical properties, c_p, c_v, k, and R for the gas being studied, it is then permissible to use any equation derived for properties, heat, and work terms.

For the normal operating ranges of pressure and temperature encountered in compressors and engines it is not necessary to use a semiperfect gas as the working substance.

PROBLEMS

8.1 An air compressor operates between air pressures of 14.7 and 120 psia without clearance. The entering air temperature is 80 F, and the value of n for the compression process is 1.3. Determine the cyclic work to compress the air.

8.2 The compressor in Prob. 8.1 has a bore of 6 in. and a stroke of 12 in. The compressor operates, single acting, at a speed of 400 rpm. Determine the horsepower required to drive the compressor.

8.3 Determine the temperature at the end of the compression process for the compressor in Prob. 8.1. If the clearance is 10%, determine the volumetric efficiency and the work of the compression process.

8.4 Determine the optimum discharge pressure for a two-stage compressor operating under the conditions of Prob. 8.1. Air is cooled to 80 F in the intercooler. Determine the cyclic work for the compressor.

8.5 The low-pressure stage of the compressor in Prob. 8.4 has a bore of 6 in. and a stroke of 12 in. and operates at a speed of 400 rpm. The clearance is 10% for both stages, and the stroke of the high-pressure stage is 12 in. $p_1 = 14.7$ psia, $p_4 = 120$ psia, and $t_1 = t_3 = 80$ F (see Fig. 8.4). Determine the mass of air delivered by the compressor, the heat removed by the cylinder jackets, the heat removed by the intercooler, the horsepower required to drive the compressor, the volumetric efficiency of both stages, and the bore of the high-pressure cylinder. $n = 1.3$ for both stages; single acting.

8.6 Determine the bore of the low-pressure cylinder in Prob. 8.5 for a double-acting compressor with a 0.75-in. rod to deliver the same quantity of air.

8.7 Given a steady-flow isentropic compressor with $p_1 = 14.7$ psia, $t_1 = 80$ F, and $p_2 = 120$ psia (see Fig. 8.7), determine the horsepower required to drive the compressor for the same quantity of air as delivered in Prob. 8.5.

8.8 A nonexpansion air engine has a bore and stroke of 4 in. The inlet air pressure is 120 psia, and the exhaust pressure is 14.7 psia. The speed of the engine is 400 rpm. Determine the HP of the engine. The inlet air temperature is 80 F.

8.9 An air expansion engine operates with the inlet and exhaust conditions given in Prob. 8.8. The engine bore, speed, and air used per cycle are also the same as in Prob. 8.8. The clearance volume is 10 in.^{3}. Determine the air used per cycle, the engine stroke, and the HP of the engine. Full expansion engine.

8.10 Given the engine in Prob. 8.9 with a stroke of 6 in., determine the HP of the engine.

8.11 Given the engine in Prob. 8.10 with a pressure at the end of the isentropic compression process of 60 psia, determine the HP of the engine.

9

EXTERNAL COMBUSTION ENGINES

External combustion engines use the open or steady-flow system where the heat addition and rejection processes may be physically separated from the processes where work is done on or by the system. The gas turbine cycle is an example of an external combustion engine where the heat is added in a combustion chamber which occupies a different location from the compressor and turbine where work is used or produced. The heat rejection process in the gas turbine is accomplished by ejecting the hot gases into the surrounding atmosphere. Another example of an external combustion device is the power plant cycle where heat is added in the boiler and heat is rejected in the condenser. Work is added or subtracted in the cycle by means of turbines or pumps.

In an internal combustion engine, such as the Otto or Diesel cycle, the burning of the fuel occurs on an intermittent basis and lasts for only a short period of time. In an external combustion device the combustion is continuous. In the internal combustion engines the maximum cycle temperature is of such short duration that the average metallic surface temperatures are well below these peak values. In an external combustion engine, the maximum temperature in the cycle is continuous and the metallic surfaces assume this maximum temperature value. In the gas turbine cycle the high temperature occurs in the first row of turbine blades. As the temperature of a metal increases, its strength decreases, requiring the use of expensive materials in the turbine. The expense of the turbine restricts its use to aircraft, some trucks, and marine transport as well as stationary power plants. While the efficiency of the modified gas turbine engine is not substantially different from that of the internal combustion engine, the high cost of the turbine has discouraged its use in the private automobile.

The above situation is unfortunate since an external combustion engine has several advantages compared to the internal combustion engine. In an Otto cycle or spark-ignited engine, combustion can occur only when the ratio of fuel to air is relatively high. In either the Otto or Diesel cycle, the time available for combustion is so short that complete combustion of the fuel is impossible. This incomplete combustion results in pollutants that must be removed from the engine exhaust. Rich mixtures also lead to high combustion temperatures which increase the variety of pollutants produced by the engine.

In an external combustion engine the combustion is continuous so that large quantities of air may be supplied to the engine. This excess air as well as a much longer time for combustion result in an almost complete burning of the fuel. Incomplete combustion not only produces carbon monoxide as a pollutant but is wasteful since all of the energy in the fuel has not been utilized. The excess air also reduces the maximum cycle temperature and lessens the tendency of the combustion products to form pollutant materials.

If the gas turbine engine could be economically constructed, an automobile would not require a transmission except for reverse operations, since a turbine produces its highest torque at low operating speeds. However, the acceleration characteristics of a gas turbine cycle engine are less than that of an internal combustion engine.

The power to weight ratio of the gas turbine is quite high, so that this device, even with the high turbine cost, is an efficient power plant for aircraft operations.

BRAYTON CYCLE

The theoretical model for the gas turbine engine is the Brayton cycle (Fig. 9.1). The air is drawn into the engine by means of an axial flow compressor and is isentropically compressed to the maximum cycle pressure. The air leaving the compressor undergoes a constant pressure process in which heat is added to the air. In the real engine, fuel is burned in the combustion chamber. After the heat addition process, in the model cycle, the gas is isentropically expanded in the turbine. Heat rejection occurs at constant pressure in the atmosphere surrounding the engine. The cycle may be analyzed with reference to the heat transfers on the TS diagram, where

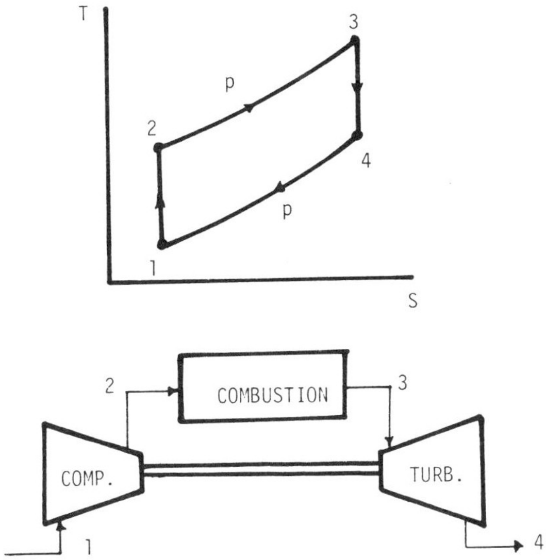

Figure 9.1

$$Q_{12} = 0, \qquad S = C$$

$$Q_{23} = mc_p(T_3 - T_2), \qquad p = C$$

$$Q_{34} = 0, \qquad S = C$$

$$Q_{41} = mc_p(T_1 - T_4), \qquad p = C$$

The net work or the net heat of the cycle is

$$\Sigma Q = \Sigma W = Q_{12} + Q_{23} + Q_{34} + Q_{41}$$

$$= mc_p(T_3 - T_2) + mc_p(T_1 - T_4)$$

$$= mc_p[(T_3 - T_2) - (T_4 - T_1)]$$

The cycle efficiency is

$$e = \Sigma W/Q_{in} = mc_p[(T_3 - T_2) - (T_4 - T_1)]/[mc_p(T_3 - T_2)]$$

$$= 1 - [(T_4 - T_1)/(T_3 - T_2)] \qquad (9.1)$$

where $Q_{in} = Q_{23} = mc_p(T_3 - T_2)$, which is the positive heat term.

For the isentropic process from state 1 to state 2 in the compressor,

$$T_2/T_1 = (p_2/p_1)^{(k-1)/k} = r_p^{(k-1)/k}$$

where $r_p = p_2/p_1$, which is the pressure ratio of the compressor. Therefore

$$T_2 = T_1 r_p^{(k-1)/k}$$

In the turbine, for the isentropic process from state 3 to state 4,

$$T_3/T_4 = (p_3/p_4)^{(k-1)/k} = (p_2/p_1)^{(k-1)/k} = r_p^{(k-1)/k}$$

since $p_3 = p_2$ and $p_1 = p_4$. Therefore

$$T_3 = T_4 r_p^{(k-1)/k}$$

Substitution of the temperature terms into Eq. (9.1) yields

$$e = 1 - [(T_4 - T_1)/(T_4 r_p^{(k-1)/k} - T_1 r_p^{(k-1)/k})]$$

$$= 1 - 1/r_p^{(k-1)/k} \tag{9.2}$$

As in the Otto and Diesel cycles the conversion from Eq. (9.1) to Eq. (9.2) removes the surrounding air temperature as a variable in the cycle efficiency and permits the evaluation of the cycle efficiency in terms of the pressure ratio of the compressor.

Since

$$r_p^{(k-1)/k} = T_2/T_1$$

Eq. (9.2) may also be written as

$$e = 1 - [1/(T_2/T_1)] = 1 - (T_1/T_2) \tag{9.3}$$

Example 9.1. Given a Brayton cycle for which $t_1 = 40$ F and $p_1 = 15$ psia, the pressure ratio of the compressor is 4.0 and the maximum temperature in the cycle is 1540 F. Determine e, Q_{in}, and ΣW.

The temperature at state 2 is

$$T_2 = T_1 r_p^{(k-1)/k} = 500(4)^{0.4/1.4} = 743 \text{ (R)}$$

and the temperature at state 4 is

$$T_4 = T_3/r_p^{(k-1)/k} = 2000/(4)^{0.4/1.4} = 1346 \text{ (R)}$$

By means of Eq. (9.1)

$$e = 1 - [(T_4 - T_1)/(T_3 - T_2)] = 1 - [(1346 - 500)/(2000 - 743)]$$

$$= 0.327 \quad \text{or} \quad 32.7\%$$

By means of Eq. (9.2)

$$e = 1 - (1/r_p^{(k-1)/k}) = 1 - [1/(4)^{0.4/1.4}] = 0.327$$

Finally, by means of Eq. (9.3)

$$e = 1 - (T_1/T_2) = 1 - (500/743) = 0.327$$

The heat added to the cycle is

$$Q_{in} = Q_{23} = c_p(T_3 - T_2) = 0.24(2000 - 743)$$

$$= 301.7 \text{ (B/lbm)}$$

The net work of the cycle may be obtained from

$$\Sigma W = eQ_{in} = 0.327(301.7) = 98.7 \text{ (B/lbm)}$$

The work of the isentropic compressor, state 1 to state 2, is

$$W_{sf} = Q - \Delta H - \Delta KE - \Delta PE$$

which, if ΔKE, $\Delta PE = 0$, reduces to

$$W_{sf} = Q - \Delta H$$

and for the isentropic compression process, $Q = 0$, so that

$$W_{12} = W_c = -\Delta h = -c_p(T_2 - T_1) = -0.24(743 - 500)$$

$$= -58.3 \text{ (B/lbm)}$$

where W_c is the work of the compressor.

In like manner, for the isentropic turbine process from state 3 to state 4, where W_t is the work of the turbine,

$$W_{34} = W_t = -c_p(T_4 - T_3) = -0.24(1346 - 2000)$$

$$= 157.0 \text{ (B/lbm)}$$

The net work of the cycle is

$$\Sigma W = W_{12} + W_{34} = W_c + W_t = -58.3 + 157.0 = 98.7 \text{ (B/lbm)}$$

which is identical to the result obtained from the efficiency equation.

Example 9.2. Rework Example 9.1 using air as a semiperfect gas. For state 1, $T_1 = 500$ R, $h_1 = 119.48$ B/lbm, and $p_{r1} = 1.0590$.

For the isentropic compression process

$$r_p = p_2/p_1 = p_{r2}/p_{r1}$$

so that

$$p_{r2} = r_p p_{r1} = 4(1.0590) = 4.236$$

From the gas tables, Appendix C,

$T = 740$ R: $h = 177.23$ B/lbm and $p_r = 4.193$

$T = 760$ R: $h = 182.08$ B/lbm and $p_r = 4.607$

and

$$f = (4.236 - 4.193)/(4.607 - 4.193) = 0.104$$

so that

$$T_2 = 740 + 0.104(760 - 740) = 742\ [743]\ (R)$$

Also

$$h_2 = 177.23 + 0.104(182.08 - 177.23)$$

$$= 177.73\ (B/lbm)$$

Perfect gas solutions are indicated by square brackets.

For state 3, $T_3 = 2000$ R, $h_3 = 504.71$ B/lbm, $p_{r3} = 174.00$, and

$$r_p = p_3/p_4 = p_{r3}/p_{r4}$$

so that

$$p_{r4} = p_{r3}/r_p = 174.00/4 = 43.50$$

By interpolation from the gas tables,

$T = 1400$ R: $h = 342.90$ B/lbm and $p_r = 42.88$

$T = 1420$ R: $h = 348.14$ B/lbm and $p_r = 45.26$

and

$$f = (43.50 - 42.88)/(45.26 - 42.88) = 0.261$$

so that

$$T_4 = 1400 + 0.261(1420 - 1400) = 1405\ [1346]\ (R)$$

Also

$$h_4 = 342.90 + 0.261(348.14 - 342.90) = 344.27\ (B/lbm)$$

The heat and work terms, ΔKE, $\Delta PE = 0$, are

$$W_{12} = W_c = -\Delta h = -(h_2 - h_1) = -(177.73 - 119.48)$$

$$= -58.2\ [-58.3]\ (B/lbm)$$

Since T_2 is only 742 R, there is no significant difference between the perfect and semiperfect gas solutions for the compressor work.

The heat added to the cycle is given by

$$Q_{23} = Q_{in} = \Delta H + W_{sf}$$

But $W_{sf} = 0$ for a constant pressure process when ΔKE, $\Delta PE = 0$. Therefore

$$Q_{23} = Q_{in} = \Delta h = h_3 - h_2 = 504.71 - 177.73$$

$$= 327.0\ [301.7]\ (B/lbm)$$

For the isentropic expansion process from state 3 to state 4

$$W_{34} = W_t = -\Delta h = -(h_4 - h_3) = -(344.27 - 504.71)$$

$$= 160.4\ [157.0]\ (B/lbm)$$

The net work of the cycle is

$$\Sigma W = W_{12} + W_{34} = W_c + W_t = -58.2 + 160.4$$

$$= 102.2\ [98.7]\ (B/lbm)$$

The efficiency of the cycle can only be determined by means of

$$e = \Sigma W/Q_{in} = 102.2/327.0$$

$$0.312\ [0.327]$$

COMPRESSOR AND TURBINE EFFICIENCY

In the previous section we have assumed that the compressor and turbine processes were isentropic. In the real gas turbine cycle the processes are irreversible and adiabatic. These irreversible processes are indicated by dashed lines in Fig. 9.2. The compressor efficiency for a semiperfect gas is defined by

$$e_c = \frac{W_{theoretical}}{W_{actual}} = \frac{W_c}{W_{c'}} = \frac{Q - \Delta h}{Q - \Delta h'} = \frac{-\Delta h}{-\Delta h'} = \frac{h_2 - h_1}{h_{2'} - h_1} \tag{9.4}$$

where it has been assumed that ΔKE, $\Delta PE = 0$.

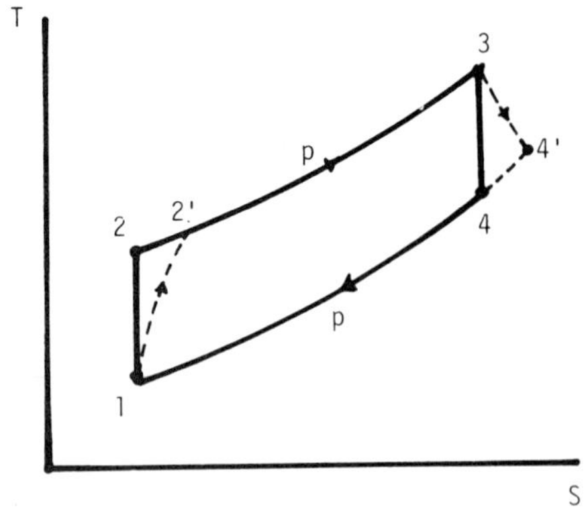

Figure 9.2

For a perfect gas

$$e_c = \frac{-\Delta h}{-\Delta h'} = \frac{-c_p(T_2 - T_1)}{-c_p(T_{2'} - T_1)} = \frac{T_2 - T_1}{T_{2'} - T_1} \tag{9.5}$$

The work done in the actual compression process, due to friction, must exceed that of the theoretical cycle.

For the gas turbine using a semiperfect gas, ΔKE, $\Delta PE = 0$,

$$e_t = \frac{W_{actual}}{W_{theoretical}} = \frac{W_t'}{W_t} = \frac{Q' - \Delta h'}{Q - \Delta h} = \frac{h_3 - h_{4'}}{h_3 - h_4} \tag{9.6}$$

and for the perfect gas

$$e_t = \frac{\Delta h'}{\Delta h} = \frac{T_3 - T_{4'}}{T_3 - T_4} \tag{9.7}$$

The actual turbine work, again due to friction, must be less than that of the theoretical process.

Example 9.3. Given $r_p = 4.0$, $p_1 = 15$ psia, $t_1 = 40$ F, and $t_3 = 1540$ F (Fig. 9.2), determine W_c', W_t', Q_{in}', and e' using air as a perfect gas. The compressor efficiency is 85%, and the turbine efficiency is 80%.

For the isentropic compression process from state 1 to state 2

$$T_2 = T_1(p_2/p_1)^{(k-1)/k} = 500(4)^{0.4/1.4}$$

$$= 743 \text{ (R)}$$

From Eq. (9.5), for the compressor efficiency,

$$e_c = (T_2 - T_1)/(T_2' - T_1) = (743 - 500)/(T_2' - 500) = 0.85$$

and

$$T_2' = 786 \text{ (R)}$$

The actual compression work, ΔKE, $\Delta PE = 0$, is

$$W_c' = \Delta h' = -c_p(T_2' - T_1) = -0.24(786 - 500)$$

$$= -68.6 \text{ (B/lbm)}$$

or

$$W_c' = W_c/e_c = [-c_p(T_2 - T_1)]/e_c$$

$$= [-0.24(743 - 500)]/0.85$$

$$= -68.6 \text{ (B/lbm)}$$

where W_c is the theoretical compressor work and W_c' is the real compressor work.

For the isentropic expansion process from state 3 to state 4

$$T_4 = T_3(p_4/p_3)^{(k-1)/k} = T_3/r_p^{(k-1)/k}$$

$$= 2000/(4.0)^{0.4/1.4} = 1346 \text{ (R)}$$

From Eq. (9.7), for the turbine efficiency,

$$e_t = (T_3 - T_4')/(T_3 - T_4) = (2000 - T_4')/(2000 - 1346) = 0.80$$

and

$$T_4' = 1477 \text{ (R)}$$

The actual turbine work is

$$W_t' = -\Delta h' = -c_p(T_4' - T_3) = -0.24(1477 - 2000) = 125.5 \text{ (B/lbm)}$$

or

$$W_t' = e_t W_t = -e_t c_p(T_4 - T_3) = -(0.8)(0.24)(1346 - 2000)$$

$$= 125.6 \text{ (B/lbm)}$$

where W_t is the theoretical turbine work and W_t' is the real turbine work.

The net work of the real cycle is

$$\Sigma W' = W_c' + W_t' = -68.6 + 125.6 = 57.0 \text{ (B/lbm)}$$

as compared to 98.7 B/lbm for the cycle where $e_c = e_t = 100\%$.

The heat added to the real cycle is

$$Q_{23}' = Q_{in}' = c_p(T_3 - T_2') = 0.24(2000 - 786)$$

$$= 291.4 \text{ (B/lbm)}$$

as compared to 301.7 B/lbm for the theoretical cycle. The heat added to the real cycle is less than that added to the theoretical cycle since $T_2' > T_2$.

The cycle efficiency is

$$e' = \Sigma W'/Q_{in}' = 57.0/291.4 = 0.196 \quad \text{or} \quad 19.6\%$$

as compared to 32.7% for the fully reversible cycle. Efficient compressor and turbine design is a critical factor in the performance of a gas turbine engine.

Example 9.4. Repeat Example 9.2 for a semiperfect gas where $W_c = -58.2$ B/lbm, $W_t = 160.4$ B/lbm, and $h_1 = 119.48$, $h_2 = 177.73$, and $h_3 = 504.71$ B/lbm.

The compression work is

$$W_c' = W_c/e_c = -58.2/0.85$$

$$= -68.5 \; [-68.6] \text{ (B/lbm)}$$

Square brackets represent perfect gas solutions of an identical problem. The turbine work is

$$W_t' = e_t W_t = 0.80(160.4)$$

$$= 128.3 \; [125.6] \text{ (B/lbm)}$$

The net work of the actual cycle is

$$\Sigma W' = W_c' + W_t' = -68.5 + 128.3$$

$$= 59.8 \; [57.0] \text{ (B/lbm)}$$

The enthalpy at the end of the compression process from Eq. (9.4) is

$$e_c = (h_2 - h_1)/(h_2' - h_1) = (177.33 - 119.48)/(h_2 - 119.48) = 0.85$$

so that

$$h_2' = 187.54 \text{ (B/lbm)}$$

The heat added to the real cycle is

$$Q_{in}' = h_3 - h_2' = 504.71 - 187.54$$

$$= 317.2 \; [291.4] \text{ (B/lbm)}$$

The cycle efficiency is

$$e' = \Sigma W'/Q_{in}' = 59.8/317.2$$

$$= 0.188 \ [0.196]$$

The difference between the perfect and semiperfect calculations of the cycle efficiency is not significant, but the heat addition and net work terms differ by 8.1% and 4.7%.

GAS TURBINE CYCLE WITH MAXIMUM NET WORK

In the gas turbine cycle, the temperature at the start of the compression process, T_1 in Fig. 9.3, is dependent on the temperature of the air entering the compressor. The maximum temperature in the cycle, T_3, occurs in the first row of the turbine blades. The maximum cycle temperature is limited by the ability of the turbine blades to maintain their strength at elevated temperatures.

For a given gas turbine cycle the temperatures T_1 and T_3 may be assumed to be fixed in value. The extremes of possible operating conditions are shown in Fig. 9.3. For high values of the pressure ratio r_p, the temperature T_{2x} approaches the value of T_3. The cyclic area on the TS diagram becomes smaller as this limit is approached, and

$$\Sigma Q = \Sigma W \rightarrow 0$$

and the gas turbine produces no useful work.

From Eq. (9.3)

$$e = 1 - (T_1/T_{2x})$$

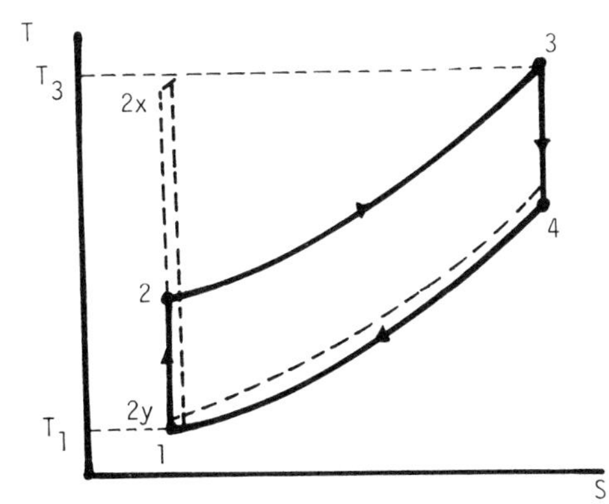

Figure 9.3

so that as T_{2x} increases, the cycle efficiency increases, and the maximum efficiency occurs when T_{2x} is nearly equal to T_3.

The other extreme occurs as the pressure ratio decreases and T_{2y} approaches the value of T_1. As before, the net heat, which is the area within the cyclic diagram, decreases with decreasing values of T_{2y}. Therefore the net work of the gas turbine becomes smaller as T_{2y} decreases.

Again, from Eq. (9.3)

$$e = 1 - (T_1/T_{2y})$$

and for

$$T_{2y} \to T_1, \qquad e \to 0$$

Since the net work at both operating extremes is zero, there must be a maximum value of the net work between these limits.

Example 9.5. Given a Brayton cycle using air as a perfect gas with $t_1 = 40$ F and $t_3 = 1540$ F, determine W_c, W_t, ΣW, Q_{in}, and e for $r_p = 8$, 10, 12, 14, and 16. The values of the compressor and turbine efficiencies are $e_c = e_t = 100\%$.

For $r_p = 10$ and an isentropic compression process,

$$T_2 = T_1(p_2/p_1)^{(k-1)/k} = T_1 r_p^{(k-1)/k} = 500(10)^{0.4/1.4}$$

$$= 965 \text{ (R)}$$

The temperature at state 4 may be obtained from

$$T_4/T_3 = (p_4/p_3)^{(k-1)/k} = (p_1/p_2)^{(k-1)/k} = T_1/T_2$$

since $p_4 = p_1$ and $p_2 = p_3$. Therefore

$$T_4 = T_1 T_3/T_2 \qquad (9.8)$$

and

$$T_4 = [500(2000)]/965 = 1036 \text{ (R)}$$

By using air as a perfect gas,

$$W_c = c_p(T_1 - T_2) = 0.24(500 - 965) = -111.6 \text{ (B/lbm)}$$

$$W_t = c_p(T_3 - T_4) = 0.24(2000 - 1036) = 231.4 \text{ (B/lbm)}$$

$$\Sigma W = W_c + W_t = -111.6 + 231.4 = 119.8 \text{ (B/lbm)}$$

$$Q_{in} = c_p(T_3 - T_2) = 0.24(2000 - 965) = 248.4 \text{ (B/lbm)}$$

and

$$e = \Sigma W/Q_{in} = 119.8/248.4 = 0.482 \quad \text{or} \quad 48.2\%$$

In like manner, using the other values of r_p, the following table may be constructed:

r_p	W_c (B/lbm)	W_t (B/lbm)	ΣW (B/lbm)	Q_{in} (B/lbm)	e (%)
8	-97.4	215.0	117.6	262.6	44.8
10	-111.6	231.4	119.8	248.4	48.2
12	-124.1	244.1	120.0	235.9	50.8
14	-135.1	254.2	119.1	224.9	53.0
16	-145.0	262.6	117.6	215.0	54.7

From the preceding table, which represents a trial and error solution, it is apparent that $r_p = 12.0$ produces the maximum net work. For r_p in excess of this value, the net work decreases, but the cycle efficiency must increase as r_p increases. Again, this presents a trade-off decision since the maximum work pressure ratio is not the most efficient pressure ratio.

For cars, trucks, or stationary power plants, the weight of the engine is not a critical factor. For aircraft engines it is of prime importance. If in the preceding problem $p_1 = 15$ psia, the value of the pressure in the combustion chamber would be

r_p:	8	10	12	14	16
p_2 (psia):	120	150	180	210	240

As the pressure in the combustion chamber increases, the strength, and thereby the weight of the combustion chamber, must increase. This weight increase adds another factor to be considered in the design of a gas turbine aircraft engine.

By the use of calculus the value of the pressure ratio that produces the maximum net work may be evaluated more precisely than by trial and error techniques.

Using a perfect gas as a working substance, the net work is

$$\Sigma W = W_c + W_t = c_p(T_1 - T_2) + c_p(T_3 - T_4) = c_p(T_1 + T_3 - T_2 - T_4)$$

Since the temperatures T_1 and T_3 are assumed to be fixed and by use of Eq. (9.8) where

$$T_4 = T_1T_3/T_2$$

the net work becomes

$$\Sigma W = c_p[T_1 + T_3 - T_2 - (T_1T_3)/T_2]$$

and the only variable in the equation is the temperature T_2. By differentiation, with respect to T_2,

$$d\ \Sigma W/dT_2 = c_p[-1 + (T_1T_3)/T_2^2]$$

and by setting the derivative equal to zero

$$-1 + (T_1T_3)/T_2^2 = 0$$

or

$$T_2 = (T_1T_3)^{0.5} \tag{9.9}$$

This value of T_2 is known to produce the maximum net work of the cycle since when $T_2 \rightarrow T_3$ or $T_2 \rightarrow T_1$, the net work approaches zero.

From Eq. (9.8)

$$T_4 = T_1T_3/T_2 = T_1T_3/(T_1T_3)^{0.5} = (T_1T_3)^{0.5} = T_2$$

For the maximum net work cycle, $T_2 = T_4$.

Example 9.6. Given the data in Example 9.5, determine r_p, W_c, W_t, ΣW, Q_{in}, and e for the maximum net cycle work.

From Eq. (9.9)

$$T_2 = (T_1T_3)^{0.5} = [500(2000)]^{0.5} = 1000\ (R)$$

and

$$T_2 = T_4 = 1000\ (R)$$

The pressure ratio that produces the maximum net work is

$$r_p = (T_2/T_1)^{k/(k-1)} = (1000/500)^{1.4/0.4} = 11.3\ [4]$$

As before

$$W_c = c_p(T_1 - T_2) = 0.24(500 - 1000) = -120.0\ [-58.3]\ (B/lbm)$$

$$W_t = c_p(T_3 - T_2) = 0.24(2000 - 1000) = 240.0 \ [157.0] \ (B/lbm)$$

$$\Sigma W = W_c + W_t = -120 + 240 = 120.0 \ [98.7] \ (B/lbm)$$

$$Q_{in} = c_p(T_3 - T_2) = 0.24(2000 - 1000) = 240.0 \ [301.7]$$

$$e = \Sigma W/Q_{in} = 120/240 = 0.500 \ [0.327]$$

Answers within square brackets are for $r_p = 4$.

Example 9.7. Assuming the working substance is air, as a semiperfect gas, determine r_p, W_c, W_t, ΣW, Q_{in}, and e for the cycle which produces the maximum net work.

Using the value of T_2 obtained from the perfect gas solution as a starting point, we shall determine the net work where T_2 = 980, 1000, 1020, and 1040 R. For T_2 = 1000 R, from the gas tables,

$$T_1 = 500 \text{ R:} \qquad h_1 = 119.48 \text{ B/lbm} \quad \text{and} \quad p_{r1} = 1.0590$$

$$T_2 = 1000 \text{ R:} \qquad h_2 = 240.98 \text{ B/lbm} \quad \text{and} \quad p_{r2} = 12.298$$

$$T_3 = 2000 \text{ R:} \qquad h_3 = 504.71 \text{ B/lbm} \quad \text{and} \quad p_{r3} = 174.00$$

The value of r_p is

$$r_p = p_2/p_1 = p_{r2}/p_{r1} = 12.298/1.0590 = 11.61$$

and

$$r_p = p_3/p_4 = p_{r3}/p_{r4} = p_2/p_1 = r_p$$

so that

$$p_{r4} = p_{r3}/r_p = 174.00/11.61 = 14.987$$

From the gas tables,

$$T = 1040 \text{ R:} \qquad h = 250.95 \text{ B/lbm} \quad \text{and} \quad p_r = 14.182$$

$$T = 1060 \text{ R:} \qquad h = 255.96 \text{ B/lbm} \quad \text{and} \quad p_r = 15.203$$

and

$$f = (14.987 - 14.182)/(15.203 - 14.182) = 0.788$$

and

$$h_4 = 250.95 + 0.788(255.96 - 250.95)$$
$$= 254.90 \ (B/lbm)$$

The compressor work is

$$W_c = h_1 - h_2 = 119.48 - 240.98 = -121.50 \text{ (B/lbm)}$$

and the turbine work is

$$W_t = h_3 - h_4 = 504.71 - 254.90 = 249.81 \text{ (B/lbm)}$$

yielding a net work of

$$\Sigma W = W_c + W_t = -121.5 + 249.81 = 128.31 \text{ (B/lbm)}$$

For the four values of T_2,

T_2 (R)	r_p	W_c (B/lbm)	W_t (B/lbm)	ΣW (B/lbm)
980	10.79	-116.54	244.47	127.93
1000	11.61	-121.50	249.81	128.31
1020	12.48	-126.49	254.99	128.50
1040	12.99	-131.47	259.94	128.47

In this case the value of T_2 that produces the maximum net work is 1000 R using air as a perfect gas and 1020 R with air as a semiperfect gas, $r_p = 12.5$ for the semiperfect gas and 11.3 for the perfect gas. For the semiperfect gas the trial and error solution must be used since there is no simple fixed relationship between the temperature and the enthalpy.

MAXIMUM NET WORK CYCLE WITH COMPRESSOR AND TURBINE EFFICIENCIES

A trial and error solution could be employed to obtain the pressure ratio which produces the maximum net work in a cycle where the compressor and turbine efficiencies are less than 100%. However, it is simpler to use calculus to obtain a simple equation that will apply for all values of the compressor efficiency e_c and the turbine efficiency e_t. By using air as a perfect gas as the working substance, from Eqs. (9.4) and (9.5) and with reference to Fig. 9.2,

$$W_c' = W_c/e_c = [c_p(T_1 - T_2)]/e_c$$

for the real or actual compressor work, and for the real or actual turbine work from Eqs. (9.6) and (9.7),

$$W_t' = e_t W_t = e_t[c_p(T_3 - T_4)]$$

The net work for the real gas turbine is

$$\Sigma W' = W_c' + W_t' = [c_p(T_1 - T_2)/e_c] + e_t[c_p(T_3 - T_4)] \tag{9.10}$$

where T_1 and T_3 are constants for the particular gas turbine installation. Since only the theoretical temperatures appear in Eq. (9.10), Eq. (9.8) may be used, where

$$T_4 = T_1T_3/T_2$$

so that

$$\Sigma W' = [c_p(T_1 - T_2)/e_c] + e_t[c_p(T_3 - T_1T_3/T_2)]$$

Taking the derivative of the actual net work with respect to the only variable temperature, T_2, yields

$$d\,(\Sigma W')/dT_2 = -c_p/e_c + e_t(c_pT_1T_3)/T_2^2$$

Setting the derivative equal to zero to obtain the maximum actual net work gives

$$c_p/e_c = e_t(c_pT_1T_3/T_2^2)$$

so that

$$T_2^2 = e_ce_tT_1T_3$$

or

$$T_2 = (e_ce_tT_1T_3)^{0.5} \tag{9.11}$$

Example 9.8. Given a Brayton cycle for which $t_1 = 40$ F, $t_3 = 1540$ F, $e_c = 85\%$, and $e_t = 80\%$, determine r_p, W_c', W_t', $\Sigma W'$, Q_{in}', and e' for a cycle that produces the maximum net cycle work (Fig. 9.2).

From Eq. (9.11)

$$T_2 = (e_ce_tT_1T_3)^{0.5} = [0.85(0.80)500(2000)]^{0.5}$$

$$= 825 \text{ (R)}$$

From Eq. (9.8)

$$T_4 = T_1T_3/T_2 = [500(2000)]/825 = 1212 \text{ (R)}$$

The optimum pressure ratio is

$$r_p = p_2/p_1 = (T_2/T_1)^{k/(k-1)} = (825/500)^{1.4/0.4}$$

$$= 5.77\ [11.3]$$

The values within the square brackets are for the cycle with e_c and e_t = 100%. The work and heat terms are

$$W_c' = [c_p(T_1 - T_2)]/e_c = [0.24(500 - 825)]/0.85$$

$$= -91.6\ [-120.0]\ (B/lbm)$$

for the compressor, and for the turbine

$$W_t' = e_t[c_p(T_3 - T_4)] = 0.80[0.24(2000 - 1212)]$$

$$= 151.3\ [240.0]\ (B/lbm)$$

The maximum net work for the actual cycle is

$$\Sigma W' = W_c' + W_t' = -91.6 + 151.3$$

$$= 59.7\ [120.0]\ (B/lbm)$$

By use of Eq. (9.5)

$$e_c = (T_2 - T_1)/(T_2' - T_1) = (825 - 500)/(T_2' - 500) = 0.85$$

so that

$$T_2' = 882\ (R)$$

The heat added is

$$Q_{in}' = c_p(T_3 - T_2') = 0.24(2000 - 882)$$

$$= 268.3\ [240]\ (B/lbm)$$

The cycle efficiency is

$$e' = \Sigma W'/Q_{in}' = 59.7/268.3$$

$$= 0.223\ [0.500]$$

As before, the severe loss in cycle efficiency due to the compression and turbine efficiency is apparent.

BRAYTON CYCLE FOR JET AIRCRAFT ENGINES

The net work produced by the gas turbine engine may be used to supply the power necessary to drive a car or truck or to drive a generator in a stationary power plant or to drive a propeller in a turboprop aircraft appli-

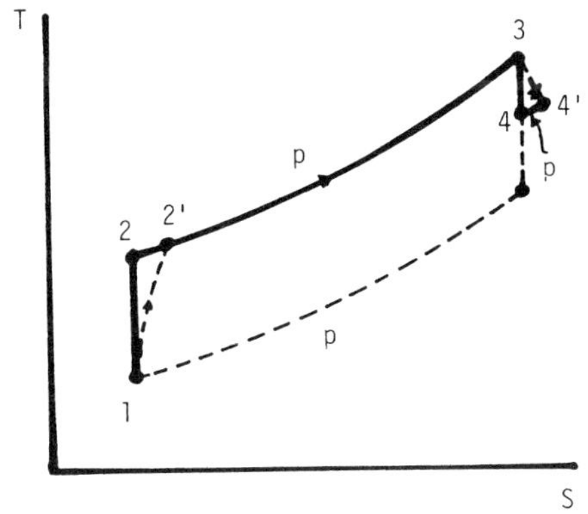

Figure 9.4

cation. In a jet aircraft engine the power developed in the turbine is used only to drive the compressor, so that the exhaust from the turbine is at a higher pressure and temperature as compared to cycles with expansion to atmospheric pressure.

Example 9.9. Determine the pressure and temperature at state 4′ for a maximum net work cycle where t_1 = 40 F, t_3 = 1540 F, e_c = 85% and e_t = 80% (Fig. 9.4). The pressure at state 1 is 15 psia.

From Example 9.8

$$W_c' = [c_p(T_1 - T_2)]/e_c = -91.6 \text{ (B/lbm)}$$

where

$$T_2 = 825 \text{ (R)}$$

The required turbine work to drive the compressor is

$$W_t' = 91.6 = e_t[c_p(T_3 - T_4)] = 0.8[0.24(2000 - T_4)]$$

so that

$$T_4 = 1523 \text{ (R)}$$

The pressure ratio is 5.77, and the pressure leaving the compressor is

$$p_2 = p_3 = r_p p_1 = 5.77(15) = 86.6 \text{ (psia)}$$

The lower value of r_p for the maximum net work cycle reduces the weight of the combustion chamber. The pressure leaving the turbine is

$$p_4 = p_4' = p_3(T_4/T_3)^{k/(k-1)}$$

$$= 86.6(1523/2000)^{1.4/0.4} = 33.4 \text{ (psia)}$$

By use of Eq. (9.7)

$$e_t = (T_3 - T_4')/(T_3 - T_4) = (2000 - T_4')/(2000 - 1523) = 0.8$$

or

$$T_4' = 1618 \text{ (R)}$$

Example 9.10. If the air in Example 9.9 is expanded in an isentropic nozzle (Fig. 9.5), determine the velocity leaving the nozzle where $\Delta PE = 0$, $W_{sf} = 0$, $V_4' = 0$, and $p_5 = 15$ psia.

For a steady-flow system

$$Q = \Delta H + \Delta KE + \Delta PE + W_{sf}$$

or

$$\Delta KE = -\Delta H = -c_p(T_5 - T_4')$$

For the isentropic expansion process in the nozzle

$$T_5 = T_4'(p_5/p_4)^{(k-1)/k} = 1618(15/33.4)^{0.4/1.4} = 1287 \text{ (R)}$$

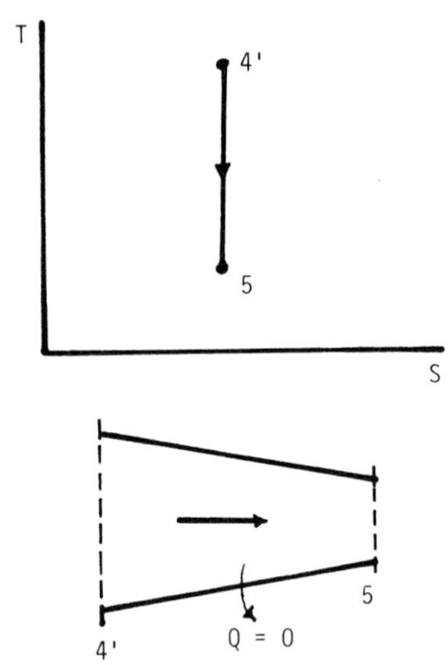

Figure 9.5

where T_5 is the temperature leaving the nozzle.

The change in kinetic energy in the nozzle is

$$\Delta KE = -c_p(T_5 - T_4') = c_p(T_4' - T_5) = 0.24(1618 - 1287)$$

$$= 79.4 \text{ (B/lbm)}$$

The relationship between the kinetic energy and the velocities is

$$(\overline{V}_5^2/2g_0J) - (\overline{V}_4'^2/2g_0J) = \Delta KE = 79.4 \text{ (B/lbm)}$$

and by assuming the velocity leaving the turbine is relatively low,

$$\overline{V}_5^2 = 2g_0J\,\Delta KE = 64.4(778)(79.4)$$

so that the velocity leaving the nozzle is

$$\overline{V}_5 = 1994 \text{ (ft/sec)} \quad \text{or} \quad 1360 \text{ (mph)}$$

AFTERBURNING

To increase the velocity of the gases leaving the nozzle and to obtain the maximum thrust required at takeoff, fuel may be injected at the turbine exit (Fig. 9.6). The injection of fuel increases the temperature in the

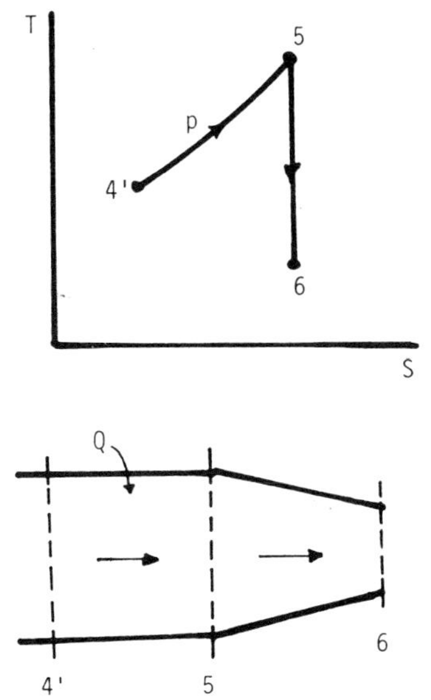

Figure 9.6

nozzle, but since the nozzle is a stationary device, the increased temperature does not pose a serious design problem with respect to the nozzle strength.

Example 9.11. Determine the velocity leaving the nozzle if 100 B/lbm is added to the air leaving the turbine. Use the data in Example 9.10.

The heat is added in a constant pressure process, where $p_5 = p_4 =$ 33.4 psia, yielding a temperature at the nozzle entrance of

$$Q = c_p(T_5 - T_4') = 0.24(T_5 - 1618) = 100$$

so that

$$T_5 = 2035 \text{ (R)}$$

The temperature leaving the nozzle is

$$T_6 = T_5(p_6/p_5)^{(k-1)/k} = 2035(15/33.4)^{0.4/1.4}$$

$$= 1619 \text{ (R)}$$

The change in kinetic energy in the nozzle is

$$\Delta KE = -c_p(T_6 - T_5) = -0.24(1619 - 2035)$$

$$= 99.8 \text{ (B/lbm)}$$

and the velocity leaving the nozzle is

$$\bar{V}_6^2 = 2g_0 J\,\Delta KE = 64.4(778)(99.8)$$

so that

$$\bar{V}_6 = 2236 \text{ (ft/sec)} \quad \text{or} \quad 1525 \text{ (mph)}$$

The heat added in the combustion chamber was 268.3 B/lbm, and increasing the heat addition by 100 B/lbm or 37.2% produces an increase in the nozzle velocity of 242 ft/sec or 12.1%.

Afterburning is not an economical mode of engine operation but may be used to produce maximum engine thrust for short periods of time.

REGENERATION

In the jet aircraft engine, the high temperature of the engine gases leaving the turbine is utilized to produce a velocity increase in the nozzle. The same effect may be obtained by the use of a nozzle in a turboprop engine.

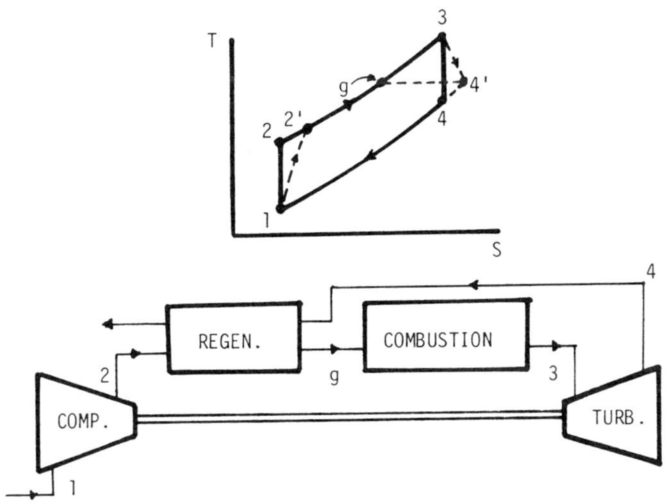

Figure 9.7

For stationary power plants or surface transportation units the energy in the high-temperature exhaust gases is wasted. To recover this energy a regenerator may be added to the cycle (Fig. 9.7). The hot gases leaving the turbine pass through a heat exchanger to increase the temperature of the air leaving the compressor. In effect, this reduces the quantity of heat that must be added to the air in the combustion chamber.

Example 9.12. Given is a Brayton cycle, with a regenerator, where $t_1 = 40$ F, $t_3 = 1540$ F, $e_c = 85\%$, and $e_t = 80\%$. The cycle is designed to produce the maximum net work. Determine Q_{in}', $\Sigma W'$, and e'.

For the maximum net work cycle

$$T_2 = (e_c e_t T_1 T_3)^{0.5} = [0.85(0.8)500(2000)]^{0.5}$$

$$= 825 \text{ (R)}$$

and

$$T_2' = T_1 + [(T_2 - T_1)/e_c] = 500 + [(825 - 500)/0.85]$$

$$= 882 \text{ (R)}$$

The temperature at state 4 is

$$T_4 = T_1 T_3/T_2 = [500(2000)]/825 = 1212 \text{ (R)}$$

and from Eq. (9.7)

$$e_t = (T_3 - T_4')/(T_3 - T_4) = (2000 - T_4')/(2000 - 1212) = 0.8$$

so that

$$T_4' = 1370 \text{ (R)}$$

For a regenerator temperature of

$$T_g = T_4' = 1370 \text{ (R)}$$

the heat added in the combustion chamber is

$$Q_{in-g} = c_p(T_3 - T_g) = 0.24(2000 - 1370)$$

$$= 151.2 \text{ (B/lbm)}$$

Without the regeneration the heat addition would be

$$Q_{in}' = c_p(T_3 - T_2') = 0.24(2000 - 882)$$

$$= 268.3 \text{ (B/lbm)}$$

The net work of the cycle, with or without a regenerator, is

$$\Sigma W' = W_c' + W_t' = c_p(T_1 - T_2') + c_p(T_3 - T_4')$$

$$= 0.24(500 - 882) + 0.24(2000 - 1370) = 59.6 \text{ (B/lbm)}$$

The cycle efficiency without a regenerator is

$$e' = \Sigma W'/Q_{in}' = 59.6/268.3 = 0.222 \text{ or } 22.2\%$$

and with a regenerator is

$$e' = \Sigma W'/Q_{in-g} = 59.6/151.2 = 0.394 \text{ or } 39.4\%$$

The regenerator increased the cycle efficiency by 17.2%.

PARTIAL REGENERATION

Many types of heat exchangers may be used as a regenerator. A disk, consisting of a metallic mesh, may be rotated between the hot gas stream leaving the turbine and the air stream from the compressor. Regardless of the type of exchanger employed in the gas turbine it is impossible to recover all of the available energy from the hot gases. The actual temperature of the air entering the combustion chamber is expressed in terms of the regenerator efficiency where

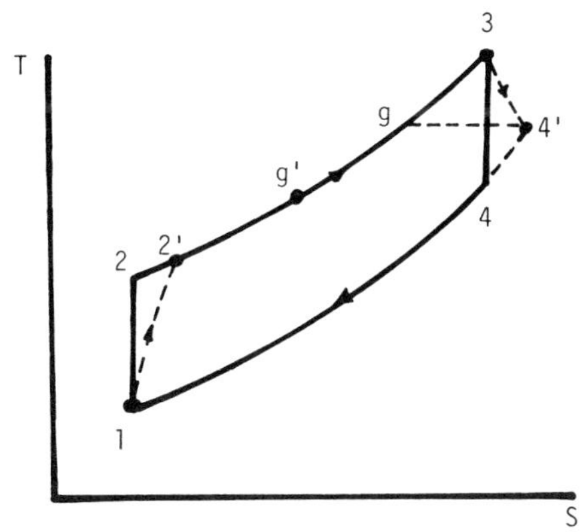

Figure 9.8

$$e_g = (h_g' - h_2')/(h_g - h_2') = (T_g' - T_2')/(T_g - T_2') \tag{9.12}$$

Using temperatures in Eq. (9.12) requires that the working substance be a perfect gas. The location of the temperatures on the TS diagram is sketched in Fig. 9.8.

Example 9.13. Rework Example 9.12 for $e_g = 55\%$.

From this example $T_2' = 882$ R and $T_4' = T_g = 1370$ R, so that from Eq. (9.12)

$$e_g = (T_g' - T_2')/(T_g - T_2') = (T_g' - 882)/(1370 - 882) = 0.55$$

which results in an inlet temperature to the combustion chamber of

$$T_g' = 1150 \text{ (R)}$$

The heat added in the combustion chamber is

$$Q_{in}'' = c_p(T_3 - T_g') = 0.24(2000 - 1150)$$

$$= 204 \text{ (B/lbm)}$$

compared to 151.2 B/lbm for a gas turbine with a 100% regenerator or 268.3 B for a gas turbine with no regenerator.

The regenerator has no effect on the net work of the cycle, so that

$$\Sigma W' = 59.6 \text{ (B/lbm)}$$

and the cycle efficiency is

$$e'' = \Sigma W'/Q_{in}'' = 59.6/204$$

$$= 0.292$$

For $e_g = 100\%$, $e = 0.394$, and for $e_g = 0$, $e = 0.222$.

VARIABLE SPEED OPERATION

Efficient compressor and turbine design is mandatory for economical operation of a gas turbine engine. Lowering the values of the compressor and turbine efficiencies results in much lower cycle efficiencies. In designing a compressor or turbine the peak efficiency occurs at a single operating speed. This means that the gas turbine is not suitable as a variable speed engine. One method of reducing the effects of variable speed operation is to allow the compressor and the high-pressure turbine that drives the compressor to operate at their design speeds (Fig. 9.9). The low-pressure turbine, whose output is the net work of the cycle, may be operated at varying speeds as required by the particular installation under consideration.

Example 9.14. Use the gas turbine in Example 9.13 where $t_1 = 40$ F, $t_3 = 1540$ F, $p_1 = 15$ psia, $e_c = 85\%$, $e_{t\text{-HP}} = 80\%$, $e_{t\text{-LP}} = 50\%$, and $e_g = 55\%$. Let $r_p = 5.77$ as in the previous examples. Determine the net work of the cycle and the cycle efficiency.

From the previous examples, $T_2' = 882$ R, and the compressor work is

$$W_c' = c_p(T_1 - T_2') = 0.24(500 - 882)$$

$$= -91.7 \text{ (B/lbm)}$$

This quantity of work must be supplied by the high-pressure turbine (HPT),

$$W_{t\text{-HP}}' = e_{t\text{-HP}}c_p(T_3 - T_4) = 0.8(0.24)(2000 - T_4) = 91.7$$

and

$$T_4 = 1522 \text{ (R)}$$

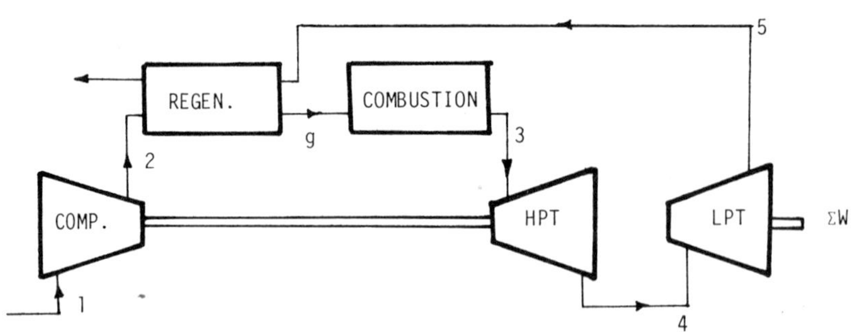

Figure 9.9

The work of the high-pressure turbine is also

$$W_{t\text{-}HP}' = c_p(T_3 - T_4') = 0.24(2000 - T_4') = 91.7$$

so that

$$T_4' = 1618 \text{ (R)}$$

The pressure leaving the high-pressure turbine is

$$p_4 = p_4' = p_3(T_4/T_3)^{k/(k-1)} = r_p p_1(T_4/T_3)^{k/(k-1)}$$

where

$$p_3 = p_2 = r_p p_1$$

so that

$$p_4 = 5.77(15)(1522/2000)^{1.4/0.4} = 33.3 \text{ (psia)}$$

The actual thermodynamic state of the air entering the low-pressure turbine is

$$p_4 = p_4' = 33.3 \text{ (psia)}$$

and

$$T_4' = 1618 \text{ (R)}$$

For a low-pressure turbine with an efficiency of 100%

$$\begin{aligned} T_5 &= T_4'(p_5/p_4)^{(k-1)/k} \\ &= 1618(15/33.3)^{0.4/1.4} \\ &= 1288 \text{ (R)} \end{aligned}$$

The net work of the cycle is

$$\begin{aligned} \Sigma W' &= W_{t\text{-}LP}' = e_{t\text{-}LP} c_p(T_4' - T_5) \\ &= 0.5(0.24)(1618 - 1288) \\ &= 39.6 \text{ (B/lbm)} \end{aligned}$$

Also

$$\Sigma W' = W_{t\text{-}LP}' = c_p(T_4' - T_5') = 0.24(1618 - T_5') = 39.6$$

and

$$T_5' = 1453 \text{ (B/lbm)}$$

The regenerator efficiency equation [Eq. (9.12)] is

$$e_g = (T_g' - T_2')/(T_g - T_2') = (T_g' - 882)/(1453 - 882) = 0.55$$

or

$$T_g' = 1196 \text{ (R)}$$

where

$$T_g = T_5' = 1453 \text{ (R)}$$

The heat added in the combustion chamber is

$$Q_{in}' = c_p(T_3 - T_g') = 0.24(2000 - 1196)$$

$$= 193.0 \text{ (B/lbm)}$$

and the cycle efficiency is

$$e' = \Sigma W'/Q_{in}' = 39.6/193.0 = 0.205 \quad \text{or} \quad 20.5\%$$

For a single turbine with an efficiency of 80% the net work of the cycle was 59.6 B/lbm, and the cycle efficiency was 29.2%. A pressure ratio of 5.77 is not the optimum value for the low- and high-pressure turbine system as given in this example.

PROBLEMS

9.1 Given a simple Brayton cycle with $r_p = 5$ and inlet and maximum cycle temperatures of 60 F and 1740 F, determine e, W_c, W_t, ΣW, and Q_{in} using air as a perfect gas.

9.2 The inlet pressure to the gas turbine in Prob. 9.1 is 14.7 psia. The volume rate of air flow into the compressor is 10,000 cfm. Determine the power of the unit.

9.3 Repeat Prob. 9.1 using air as a semiperfect gas.

9.4 Assume that the gas turbine in Prob. 9.1 has a compressor efficiency of 80% and a turbine efficiency of 75%. Determine W_c', W_t', $\Sigma W'$, Q_{in}', and e'.

9.5 Repeat Prob. 9.4 for a semiperfect gas.

9.6 Determine the pressure ratio, e, W_c, W_t, ΣW, and Q_{in} for the gas turbine in Prob. 9.1 operating on the maximum net work cycle.

9.7 Determine r_p, W_c', W_t', $\Sigma W'$, Q_{in}', and e for the gas turbine in Prob. 9.4 operating on the maximum net work cycle.

9.8 Assume that the gas turbine in Prob. 9.7 is used in a jet aircraft engine. Determine the pressure and temperature leaving the turbine and the velocity of the air leaving a nozzle which has an exhaust pressure of 14.7 psia.

9.9 The engine in Prob. 9.8 includes an afterburner which adds 150 B/lbm to the air leaving the turbine. Determine the exit velocity.

9.10 Repeat Prob. 9.7 for a gas turbine having a regenerator efficiency of 60%.

9.11 The gas turbine in Prob. 9.10 is used to drive an automobile. The low-pressure turbine has an efficiency of 60%. Determine the net work and the efficiency of this variable speed unit. The exhaust pressure is 14.7 psia.

10

VAPORS

In the previous chapters the perfect gas and the semiperfect gas have been defined and used as thermodynamic working substances. In this chapter we shall study vapors and how they differ from the perfect and semiperfect gases. The changes in properties and in the heat and work terms using a vapor as the working substance will also be studied.

VAPORS AND THE PERFECT GAS

A vapor is a substance in the gaseous phase that does not obey the perfect gas law. Gases, as they approach the temperature at which they will condense, deviate from the perfect gas law. Nitrogen, oxygen, hydrogen, and many other gases condense at such a low absolute temperature that for most engineering problems they may be treated as perfect or semiperfect gases.

Other gases, such as water and many refrigerants, condense at temperatures varying from -120 to 700 F. The very fact that these materials condense at these temperatures is the principal reason for their use in the steam power plant cycle or in the refrigeration cycle.

The steam tables, Appendix D, give the tabular values for the thermodynamic properties of water. In Appendix E the properties of ammonia are listed, and in Appendix F the properties of Freon-12 are given. For water, in the gaseous phase, at three selected states, the values of the pressure, temperature, and specific volume, from Table 2 in Appendix D, are

p (psia)	t (F)	v_g (ft^3/lbm)
1.0	101.7	333.6
14.7	212.0	26.80
500.0	467.0	0.928

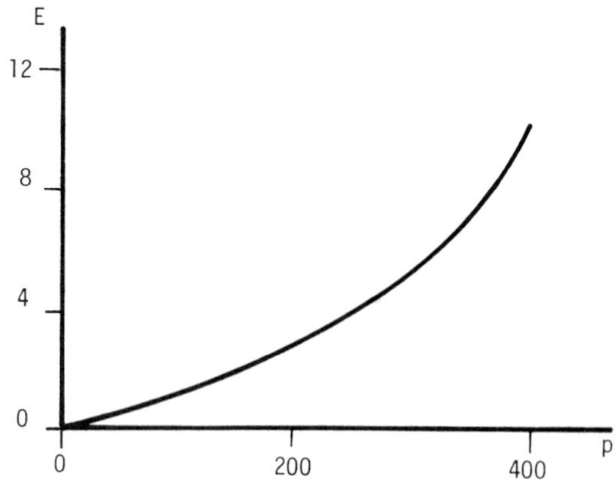

Figure 10.1

Example 10.1. Determine the value of the specific volume of water at the three pressures by use of the perfect gas law. What is the error caused by assuming that the perfect gas law may be applied?

For p = 1 psia where R for water is 85.8 ft-lbf/lbm-R, from Appendix B

$$v = RT/p = [85.8(562)]/[1(144)]$$

$$= 334.9 \ (ft^3/lbm)$$

and the error is

$$E = (334.9 - 333.6)/333.6 = 0.0039$$

$$= 0.39\%$$

For p = 14.7 psia, v = 27.23, and E = 1.60%, and for p = 500 psia, v = 1.1046, and E = 19.1%.

As the pressure or temperature at which the water will condense increases, the error caused by assuming a perfect gas relationship increases (Fig. 10.1).

SATURATED VAPOR

A working substance that is completely vaporous at its condensation pressure and temperature is a saturated vapor. In the steam tables the subscript g is used to designate a saturated vapor. If heat, at constant pressure, is added to a saturated vapor, the temperature will rise. If heat is removed from a saturated vapor at constant pressure, the temperature will remain constant, but some of the saturated vapor will be condensed into its liquid phase.

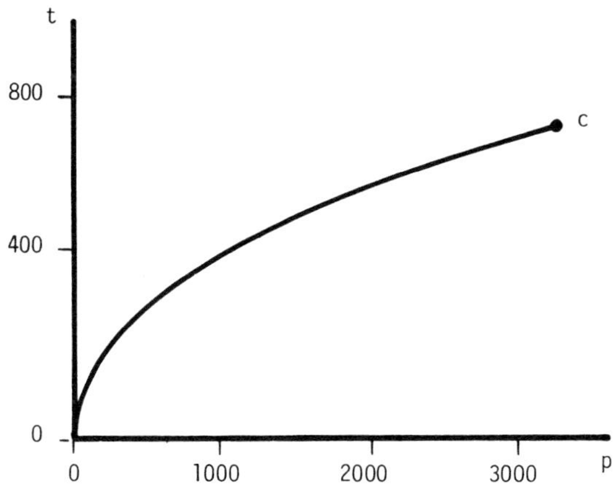

Figure 10.2

A plot of the pressures and temperatures of saturated vapors is sketched in Fig. 10.2. The critical point c is the highest value of the temperature and pressure for which the saturated vapor can exist. For water these values are $t_c = 705.4$ F and $p_c = 3206$ psia.

SATURATED LIQUID

A saturated liquid consists of only the liquid phase of the working substance at its saturation pressure and temperature. The plot in Fig. 10.2 relates the pressures and temperatures for both saturated vapors and saturated liquids. At constant pressure, removal of heat from the saturated liquid results in a temperature drop; the addition of heat results in some of the liquid becoming vapor while the temperature remains constant. In the steam tables the subscript f is used to designate the saturated liquid.

WET REGION

From the steam tables the following values are obtained for water at 14.696 psia at 212 F:

$$v_f = 0.01672 \qquad v_g = 26.80 \qquad (\text{ft}^3/\text{lbm})$$
$$h_f = 180.1 \qquad h_g = 1150.4 \qquad (\text{B/lbm})$$
$$s_f = 0.3120 \qquad s_g = 1.7566 \qquad (\text{B/lbm-R})$$

Values of the internal energy may be listed in the tables or may be calculated from

$$u = h - (pv/J)$$

for either the saturated vapor or saturated liquid.

Values which are subscripted fg represent the results of the calculation

$$h_{fg} = h_g - h_f$$

which is the result of subtracting the saturated liquid value of a property from the saturated vapor value.

Table 1 in Appendix D lists the property values of integer values of the temperature, and Table 2 lists the property values for integer values of the pressure.

In Fig. 10.3, the saturated liquid and vapor lines are plotted on the pV, TS, hs, and ph planes. The saturated liquid line lies to the left of the critical state on all curves. The wet region lies between the saturated liquid and saturated vapor lines. Figure 10.2, which relates the pressures and temperatures for the saturated vapor and saturated liquid, is a single line. For a pressure of 14.7 psia the temperature is 212 F for all points in the wet region for water. Since stating the value of the pressure defines the value of the temperature in the wet region, pressure and temperature are dependent properties. To define the thermodynamic state in the wet region either the pressure or temperature represents only one of the

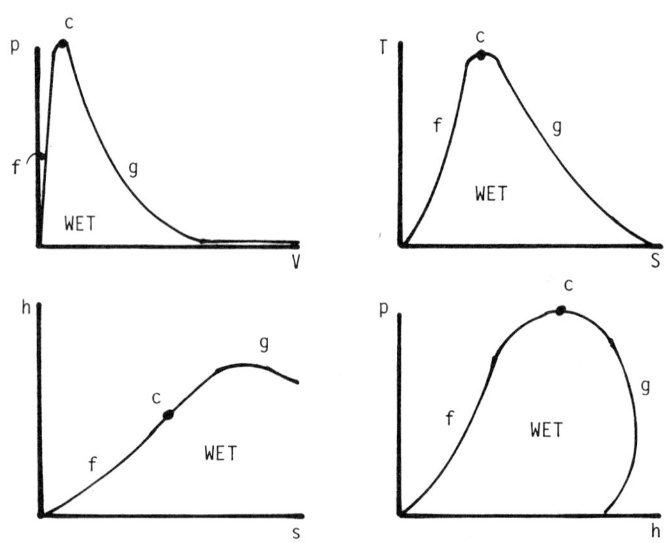

Figure 10.3

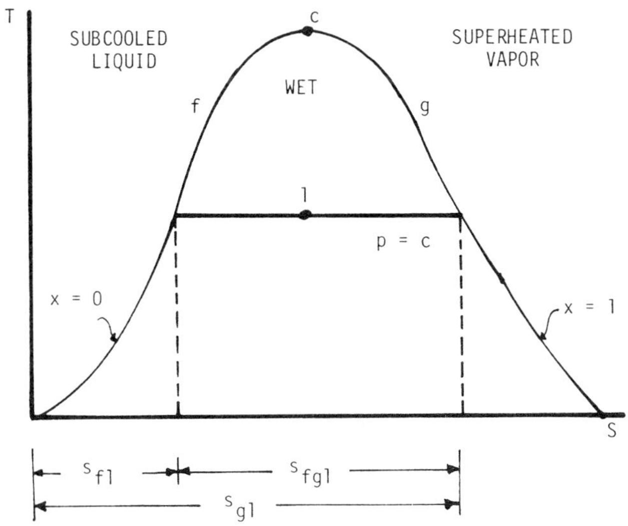

Figure 10.4

two required thermodynamic properties. The other property could be the specific volume, the enthalpy, the entropy, or the internal energy.

Assume that we know the temperature and the entropy of water in the wet region. These two independent properties determine a single state point, 1, as shown in Fig. 10.4. The value of s_{fg}, which is $s_g - s_f$, is represented by the length of the horizontal line from s_f to s_g. This horizontal line depicts both a constant pressure and constant temperature process. The distance from s_f to state point is equal to $s_1 - s_f$. We shall define a new term, the quality, which is defined by

$$x_1 = (s_1 - s_{f1})/(s_{g1} - s_{f1}) \tag{10.1}$$

The quality is equal to the fraction of vapor present in a liquid-vapor mixture. The moisture or liquid component in the liquid-vapor mixture is given by

$$m = 1 - x$$

For a saturated liquid, x = 0 or 0% and m = 1 or 100%, and for a saturated vapor, x = 1 or 100% and m = 0 or 0%.

The quality or moisture content has a numerical value only on the saturated liquid line, on the saturated vapor line, or in the wet region.

Rewriting Eq. (10.1) so that

$$s = s_f + x(s_g - s_f) = s_f + xs_{fg} \tag{10.2}$$

the specific volume, enthalpy, and internal energy may also be expressed in terms of the quality so that

$$v = v_f + xv_{fg} \tag{10.3}$$

$$h = h_f + xh_{fg} \tag{10.4}$$

and

$$u = u_f + xu_{fg} \tag{10.5}$$

The subscript 1 has been omitted in the above four equations.

Example 10.2. Given water at a temperature of 300 F, determine p, v, h, u, and s for a quality of 0%, 100%, and 50%.

For a quality of 0%, x = 0 and m = 1.0, and the thermodynamic properties are the f properties in Table 1 of Appendix D, so that p = 67.01 psia, $v = v_f = 0.01745$ ft^3/lbm, $h = h_f = 269.6$ B/lbm, and $s = s_f = 0.4369$ B/lbm-R.

The internal energy may be calculated from

$$u = u_f = h_f - (pv_f/J) = 269.6 - (144/778)[67.013(0.01745)]$$

$$= 269.4 \text{ (B/lbm)}$$

For a quality of 100%, x = 1.0 and m = 0.0, and the thermodynamic properties are the g properties in Table 1, so that p = 67.01 psia, $v = v_g = 6.466$ ft^3/lbm, $h = h_g = 1179.7$ B/lbm, and $s = s_g = 1.6350$ B/lbm-R.

The internal energy is

$$u = u_g = h_g - (pv_g/J) = 1179.7 - (144/778)[67.01(6.466)]$$

$$= 1099.5 \text{ (B/lbm)}$$

For a quality of 50%, x = 0.5 and m = 0.5, and by means of Eq. (10.3)

$$v = v_f + xv_{fg} = 0.01745 + 0.5(6.449)$$

$$= 3.242 \text{ (ft}^3\text{/lbm)} \quad \text{or} \quad 0.202 \text{ (m}^3\text{/kg)}$$

By means of Eq. (10.4)

$$h = h_f + xh_{fg} = 269.6 + 0.5(910.1)$$

$$= 724.6 \text{ (B/lbm)} \quad \text{or} \quad 1685.4 \text{ (kJ/kg)}$$

and by means of Eq. (10.2)

$$s = s_f + xs_{fg} = 0.4369 + 0.5(1.1980)$$

$$= 1.0359 \text{ (B/lbm-R)} \quad \text{or} \quad 4.3342 \text{ (kJ/kg-K)}$$

The internal energy may be calculated from

$$u = h - (pv/J) = 724.6 - (144/778)[67.01(3.242)]$$

$$= 684.4 \text{ (B/lbm)} \quad \text{or} \quad 1591.9 \text{ (kJ/kg)}$$

where p = 67.013 psia for a temperature of 300 F.

Example 10.3. Determine p, v, h, s, and u for ammonia having a quality of 30% and a temperature of -30 F.

From Table 1, Appendix E, p = 13.90 psia or 95.8 kPa-a, so that

$$v = v_f + x(v_g - v_f) = 0.0234 + 0.3(18.97 - 0.0234)$$

$$= 5.71 \text{ (ft}^3\text{/lbm)} \quad \text{or} \quad 0.356 \text{ (m}^3\text{/kg)}$$

$$h = h_f + x(h_g - h_f) = 10.7 + 0.3(601.4 - 10.7)$$

$$= 187.9 \text{ (B/lbm)} \quad \text{or} \quad 437.1 \text{ (kJ/kg)}$$

and

$$s = s_f + x(s_g - s_f) = 0.0250 + 0.3(1.4001 - 0.0250)$$

$$= 0.4375 \text{ (B/lbm-R)} \quad \text{or} \quad 1.8306 \text{ (kJ/kg-K)}$$

The internal energy is

$$u = h - (pv/J) = 187.9 - (144/778)[13.90(5.71)]$$

$$= 173.2 \text{ (B/lbm)} \quad \text{or} \quad 402.9 \text{ (kJ/kg)}$$

Example 10.4. Determine p, v, h, s, and u for Freon-12 at a temperature of 100 F and a quality of 80%.

From Table 1, Appendix F, p = 131.86 psia, and

$$v = v_f + x(v_g - v_f) = 0.01269 + 0.8(0.30794 - 0.01269)$$

$$= 0.24889 \text{ (ft}^3\text{/lbm)}$$

$$h = h_f + x(h_g - h_f) = 31.100 + 0.8(87.029 - 31.100)$$

$$= 75.843 \text{ (B/lbm)}$$

and

$$s = s_f + x(s_g - s_f) = 0.063227 + 0.8(0.16315 - 0.063227)$$

$$= 0.14317 \text{ (B/lbm-R)}$$

Finally by calculation

$$u = h - (pv/J) = 75.843 - (144/778)[131.86(0.24889)]$$

$$= 69.769 \text{ (B/lbm)}$$

Note that in the ammonia and Freon-12 tables values of fg are not listed and that it is necessary to subtract the saturated liquid property value from the saturated vapor property value.

COMPRESSED OR SUBCOOLED LIQUIDS

The compressed or subcooled liquid region lies to the left of the saturated liquid line as shown in Fig. 10.4. For low pressures the properties are those of the saturated liquid at the temperature of the liquid.

Example 10.5. Determine v, h, s, and u for water at a pressure of 200 psia and a temperature of 300 F. To determine if the water is in the subcooled liquid, note that the temperature for a saturated liquid at 200 psia from Table 2, Appendix D, is

$$t = 381.8 \text{ (F)}$$

Since the given value of the temperature is below this value, the water is subcooled.

From Table 1, Appendix D, for a temperature of 300 F, $v_f = 0.01745$ ft³/lbm, $h_f = 269.6$ B/lbm, and $s_f = 0.4369$ B/lbm-R.

The value of the internal energy is

$$u = h - (pv/J) = 269.6 - (144/778)[200(0.01745)] = 268.94 \text{ (B/lbm)}$$

Example 10.6. Repeat Example 10.5 using the compressed liquid tables, Table 4, Appendix D.

From the tables

$$v - v_f = -1.1 \times 10^{-5} \quad \text{or} \quad v = 0.01745 - 0.00001 = 0.01744 \text{ (ft}^3\text{/lbm)}$$

The error in v by use of the saturated liquid values is 0.06%. Also from the tables,

$$h - h_f = +0.23 \quad \text{or} \quad h = 269.6 + 0.2 = 269.8 \text{ (B/lbm)}$$

for an error of -0.09%, and

$$s - s_f = -0.21 \times 10^{-3} \quad \text{or} \quad 0.4369 - 0.0002 = 0.4367 \text{ (B/lbm-R)}$$

for an error of 0.05%.

The value of the internal energy u is

$$u = h - (pv/J) = 269.8 - (144/778)(200)(0.01744)$$
$$= 269.2 \text{ (B/lbm)}$$

compared to 268.9 B/lbm using the saturated liquid tables, which is an error of -0.11%.

Example 10.7. Determine the error caused by using the saturated liquid tables for a pressure of 3000 psia and a temperature of 600 F.

From Table 1, Appendix D, for t = 600 F, $v_f = 0.0236$ ft^3/lbm, $h_f =$ 617.0 B/lbm, and $s_f = 0.8131$ B/lbm-R.

The value of the internal energy is

$$u = h - (pv/J) = 617.0 - (144/778)(3000)(0.0236)$$
$$= 603.9 \text{ (B/lbm)}$$

From Table 4, $v - v_f = -87.9 \times 10^{-5}$, $h - h_f = -6.9$, and $s - s_f = -12.4 \times 10^{-3}$.

The error for the specific volume is

$$E = (87.9 \times 10^{-5})/(0.0236 - 87.9 \times 10^{-5}) = 0.039 \quad \text{or} \quad 3.9\%$$

For the enthalpy the error is

$$E = 6.9/(617.0 - 6.9) = 0.011 \quad \text{or} \quad 1.1\%$$

and for the entropy

$$E = (12.4 \times 10^{-3})/(0.8131 - 12.4 \times 10^{-3}) = 0.015 \quad \text{or} \quad 1.5\%$$

The exact value of u is

$$u = h - (pv/J) = (617.0 - 6.9) - (144/778)(3000)(0.0236 - 87.9 \times 10^{-5})$$
$$= 597.5$$

so that

$$E = (603.9 - 597.5)/597.5 = 0.011 \text{ or } 1.1\%$$

Even at a pressure of 3000 psia, except for the volume, the error in the properties is within normal engineering accuracy so that the saturated liquid values may be used.

SUPERHEATED VAPOR

When the temperature exceeds the value of the temperature at the condensation or saturation pressure the water is in the superheated region. In the superheated region the temperature and the pressure are independent properties, and the remaining property values may be obtained by a cross-index of these properties in Table 3, Appendix D. The superheated vapor lies to the right of the saturated vapor line as shown in Fig. 10.4. Table 2 in Appendix E and Table 2 in Appendix F present the same data for ammonia and Freon 12.

Example 10.8. Determine the properties for superheated steam at a pressure of 100 psia and a temperature of 600 F. This must be in the superheated region since at a pressure of 100 psia the saturation temperature is 327.8 F.

From the superheated tables, $v = 6.218\ ft^3/lbm$, $h = 1329.1$ B/lbm, and $s = 1.7581$ B/lbm-R. By calculation, the internal energy is

$$u = h - (pv/J) = 1329.1 - (144/778)(100)(6.218)$$

$$= 1214.0\ (B/lbm)$$

Example 10.9. Determine the properties of superheated steam at a pressure of 100 psia and a temperature of 630 F from the superheated tables.

t = 600	t = 700	(F)
v = 6.218	v = 6.835	(ft^3/lbm)
h = 1329.1	h = 1378.9	(B/lbm)
s = 1.7581	s = 1.8029	(B/lbm-R)

The properties at 630 F require a single interpolation between 600 and 700 F so that the linear fraction is

$$f = (630 - 600)/(700 - 600) = 0.3$$

and

$$v = 6.218 + 0.3(6.835 - 6.218) = 6.403\ (ft^3/lbm)$$

$$h = 1329.1 + 0.3(1378.9 - 1329.1) = 1344.0\ (B/lbm)$$

and

$$s = 1.7581 + 0.3(1.8029 - 1.7581) = 1.7715\ (B/lbm\text{-}R)$$

The value of the internal energy is

$$u = h - (pv/J) = 1344.0 - (144/778)(100)(6.403)$$

$$= 1222.5 \text{ (B/lbm)}$$

Example 10.10. Determine the properties of superheated steam at 105 psia and 630 F. The evaluation of the properties requires a double interpolation in the superheated tables.

From Table 3, Appendix , at 120 psia,

t = 600	t = 700	(F)
v = 5.165	v = 5.683	(ft^3/lbm)
h = 1327.7	h = 1377.8	(B/lbm)
s = 1.7370	s = 1.7822	(B/lbm-R)

Again f = 0.3, so that at a pressure of 120 psia and a temperature of 630 F

$$v = 5.165 + 0.3(5.683 - 5.165) = 5.320 \text{ (ft}^3\text{/lbm)}$$

$$h = 1327.7 + 0.3(1377.8 - 1327.7) = 1342.7 \text{ (B/lbm)}$$

and

$$s = 1.7370 + 0.3(1.7822 - 1.7370) = 1.7506 \text{ (B/lbm)}$$

We now have values for the properties at 630 F and pressures of 100 and 120 psia, so that

p = 100	p = 120	(psia)
v = 6.403	v = 5.320	(ft^3/lbm)
h = 1344.0	h = 1342.7	(B/lbm)
s = 1.7715	s = 1.7506	(B/lbm-R)

For a pressure of 105 psia

$$f = (105 - 100)/(120 - 100) = 0.250$$

The properties are

$$v = 6.403 - 0.25(6.403 - 5.320) = 6.132 \text{ (ft}^3\text{/lbm)}$$

$$h = 1344.0 - 0.25(1344.0 - 1342.7) = 1343.7 \text{ (B/lbm)}$$

and

$$s = 1.7715 - 0.25(1.7715 - 1.7506) = 1.7663 \text{ (B/lbm-R)}$$

The internal energy is

$$u = h - (pv/J) = 1343.7 - (144/778)(105)(6.132)$$
$$= 1224.5 \text{ (B/lbm)}$$

Example 10.11. Determine the property values of superheated steam at a pressure of 100 psia and a temperature of 360 F.

From Table 3, Appendix D, the property values are listed at a pressure of 100 psia and a temperature of 400 F. The properties at a lower temperature become those of a saturated vapor at 100 psia as listed in Table 2.

For a pressure of 100 psia

$t = 327.8$	$t = 400$	(F)
$v_g = 4.432$	$v = 4.937$	(ft^3/lbm)
$h_g = 1187.2$	$h = 1227.6$	(B/lbm)
$s_g = 1.6026$	$s = 1.6518$	(B/lbm-R)

For a temperature of 360 F

$$f = (360 - 327.8)/(400 - 327.8) = 0.446$$

and the property values are

$$v = 4.432 + 0.446(4.937 - 4.432) = 4.657 \text{ (ft}^3\text{/lbm)}$$

$$h = 1187.2 + 0.446(1227.6 - 1187.2) = 1205.2 \text{ (B/lbm)}$$

$$s = 1.6026 + 0.446(1.6518 - 1.6026) = 1.6245 \text{ (B/lbm-R)}$$

and

$$u = h - (pv/J) = 1205.2 - (144/778)(100)(4.657)$$
$$= 1119.0 \text{ (B/lbm)}$$

PROCESSES USING VAPORS

The techniques required to determine the property values for subcooled liquids, saturated liquids, vapor-liquid mixtures, saturated vapors, and superheated vapors have now been established. As in our discussion of perfect gases, we shall consider the various processes using vapors and from these processes form theoretical cycles that will simulate real engineering devices.

For vapor processes, the following equations may be used:

Nonflow energy equation: $Q = \Delta U + W_{nf}$

Steady-flow energy equation: $Q = \Delta H + \Delta KE + \Delta PE + W_{sf}$

Entropy change: $\Delta S = Q/T \quad \text{for } T = C$

Nonflow work: $W_{nf} = \int_{V_1}^{V_2} p\, dV$

Steady-flow work: $W_{sf} = \int_{p_1}^{p_2} -V\, dp \qquad (\Delta KE, \Delta PE = 0)$

The equations for ΔS, W_{nf}, and W_{sf} are valid only for reversible processes.

CONSTANT VOLUME PROCESS

A reversible constant volume process for which $p_2 > p_1$ is sketched on the pV, TS, hs, and ph planes in Fig. 10.5.

Example 10.12. Given is 5 lbm of water at a pressure of 100 psia and a volume of 15.30 ft^3. The water undergoes a constant volume reversible

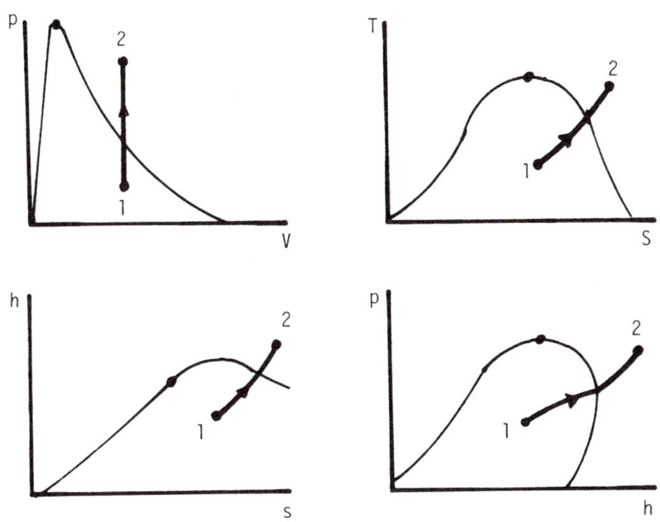

Figure 10.5

process to a pressure of 200 psia. Determine t_1, t_2, ΔU, ΔH, ΔS, Q, W_{nf}, and W_{sf} (ΔKE, ΔPE = 0). Determine W_{sf} if ΔKE = -50 B and ΔPE = 0.

The specific volume for the water is

$$v_1 = v_2 = V_1/m = 15.30/5 = 3.060 \text{ (ft}^3\text{/lbm)} \quad \text{or} \quad 0.191 \text{ (m}^3\text{/kg)}$$

From Table 2, Appendix D, for a pressure of 100 psia

$$v_f = 0.01774 \quad \text{and} \quad v_g = 4.432 \text{ (ft}^3\text{/lbm)}$$

Since the value of v_1 is between the values of v_f and v_g, the state of the water is in the wet region. If the value of a property is less than the saturated liquid property value at a given pressure, then the substance is a subcooled liquid. If the value of a property is between the saturated liquid and saturated vapor property values at a given pressure or temperature, the substance is in the wet region. If the property value exceeds the saturated vapor property value for a given pressure, the substance is in the superheated region.

Since the initial state of the water is in the wet region, the saturation temperature for a saturation pressure of 100 psia is

$$t_1 = 327.8 \text{ (F)}$$

From Eq. (10.3)

$$v_1 = v_{f1} + x_1 v_{fg1} = v_{f1} + x_1(v_{g1} - v_{f1})$$

$$= 0.01774 + x_1(4.432 - 0.018) = 3.060$$

and

$$x_1 = 0.689 \quad \text{or} \quad 68.9\%$$

From Eq. (10.4)

$$h_1 = h_{f1} + x_1 h_{fg1} = 298.4 + (0.689)(888.8)$$

$$= 910.8 \text{ (B/lbm)} \quad \text{or} \quad 2118 \text{ (kJ/kg)}$$

where h_f = 298.4 and h_{fg} = 888.8 B/lbm at a pressure of 100 psia.

From Eq. (10.2)

$$s_1 = s_{f1} + x_1 s_{fg1} = 0.4740 + 0.689(1.1286)$$

$$= 1.2516 \text{ (B/lbm-R)} \quad \text{or} \quad 5.2367 \text{ (kJ/kg-K)}$$

where s_f = 0.4740 and s_{fg} = 1.1286 B/lbm-R at a pressure of 100 psia.

The value of the internal energy at the initial state is

$$u_1 = h_1 - (p_1 v_1/J) = 910.8 - (144/778)(100)(3.060)$$

$$= 854.2 \text{ (B/lbm)} \quad \text{or} \quad 1987 \text{ (kJ/kg)}$$

For the final state of the process, p_2 = 200 psia and $v_2 = v_1 = 3.060$ ft^3/lbm. At a pressure of 200 psia the value of the specific volume for the saturated vapor is

$$v_{g2} = 2.288 \text{ (ft}^3\text{/lbm)}$$

Since $v_2 > v_{g2}$, the water or steam is in the superheated region.

From Table 3, Appendix D, for the given values of p_2 and v_2, t_2 = 600 F, h_2 = 1322.1 B/lbm, and s_2 = 1.6767 B/lbm-R, and

$$u_2 = h_2 - (p_2 v_2/J) = 1322.1 - (144/778)(200)(3.060)$$

$$= 1208.8 \text{ (B/lbm)} \quad \text{or} \quad 2812 \text{ (kJ/kg)}$$

The change in internal energy is

$$\Delta U = m\,\Delta u = m(u_2 - u_1) = 5(1208.8 - 854.2)$$

$$= 1773.0 \text{ (B)} \quad \text{or} \quad 1869 \text{ (kJ)}$$

The change in enthalpy is

$$\Delta H = m\,\Delta h = m(h_2 - h_1) = 5(1322.1 - 910.8)$$

$$= 2056.5 \text{ (B)} \quad \text{or} \quad 2168 \text{ (kJ)}$$

The change in entropy is

$$\Delta S = m\,\Delta s = m(s_2 - s_1) = 5(1.6767 - 1.2516)$$

$$= 2.1255 \text{ (B/R)} \quad \text{or} \quad 4.0325 \text{ (kJ/K)}$$

For a reversible process, the nonflow work is $W_{nf} = 0$, V = C, so that the heat transfer is

$$Q = \Delta U + W_{nf} = 1773.0 \text{ (B)}$$

and the steady-flow work is

$$W_{sf} = Q - \Delta H - \Delta KE - \Delta PE$$

$$= 1773.0 - 2056.5 = -283.5 \text{ (B)}$$

when $\Delta KE, \Delta PE = 0$.

The steady-flow work for a reversible process with no changes in kinetic or potential energy may also be calculated by means of

$$W_{sf} = \int_{p_1}^{p_2} -V\,dp = -V(p_2 - p_1) = -mv(p_2 - p_1)$$

$$= -(144/778)(5)(3.060)(200 - 100)$$

$$= -283.2 \text{ (B)}$$

This is the negative of the area in back of the curve on the pV plane.

For $\Delta KE = -50$ B

$$W_{sf} = Q - \Delta H - \Delta KE = 1773.0 - 2056.5 + 50.0$$

$$= -233.5 \text{ (B)}$$

which is not the area in back of the curve.

The presence of kinetic or potential energy terms produces a different value of the steady-flow work except when $\Delta KE = -\Delta PE$.

CONSTANT PRESSURE PROCESS

A reversible constant pressure process for which $V_2 > V_1$ is sketched in Fig. 10.6.

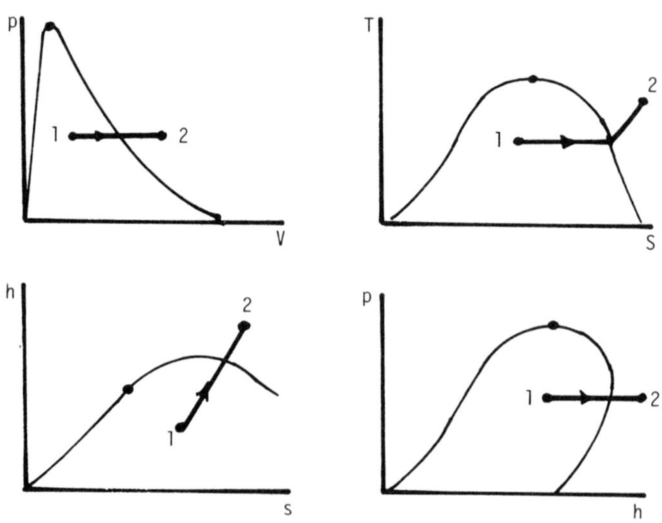

Figure 10.6

Example 10.13. Given the initial property values in Example 10.12 and t_2 = 600 F for a reversible constant pressure process, determine W_{nf}, Q, and W_{sf} (ΔKE, ΔPE = 0).

The properties at the initial state are v_1 = 3.060 ft^3/lbm, u_1 = 854.2 B/lbm, h_1 = 910.8 B/lbm, and s_1 = 1.2516 B/lbm-R.

The final state is defined by $p_2 = p_1$ = 100 psia and t_2 = 600 F. For a saturation pressure of 100 psia, the saturation temperature is 327.8 F, and since t_2 = 600 F, the water lies in the superheated region. From Table 3, Appendix D, for the given final state, v_2 = 6.218 ft^3/lbm, h_2 = 1329.1 B/lbm, s_2 = 1.7581 B/lbm-R, and

$$u_2 = h_2 - (p_2v_2/J) = 1329.1 - (144/778)(100)(6.218)$$

$$= 1214.0 \text{ (B/lbm)}$$

Again, only for a reversible process, the nonflow work is

$$W_{nf} = \int_{V_1}^{V_2} p\, dV = p(V_2 - V_1) = mp(v_2 - v_1), \qquad P = C$$

$$= 5(144/778)(100)(6.218 - 3.060)$$

$$= 292.3 \text{ (B)}$$

which is the area below the process curve on the pV plane.

The heat transfer is

$$Q = \Delta U + W_{nf} = m(u_2 - u_1) + W_{nf}$$

$$= 5(1214.0 - 854.2) + 292.3$$

$$= 2091.3 \text{ (B)}$$

and the steady-flow work is

$$W_{sf} = Q - \Delta H = Q - m(h_2 - h_1) \qquad (\Delta KE, \Delta PE = 0)$$

$$= 2091.3 - 5(1329.1 - 910.8)$$

$$= -0.2 \text{ (B)}$$

The steady-flow work could also be calculated from

$$W_{sf} = -\int_{V_1}^{V_2} V\, dp = 0 \qquad \text{for } p = c$$

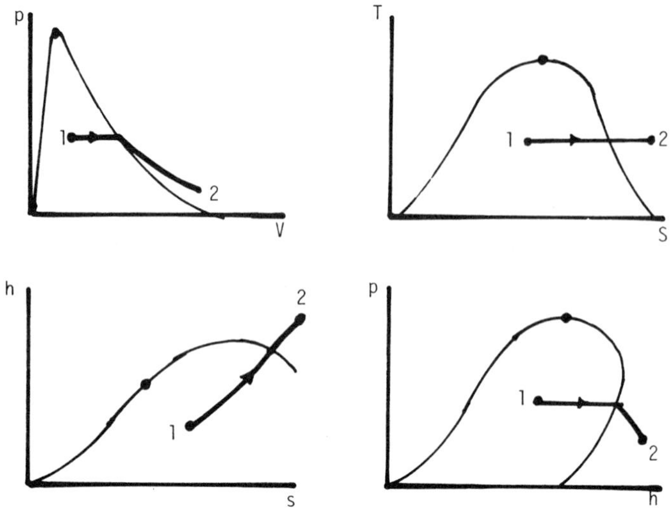

Figure 10.7

CONSTANT TEMPERATURE PROCESS

A reversible constant temperature process for which $V_2 > V_1$ is sketched in Fig. 10.7. Note that in the wet region a constant temperature process is also a constant pressure process.

Example 10.14. Given the initial state in Example 10.12 and p_2 = 14.696 psia, determine v_2, Δu, Δh, Δs, Q, W_{nf}, and W_{sf} (ΔKE, ΔPE = 0).

The final state is defined by p_2 = 14.696 psia and $t_2 = t_1$ = 327.8 F. For p_2 = 14.696 psia, the saturation temperature is 212 F. Since t_2 is 327.8 F, the final state is in the superheated region. To determine the final properties it is necessary to perform a single interpolation in Table 3, Appendix D.

For p_2 = 14.696 the values are

t = 300	t = 400	(F)
v = 30.53	v = 34.68	(ft^3/lbm)
h = 1192.8	h = 1239.9	(B/lbm)
s = 1.8160	s = 1.8743	(B/lbm-R)

The linear fraction is

f = (327.8 - 300)/(400 - 300)

= 0.278

and the property values are

$$v_2 = 30.53 + 0.278(34.68 - 30.53) = 31.68 \ (\text{ft}^3/\text{lbm})$$

$$h_2 = 1192.8 + 0.278(1239.9 - 1192.8) = 1205.9 \ (\text{B/lbm})$$

and

$$s_2 = 1.8160 + 0.278(1.8743 - 1.8160) = 1.8322 \ (\text{B/lbm-R})$$

The internal energy is

$$u_2 = h_2 - (p_2 v_2/J) = 1205.9 - (144/778)(14.696)(31.68)$$

$$= 1119.7 \ (\text{B/lbm})$$

The changes in property values are

$$\Delta u = u_2 - u_1 = 1119.7 - 854.2 = 265.5 \ (\text{B/lbm})$$

$$\Delta h = h_2 - h_1 = 1205.9 - 910.8 = 295.1 \ (\text{B/lbm})$$

and

$$\Delta s = s_2 - s_1 = 1.8322 - 1.2516 = 0.5806 \ (\text{B/lbm-R})$$

The heat transfer for the reversible process is

$$Q = T\,\Delta s = (327.8 + 460)0.5806$$

$$= 457.4 \ (\text{B/lbm})$$

which is the area below the process curve on the TS plane.

The nonflow work is

$$W_{nf} = Q - \Delta u = 457.4 - 265.5 = 191.9 \ (\text{B/lbm})$$

and the steady-flow work is

$$W_{sf} = Q - \Delta h = 457.4 - 295.1 \qquad (\Delta KE, \Delta PE = 0)$$

$$= 162.3 \ (\text{B/lbm})$$

For a perfect gas where ΔKE, $\Delta PE = 0$,

$$Q = W_{nf} = W_{sf}$$

and for a perfect gas undergoing an isothermal process,

$$\Delta u = 0 \quad \text{and} \quad \Delta h = 0$$

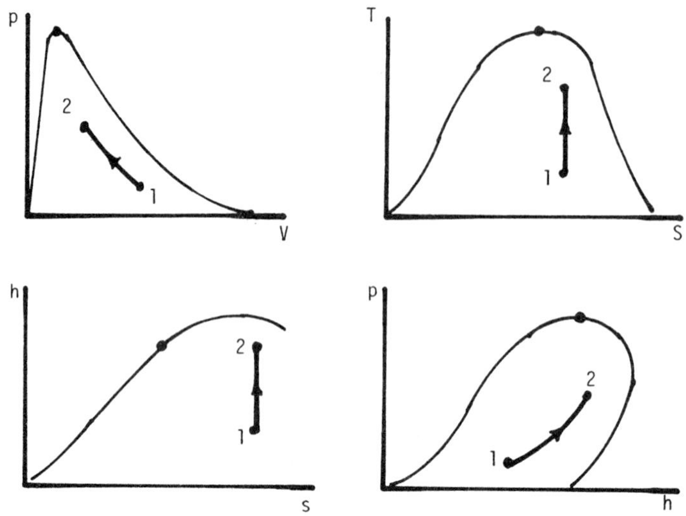

Figure 10.8

CONSTANT ENTROPY PROCESS

An isentropic process for which $T_2 > T_1$ is sketched in Fig. 10.8.

Example 10.15. Given is the initial state point from Example 10.12. The water undergoes a constant entropy process to a pressure of 200 psia. Determine t_2, v_2, Δu, Δh, Δs, Q, W_{nf}, and W_{sf} (ΔKE, $\Delta PE = 0$).

The known properties at the final state are $s_2 = s_1 = 1.2516$ B/lbm-R and $p_2 = 200$ psia.

From Table 2, Appendix D, at 200 psia, $s_{f2} = 0.5435$ and $s_{g2} = 1.5453$ B/lbm-R. Since $s_{g2} > s_2 > s_{f2}$, the water is in the wet region. From Eq. (10.2)

$$s_2 = s_{f2} + x_2 s_{fg2}$$

$$= 0.5435 + x_2(1.0018) = 1.2516$$

so that

$$x_2 = 0.707 \quad \text{or} \quad 70.7\%$$

Since the water is in the wet region, the temperature is the saturation temperature for a saturation pressure of 200 psia or

$$t_2 = 381.8 \text{ (F)}$$

The specific volume and enthalpy are

$$v_2 = v_{f2} + x_2(v_{g2} - v_{f2}) = 0.01839 + 0.707(2.288 - 0.018)$$

$$= 1.623\ (\mathrm{ft^3/lbm})$$

and

$$h_2 = h_{f2} + x_2(h_{fg2}) = 355.4 + 0.707(843.0)$$

$$= 951.4\ (\mathrm{B/lbm})$$

The internal energy at the final state is

$$u_2 = h_2 - (p_2v_2/J) = 951.4 - (144/778)(200)(1.623)$$

$$= 891.3\ (\mathrm{B/lbm})$$

The property changes are

$$\Delta u = u_2 - u_1 = 891.3 - 854.2 = 37.1\ (\mathrm{B/lbm})$$

$$\Delta h = h_2 - h_1 = 951.4 - 910.8 = 40.6\ (\mathrm{B/lbm})$$

and

$$\Delta s = s_2 - s_1 = 0.0$$

The heat and work terms are

$$Q = 0$$

$$W_{nf} = Q - \Delta u = -37.1\ (\mathrm{B/lbm})$$

and

$$W_{sf} = Q - \Delta h = -40.6\ (\mathrm{B/lbm})$$

POLYTROPIC PROCESS

The property changes as well as the heat and work terms may be evaluated for a reversible polytropic process using a vapor as a working substance if

$$pV^n = C$$

is a valid representation of the relationship between the pressures and volumes.

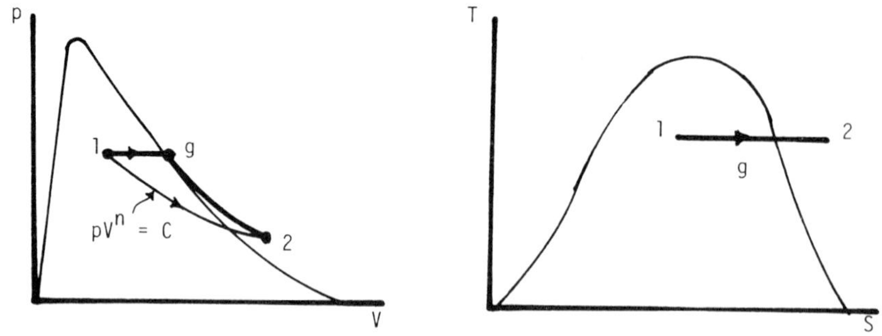

Figure 10.9

Example 10.16. Rework the reversible isothermal process (Example 10.14) assuming that the process in the superheat region follows a path defined by $pV^n = C$.

From the isothermal example the properties at state points 1, g, and 2 (Fig. 10.9) are

State point	1	g	2
p (psia)	100	100	14.696
t (F)	327.8	327.8	327.8
v (ft^3/lbm)	3.060	4.432	31.68
h (B/lbm)	910.8	1187.2	1205.9
s (B/lbm-R)	1.2516	1.6026	1.8322
u (B/lbm)	854.2	1105.1	1119.7

The heat and work terms for the isothermal process were $Q = 457.4$ B/lbm, $W_{nf} = 191.9$ B/lbm, and $W_{sf} = 162.3$ B/lbm, where $\Delta KE, \Delta PE = 0$. Since properties are point functions Δv, Δh, Δs, and Δu are independent of the process path between the initial and final thermodynamic states—the property value changes are the same for the polytropic and isothermal processes.

For the process from state 1 to state g, which is a constant pressure and constant temperature process,

$$Q_{1g} = T_1(s_g - s_1) = (328 + 460)(1.6026 - 1.2516)$$

$$= 276.6 \text{ (B/lbm)}$$

The nonflow work is

$$W_{nf-1g} = Q_{1g} - \Delta u_{g1} = Q_{1g} - (u_g - u_1)$$

$$= 276.6 - (1105.1 - 854.2) = 25.7 \text{ (B/lbm)}$$

or for a reversible constant pressure process

$$W_{nf-1g} = \int_{v_1}^{v_g} p\ dv = [p_1(v_g - v_1)]/J$$

$$= (144/778)(100)(4.432 - 3.060) = 25.4\ (B/lbm)$$

The steady-flow work is

$$W_{sf-1g} = Q_{1g} - \Delta h_{g1} = Q_{1g} - (h_g - h_1)$$

$$= 276.6 - (1187.2 - 910.8) = 0.2\ (B/lbm)$$

or for the constant pressure process ($\Delta KE, \Delta PE = 0$)

$$W_{sf-1g} = -\int_{p_1}^{p_g} V\ dp = 0.0$$

Slight differences in W_{sf} are caused by expressing the absolute temperature to the nearest degree in the heat transfer equation.

For the process from state g to state 2, let

$$p_g v_g^n = p_2 v_2^n$$

or

$$100(4.432)^n = 14.696(31.68)^n$$

so that

$$7.148^n = 6.804$$

and

$$n = 0.975$$

For a perfect gas undergoing an isothermal process, n = 1.

The nonflow work for n = 0.975 is

$$W_{nf-g2} = (p_2 v_2 - p_g v_g)/[J(1 - n)]$$

$$= (144/778)[14.696(31.68) - 100(4.432)][1/(1 - 0.975)]$$

$$= 165.6\ \ (B/lbm)$$

The nonflow work from state 1 to state 2 is

$$W_{nf-12} = W_{nf-1g} + W_{nf-g2} = 25.7 + 165.6 = 191.3\ (B/lbm)$$

as compared to 191.9 B/lbm for the isothermal process.

The heat transfer from state g to state 2 is

$$Q_{g2} = \Delta u_{2g} + W_{nf-g2} = (1119.7 - 1105.1) + 165.6$$

$$= 180.2 \text{ (B/lbm)}$$

and the heat transfer is

$$Q_{12} = Q_{1g} + Q_{g2} = 276.6 + 180.2 = 456.8 \text{ (B)}$$

as compared to 457.4 B/lbm for the isothermal solution.

The steady-flow work, ΔKE, $\Delta PE = 0$, is

$$W_{sf-g2} = Q_{g2} - \Delta h_{2g} = 180.2 - (1205.9 - 1187.2)$$

$$= 161.5 \text{ (B/lbm)}$$

and for the complete process

$$W_{sf-12} = W_{sf1-g} + W_{sf-g2} = 0.0 + 161.5$$

$$= 161.5 \text{ (B/lbm)}$$

which was 162.3 B/lbm for the isothermal process.

The assumption that $pV^n = C$ is valid for the process from state g to state 2 yields values for the reversible process heat and work terms that are reasonably close to the values obtained for the isothermal process.

Example 10.17. Rework Example 10.16 assuming that

$$p_1 v_1^{\ n} = p_2 v_2^{\ n}$$

Compare the heat and work terms for the above relationship and determine the error in the solution as compared to the isothermal solution. Figure 10.9 illustrates the continuous process from state 1 to state 2.

From the state point data

$$100(3.060)^n = 14.696(31.68)^n$$

so that

$$n = 0.820$$

The nonflow work is

$$W_{nf} = (p_2v_2 - p_1v_1)/[J(1 - n)]$$
$$= (144/778)[14.696(31.68) - 100(3.060)]/(1 - 0.820)$$
$$= 164.1 \text{ (B/lbm)}$$

compared to 191.9 B/lbm for the isothermal process, and the error is -14.5%.

The heat transfer is

$$Q_{12} = \Delta u_{21} + W_{nf-12} = (1119.7 - 854.2) + 164.1$$
$$= 429.6 \text{ (B/lbm)}$$

which when compared to the isothermal value of 457.4 B/lbm results in an error of -6.1%.

The steady-flow work is

$$W_{sf} = Q - \Delta h = 429.6 - (1205.9 - 910.8)$$
$$= 134.5 \text{ (B/lbm)}$$

which is -17.1% in error compared to the isothermal value of 162.3 B/lbm.

The error in Example 10.17 is a result of assuming that the isothermal process may be replaced by a polytropic process with a single value of n. Since heat and work terms are path functions, the path must be specified before these terms may be evaluated. From Fig. 10.9, the source of the errors is a result of different process paths between the initial and final states.

IRREVERSIBLE ADIABATIC PROCESS

The reversible adiabatic or isentropic process is the ideal compression or expansion process for a pump or turbine. Real expansion or compression processes are represented by an irreversible adiabatic process. For a steady-flow steam turbine, which is an expansion device, the turbine efficiency is given by

$$e_t = W_{actual}/W_{ideal}$$

and for an adiabatic process where ΔKE, $\Delta PE = 0$, the work is given by

$$W_{sf} = Q - \Delta H = -\Delta H = -(H_2 - H_1) = H_1 - H_2$$

In Fig. 10.10, a reversible adiabatic expansion process occurs between states 1 and 2. Irreversible adiabatic processes are indicated by final state points 2′ and 2″, where $p_2 = p_2' = p_2''$.

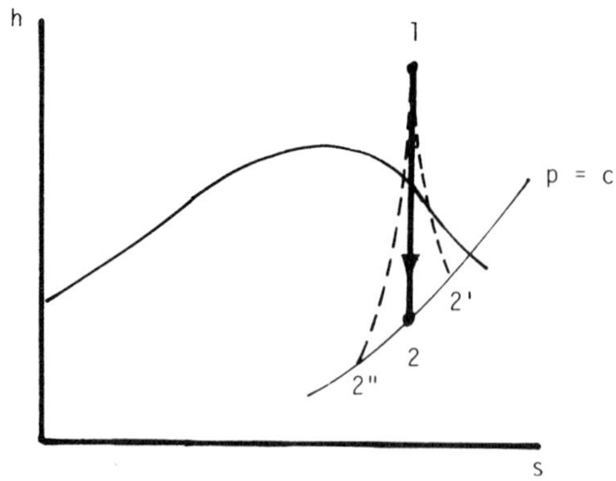

Figure 10.10

Due to friction and fluid turbulence the actual turbine work must be less than the ideal turbine work so that

$$e_t < 100\%$$

For the process from state 1 to state 2′ the turbine efficiency would be

$$e_t = (h_1 - h_2')/(h_1 - h_2)$$

and from Fig. 10.10, since $h_2' > h_2$, then $e_t < 100\%$. This is a logical result and reflects the loss of work in the real or actual process.

For the expansion process from state 1 to state 2″ the turbine efficiency would be

$$e_t = (h_1 - h_2'')/(h_1 - h_2)$$

and from Fig. 10.10, since $h_2 > h_2''$, then $e_t > 100\%$.

This is impossible since the actual output of the turbine, which includes the losses due to friction, would exceed that of the ideal turbine, which is represented by a reversible process which is frictionless.

Example 10.18. In Fig. 10.10, assume p_1 = 200 psia and t_1 = 600 F. The pressure at the end of the expansion process is 10 psia. Assume $s_2' - s_1$ = 0.05 B/lbm-R and $s_2'' - s_1$ = -0.05 B/lbm-R. Determine the efficiency of a turbine connecting these state points.

From Table 3, Appendix D, for p_1 = 200 psia and t_1 = 600 F, h_1 = 1322.1 B/lbm and s_1 = 1.6767 B/lbm-R. At a pressure of 10 psia, s_g = 1.7876 B/lbm-R so that for the ideal process the final state point is in the wet region or

$$s_2 = s_{f2} + x_2 s_{fg2}$$

$$= 0.2835 + x_2(1.5041) = 1.6767$$

or

$$x_2 = 0.926$$

The final enthalpy is given by

$$h_2 = h_{f2} + x_2 h_{fg2} = 161.2 + (0.926)(982.1)$$

$$= 1070.6 \text{ (B/lbm)}$$

The ideal work is

$$W_{sf} = h_1 - h_2 = 1322.1 - 1070.6 = 251.5 \text{ (B/lbm)}$$

For the real process from state 1 to state 2′

$$s_2' - s_1 = 0.05$$

or

$$s_2' = s_1 + 0.05 = 1.6767 + 0.05 = 1.7267 \text{ (B/lbm-R)}$$

The quality x_2', for $p_2 = 10$ psia, is

$$s_2' = s_{f2}' + x_2' s_{fg2} = 0.2835 + x_2'(1.5401) = 1.7267$$

or

$$x_2' = 0.937$$

The final enthalpy is

$$h_2' = h_{f2}' + x_2' h_{fg2}' = 161.2 + (0.937)(982.1)$$

$$= 1081.4 \text{ (B/lbm)}$$

and the actual work is

$$W_{sf}' = h_1 - h_2' = 1322.1 - 1081.4 = 240.7 \text{ (B/lbm)}$$

This yields a turbine efficiency of

$$e_t' = W_{actual}/W_{ideal} = 240.7/251.5$$

$$= 0.957$$

If the final state point is 2″,

$$s_2'' - s_1 = -0.05$$

and

$$s_2'' = s_1 - 0.05 = 1.6767 - 0.05 = 1.6267$$

The final quality is

$$s_2'' = s_{f2}'' + x_2'' s_{fg2}'' = 0.2835 + x_2''(1.5041) = 1.6267$$

or

$$x_2'' = 0.893$$

The final enthalpy is

$$h_2'' = h_{f2}'' + x_2'' h_{fg2}'' = 161.2 + (0.893)(982.1)$$

or

$$h_2'' = 1038.2 \text{ (B/lbm)}$$

and the work for the assumed process is

$$W_{sf}'' = h_1 - h_2'' = 1322.1 - 1038.2$$

$$= 283.9 \text{ (B/lbm)}$$

giving an apparent turbine efficiency of

$$e_t'' = W_{actual}/W_{ideal} = 283.9/251.5 = 1.129$$

or

$$e_t'' = 112.9\%$$

which is impossible.

For a steady-flow compression process (Fig. 10.11)

$$e_c = W_{ideal}/W_{actual} = -\Delta h/-\Delta h' = \Delta h/\Delta h'$$

and for an irreversible process where $\Delta KE, \Delta PE = 0$,

$$e_c = (h_2 - h_1)/(h_2' - h_1)$$

where the final state point is 2′.

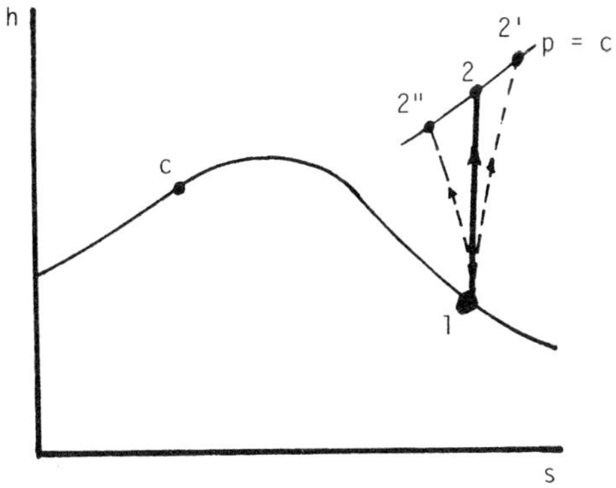

Figure 10.11

The actual work of the compression process exceeds the ideal or theoretical work, again due to friction and fluid turbulence. Most real compression processes begin either on the saturated liquid or saturated vapor lines.

Example 10.19. Given an initial state for water of 14.696 psia and a quality of 100%, determine the compressor efficiency for $s_2' - s_2 = 0.05$ and $s_2'' - s_2 = -0.05$ B/lbm (Fig. 10.11) and a final pressure of 60 psia.

For the given initial conditions of $p_1 = 14.696$ psia and $x_1 = 1.00$, the entropy and enthalpy are $s_1 = s_{g1} = 1.7566$ B/lbm-R and $h_1 = h_{g1} = 1150.4$ B/lbm. For the reversible process, $s_2 = s_1 = 1.7566$ B/lbm-R and $p_2 = 60$ psia.

By interpolation from Table 3, Appendix D, where

p = 60	p = 60	(psia)
t = 400	t = 500	(F)
s = 1.7135	s = 1.7678	(B/lbm-R)
h = 1233.6	h = 1283.0	(B/lbm)

we see that

$$f = (1.7566 - 1.7135)/(1.7678 - 1.7135) = 0.794$$

and

$$h_2 = 1233.6 + 0.794(1283.0 - 1233.6)$$

$$= 1272.8 \text{ (B/lbm)}$$

The work for the reversible steady-flow compression process is

$$W_{sf} = h_1 - h_2 = 1150.4 - 1272.8$$

$$= -122.4 \text{ (B/lbm)}$$

For the irreversible adiabatic process where $s_2' - s_2 = 0.05$,

$$s_2' = 1.7566 + 0.05 = 1.8066 \text{ (B/lbm-R)}$$

To calculate h_2' it is necessary to interpolate in the superheat tables between 500 and 600 F at a pressure of 60 psia. The enthalpy is

$$h_2' = 1322.1 \text{ (B/lbm)}$$

and the work for the irreversible adiabatic process is

$$W_{sf}' = h_1 - h_2' = 1150.4 - 1322.1$$

$$= -171.7 \text{ (B/lbm)}$$

The compressor efficiency is

$$e_c' = W_{sf}/W_{sf}' = -122.4/-171.7$$

$$= 0.713 \quad \text{or} \quad 71.3\%$$

For the assumed irreversible adiabatic process from state 1 to state 2″

$$s_2'' = 1.7066 \text{ (B/lbm-R)}$$

Again, by interpolation,

$$h_2'' = 1228.0 \text{ (B/lbm)}$$

For the assumed process

$$W_{sf}'' = h_1 - h_2'' = 1150.4 - 1228.0$$

$$= -77.6 \text{ (B/lbm)}$$

and the compressor efficiency is

$$e_c'' = W_{sf}/W_{sf}'' = -122.4/-77.6$$

$$= 1.577 \quad \text{or} \quad 157.7\%$$

which is impossible.

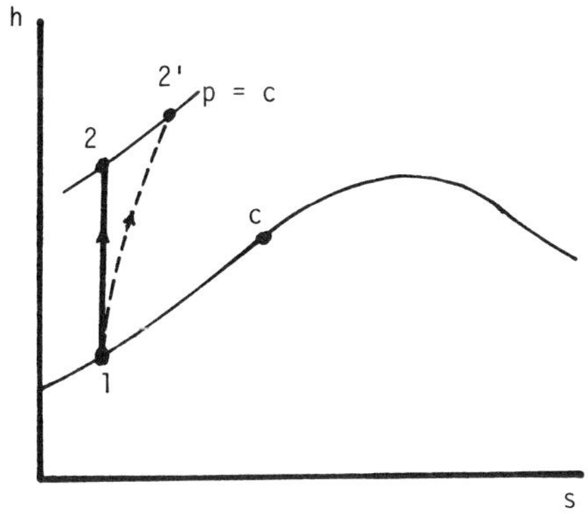

Figure 10.12

Example 10.20. Water at a pressure of 14.696 psia and a quality of 0% is compressed to a pressure of 200 psia. The pump efficiency is 80%. Determine the work of the irreversible adiabatic process (Fig. 10.12).

Since the compression process occurs solely in the liquid region where the fluid is incompressible, $v_1 = v_2 = c$. The steady-flow work for the reversible process is

$$W_{sf} = -\int_{p_1}^{p_2} v\,dp = [-v(p_2 - p_1)]/J$$

$$= [0.01672(200 - 14.696)](144/778)$$

$$= 0.57 \text{ (B/lbm)}$$

where $v = v_f$ at the saturation pressure of 14.696 psia.

The work for the irreversible adiabatic process W_{sf}' is

$$e_p = W_{sf}/W_{sf}'$$

or

$$W_{sf}' = W_{sf}/e_p = -0.57/0.80$$

$$= -0.71 \text{ (B/lbm)}$$

The enthalpy after the irreversible adiabatic process is

$$W_{sf} = h_1 - h_2' = 180.1 - h_2' = -0.71$$

or

$$h_2' = 180.8$$

where $h_1 = h_{f1}$ at the saturation pressure.

THROTTLING CALORIMETERS

A throttling process is a free expansion process from a high-pressure to a low-pressure region (Fig. 10.13). The process is an irreversible adiabatic process for which $W_{sf} = 0$ and $Q = 0$.

From the steady-flow process equation

$$Q = \Delta H + \Delta KE + \Delta PE + W_{sf}$$

and for $\Delta KE, \Delta PE = 0$

$$\Delta H = 0$$

or

$$h_1 = h_2$$

If the water is in the wet region and has a relatively high quality, it will flash into superheated steam when the pressure is decreased.

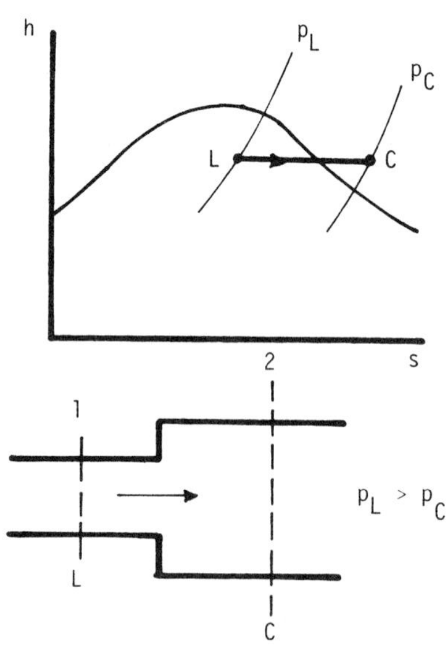

Figure 10.13

In the superheat region, the pressure and temperature are independent properties which may be used to determine the enthalpy in the calorimeter. Since the enthalpy is constant, the enthalpy in the wet region is known. If the pressure or temperature in the wet region is also known, the quality and the other thermodynamic properties may be calculated.

Example 10.21. Given are a line pressure of 300 psia and a calorimeter pressure of 14.696 psia. The temperature in the calorimeter is 300 F. Determine the enthalpy and the quality in the line.

For a pressure of 14.696 psia and a temperature of 300 F, which is in the superheated region,

$$h_c = h_L = 1192.8 \text{ (B/lbm)}$$

For the throttling process this is the enthalpy in the steam line.

At a pressure of 300 psia, $h_f = 393.8$, $h_{fg} = 809.0$, and $h_g = 1202.8$ B/lbm. Since $h_g > h_L$, the steam in the line is in the wet region, and the quality is

$$h_L = h_c = h_{fL} + x_L h_{fgL}$$

$$= 393.8 + x_L(809.0) = 1192.8$$

and

$$x_L = 0.988 \quad \text{or} \quad 98.8\%$$

A disadvantage of the throttling calorimeter is that if the moisture content in the line is excessive, the steam condition in the calorimeter may remain in the wet region. When this occurs, the calorimeter pressure and temperature will not establish the enthalpy in the line or in the calorimeter.

Example 10.22. For a line pressure of 300 psia and a calorimeter pressure of 14.696 psia, determine the minimum possible quality in the line that flashes the steam into the superheat region. Determine the entropy change.

The throttling process is sketched on the hs diagram in Fig. 10.14. The limiting value for calorimeter conditions that are in the superheat region are, at 14.696 psia,

$$h_c = h_g = 1150.4 \text{ (B/lbm)} \quad \text{and} \quad s_c = s_g = 1.7566 \text{ (B/lbm-R)}$$

Actually h_c would have to be slightly greater than h_g to be in the superheat

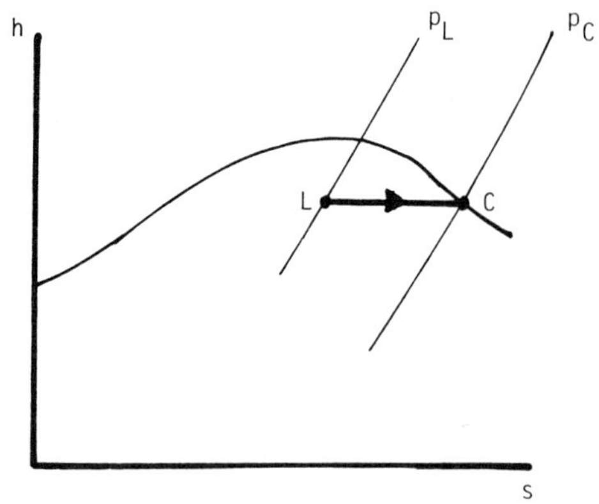

Figure 10.14

region. At a pressure of 300 psia the minimum quality that could be obtained from a throttling calorimeter is

$$h_L = h_c = h_{fL} + x_L(h_{fgL})$$

$$= 393.8 + x_L(809.0) = 1150.4$$

so that

$$x_L = 0.935 \quad \text{or} \quad 93.5$$

The entropy in the line at 300 psia is

$$s_L = s_{fL} + x_L(s_{fgL})$$

$$= 0.5879 + 0.935(0.9225)$$

$$= 1.4504 \text{ (B/lbm-R)}$$

The change in entropy for the throttling process is

$$s = s_c - s_L = 1.7566 - 1.4504 = 0.3062 \text{ (B/lbm-R)}$$

where $s_c = s_g$ at 14.696 psia.

Note that for all irreversible adiabatic processes the entropy must increase.

SUPERHEATING CALORIMETERS

If the quality of the steam in the line is too low, heat may be added to the steam in the calorimeter so that the final state is in the superheat region. One difficulty with this procedure is that the rate of steam flow through the calorimeter must be established.

Example 10.23. Given are a line pressure of 300 psia and a quality of 50%. The calorimeter pressure and temperature are 14.696 psia and 300 F. How much heat was added to the steam in the calorimeter?

For p = 300 psia and x = 50%,

$$h_L = h_{fL} + x_L(h_{fgL}) = 393.8 + 0.5(809.0)$$

$$= 798.3 \text{ (B/lbm)}$$

The enthalpy in the calorimeter at p = 14.696 psia and a temperature of 300 F is

$$h_c = 1192.8 \text{ (B/lbm)}$$

The heat that must be added to the steam in the calorimeter is

$$Q = \Delta h = h_c - h_L = 1192.8 - 798.3$$

$$= 394.5 \text{ (B/lbm)}$$

SEPARATING CALORIMETERS

Another technique that may be employed for very wet steam mixtures is the separating calorimeter. The wet mixture is passed through a curved section so that the liquid particles, which are heavier than the vapor particles,

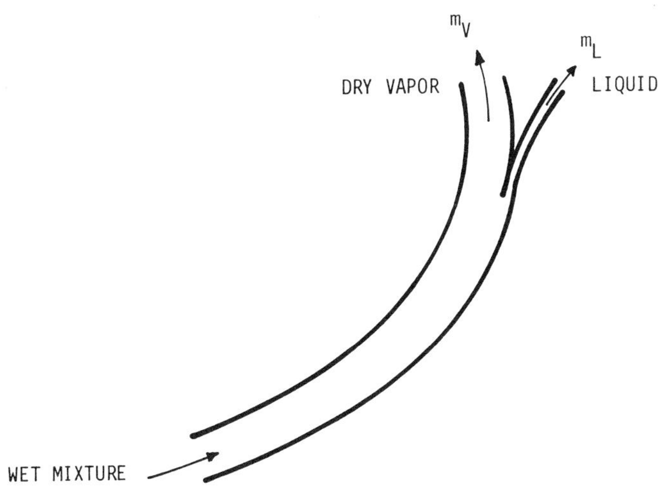

Figure 10.15

are thrown by centrifugal force to the outside of the separator (Fig. 10.15). If the flow of the vapor is measured, then the quality is given by

$$x = m_V/(m_V + m_L)$$

where m_V is the mass flow of the vapor and m_L is the amount of liquid collected in a given period of time.

MOLLIER DIAGRAM

Properties of steam or water in the subcooled liquid, wet, and superheated regions may be obtained from the steam tables. Quite often this requires either single or double interpolation and is a somewhat tedious process. For many engineering calculations, the Mollier chart or diagram (Fig. 10.16) is sufficiently accurate and permits rapid solutions of complex cycle problems.

A more detailed Mollier diagram, as supplied by Combustion Engineering Corporation, is presented in Appendix G. On the Mollier diagram the vertical axis represents the enthalpy, and the horizontal axis gives entropy values. Lines of constant pressure are inclined vertically upward from left to right in the wet and superheat regions. Lines of constant temperature are presented in the superheat region and become nearly horizontal as the entropy increases or the pressure decreases. Horizontal lines for a constant temperature process indicate that the vapor is behaving as a perfect gas.

Lines of constant moisture content are plotted in the wet region, and the percentage value of the quality may be calculated from

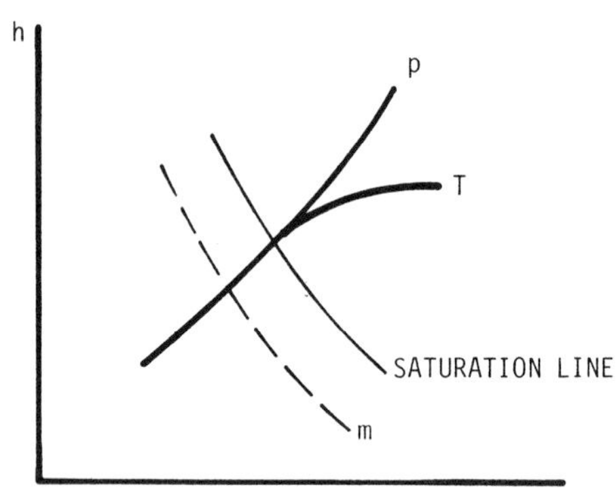

Figure 10.16

$$x = 100 - m$$

Lines of constant superheat are also indicated in the superheat region. For a given pressure the degree of superheat is the difference between the vapor temperature and the temperature of the saturated vapor.

<u>Example 10.24</u>. Steam at a pressure and temperature of 50 psia and 360 F undergoes an isentropic process to a pressure of 1 psia. Determine the change in enthalpy.

The process is sketched in the hs diagram in Fig. 10.17.

From the Mollier chart, Appendix G,

$$h_1 = 1215 \text{ (B/lbm)}$$

Dropping vertically to a pressure of 1 psia gives

$$h_2 = 956 \text{ (B/lbm)}$$

Note that at low pressure lines of constant pressure in psia are solid lines and lines of constant pressure in in.-Hg-abs are represented by dashed lines.

The change in enthalpy is

$$\Delta h = h_2 - h_1 = 956 - 1215 = -259 \text{ (B/lbm)}$$

When the Mollier diagram is used for cycle analysis, values of the enthalpy in the superheat and wet regions are obtained from the diagram. The enthalpy along the saturated liquid line must be obtained from the steam tables since the minimum enthalpy, for a saturated liquid, on the Mollier chart is 835 B/lbm.

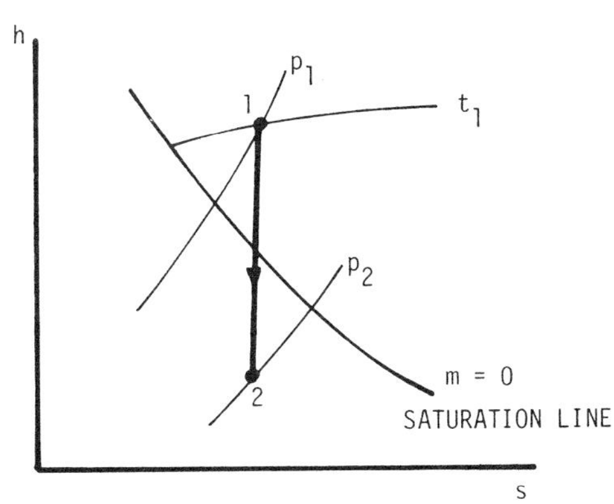

Figure 10.17

PROBLEMS

10.1 For steam or water, complete the following chart where the units are p (psia), t (F), v (ft^3/lbm), s (B/lbm-R), u (B/lbm), h (B/lbm), and x (%). If a particular value is not defined in a particular region, insert NV, and if there is no solution for the given properties, insert NS.

p	t	v	u	h	s	x
14.7	60	–	–	–	–	–
100	–	–	–	168.0	–	–
150	358.4	–	–	–	–	–
–	300	–	–	–	–	60
60	–	3.0	–	–	–	–
–	400	–	–	–	1.0	–
3206	705.4	–	–	–	–	–
120	600	–	–	–	–	–
900	–	–	–	1312.9	–	–
–	900	4.452	–	–	–	–
140	–	–	–	–	1.5751	–

10.2 Five lbm of steam at a pressure of 80 psia undergoes a constant pressure process from a temperature of 60 to 600 F. Determine V_1, V_2, ΔU, ΔH, ΔS, Q, W_{nf}, and W_{sf} (ΔKE = -50 B). Sketch the process on the pV and TS planes.

10.3 Steam at a pressure of 14.696 psia and a quality of 100% undergoes a constant volume process to a temperature of 800 F. Determine v, t_1, p_2, Δu, Δh, Δs, Q, W_{nf}, and W_{sf} (ΔKE, ΔPE = 0). Sketch the process on the pV and TS planes.

10.4 Steam at a pressure and temperature of 160 psia and 500 F undergoes an isothermal process to an enthalpy of 800 B/lbm. Determine v_1, p_2, v_2, x_2, Δu, Δh, Δs, Q, W_{nf}, and W_{sf} (ΔKE, ΔPE = 0). Sketch the process on the pV and TS planes.

10.5 Steam at a pressure of 200 psia and a temperature of 800 F undergoes an isentropic process to a pressure of 1 psia. Determine t_2, x_2, v_1, v_2, Δu, Δh, Δs, Q, W_{nf}, and W_{sf} (ΔKE, ΔPE = 0). Sketch the process on the pV, TS, and hs planes.

10.6 Steam at a pressure and temperature of 10 psia and 300 F undergoes a polytropic process, for which $pV^n = C$, to a pressure and temperature of 300 psia and 1000 F. Determine v_1, v_2, n, Δu, Δh, Δs, W_{nf}, Q, and W_{sf} (ΔKE, $\Delta PE = 0$).

10.7 Given the data in Prob. 10.5, determine Δh and W_{sf} (ΔKE, $\Delta PE = 0$) by use of the Mollier chart. A turbine operating between the state points in this problem has an efficiency of 85%. Determine the enthalpy and entropy after an irreversible adiabatic process.

10.8 A boiler pump received water as a saturated liquid at a temperature of 70 F and discharges the water at a pressure of 300 psia. Determine the pump work and final enthalpy for a pump efficiency of 100% and 80%.

10.9 Steam has a line pressure of 200 psia and a throttling calorimeter pressure and temperature of 14.696 psia and 240 F. Determine the quality in the line by use of the steam tables and the Mollier chart.

10.10 A superheating calorimeter has a pressure and temperature of 20 psia and 300 F. A DC heating coil, in the calorimeter, has a voltage of 120 V and a current flow of 10 A. The steam flow through the calorimeter is 0.125 lbm/min. For a line pressure of 400 psia, determine the quality.

11

STEAM POWER PLANT CYCLE

The theoretical model for the steam power plant is the Rankine cycle. The component parts of the basic Rankine cycle are illustrated in Fig. 11.1. The heat is added to the cycle, in the boiler, during a constant pressure process, 41. The heat addition process may be the result of the combustion of coal, oil, or natural gas. It may also, without any alteration of the basic cycle, be the result of a nuclear reaction.

The process from state 1 to state 2 is represented by an isentropic expansion process in a steam turbine. This process produces the output work for the cycle. The condenser, which is a constant pressure device, is used to convert the steam leaving the turbine to a saturated liquid. This pro-

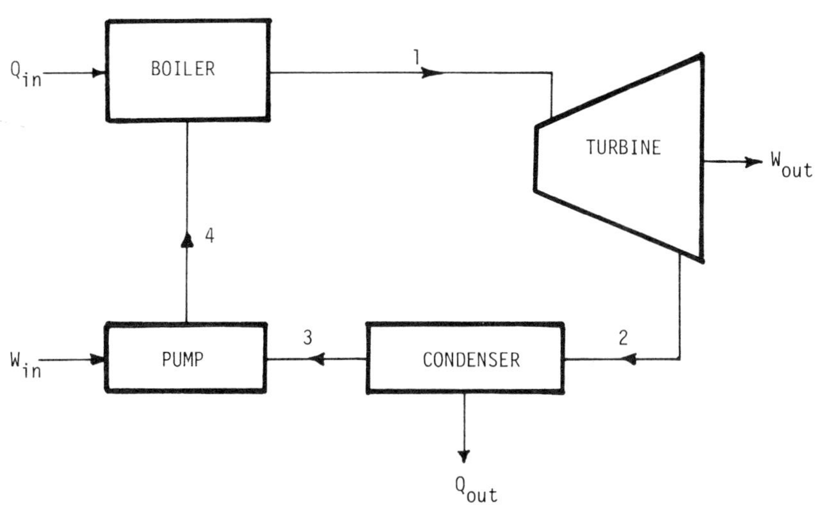

Figure 11.1

cess, from state 2 to state 3, represents the heat rejection component of the Rankine cycle.

The pressure of the water leaving the condenser is raised to the boiler pressure by means of the boiler feed pump. This process, from state 3 to state 4, requires that work be done on the system during an isentropic process.

RANKINE CYCLE PERFORMANCE

The general energy equation for the steady-flow processes in the Rankine cycle is

$$Q = \Delta H + \Delta KE + \Delta PE + W_{sf}$$

Throughout our discussion of the Rankine cycle we shall assume that

$$\Delta KE, \Delta PE = 0$$

so that

$$Q = \Delta H + W_{sf}$$

For the boiler feed pump we shall also use the steady-flow equation,

$$W_{sf} = -\int_{p_3}^{p_4} v\, dp$$

which also requires that $\Delta KE, \Delta PE = 0$.

The basic Rankine cycle is sketched on the TS diagram in Fig. 11.2.

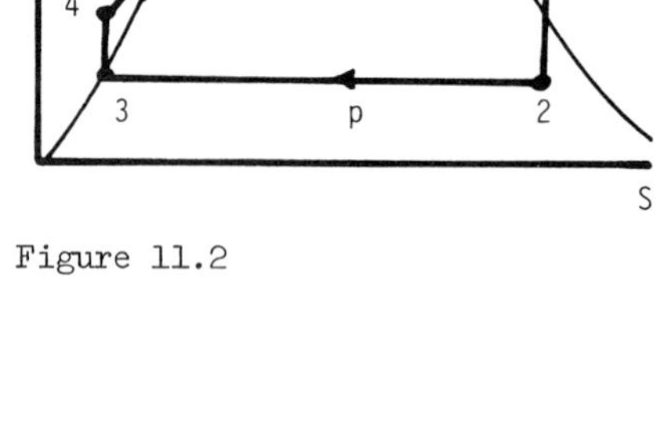

Figure 11.2

For the boiler, $W_{41} = 0$ since p = c, and

$$Q_{41} = \Delta H = H_1 - H_4$$

or for 1 lbm of steam

$$Q_{41} = h_1 - h_4 \tag{11.1}$$

For the isentropic turbine, $Q_{12} = 0$ since s = c, and

$$W_{12} = -\Delta H = -(H_2 - H_1)$$

and for 1 lbm of steam

$$W_{12} = h_1 - h_2 \tag{11.2}$$

For the condenser, $W_{23} = 0$ since p = c, and

$$Q_{23} = \Delta H = H_3 - H_2$$

also for 1 lbm of steam

$$Q_{23} = h_3 - h_2 \tag{11.3}$$

For the isentropic pump, since water is practically incompressible,

$$v_4 = v_3 = C$$

or

$$W_{34} = -\int_{p_3}^{p_4} v\, dp = -v_3(p_4 - p_3) \tag{11.4}$$

For the isentropic process, Q = 0, so that

$$W_{34} = -\Delta H = -(H_4 - H_3)$$

and for 1 lbm of steam

$$W_{34} = h_3 - h_4 \tag{11.5}$$

Example 11.1. Given is the basic Rankine cycle illustrated in Fig. 11.3 where $p_4 = p_1 = 100$ psia and $p_2 = p_3 = 14.696$ psia. The quality of the steam at state 1 is 100%, and the quality leaving the condenser at state 3 is always fixed at 0%. Determine Q_{in}, Q_{out}, W_{in}, W_{out}, and the cycle efficiency e.

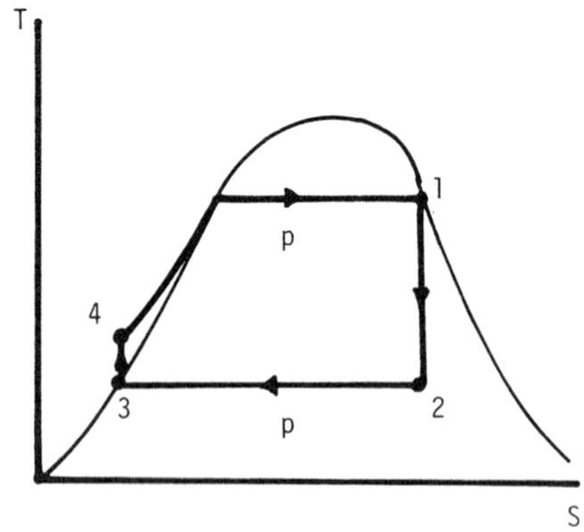

Figure 11.3

For state 1, $p_1 = 100$ psia and $x_1 = 100\%$. By use of Table 2, Appendix D,

$$h_1 = h_{g1} = 1187.2 \text{ (B/lbm)}$$

and

$$s_1 = s_{g1} = 1.6026 \text{ (B/lbm-R)}$$

For state 2, $p_2 = 14.696$ psia and $s_2 = s_1 = 1.6026$ B/lbm-R, and by use of Table 2,

$$s_{f2} = 0.3120 \quad \text{and} \quad s_{fg2} = 1.4446 \text{ (B/lbm-R)}$$

Also

$$h_{f2} = 180.1 \quad \text{and} \quad h_{fg2} = 970.3 \text{ (B/lbm)}$$

The quality at state 2 may be obtained from

$$s_2 = s_{f2} + x_2 s_{fg2}$$

or

$$1.6026 = 0.3120 + x_2(1.4446)$$

so that

$$x_2 = 0.893$$

The enthalpy at state 2 is

$$h_2 = h_{f2} + x_2 h_{fg2} = 180.1 + 0.893(970.3)$$
$$= 1046.6 \text{ (B/lbm)}$$

The enthalpy at state 3, for $p_3 = 14.696$ psia and $x_3 = 0\%$, is

$$h_3 = h_{f3} = h_{f2} = 180.1 \text{ (B/lbm)}$$

By use of the steady-flow work equation [Eq. (11.4)]

$$W_{34} = W_{in} = -v_3(p_4 - p_3)/J, \qquad \text{where } v_3 = v_{f3}$$

$$= -(144/778)(0.0167)(100 - 14.696)$$

$$= -0.3 \text{ (B/lbm)}$$

so that

$$W_{34} = h_3 - h_4$$

and

$$h_4 = h_3 - W_{34} = 180.1 + 0.3 = 180.4 \text{ (B/lbm)}$$

From Eq. (11.1), the heat added to the cycle is

$$Q_{in} = Q_{41} = h_1 - h_4 = 1187.2 - 180.4$$

$$= 1006.8 \text{ (B/lbm)}$$

From Eq. (11.2), the work output of the cycle is

$$W_{out} = W_{12} = h_1 - h_2 = 1187.2 - 1046.6$$

$$= 140.6 \text{ (B/lbm)}$$

From Eq. (11.3), the heat rejected by the cycle is

$$Q_{out} = Q_{23} = h_3 - h_2 = 180.1 - 1046.6$$

$$= -866.5 \text{ (B/lbm)}$$

The work added to the cycle has previously been determined by Eq. (11.4), so that

$$W_{in} = W_{34} = -0.3 \text{ (B/lbm)}$$

The net heat of the cycle is

$$\Sigma Q = Q_{in} + Q_{out} = 1006.8 - 866.5 = 140.3 \text{ (B/lbm)}$$

and the net work is

$$\Sigma W = W_{in} + W_{out} = -0.3 + 140.6 = 140.3 \text{ (B/lbm)}$$

Therefore

$$\Sigma Q = \Sigma W$$

which satisfies the First Law of Thermodynamics.

The cycle efficiency is

$$e = \Sigma W/Q_{in} = \Sigma Q/Q_{in} = 140.3/1006.8$$

$$= 0.139 \quad \text{or } 13.9\%$$

Only 13.9% of the heat added to the cycle is converted to work. This is a result of the relatively low value of the enthalpy at state 1 and the high value of the enthalpy at state 2.

The moisture content leaving the turbine is

$$m_2 = 1 - x_2 = 1 - 0.893 = 0.107 \quad \text{or} \quad 10.7\%$$

Liquid water particles, traveling at high velocities, will erode the blades in the latter stages of the turbine. To prevent erosion, the moisture content at state 2 must be lowered or eliminated.

EFFECT OF NEGLECTING PUMP WORK

The work of the boiler feed pump is such a small quantity compared to the other heat and work terms that it may be neglected from cycle efficiency calculations. When the pump work is neglected the heat supplied to the Rankine cycle is approximated by

$$Q_{in} = h_1 - h_3$$

which is slightly higher than the value obtained for Eq. (11.1).

The net work of the cycle is approximated by

$$\Sigma W = h_1 - h_2$$

which is also slightly higher than the value obtained from

$$\Sigma W = W_{in} + W_{out}$$

The cycle efficiency, using these approximations, is

$$e = \Sigma W/Q_{in} = (h_1 - h_2)/(h_1 - h_3) \tag{11.6}$$

Example 11.2. Determine the cycle efficiency for Example 11.1 by means of Eq. (11.6).

From Eq. (11.6)

$$e = (h_1 - h_2)/(h_1 - h_3) = (1187.2 - 1046.6)/(1187.2 - 180.1)$$

$$= 0.140 \quad \text{or} \quad 14.0\%$$

compared to 13.9% as calculated in Example 11.1.

EFFECT OF SUPERHEATING STEAM IN THE BOILER

A separate coil of tubes may be inserted at the hottest location in the furnace to superheat the steam before it enters the turbine (Fig. 11.4). Superheating the steam serves to raise the cycle efficiency and to reduce the moisture content in the turbine.

Example 11.3. Given is the basic Rankine cycle where $p_1 = p_4 = 100$ psia, $p_2 = p_3 = 14.696$ psia, $x_3 = 0\%$, and $t_1 = 800$ F. Neglect pump work.

From the superheated steam tables, Appendix D,

$$h_1 = 1428.9 \text{ (B/lbm)} \quad \text{and} \quad s_1 = 1.8443 \text{ (B/lbm-R)}$$

where

$$p_1 = 100 \text{ (psia)} \quad \text{and } t_1 = 800 \text{ (F)}$$

For a condenser pressure of 14.696 psia,

$$s_g = 1.7566 \text{ (B/lbm-R)}$$

and since

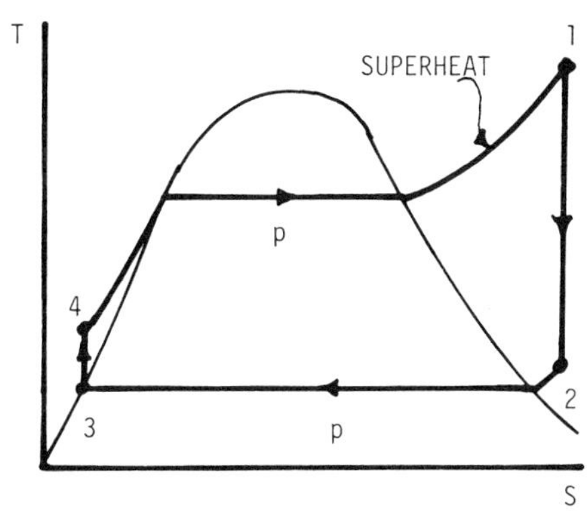

Figure 11.4

$$s_2 = s_1 = 1.8443 \text{ (B/lbm-R)}$$

the steam leaving the condenser is in the superheated region.

From the superheat tables for a pressure of 14.696 psia

$t = 300$	$t = 400$	(F)
$s = 1.8160$	$s = 1.8743$	(B/lbm-R)
$h = 1192.8$	$h = 1239.9$	(B/lbm)

so that

$$f = (1.8443 - 1.8160)/(1.8743 - 1.8160) = 0.485$$

and

$$h_2 = 1192.8 + 0.485(1239.9 - 1192.8)$$

$$= 1215.6 \text{ (B/lbm)}$$

The steam leaving the condenser at a pressure of 14.696 psia and a quality of 0% has an enthalpy of

$$h_3 = h_{f3} = 180.1 \text{ (B/lbm)}$$

From Eq. (11.6) the cycle efficiency is

$$e = (h_1 - h_2)/(h_1 - h_3) = (1428.9 - 1215.6)/(1428.6 - 180.1)$$

$$= 0.171 \quad \text{or} \quad 17.1\%$$

which represents an efficiency increase of 3.1% as compared to a cycle with saturated steam entering the turbine. The steam leaving the turbine is in the superheat region so that erosion of the turbine blades due to water droplets is eliminated.

Example 11.4. Use the same data as for Example 11.3 except $p_1 = 1000$ psia and $t_1 = 800$ F, for which $h_1 = 1389.2$ B/lbm and $s_1 = 1.5670$ B/lbm-R. For $p_2 = 14.696$ psia and $s_{f2} = 0.3120$ and $s_{fg2} = 1.4446$ B/lbm-R, the quality and moisture content leaving the turbine are

$$s_2 = s_1 = s_{f2} + x_2 s_{fg2} = 0.3120 + 1.4446x_2 = 1.5670$$

or

$$x_2 = 0.868 \quad \text{or} \quad m_2 = 1 - x_2 = 0.132 \quad \text{or} \quad 13.2\%$$

Raising the boiler pressure results in a high moisture content in the later stages of the turbine.

The enthalpy leaving the turbine at 14.696 psia is

$$h_2 = h_{f2} + x_2 h_{fg2} = 180.1 + 0.868(970.3)$$

$$= 1022.3 \text{ (B/lbm)}$$

As before, the enthalpy leaving the condenser is

$$h_3 = h_{f3} = h_{f2} = 180.1 \text{ (B/lbm)}$$

The cycle efficiency, by means of Eq. (11.6), is

$$e = (h_1 - h_2)/(h_1 - h_3) = (1389.2 - 1022.3)/(1389.2 - 180.1)$$

$$= 0.303 \quad \text{or} \quad 30.3\%$$

Raising the boiler efficiency to 1000 psia has increased the cycle efficiency by 13.2% as compared to the superheat cycle operating at 100 psia and 800 F.

EFFECT OF REDUCING THE CONDENSER PRESSURE

The condenser pressure is controlled by the availability of cooling water at a low temperature. The lower the temperature or pressure of the steam in the condenser, the higher the cycle efficiency.

Example 11.5. Given is a Rankine cycle with p_1 = 1000 psia and t_1 = 800 F; also $p_2 = p_3$ = 0.5069 psia, which corresponds to a condenser temperature of 80 F. Properties leaving the boiler are h_1 = 1389.2 B/lbm, and s_1 = 1.5670 B/lbm-R. For a condenser temperature of 80 F, s_{f2} = 0.0932 and s_{fg2} = 1.9428 B/lbm-R and h_{f2} = 48.0 and h_{fg2} = 1048.6 B/lbm.

The quality of the steam leaving the turbine is

$$s_2 = s_1 = s_{f2} + x_2 s_{fg2} = 0.0932 + 1.9428x_2 = 1.5670$$

or

$$x_2 = 0.759 \quad \text{and } m_2 = 0.241 \quad \text{or} \quad 24.1\%$$

and

$$h_2 = h_{f2} + x_2 h_{fg2} = 48.0 + 0.759(1048.6)$$

$$= 843.8 \text{ (B/lbm)}$$

The cycle efficiency is

$$e = (h_1 - h_2)/(h_1 - h_3) = (1389.2 - 843.8)/(1389.2 - 48.0)$$

$$= 0.407 \quad \text{or} \quad 40.7\%$$

A reduction of the condenser pressure to 0.5069 psia has resulted in an efficiency increase of 10.4% compared to a similar cycle with a condenser pressure of 14.696 psia.

Unfortunately, the moisture content in the turbine has been increased from 13.2% to 24.1% by reducing the condenser pressure.

REHEAT

If the steam is returned to the boiler when the quality drops below 100%, it may be reheated so that later expansion in the turbine results in a lower moisture content (Fig. 11.5). This cycle is sketched in the TS diagram in Fig. 11.6.

Example 11.6. Given is the Rankine cycle in Example 11.5 where p_1 = 1000 psia, t_1 = 800 F, and $p_4 = p_5$ = 0.5069 psia. The steam is reheated to 800 F when the steam in the turbine reaches the saturated state for the first time.

To prevent the necessity of interpolation in the steam tables the Mollier chart may be used to determine the enthalpy at all state points except that of the saturated liquid leaving the condenser, state point 5.

From the Mollier chart (Fig. 11.7), h_1 = 1387, h_2 = 1195, h_3 = 1424, and h_4 = 962 B/lbm, and m_4 = 13.4%. For a condenser temperature of 80 F, $h_5 = h_{f5}$ = 48.0 B/lbm.

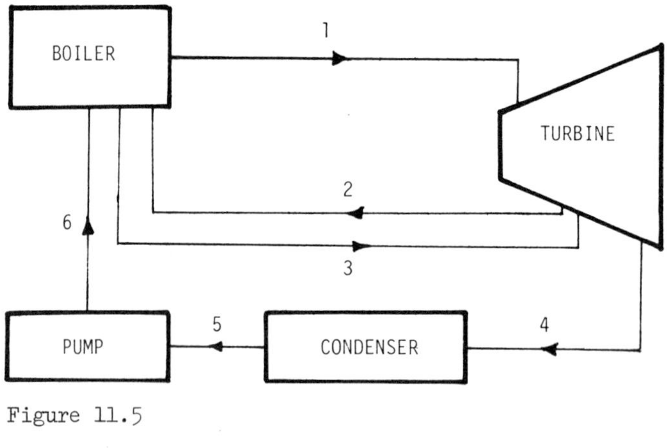

Figure 11.5

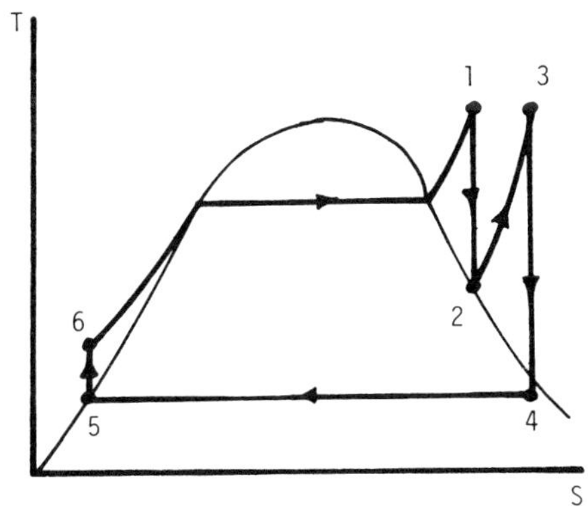

Figure 11.6

The heat added to the cycle is

$$Q_{in} = (h_1 - h_5) + (h_3 - h_2)$$

$$= (1387 - 48) + (1424 - 1195) = 1568 \text{ (B/lbm)}$$

and the work of the turbine is

$$W_{out} = (h_1 - h_2) + (h_3 - h_4) = (1387 - 1195) + (1424 - 962)$$

$$= 654 \text{ (B/lbm)}$$

which assumes that the pump work has been neglected.

The cycle efficiency is

$$e = \Sigma W/Q_{in} = 654/1568 = 0.417 \quad \text{or} \quad 41.7\%$$

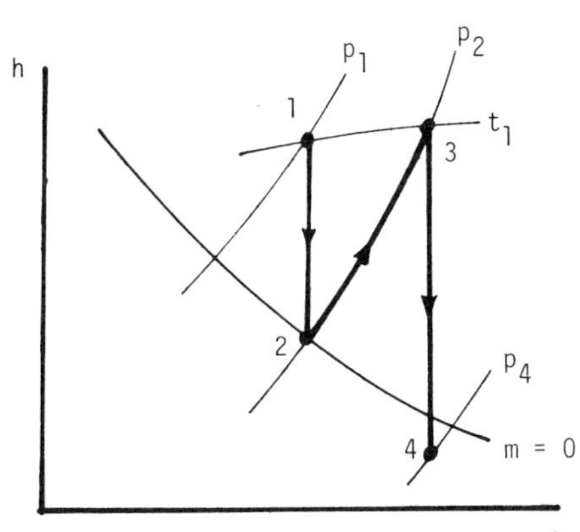

Figure 11.7

which is 1.0% more than the cycle without reheat. By using reheat the moisture content has been reduced from 24.1% to 13.4%.

Example 11.7. Assume that a second reheat occurs when the steam in the turbine becomes saturated for the second time. The process is sketched on the Mollier diagram in Fig. 11.8. The reheat temperature for first state is 800 F. As before, $h_1 = 1387$, $h_2 = 1195$, and $h_3 = 1424$ B/lbm. From the Mollier chart, $h_4 = 1143$ and $h_5 = 1357$ B/lbm from tables for $p_5 = 10$ psia and $s_5 = s_6 = s_{g6} = 2.0360$ B/lbm-R. The pressure and temperature at state 5 are $p_5 = 10$ psia and $t_5 = 646$ F.

The enthalpy at state 6 is $h_6 = 1097$ B/lbm. As before, for the saturated liquid at 80 F, $h_7 = 48.0$ B/lbm.

The heat added to the cycle is

$$Q_{in} = (h_1 - h_7) + (h_3 - h_2) + (h_5 - h_4)$$

$$= (1387 - 48) + (1424 - 1195) + (1357 - 1143)$$

$$= 1782 \text{ (B/lbm)}$$

and the net work, neglecting pump work, is

$$\Sigma W = (h_1 - h_2) + (h_3 - h_4) + (h_5 - h_6)$$

$$= (1387 - 1195) + (1424 - 1143) + (1357 - 1097)$$

$$= 733 \text{ (B/lbm)}$$

The cycle efficiency is

$$e = \Sigma W/Q_{in} = 732/1782 = 0.411 \quad \text{or} \quad 41.1\%$$

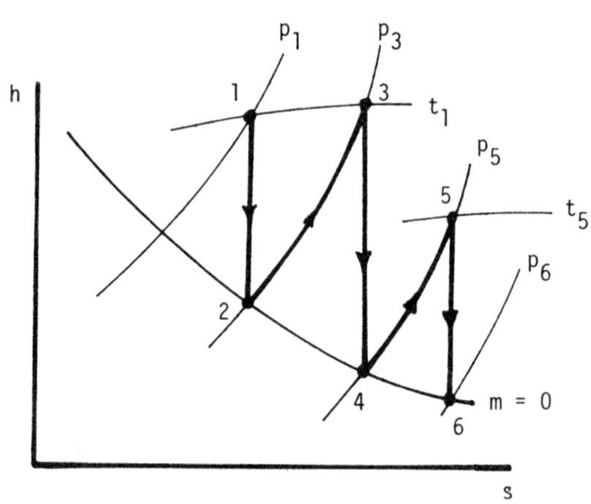

Figure 11.8

The second stage of reheat has reduced the efficiency by

$$\Delta e = 0.411 - 0.417 = -0.006 \quad \text{or} \quad -0.6\%$$

but there are no water droplets passing through the turbine.

REGENERATION

Steam may be extracted at any point in the turbine and used to heat the water leaving the condenser (Fig. 11.9). If this heating process occurs in an open heater (Fig. 11.10), the energy balance equation is

$$m_e h_e + (1 - m_e)h_7 = h_8$$

The enthalpy of the steam entering the boiler has been increased so that less heat must be added in the boiler.

Example 11.8. Given is Example 11.7, with extraction at 14.696 psia. The enthalpy of the extracted steam, from the Mollier chart, is

$$h_e = 1172 \text{ (B/lbm)}$$

The required amount of extracted steam may be determined by means of an energy balance (Fig. 11.10), where $Q = 0$, so that

$$1172m_e + 48(1 - m_e) = 180$$

or $\quad m_e = 0.117 \text{ (lbm/lbm)}$

where $h_8 = h_{f8} = 180$ B/lbm at 14.696 psia.

The heat added in the cycle is

$$\begin{aligned} Q_{in} &= (h_1 - h_8) + (h_3 - h_2) + (1 - m_e)(h_5 - h_4) \\ &= (1387 - 180) + (1424 - 1195) + (1 - 0.117)(1357 - 1143) \\ &= 1625 \text{ (B/lbm)} \end{aligned}$$

Note that for the process from state 4 to state 5 the amount of steam to be heated is reduced by the amount of extracted steam.

The net work of the cycle, neglecting the condenser and the boiler pump, is

$$\begin{aligned} \Sigma W &= (h_1 - h_2) + (h_3 - h_e) + (1 - m_e)(h_e - h_4) + (1 - m_e)(h_5 - h_6) \\ &= (1387 - 1195) + (1424 - 1172) + 0.883(1172 - 1143) + 0.883(1357 - 1097) \\ &= 699 \text{ (B/lbm)} \end{aligned}$$

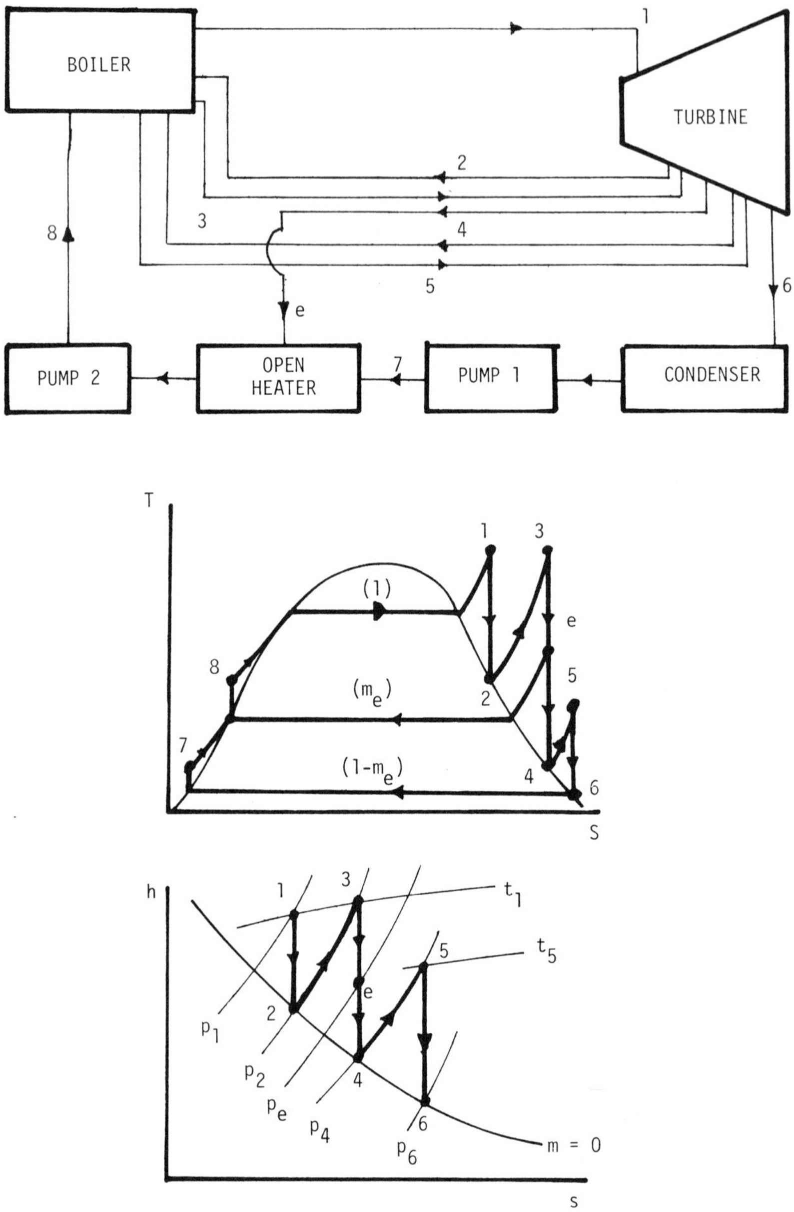
BOILER
TURBINE
PUMP 2
OPEN HEATER
PUMP 1
CONDENSER
1
2
3
4
5
6
7
8
e
T
S
(1)
(m_e)
$(1-m_e)$
h
s
t_1
t_5
p_1
p_2
p_e
p_4
p_6
m = 0

Figure 11.9

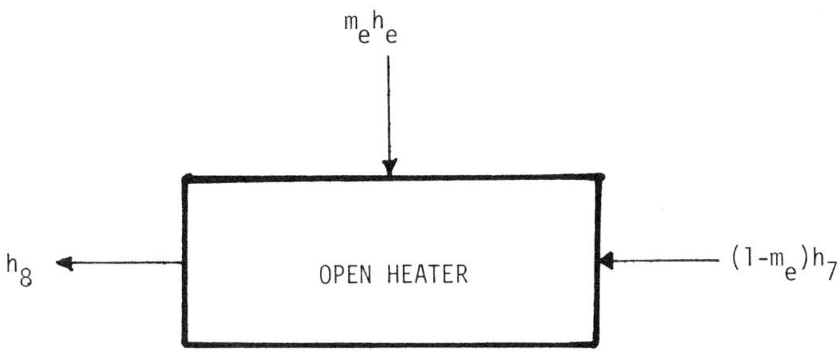

Figure 11.10

where

$$1 - m_e = 0.883$$

The cycle efficiency is

$$e = \Sigma W/Q_{in} = 699/1625 = 0.430$$

By use of a single regeneration stage the cycle efficiency has been increased from 41.1% to 43.0% or by 1.9%.

Example 11.9. Assume a second stage of regeneration is used in Example 11.8 acting at a pressure of 200 psia. As shown in Fig. 11.11, the high-pressure extraction requires a steam flow to the open heater of

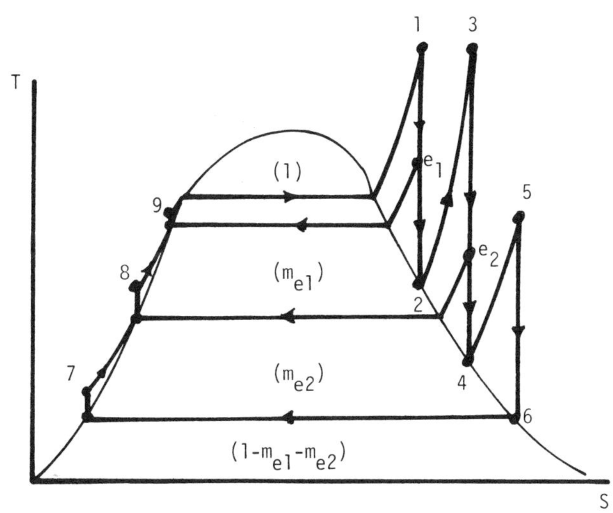

Figure 11.11

$$m_{e1}h_{e1} + (1 - m_{e1})h_8 = h_9$$

where $h_{e1} = 1214$ and $h_9 = h_{f9} = 355$ B/lbm at $p_1 = 200$ psia, so that

$$m_{e1}(1214) + (1 - m_{e1})(180) = 355$$

and

$$m_{e1} = 0.169$$

For the low-pressure extraction stage

$$m_{e2}h_{e2} + (1 - m_{e2} - m_{e1})h_7 = (1 - m_{e1})h_8$$

$$1172m_{e2} + (0.831 - m_{e2})(48) = 0.831(180)$$

or

$$m_{e2} = 0.098$$

where h_{e2} is equal to h_e in Example 11.8.

The heat added to the cycle is

$$Q_{in} = (h_1 - h_9) + (1 - m_{e1})(h_3 - h_2) + (1 - m_{e1} - m_{e2})(h_5 - h_4)$$

$$= (1387 - 355) + 0.831(1424 - 1195) + 0.733(1357 - 1143)$$

$$= 1379 \text{ (B/lbm)}$$

The net work of the cycle, neglecting pump work, is

$$\Sigma W = (h_1 - h_{e1}) + (1 - m_{e1})(h_{e1} - h_2) + (1 - m_{e1})(h_3 - h_{e2})$$

$$+ (1 - m_{e1} - m_{e2})(h_{e2} - h_4) + (1 - m_{e1} - m_{e2})(h_5 - h_6)$$

$$= (1387 - 1214) + 0.831(1214 - 1195) + 0.831(1424 - 1172)$$

$$+ 0.733(1172 - 1143) + 0.733(1357 - 1097)$$

$$= 610 \text{ (B/lbm)}$$

The cycle efficiency is

$$e = \Sigma W/Q_{in} = 610/1379 = 0.442 \quad \text{or} \quad 44.2\%$$

The second open heater has increased the efficiency from 43.0% to 44.2% or by 1.2%.

No attempt has been made to determine if the extraction pressures in this example represent the optimum values that would produce the maximum cycle efficiency. Open heaters are generally operated above atmospheric pressure so that any leakage would be from the heater, instead of into the heater, and is thereby easily detected.

TURBINE EFFICIENCY

The turbine efficiency is defined by the equation

$$e_t = \text{actual work/ideal work}$$

$$= (h_1 - h_2')/(h_1 - h_2)$$

as shown in Fig. 11.12.

When an extraction point occurs in the turbine, the enthalpy at the extraction point is increased and is equal to

$$h_e' = h_1 - e_t(h_1 - h_e)$$

as shown in Fig. 11.13.

Example 11.10. Determine the cycle efficiency in Example 11.9 for a turbine efficiency of 80% (Fig. 11.11).

The enthalpy at h_2', leaving the high-pressure turbine stage, is

$$e_t = (h_1 - h_2')/(h_1 - h_2) = (1387 - h_2')/(1387 - 1195) = 0.8$$

or

$$h_2' = 1233 \text{ (B/lbm)}$$

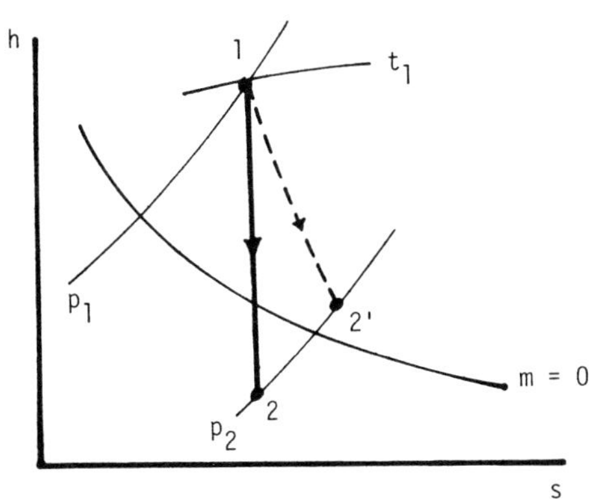

Figure 11.12

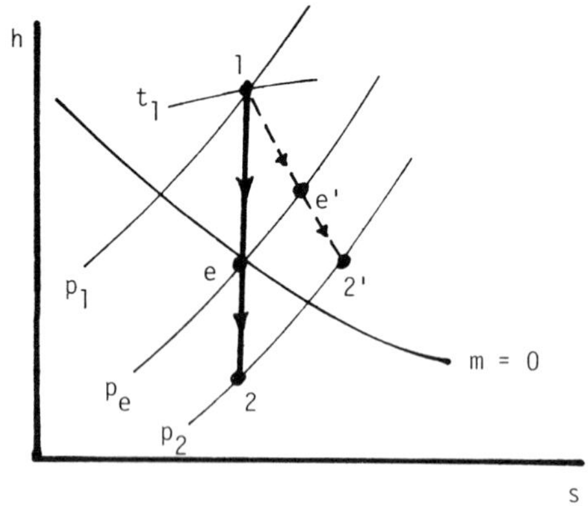

Figure 11.13

The enthalpy of the steam entering the high-pressure open heater is

$$h_{e1}' = h_1 - e_t(h_1 - h_{e1}) = 1387 - 0.8(1387 - 1214)$$

$$= 1249 \text{ (B/lbm)}$$

The enthalpy at h_4', leaving the middle-pressure turbine stage, is

$$e_t = (h_3 - h_4')/(h_3 - h_4) = (1424 - h_4')/(1424 - 1143) = 0.8$$

or

$$h_4' = 1199 \text{ (B/lbm)}$$

The enthalpy of the steam entering the low-pressure open heater is

$$h_{e2}' = h_3 - e_t(h_3 - h_{e2}) = 1424 - 0.8(1424 - 1172)$$

$$= 1222 \text{ (B/lbm)}$$

The enthalpy at state h_6', leaving the low-pressure turbine stage, is

$$e_t = (h_5 - h_6')/(h_5 - h_6) = (1357 - h_6')/(1357 - 1097) = 0.8$$

$$h_6' = 1149 \text{ (B/lbm)}$$

The mass of steam extracted in the high-pressure heater is

$$m_{e1}'h_{e1}' + (1 - m_{e1}')h_8 = h_9$$

$$1249m_{e1}' + (1 - m_{e1}')(180) = 355$$

and

$$m_{e1}' = 0.164$$

The amount of steam extracted in the low-pressure heater is

$$m_{e2}'h_{e2}' + (1 - m_{e2}' - m_{e1}')h_7 = (1 - m_{e1}')h_8$$

$$m_{e2}'(1222) + (1 - m_{e2}' - 0.164)(48) = (1 - 0.164)(180)$$

and

$$m_{e2}' = 0.094$$

The heat addition in the cycle is

$$Q_{in} = (h_1 - h_9) + (1 - m_{e1}')\ (h_3 - h_2') + (1 - m_{e1}' - m_{e2}')(h_5 - h_4')$$

$$= (1387 - 355) + 0.836(1424 - 1233) + 0.742(1357 - 1199)$$

$$= 1309\ (\text{B/lbm})$$

The net work of the cycle is

$$\Sigma W = (h_1 - h_{e1}') + (1 - m_{e1}')(h_{e1}' - h_2') + (1 - m_{e1}')(h_3 - h_{e2}')$$

$$+ (1 - m_{e1}' - m_{e2}')(h_{e2}' - h_4') + (1 - m_{e1}' - m_{e2}')(h_5 - h_6')$$

$$= (1387 - 1249) + 0.836(1249 - 1233) + 0.836(1424 - 1222)$$

$$+ 0.742(1222 - 1199) + 0.742(1357 - 1149)$$

$$= 492\ (\text{B/lbm})$$

The cycle efficiency is

$$e = \Sigma W/Q_{in} = 492/1309 = 0.376 \quad \text{or} \quad 37.6\%$$

and the cycle efficiency is reduced by 6.6% when the turbine efficiency is 80%.

POWER PLANT ECONOMICS

Assume that the cycles discussed in the previous sections are used to generate 250,000 kW of electrical power with a boiler efficiency of 85% and an electric generator efficiency of 95%. Coal will be used to fire the boiler, which has a heating value of HV = 14,000 B/lbm and a cost of $20 per ton.

Since

$1 \text{ kW} = 3412 \text{ B/hr}$

then

$250{,}000 \text{ kW} = 853 \times 10^6 \text{ B/hr}$

For a generator efficiency of 95%, the work to be produced by the turbine is

$\Sigma W = (853 \times 10^6)/0.95 = 898 \times 10^6 \text{ (B/hr)}$

From Example 11.10, for a turbine efficiency of 80%, the steam rate was

$m = (898 \times 10^6)/492 = 1.825 \times 10^6 \text{ (lbm/hr)}$

where ΣW for the cycle was 492 B/lbm.

The quantity of the heat added in the boiler is

$Q_{in} = (1.825 \times 10^6)(1309) = 2.388 \times 10^9 \text{ (B/hr)}$

where the heat added per pound of steam was 1309 B/lbm.

For a boiler efficiency of 85%, the heat to be supplied by the coal is

$Q_{coal} = (2.388 \times 10^9)/0.85 = 2.809 \times 10^9 \text{ (B/hr)}$

The mass of coal required is

$m_{coal} = Q_{coal}/HV = (2.809 \times 10^9)/14{,}000$

$= 2.006 \times 10^5 \text{ (lbm/hr)}$

$= (2.006 \times 10^5)/2000 = 100.3 \text{ (tons/hr)}$

The fuel costs are

Cost = 100.3(20) = \$2006/hr

The unit power costs are

Cost = 2006/250,000 = 0.00803 (dollars/kWh)

= 0.803 (cents/kWh)

If the plant operates 24 hr a day for 365 days per year, the fuel costs are

Total cost = 2006(24)(365) = \$17,572,000

In 1970 the average cost of electric power was about 2.0 cents/kWh, but this cost includes plant value, land, labor, etc.

Calculations for the previous examples are listed in the following table:

Boiler Press. (psia)	Turbine Temp. (%)	Condenser Pressure (psia)	Reheat	Regeneration	Efficiency (%)	Cost (¢/kWh)	Total Cost ($\$^{6}$)
100	327.8	14.7	No	No	14.0	2.16	47.35
100	800	14.7	No	No	17.1	1.17	38.69
1000	800	14.7	No	No	30.3	0.994	21.78
1000	800	0.507	No	No	40.7	0.742	16.26
1000	800	0.507	p = 160	No	41.7	0.724	15.85
1000	800	0.507	p = 160, 10	No	41.1	0.734	16.07
1000	800	0.507	p = 160, 10	p = 14.7	43.0	0.702	15.37
1000	800	0.507	p = 160, 10	p = 200, 14.7	44.2	0.682	14.94
1000	800	0.507	p = 160, 10	p = 200, 14.7	37.6	0.803	17.57

The last line in the table is for the system with an 80% turbine efficiency. The pump work has been neglected in all calculations. The calculations also assume an electric generator efficiency of 95% and a boiler efficiency of 85%.

From the total cost data, increasing the boiler temperature from 327.8 to 800 F results in a savings of \$8.7 million per year. Raising the boiler pressure from 100 to 1000 psia saves \$16.9 million per year. Reduction of the condenser pressure from 14.7 to 0.51 psia saves \$5.5 million per year. Removing the moisture from the turbine by two stages of reheat saves \$0.2 million per year. Two stages of regeneration added to the moisture-free system results in a savings of \$1.2 million per year. If the turbine efficiency is 80% instead of 100%, the increased costs are \$2.6 million per year. For the plant, with a turbine efficiency of 80%, increasing the plant efficiency by 1% will save \$0.8 million per year.

BOILER PUMP

While the pump work has a negligible effect on the cycle efficiency, the power requirement to drive the pump in a large plant is not insignificant.

Example 11.11. Determine the power required to drive a boiler pump which has an efficiency of 90% in a 250,000-kW plant. Water enters the pump as a saturated liquid at 200 psia and leaves at 1000 psia.

For water as an incompressible fluid,

$$W_{sf} = -\int_{p_1}^{p_2} v\ dp = -v(p_2 - p_1)$$

$$= (144/778)(0.01839)(1000 - 200)$$

$$= -2.72\ (B/lbm)$$

where

$$v_f = 0.01839\ (ft^3/lbm) \qquad \text{at } 200\ (psia)$$

For a steam flow of 1.825×10^6 lbm/hr the work is

$$W_{sf} = -(1.825 \times 10^6)(2.72) = -4.970 \times 10^6\ (B/hr)$$

$$= -8.283 \times 10^4\ (B/min)$$

$$= -1952\ (HP)$$

For a pump with an efficiency of 90%, the horsepower requirement is

$$W_{sf} = -(1952/0.9) = -2170\ (HP)$$

$$= -1617\ (kW)$$

which is 0.6% of the work output of the 250,000-kW plant. The cycle efficiency, including pump work, is 35.6% or 0.1% lower than the efficiency when the pump work is neglected.

STEAM AND HEAT RATE

The steam rate is a value which is often used to compare the operating characteristics of different steam power plants. The steam rate is defined by

Steam rate = mass of steam per hour/net work

The heat rate is defined by

Heat rate = Q_{in}/net work

Modern power plants have a heat rate of approximately 9000 B/kWh.

Example 11.12. Determine the steam and heat rates for the cycle in Example 11.10 where the net work = 250,000 kW, the steam supplied = 1.825×10^6 lbm/hr, and the heat in = 2.388×10^9 B/hr.

Steam rate = $(1.825 \times 10^6)/250{,}000 = 7.30$ (lbm/kWh)

Heat rate = $(2.388 \times 10^9)/250{,}000 = 9552$ (B/kWh)

PROBLEMS

11.1 A basic Rankine cycle has a boiler pressure of 200 psia and a condenser temperature of 180 F. The steam leaves the boiler as a saturated vapor. By use of the steam tables, Appendix D, determine Q_{in}, Q_{out}, W_{in}, W_{out}, x_2, and e.

11.2 Determine the efficiency of the cycle in Prob. 11.1 when pump work is omitted.

In the following problems, neglect pump work and use the Mollier chart where possible.

11.3 The steam leaves the boiler in Prob. 11.1 at a temperature of 700 F and a pressure of 200 psia. Determine e.

11.4 The steam leaves the boiler in Prob. 11.1 at a temperature of 700 F and a pressure of 500 psia. Determine e and m_2.

11.5 The steam leaves the boiler at the condition in Prob. 11.4 and has a condenser pressure of 2.5 in.-Hg-abs. Determine e and m_2. Sketch the cycle on the TS and hs planes.

11.6 The cycle in Prob. 11.5 is modified so that the steam is reheated to 700 F when the steam reaches the saturation state for the first time in the turbine. Determine the pressure in the reheat line, e, and the moisture content leaving the turbine. Sketch the process on the TS and hs planes.

11.7 A second reheat stage is added to the cycle in Prob. 11.6 so that the steam leaves the turbine as a saturated vapor. Determine the pressure and temperature after the second reheat and e. Sketch the cycle on the TS and hs planes.

11.8 The cycle in Prob. 11.7 has a single stage of regeneration at 14.696 psia. Determine the quantity of extracted steam, ΣW, Q_{in}, and e. Sketch the cycle on the TS and hs planes.

11.9 Determine the cost in cents/kWh to operate the theoretical cycle in Prob. 11.8. The fuel cost is $15 per ton of a fuel having a heating value of 14,000 B/lbm. Boiler and generator efficiencies are 100%.

11.10 Repeat Prob. 11.8 for a turbine efficiency of 75%.

11.11 Using the data in Probs. 11.9 and 11.10 for a cycle having a turbine efficiency of 75%, an electric generator efficiency of 90%, and a boiler efficiency of 70%, determine the unit cost.

11.12 Determine the steam rate and the heat rate for the cycle in Prob. 11.11.

HEAT PUMP AND REFRIGERATION CYCLES

Assume that the basic Rankine cycle introduced in Chap. 11 is reversed, so that the cycle proceeds in the counterclockwise direction on the TS plane (Fig. 12.1). For the ideal reversed cycle, the processes and components consist of

1. Process 1-2. Isentropic compression by use of a steady-flow compressor. State 1 is assumed to lie on the saturated vapor line.
2. Process 2-3. Constant pressure heat rejection by use of a condenser.

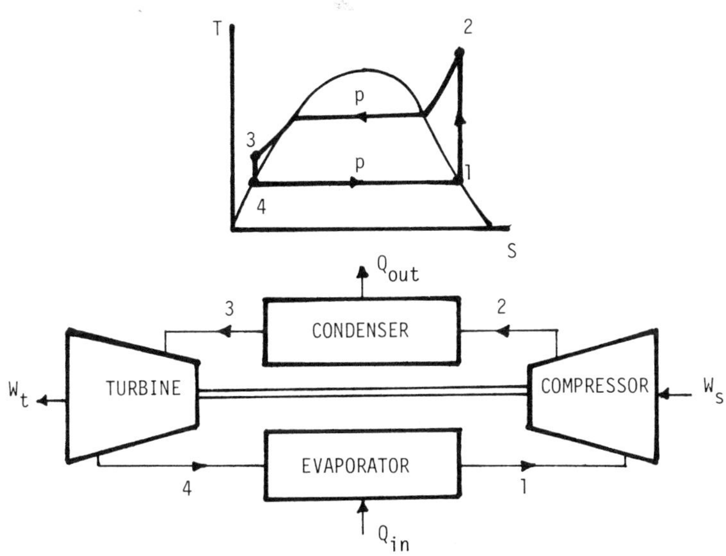

Figure 12.1

3. Process 3-4. Isentropic expansion by use of a steady-flow turbine. State 4 is assumed to lie on the saturated liquid line.

4. Process 4-1. Constant pressure heat addition by use of an evaporator.

The reversed cycle has two possible applications. The heat rejected by the condenser could be used for heating purposes. In this case, the reversed cycle is described as a heat pump cycle. The heat input, at a low evaporator temperature, could be used for cooling purposes. In this case, the reversed cycle is designated a refrigeration cycle.

The efficiency expression

$$e = \Sigma W/Q_{in}$$

has no meaning for a reversed cycle. The parameter used to evaluate either the heat pump or refrigeration cycle is the coefficient of performance, COP, or γ.

For the heat pump cycle

$$COP_P = \gamma_P = |Q_{out}|/|\Sigma W| \qquad (12.1)$$

The vertical brackets are used to indicate the magnitude or positive value of the heat and work terms and are used in the calculation of the coefficient of performance.

For the refrigeration cycle

$$COP_R = \gamma_R = |Q_{in}|/|\Sigma W| \qquad (12.2)$$

For any reversed cycle, ideal or actual,

$$\begin{aligned}\gamma_P - \gamma_R &= (|Q_{out}|/|\Sigma W|) - |Q_{in}|/|\Sigma W|)\\ &= (1/|\Sigma W|)(|Q_{out}| - |Q_{in}|) = (1/|\Sigma W|)|\Sigma W|\\ &= 1\end{aligned} \qquad (12.3)$$

Example 12.1. Given water as the working substance in a reversed cycle, the evaporator temperature is 35 F and the condenser pressure is 14.696 psia. Determine the coefficients of performance.

From the steam tables for state 1, t_1 = 35 F; p_1 = 0.1000 psia, where x_1 = 100%, so that $s_1 = s_{g1}$ = 2.1770 B/lbm-R; and $h_1 = h_{g1}$ = 1077.1 B/lbm.

The properties at state 2, after the isentropic compression process, are $s_2 = s_1$ = 2.1770 B/lbm-R and p_2 = 14.696 psia.

The determination of the value of the enthalpy at state 2 requires an interpolation using values from the superheated steam tables, where for $p = 14.696$ psia,

$t = 1000$	$t = 1200$	(F)
$s = 2.1319$	$s = 2.1989$	(B/lbm-R)
$h = 1533.1$	$h = 1637.5$	(B/lbm)

so that

$$f = (2.1770 - 2.1319)/(2.1989 - 2.1319) = 0.673$$

and

$$h_2 = 1533.1 + 0.673(1637.5 - 1533.1)$$

$$= 1603.4 \text{ (B/lbm)}$$

For state 4, $t_4 = 35$ F, $p_4 = 0.1000$ psia, and $v_4 = v_{f4} = 0.01602$ ft^3/lbm, where $h_4 = h_{f4} = 3.02$ B/lbm since $x_4 = 0\%$.

The work of the turbine, using water as an incompressible liquid, is

$$W_{34} = W_t = [-v_4(p_4 - p_3)]/J$$

$$= -(144/778)(0.01602)(0.1000 - 14.696)$$

$$= 0.04 \text{ (B/lbm)}$$

For state 3

$$W_{34} = W_t = h_3 - h_4$$

or

$$h_3 = W_t + h_4 = 0.04 + 3.02 = 3.06 \text{ (B/lbm)}$$

The work of the compressor is

$$W_c = h_1 - h_2 = 1077.1 - 1603.4 = -526.3 \text{ (B/lbm)}$$

The net work for the cycle, neglecting the small amount of turbine work, is

$$\Sigma_W = W_c = -526.3 \text{ (B/lbm)}$$

The heat added in the evaporator is

$$Q_{41} = Q_{in} = h_1 - h_4 = 1077.1 - 3.0 = 1074.1 \text{ (B/lbm)}$$

and the heat rejected in the condenser is

$$Q_{23} = Q_{out} = h_3 - h_2 = 3.0 - 1603.4 = -1600.4 \text{ (B/lbm)}$$

The net heat for the cycle is

$$\Sigma Q = Q_{in} + Q_{out} = 1074.1 - 1600.4$$

$$= -526.3 \text{ (B/lbm)}$$

For the heat addition and heat rejection constant pressure processes, $W_{sf} = 0$ and ΔKE, $\Delta PE = 0$.

For the isentropic work processes, $Q = 0$, and it has also been assumed that ΔKE, $\Delta PE = 0$.

From Eq. 12.1, the coefficient of performance of the heat pump cycle is

$$\gamma_P = |Q_{out}|/|\Sigma W| = 1600.4/526.3 = 3.04$$

From Eq. (12.2), the coefficient of performance of the refrigeration cycle is

$$\gamma_R = |Q_{in}|/|\Sigma W| = 1074.1/526.3 = 2.04$$

Eq. (12.3) is verified since

$$\gamma_P - \gamma_R = 3.04 - 2.04 = 1.0$$

Water is not a satisfactory refrigerant since the evaporator cannot operate below 32 F, the freezing point of water. For a modest condenser pressure of 14.696 psia, the temperature at the end of the compression process is over 1100 F. This superheat temperature is entirely too high for a reasonable heat pump or refrigeration cycle and would lead to serious lubrication problems.

Example 12.2. Assume a reversed cycle uses ammonia as a working substance. The evaporator temperature is -50 F, and the condenser pressure is 70 psia. Determine the coefficients of performance.

From Appendix E for ammonia at -50 F, $s_1 = s_{g1} = 1.4497$ B/lbm-R and $h_1 = h_{g1} = 593.7$ B/lbm. For state 2 after the isentropic compression process, $s_2 = s_1 = 1.4497$ B/lbm-R and $p_2 = 70$ psia. From the superheat tables for ammonia, Table 2, where $p_2 = 70$ psia,

t = 220	t = 240	(F)
s = 1.4469	s = 1.4631	(B/lbm-R)
h = 727.7	h = 738.9	(B/lbm)

and

$$f = (1.4497 - 1.4469)/(1.4631 - 1.4469) = 0.1728$$

so that

$$h_2 = 727.7 + 0.1728(738.9 - 727.7)$$

$$= 729.6 \text{ (B/lbm)}$$

and

$$t_2 = 220 + 0.1728(240 - 220) = 223 \text{ (F)}$$

which is much more realistic than the temperature entering the condenser when water is used.

The properties at state 4, at the entrance to the evaporator, are $t_4 = -50$ F, $p_4 = 7.67$ psia, $v_4 = v_{f4} = 0.0230$ ft^3/lbm and $h_4 = h_{f4} = -10.6$ B/lbm.

The work of the isentropic turbine is

$$W_{34} = W_t = -v_4(p_4 - p_3) = -(144/778)(0.0230)(7.67 - 70)$$

$$= 0.3 \text{ (B/lbm)}$$

The enthalpy at state 3 is

$$h_3 = W_t + h_4 = 0.3 - 10.6 = -10.3 \text{ (B/lbm)}$$

The work of the isentropic compressor is

$$W_{12} = W_c = h_1 - h_2 = 593.7 - 729.6 = -135.9 \text{ (B/lbm)}$$

and the net work of the cycle is

$$\Sigma W = W_c + W_t = -135.9 + 0.3 = -135.6 \text{ (B/lbm)}$$

The heat added in the evaporator is

$$Q_{41} = Q_{in} = h_1 - h_4 = 593.7 + 10.6 = 604.3 \text{ (B/lbm)}$$

and the heat rejected in the condenser is

$$Q_{23} = Q_{out} = h_3 - h_2 = -10.3 - 729.6 = -739.9 \text{ (B/lbm)}$$

The net heat of the cycle is

$$\Sigma Q = Q_{in} + Q_{out} = 604.3 - 739.9 = -135.6 \text{ (B/lbm)}$$

The coefficients of performance are as follows: As a heat pump

$$\gamma_P = |Q_{out}|/|\Sigma W| = 739.9/135.6 = 5.46$$

and as a refrigerator

$$\gamma_R = |Q_{in}|/|\Sigma W| = 604.3/135.6 = 4.46$$

VAPOR COMPRESSION REFRIGERATION CYCLE

The principal difference between the vapor compression cycle and the reversed Rankine cycle is the replacement of the turbine with an expansion valve. The turbine produces very little work and is not justified economically in most heat pump or refrigeration applications.

The vapor compression cycle is sketched in Fig. 12.2. For the expansion valve

$$Q = \Delta H + \Delta KE + \Delta PE + W_{sf}$$

The process in the expansion valve is a throttling process where Q, ΔKE, ΔPE, $W_{sf} = 0$, so that

$$h_3 = h_4$$

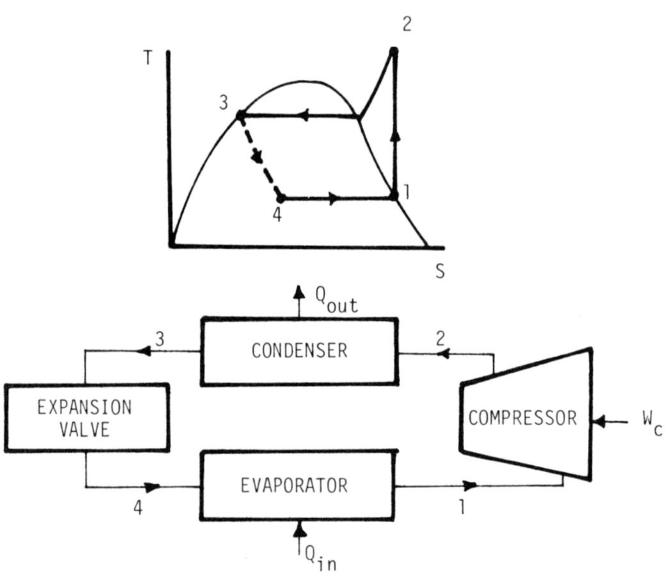

Figure 12.2

The replacement of the turbine by an expansion valve does not result in an appreciable loss of work. The use of an expansion valve, which requires an irreversible adiabatic process, will substantially lower the coefficients of performance, due to decreased heat transfer.

Example 12.3. Rework Example 12.2 using ammonia as the working substance and assuming a vapor compression cycle replaces the reversed Rankine cycle.

As in Example 12.2, for ammonia, at $t_1 = -50$ and $x_1 = 100\%$,

$$h_1 = 593.7 \text{ (B/lbm)}$$

After the isentropic compression process to a pressure of 70 psia,

$$h_2 = 729.6 \text{ (B/lbm)}$$

and the enthalpy leaving the condenser at a pressure of 70 psia and $x = 0\%$, by interpolation, is

$t = 35$	$t = 40$	(F)
$p = 66.26$	$p = 73.72$	(psia)
$h_f = 81.2$	$h_f = 86.8$	(B/lbm)

where

$$f = (70 - 66.26)/(73.72 - 66.26) = 0.501$$

and

$$h_3 = h_4 = 81.2 + 0.501(86.8 - 81.2) = 84.0 \text{ (B/lbm)}$$

The net work for the cycle is the same as the compressor work,

$$\Sigma W = W_c = h_1 - h_2 = 593.7 - 729.6$$

$$= -135.9 \ [-135.6] \text{ (B/lbm)}$$

where square brackets are the values for the reversed Rankine cycle. The difference between these two values is a result of the omission of the turbine work in the reversed Rankine cycle.

The heat transferred from the condenser is

$$Q_{23} = Q_{out} = h_3 - h_2 = 84.0 - 729.6$$

$$= -645.6 \ [-739.9] \text{ (B/lbm)}$$

The decrease in the heat transferred from the system is due to the higher temperature leaving the condenser.

The heat transfer into the system in the evaporator is

$$Q_{41} = Q_{in} = h_1 - h_4 = 593.7 - 84.0$$

$$= 509.7\ [604.3]\ (\text{B/lbm})$$

The loss of the refrigerating capability is due to the entropy increase that occurs in the expansion valve. This irreversible adiabatic process will also result in a considerable lowering of the coefficients of performance, where

$$\gamma_P = |Q_{out}|/|\Sigma W| = 645.6/135.9$$

$$= 4.75\ [5.46]$$

and

$$\gamma_R = |Q_{in}|/|\Sigma W| = 509.7/135.9$$

$$= 3.75\ [4.46]$$

Even though the use of an expansion valve results in a substantial decrease in the performance of the heat pump and refrigeration cycles, the simplicity of the vapor compressor cycle indicates its use for most engineering applications.

REFRIGERANT PROPERTIES

We have established that water is not a satisfactory refrigerant due to the high temperatures leaving the condenser and the inability to use water for temperatures below 32 F. The properties required for a satisfactory refrigerant may be divided into thermodynamic and nonthermodynamic considerations. The thermodynamic properties of importance include a sufficiently high critical temperature to permit efficient condenser operation, high values of the enthalpy change when converting the saturated liquid to a saturated vapor (latent heat), low specific volumes in the vapor phase to reduce compressor size, and reasonable temperature values for the pressures encountered in the cycle.

Nonthermodynamic properties include the cost of the working substance, its lubricating properties in the compressor, how it reacts with the materials used in the system, whether or not it is explosive in air, its characteristics with regard to its being toxic or poisonous, and if there is leakage, whether it is easily detected.

Many materials satisfy the above requirements. Good commercial refrigerants include ammonia and sulfur dioxide. Ammonia can be explosive and both materials are toxic, but they have desirable thermodynamic properties that permit their use in carefully monitored industrial systems.

For home use, Freon-12 and other materials of a like nature are desirable not only for their thermodynamic properties but because they are nontoxic and their leakage from the system is not difficult to detect.

Example 12.4. Determine the coefficients of performance when Freon-12 replaces ammonia in the vapor compression cycle where $t_1 = -50$ F, $x_1 = 100\%$, and $p_2 = p_3 = 70$ psia.

From Appendix F for Freon-12, $h_1 = h_{g1} = 71.805$ B/lbm and $s_1 = s_{g1} = 0.17533$ B/lbm-R, where $x_1 = 100\%$.

From the superheat tables at p = 60 psia,

t = 80	t = 120	(F)
h = 87.330	h = 93.731	(B/lbm)
s = 0.17497	s = 0.18461	(B/lbm-R)

so that for $s_1 = s_2 = 0.17533$ B/lbm-R,

$$f = (0.17533 - 0.17497)/(0.18461 - 0.17497) = 0.037$$

and

$$h = 87.330 + 0.037(93.731 - 87.330) = 87.566 \text{ (B/lbm)}$$

For a pressure of 80 psia,

t = 80	t = 120	(F)
h = 86.316	h = 92.945	(B/lbm)
s = 0.16885	s = 0.18070	(B/lbm-R)

and, as before, f = 0.546 and h = 89.942 B/lbm.

Interpolating to a pressure of 70 psia gives

p = 60	p = 80	(psia)
h = 87.566	h = 89.942	(B/lbm)

and f = 0.5, so that

$$h_2 = 87.566 + 0.5(89.942 - 87.566)$$

$$= 88.754 \text{ (B/lbm)}$$

for $p_2 = 70$ psia and $s_2 = 0.17533$ B/lbm-R.

Leaving the condenser, the properties of Freon-12 as a saturated liquid at a pressure of 70 psia are calculated from

$t = 55$	$t = 60$	(F)
$h_f = 20.634$	$h_f = 21.766$	(B/lbm)
$p = 66.743$	$p = 72.433$	(psia)

so that

$$f = (70 - 66.743)/(72.433 - 66.743) = 0.572$$

and

$$h_3 = h_4 = 20.634 + 0.572(21.766 - 20.634)$$

$$= 21.282 \text{ (B/lbm)}$$

The work of the compressor is

$$\Sigma W = W_c = h_1 - h_2 = 71.805 - 88.754$$

$$= -16.949 \text{ (B/lbm)}$$

Also

$$Q_{in} = h_1 - h_4 = 71.805 - 21.282$$

$$= 50.523 \text{ (B/lbm)}$$

and

$$Q_{out} = h_3 - h_2 = 21.282 - 88.754$$

$$= -67.472 \text{ (B/lbm)}$$

The coefficients of performance are

$$\gamma_P = |Q_{out}|/|\Sigma W| = 67.472/16.949$$

$$= 3.98$$

and

$$\gamma_R = |Q_{in}|/|\Sigma W| = 50.523/16.949$$

$$= 2.98$$

Comparable values using ammonia as the refrigerant were 4.75 and 3.75.

COMPRESSOR EFFICIENCY

The efficiency of a compressor is

$$e_c = W_{ideal}/W_{actual} = (h_2 - h_1)/(h_2' - h_1)$$

which assumes that $\Delta KE, \Delta PE = 0$ (Fig. 12.3).

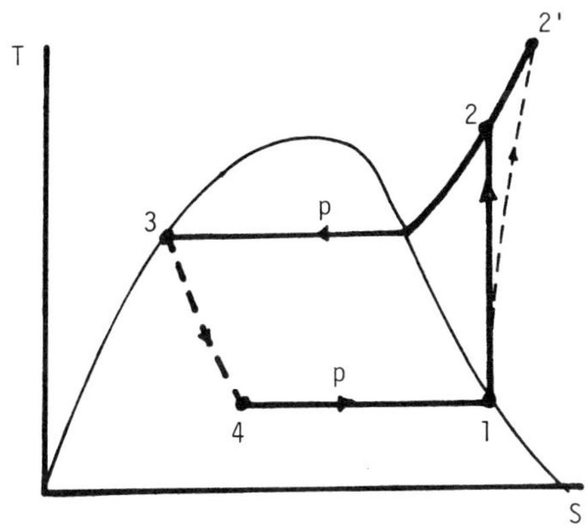

Figure 12.3

Example 12.5. Determine Q_{in}', Q_{out}', $\Sigma W'$, γ_P', and γ_R' using the data in the previous two examples and using ammonia and Freon-12 as the refrigerants. The compressor efficiency is 80%. For these examples, t_1 = -50 F, x_1 = 100%, and $p_2 = p_3$ = 70 psia.

When ammonia was used as the working substance, $h_1 = 593.7$, $h_2 = 729.6$, and $h_3 = h_4 = 84.0$ B/lbm.

The heat added to the system is not affected by the compressor efficiency so that

$$Q_{in}' = Q_{in} = h_1 - h_4 = 593.7 - 84.0 = 509.7 \text{ (B/lbm)}$$

The actual work of the cycle is

$$\Sigma W' = \Sigma W/e_c = (h_1 - h_2)/e_c = (593.7 - 729.6)/0.8$$

$$= -169.9 \ [-135.9] \text{ (B/lbm)}$$

The actual work of the cycle is also given by

$$\Sigma W' = h_1 - h_2' = 593.7 - h_2' = -169.9$$

so that

$$h_2' = 763.6 \text{ (B/lbm)}$$

The heat rejected in the condenser is

$$Q_{out}' = h_3 - h_2' = 84.0 - 763.6 = -679.6 \ [-645.6] \text{ (B/lbm)}$$

The coefficients of performance are

$$\gamma_P' = |Q_{out}'|/|\Sigma W'| = 679.6/169.9 = 4.00 \ [4.75]$$

and

$$\gamma_R{}' = |Q_{in}|/|\Sigma W'| = 509.7/169.9 = 3.00\ [3.75]$$

Values within the square brackets are for the cycle with a compressor efficiency of 100%.

For Freon-12, $h_1 = 71.805$, $h_2 = 88.754$, and $h_3 = h_4 = 21.282$ (B/lbm). For a compressor efficiency of 80%,

$$e_c = (h_2 - h_1)/(h_2{}' - h_1) = (88.754 - 71.805)/(h_2{}' - 71.805) = 0.8$$

so that

$$h_2{}' = 92.991\ (\text{B/lbm})$$

The heat and work terms are

$$Q_{in} = Q_{in}{}' = h_1 - h_4 = 71.805 - 21.282 = 50.523\ (\text{B/lbm})$$

$$\Sigma W' = W_c{}' = h_1 - h_2{}' = 71.805 - 92.991 = -21.186\ [-16.945]\ (\text{B/lbm})$$

and

$$Q_{out} = h_3 - h_2{}' = 21.282 - 92.991 = -71.709\ [67.463]\ (\text{B/lbm})$$

The coefficients of performance are

$$\gamma_P{}' = |Q_{out}{}'|/|\Sigma W'| = 71.709/21.186 = 3.38\ [3.98]$$

$$\gamma_R{}' = |Q_{in}{}'|/|\Sigma W'| = 50.523/21.186 = 2.38\ [2.98]$$

SYSTEM PERFORMANCE

Refrigeration units are usually classified in terms of tons of refrigeration, where

1 ton of refrigeration = 200 B/min = 3.51 kW

Example 12.6. Determine the horsepower required to drive the compressor in Example 12.5 where the capacity of the system is 5 tons.

Using ammonia as the working substance, the mass flow is

$$m' = \text{refrigeration load}/Q_{in}{}' = [5(200)]/509.7$$

$$= 1.962\ (\text{lbm/min})$$

The total work required to drive the compressor is

$$\Sigma W_t{}' = m'\Sigma W' = 1.962(-169.9) = -333.3\ (\text{B/min})$$

and the horsepower required to drive the compressor is

$$\Sigma W_t' = [-333.3(778)]/33{,}000 = -7.86 \text{ (HP)}$$

By using Freon-12 as the working substance,

$$m' = \text{refrigeration load}/Q_{in}' = [5(200)]/50.523$$
$$= 19.79 \text{ (lbm/min)}$$

The total work required to drive the compressor is

$$\Sigma W_t' = m'\Sigma W' = 19.79(-21.186) = -419.3 \text{ (B/lbm)}$$

which is equivalent to -9.89 HP.

The power savings where ammonia is used as the working substance are

$$\Delta HP = (9.89 - 7.86)/7.86 = 0.258 \quad \text{or} \quad 25.8\%$$

Example 12.7. If the compressor in Example 12.6 has a stroke of 8 in. and operates at a speed of 400 rpm, determine the compressor diameter for ammonia and Freon.

For ammonia at $t_1 = -50$ F and $x_1 = 100\%$,

$$v_1 = v_{g1} = 33.08 \text{ (ft}^3\text{/lbm)}$$

The displacement of the compressor is

$$V_D = m'v_1 = 1.962(33.08) = 64.90 \text{ (ft}^3\text{/min)}$$

and the piston diameter is

$$V_D = (\Pi/4)D^2LN$$

so that

$$64.90 = (\Pi/4)D^2(8/12)(400)$$

and

$$D = 0.557 \text{ (ft)} = 6.68 \text{ (in.)}$$

For Freon-12, at the same inlet conditions to the compressor,

$$v_1 = v_{g1} = 4.9742 \text{ (ft}^3\text{/lbm)}$$

The displacement of the compressor is

$$V_D = m'v_1 = 19.79(4.9742) = 98.44 \text{ (ft}^3\text{/min)}$$

and the piston diameter is

$$V_D = (\Pi/4)D^2LN$$

so that

$$98.44 = (\Pi/4)D^2(8/12)(400)$$

and

$$D_2 = 0.686 \text{ (ft)} = 8.24 \text{ (in.)}$$

The use of ammonia not only saves power, but the compressor is significantly smaller. For home units, Freon is still the preferred refrigerant since it is safer to use than ammonia. The ammonia system has a flow rate of 1.962 lbm/min and the Freon-12 a flow rate of 19.79 lbm/min.

pH DIAGRAM

The Mollier chart was used with the Rankine cycle to eliminate the need to interpolate enthalpy values within the steam tables. The pH diagram serves the same purpose with respect to the refrigeration or heat pump cycle (Fig. 12.4). A pH diagram for Freon-12 is presented in Appendix H. The vertical axis lists the value of the pressure plotted to a logarithmic scale. The horizontal axis gives values of the enthalpy plotted on a linear scale with a change of scale located at 60 B/lbm. Within the wet region, values of the quality x are plotted. In the superheat region, constant temperature, constant entropy, and constant volume lines are plotted. The constant temperature lines are nearly vertical, especially at low pressure, and the

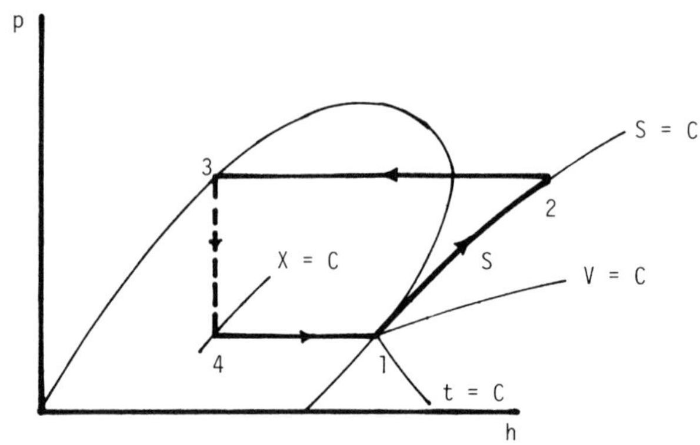

Figure 12.4

lines of constant volume are almost horizontal at high values of the enthalpy.

Since the pressures within the evaporator and condenser are constant, these processes, 41 and 23, are horizontal lines, as shown in Fig. 12.4. For an isentropic compressor the constant entropy path is plotted from state 1 to state 2. The expansion valve is represented by a constant enthalpy or throttling process, from state 3 to state 4, which is a vertical line. The specific volume at state 1, which may be needed to size the compressor, may also be obtained from the pH diagram.

Example 12.8. By use of the pH diagram, determine the coefficients of performance for a Freon-12 system where t_1 = -50 F, x_1 = 100%, and p_2 = 70 psia. The compressor efficiency is 100%. Determine the specific volume at state 1.

From the pH diagram in Appendix H, h_1 = 71.5, h_2 = 88, and $h_3 = h_4$ = 21 B/lbm.

The coefficients of performance are

$$\gamma_P = |Q_{out}|/|\Sigma W| = (h_2 - h_3)/(h_2 - h_1)$$

$$= (88 - 21)/(88 - 72) = 4.2 \ [4.0]$$

and

$$\gamma_R = |Q_{in}|/|\Sigma W| = (h_1 - h_4)/(h_2 - h_1)$$

$$= (71.5 - 21)/(88 - 72) = 3.2 \ [3.0]$$

Values within the square brackets are based on tabular calculations.

From the pH diagram, the specific volume at state 1 is

$$v_1 = 5.0 \ [4.97] \ (ft^3/lbm)$$

VAPOR COMPRESSION CASCADE SYSTEM

If two-stage compression is used with a closed heat exchanger (Figs. 12.5 and 12.6), the coefficients of performance of the cycle will be increased. The net work of the cycle is

$$\Sigma W = W_{LP} + W_{HP}$$

and the coefficients of performance are

$$\gamma_P = |Q_{in}|/|\Sigma W| \quad \text{and} \quad \gamma_R = |Q_{out}|/|\Sigma W|$$

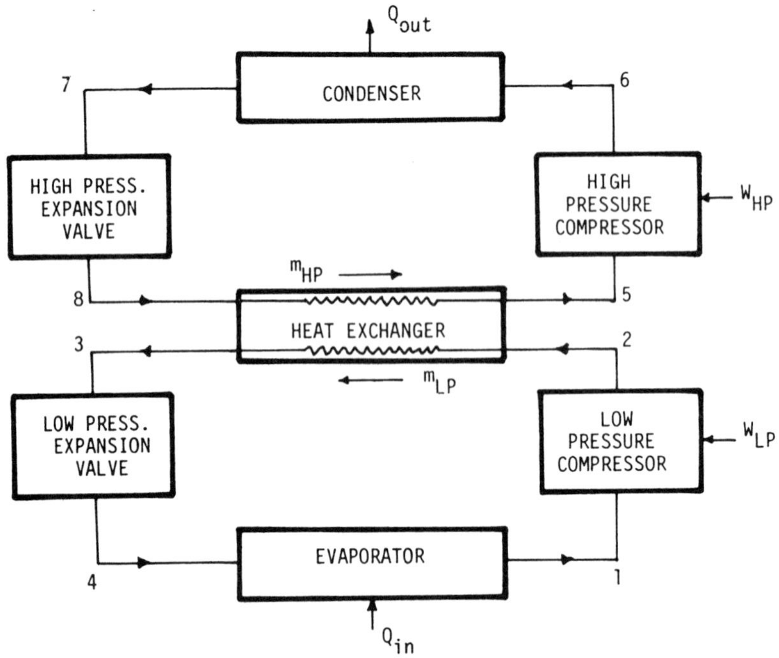

Figure 12.5

Example 12.9. Given is a Freon-12 system where $t_1 = -50$ F, $t_3 = t_5 = 0$ F, $x_1 = x_5 = 100\%$, and $p_7 = 70$ psia. Compressor efficiencies are 100%. Determine γ_R and γ_P for the cascade system.

For $t_1 = -50$ F and $x_1 = 100\%$,

$$h_1 = h_{g1} = 71.805 \text{ (B/lbm)} \quad \text{and} \quad s_1 = s_{g1} = 0.17533$$

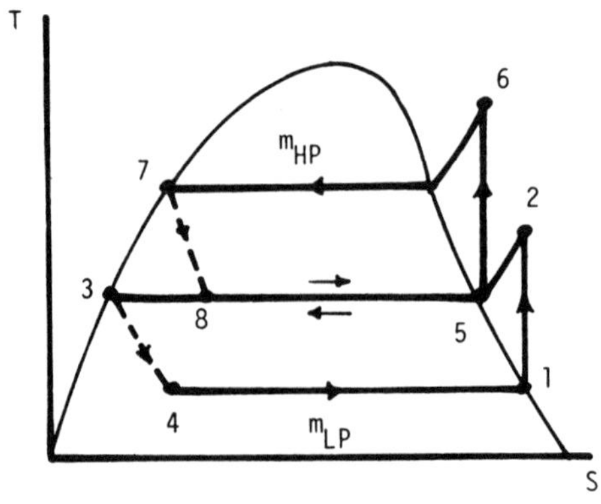

Figure 12.6

For state 2, where $t_3 = 0$ F and $p_2 = p_3 = 23.849$ psia,

$$s_2 = s_1 = 0.17533 \text{ (B/lbm-R)}$$

and by double interpolation

$$h_2 = 80.137 \text{ (B/lbm)}$$

For $t_3 = 0$ F and $x_3 = 0\%$,

$$h_3 = h_4 = h_{f3} = 8.521 \text{ (B/lbm)}$$

At state 5 for $t_5 = 0$ F and $x_5 = 100\%$,

$$h_5 = h_{g5} = 77.271 \text{ (B/lbm)} \quad \text{and} \quad s_5 = s_{g5} = 0.16888 \text{ (B/lbm)}$$

For $p_6 = 70$ psia,

$$s_6 = s_5 = 0.16888 \text{ (B/lbm-R)}$$

so that

$$h_6 = 85.236 \text{ (B/lbm)}$$

For $p_7 = 70$ psia and $x_7 = 0$,

$$h_7 = h_8 = 21.282 \text{ (B/lbm)}$$

The heat balance for the heat exchanger is

$$m_{LP}(h_2 - h_3) = m_{HP}(h_5 - h_8)$$

and for a mass flow of $m_{LP} = 1$ lbm in the low-pressure loop,

$$\begin{aligned} m_{HP} &= (h_2 - h_3)/(h_5 - h_8) \\ &= (80.137 - 8.521)/(77.271 - 21.282) \\ &= 1.279 \end{aligned}$$

The net work of the cycle is

$$\begin{aligned} \Sigma W = W_{LP} + W_{HP} &= m_{LP}(h_1 - h_2) + m_{HP}(h_5 - h_6) \\ &= 1(71.805 - 80.137) + 1.279(77.271 - 85.236) \\ &= -18.52\ [-16.94] \text{ (B/lbm)} \end{aligned}$$

The heat transferred in the evaporator is

$$Q_{in} = h_1 - h_4 = 71.805 - 8.521$$

$$= 63.284\ [50.523]\ (B/lbm)$$

The coefficient of performance, as a refrigerator, is

$$\gamma_R = |Q_{in}|/|\Sigma W| = 63.284/18.52$$

$$= 3.42\ [2.98]$$

Values within the square brackets are for a cycle using only one stage of compression.

The heat transfer in the condenser is

$$Q_{out} = m_{HP}(h_7 - h_6) = 1.279(21.282 - 85.236)$$

$$= -81.797$$

so that

$$\gamma_P = |Q_{out}|/|\Sigma W| = 81.797/18.52 = 4.42\ [3.98]$$

Note that

$$\gamma_P - \gamma_R = 1.0$$

CASCADE SYSTEM, OPEN HEAT EXCHANGER

Instead of a closed heat exchanger, assume that an open heat exchanger is used as in Fig. 12.7.

The energy balance for the heat exchanger, assuming a mass flow of m_{LP} = 1 lbm in the low-pressure loop, is

$$h_2 + mh_{g3} = (1 + m)h_5$$

where h_5 is the enthalpy entering the high-pressure compressor and h_{g3} is the enthalpy of a saturated vapor at the open heating pressure p_2. For this cycle it is necessary to include a separator between the two expansion valves. Only the saturated vapor, m, is supplied to the open heater, and only saturated liquid is admitted to the low-pressure expansion valve. See Fig. 12.6 for the TS diagram; state 5 is in the superheat region.

The heat balance of the separator is

$$(1 + m)h_8 = h_3 + mh_{g3}$$

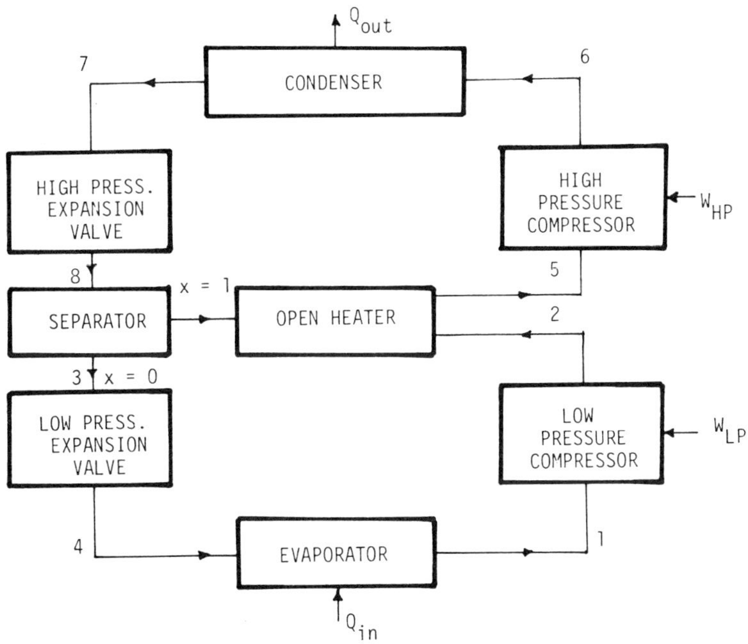

Figure 12.7

The work and heat terms are

$$Q_{in} = h_1 - h_4$$

$$Q_{out} = (1 + m)(h_6 - h_7)$$

and

$$\Sigma W = W_{LP} + W_{HP} = (h_1 - h_2) + (1 + m)(h_5 - h_6)$$

Example 12.10. Given the pressures and temperatures used in Example 12.9, determine the coefficients of performance for the cycle with an open heater and separator.

From Example 12.9, $t_1 = -50$ F, $x_1 = 100\%$, $h_1 = 71.805$ B/lbm, $t_3 = 0$ F, $p_2 = p_3 = 23.849$ psia, $h_2 = 80.137$ B/lbm, and $h_3 = h_4 = 8.521$ B/lbm. $p_7 = p_6 = 70$ psia, and $h_7 = h_8 = 21.282$ B/lbm. For $t_3 = 0$ F, $h_{g3} = 77.271$ B/lbm.

Substituting the known enthalpy values in the heater and separator energy balance equations yields

$$h_2 + mh_{g3} = (1 + m)h_5$$

$$80.137 + m(77.271) = (1 + m)h_5$$

and

$$(1 + m)h_8 = h_3 + mh_{g3}$$

$$(1 + m)(21.282) = 8.521 + m(77.271)$$

The value of m from the separator equation is

$$m = 0.228$$

The value of h_5 is

$$h_5 = [1/(1 + 0.228)][80.137 + 0.228(77.27)]$$

$$= 79.605 \text{ (B/lbm)}$$

The pressure and enthalpy at state 5 are $p_5 = p_2 = 23.849$ psia and $h_5 =$ 79.605 B/lbm, which is in the superheat region.

By double interpolation from the superheat tables

$$s_5 = 0.17502 \text{ (B/lbm-R)}$$

The properties, leaving the high-pressure compressor, are $p_6 = 70$ psia and $s_5 = s_6 = 0.17502$ B/lbm-R. Again by double interpolation

$$h_6 = 88.562 \text{ (B/lbm)}$$

The heat added to the cycle is

$$Q_{in} = h_1 - h_4 = 71.805 - 8.521 = 63.284 \text{ (B/lbm)}$$

The heat rejected from the cycle is

$$Q_{out} = (1 + m)(h_7 - h_6) = (1 + 0.228)(21.282 - 88.562)$$

$$= -82.620 \text{ (B/lbm)}$$

The net work is

$$\Sigma W = (h_1 - h_2) + (1 + m)(h_5 - h_6)$$

$$= (71.805 - 80.137) + (1 + 0.228)(79.605 - 88.562)$$

$$= -19.331$$

The coefficients of performance are

$$\gamma_R = |Q_{in}|/|\Sigma W| = 63.284/19.331 = 3.27$$

and

$$\gamma_P = |Q_{out}|/|\Sigma W| = 82.620/19.331 = 4.27$$

The open heater-separator system has coefficients of performance that are nearly equal to those of the closed heater system, 3.42 and 4.42, and both are superior to the single-stage compressor system, where $\gamma_R = 2.98$ and $\gamma_P = 3.98$.

EVAPORATIVE COOLING

The evaporative cooling system, using water as a working substance, is sketched in Fig. 12.8. The principal use of this system is for the production of cold liquid water. The components consist of a steam ejector, which is a device capable of creating a low-pressure zone at the narrowest section of a converging-diverging nozzle; a pump to circulate the chilled water; a heat exchanger which transfers the heat from the water to the system being cooled; and a spray system which breaks the returning water into small droplets.

The mass balance equation for the system is

$$m_{in} = m_{out}$$

$$m_2 + m_3 = m_1 + m_4$$

where m_1, m_2, and m_3 are subcooled liquids and m_4 is a saturated vapor. The mass flows into and out of the heat exchanger are identical so that

$$m_1 = m_2$$

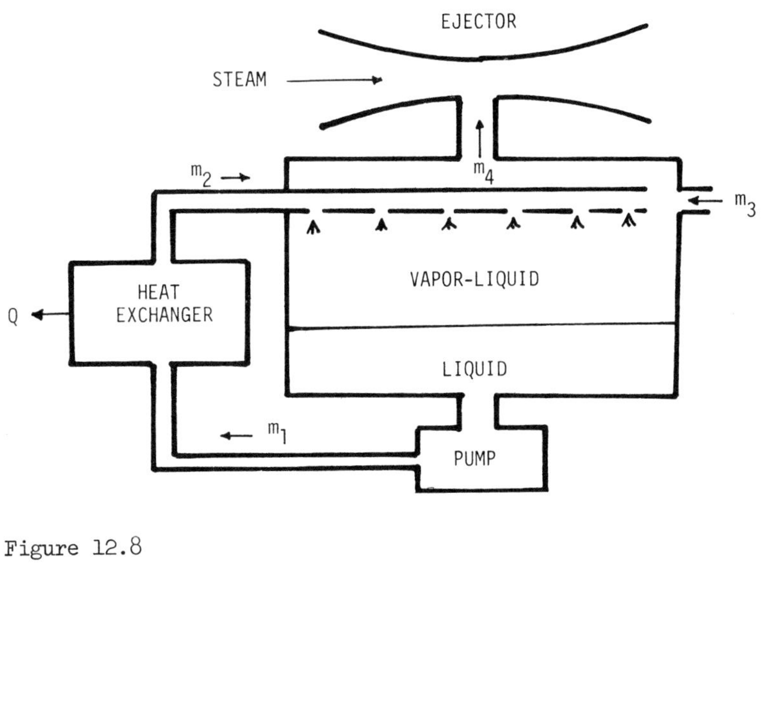

Figure 12.8

The vapor drawn off by the ejector m_4 must be replaced by the introduction of new liquid water m_3, so that

$$m_3 = m_4$$

The energy balance equation is

$$m_2h_2 + m_3h_3 = m_1h_1 + m_4h_4$$

The mass of vapor leaving the system is

$$m_4 = xm_2 = m_3$$

where x is the quality of the vapor-liquid mixture in the evaporator. In effect, the vapor leaving the system is the result of introducing the water droplets into a low-pressure area so that a portion of the water "flashes" into vapor.

The energy balance equation is now

$$m_2h_2 + xm_2h_3 = m_2h_1 + xm_2h_4$$

so that

$$h_2 + xh_3 = h_1 + xh_4$$

For the subcooled liquids m_1, m_2, and m_3,

$$h_1 = h_{f1} \qquad \text{at } t_1$$

$$h_2 = h_{f2} \qquad \text{at } t_2$$

$$h_3 = h_{f3} \qquad \text{at } t_3$$

and

$$h_4 = h_{g4} \qquad \text{at } t_1$$

Example 12.11. Given is the evaporative cooling system in Fig. 12.8, where $t_1 = 45$, $t_2 = 70$, and $t_3 = 65$ F. The heat to be transferred from the chilled water is 2000 B/min or 10 tons. Determine the quantity of water to be circulated, the amount of makeup water, and the pressure required in the evaporator.

The quantity of water to be circulated is, c = 1 B/lbm-R,

$$Q = m_2(h_2 - h_1) = m_2c(t_2 - t_1)$$

$$= m_2(1)(70 - 45) = 2000$$

or

$$m_2 = 80.0 \text{ (lbm/min)}$$

For an evaporator temperature of 45 F, the required evaporator pressure is 0.1475 psia or 0.3 in.-Hg-abs. The ejector also serves to draw off any air that may leak into the system. Air leaking into an evaporator or condenser reduces the performance of either device, and an ejector is the standard technique for air removal.

The enthalpies for the water are

$$h_{f1} = 13.1 \text{ (B/lbm)} \qquad \text{at } t_1 = 45 \text{ (F)}$$

$$h_{f2} = 38.0 \text{ (B/lbm)} \qquad \text{at } t_2 = 70 \text{ (F)}$$

$$h_{f3} = 33.0 \text{ (B/lbm)} \qquad \text{at } t_3 = 65 \text{ (F)}$$

and

$$h_{g4} = 1081.5 \text{ (B/lbm)} \qquad \text{at } t_1 = 45 \text{ (F)}$$

From Eq. (12.4), the quality is

$$h_{f2} + xh_{f3} = h_{f1} + xh_{g4}$$

$$38.0 + 33.0x = 13.1 + 1081.5x$$

so that

$$x = 0.0237$$

The mass of vapor leaving the system or the amount of makeup water is

$$m_4 = m_3 = xm_2 = 0.0237(80.0)$$

$$= 1.896 \text{ (lbm/min)}$$

$$= 0.227 \text{ (gal/min)} \quad \text{or} \quad 13.7 \text{ (gal/hr)}$$

The liquid water in the bottom of the evaporator is in equilibrium with the liquid-vapor mixture and serves as a reserve supply of chilled water in case the heat load should be changed. The water level in the evaporator may be maintained by means of a float device that opens a valve in the makeup water line if the demand for chilled water suddenly increases.

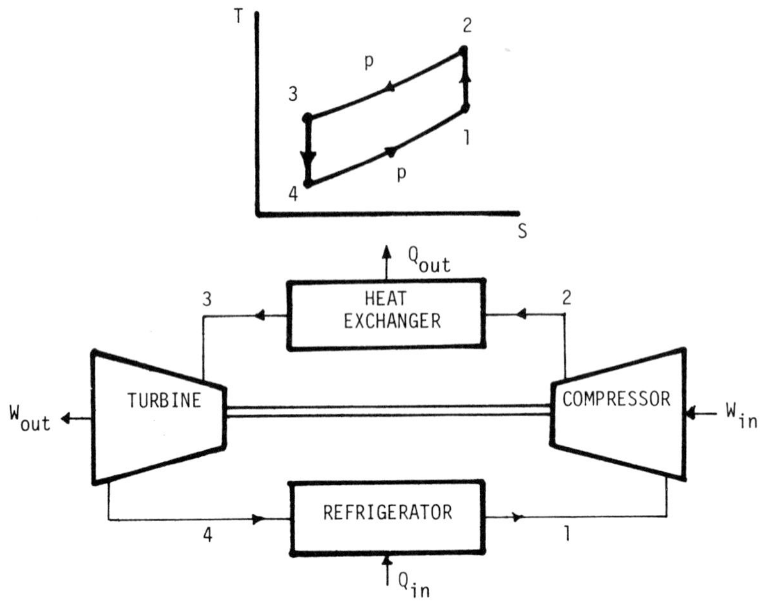

Figure 12.9

GAS REFRIGERATION

As noted in our discussion of air engines in Chap. 8, the isentropic expansion of air produces an extremely low temperature for the air leaving the engine. Instead of an air engine, we may also produce cold air by means of the reversed gas turbine or Brayton cycle (Fig. 12.9).

The work added to the system, in the compressor, is

$$W_{in} = W_{12} = (h_1 - h_2)$$

and the work supplied by the system, in the turbine, is

$$W_{out} = W_{34} = h_3 - h_4$$

The heat added in the refrigerator is

$$Q_{in} = Q_{41} = h_1 - h_4$$

and the heat transfer from the heat exchanger is

$$Q_{out} = Q_{23} = h_3 - h_2$$

The coefficients of performance tend to be relatively low, and use is generally restricted to systems that have a large cooling load and also require equipment that meets low size or weight requirements. The cooling of aircraft cockpits has been accomplished by means of the gas refrigeration system.

It is mandatory that the gas used in this system be absolutely dry, since moisture will freeze below 32 F and block the air lines.

Example 12.12. Given are p_1 = 15 psia and t_1 = 60 F. The pressure ratio in the air compressor is r_p = 5. The air is cooled to 80 F in the heat exchanger. Determine the coefficients of performance as a refrigerator.

The temperature leaving the compressor, for e_c = 100%, is

$$T_2 = T_1(p_2/p_1)^{(k-1)/k} = T_1 r_p^{(k-1)/k}$$

$$= 520(5)^{0.286} = 824 \text{ (R)}$$

The temperature leaving the turbine, for e_t = 100%, is

$$T_4 = T_3(p_4/p_3)^{(k-1)/k} = T_3(p_1/p_2)^{(k-1)/k} = T_3/r_p^{(k-1)/k}$$

$$= 540/(5)^{0.286} = 341 \text{ (R)}$$

The work required to drive the compressor is

$$W_{in} = W_{12} = h_1 - h_2 = c_p(T_1 - T_2)$$

$$= 0.24(520 - 824) = -73.0 \text{ (B/lbm)}$$

and the work obtained from the turbine is

$$W_{out} = W_{34} = h_3 - h_4 = c_p(T_3 - T_4)$$

$$= 0.24(540 - 341) = 47.8 \text{ (B/lbm)}$$

The net work of the system is

$$\Sigma W = W_{in} + W_{out} = -73.0 + 47.8 = -25.2 \text{ (B/lbm)}$$

The heat added in the refrigerator is

$$Q_{in} = h_1 - h_4 = c_p(T_1 - T_4)$$

$$= 0.24(520 - 341) = 43.0 \text{ (B/lbm)}$$

and the coefficient of performance is

$$\gamma_R = |Q_{in}|/|\Sigma W| = 43.0/25.2$$

$$= 1.71$$

This value is well below the coefficients of performance obtained from the vapor compressor refrigerator system.

ABSORPTION REFRIGERATION SYSTEM

An absorption refrigeration system is sketched in Fig. 12.10. From the condenser, through the expansion valve, and leaving the evaporator, the system is identical to the vapor compressor refrigeration system. The absorber-generator replaces the compressor in the vapor compression cycle. This component of the apparatus is shown in greater detail in Fig. 12.11. Ammonia leaves the evaporator as a saturated vapor and is mixed with water in the absorber. In the absorption process, heat is removed by means of a cooling water coil in the absorber. Liquid that has a high ammonia content is pumped through a heat exchanger to the separator where the vaporous ammonia is sent to the condenser. The ammonia-rich liquid enters the generator where, by the addition of heat, more ammonia vapor is formed and sent to the condenser. The weak ammonia solution at the bottom of the generator is drained through the heat exchanger and in a regeneration process heats the fluid that is going to the separator. The weak ammonia solution is then returned to the absorber.

As compared to the vapor compression system the circulating pump between the absorber and the heat exchanger requires very little work. The

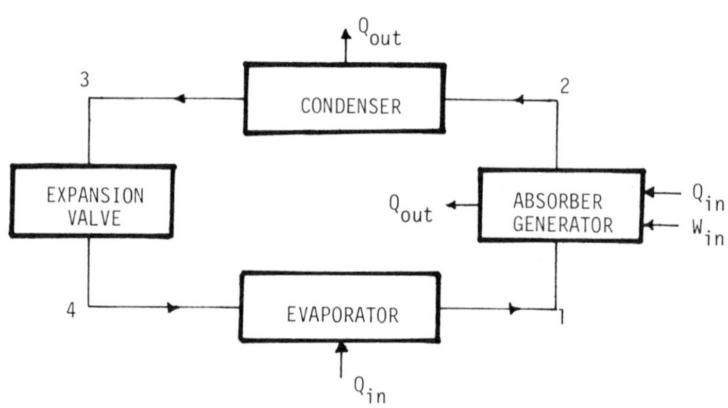

Figure 12.10

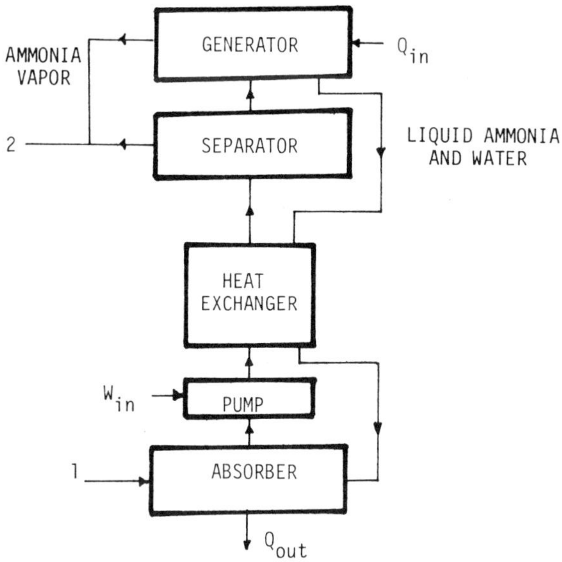

Figure 12.11

major energy input is in the generator, which may be any heating source, such as steam or a gas flame.

The Servel system replaces the circulating pump with a series of liquid seals using hydrogen to produce essentially equal pressure on both sides of the seal. This system permits the ammonia to move from the low pressure leaving the evaporator to the high pressure entering the condenser, without any working parts in the system. The Servel system relies on a gas flame to vaporize the ammonia-water mixture.

CRYOGENICS, GAS LIQUEFACTION

The low temperature required to liquefy gases requires a different system from that used in normal heating or cooling installations. The Linde cycle is sketched in Fig. 12.12. The compressor is actually a series of pressure stages to minimize the work. When leaving the compressor the vapor is cooled to as near an ambient temperature as possible. The gas then passes through a heat exchanger and through an expansion valve. On starting the unit, there is no liquid gas in the receiver, so that gas passes back through the heat exchanger and lowers the gas temperature before the expansion valve. This procedure continues until the temperatures entering the expansion valve have been reduced to such a point that liquid begins

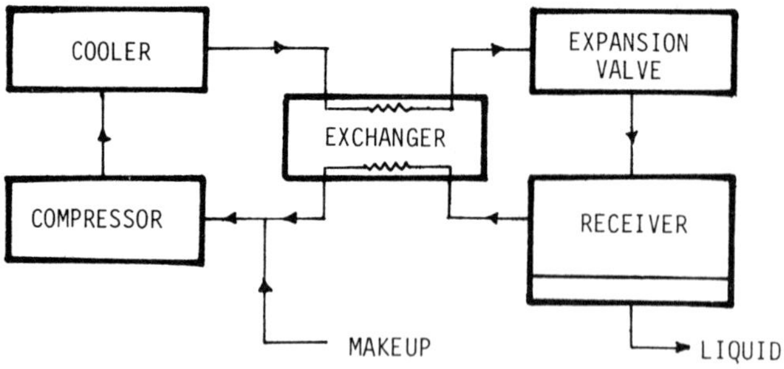

Figure 12.12

to be produced in the receiver. When steady state conditions are reached, the liquid gas may be drawn from the system and makeup gas supplied to the system.

PROBLEMS

12.1 A vapor compression system using ammonia operates with an evaporator temperature of 0 F and a condenser pressure of 100 psia. Determine Q_{in}, Q_{out}, ΣW, γ_P, and γ_R.

12.2 Repeat Prob. 12.1 using Freon-12 as the working substance.

12.3 Repeat Prob. 12.1 for a compressor with an efficiency of 80%.

12.4 Repeat Prob. 12.2 for a compressor efficiency of 80%.

12.5 Determine the horsepower required for a 2-ton refrigeration unit using the system in Prob. 12.3.

12.6 Determine the horsepower required for a 2-ton refrigeration unit using the system in Prob. 12.4.

12.7 The system in Prob. 12.5 operates at a speed of 220 rpm and has a piston stroke that is equal to the piston diameter. Determine the piston stroke.

12.8 Repeat Prob. 12.2 by use of the pH diagram.

12.9 Repeat Prob. 12.1 for an evaporator temperature of -40 F.

12.10 Given a cascade ammonia vapor compression system where the evaporator operates at -40 F, a heat exchanger temperature of 0 F, and a condenser pressure of 100 psia, determine the coefficients of performance.

12.11 Repeat Prob. 12.10 using an open heater. The open heater pressure is equivalent to a saturation temperature of 0 F.

12.12 An evaporative cooling system is to produce 20 gpm of chilled water at 50 F. Makeup water is supplied at 70 F, and the returning water temperature is 80 F. Determine the tons of refrigeration and the quantity of makeup water in gpm.

12.13 A reversed Brayton cycle is used in a refrigerator. p_1 = 15 psia, t_1 = 40 F, and r_p = 10. The air is cooled to 100 F in the heat exchanger. Determine the coefficient of performance γ_R.

12.14 Repeat Prob. 12.13 for e_c = 85% and e_t = 80%.

13

MIXTURES

In previous chapters we have defined the perfect and semiperfect gases as the working substances for use in internal and external combustion engines as well as air compressors and air engines.

In our discussion of the steam power plant and the refrigeration cycle we added vapors to our collection of working substances.

In this chapter we shall study mixtures of perfect gases and mixtures of perfect gases and vapors.

Processes and cycles may be evaluated for perfect gases if the values of the gas constant and specific heats are known. For pure gases or common gas mixtures – air, for example – these properties are generally available (Appendix B). We now wish to extend our capability in working with perfect gases so that any mixture of gases may be used as the working substance in a thermodynamic problem.

GRAVIMETRIC ANALYSIS

To determine the properties of a gas mixture the composition of the sample must be known. This information may be obtained by chemical separation of the components or by instruments that detect and measure the quantity of a particular gas in a gaseous mixture. We shall assume that, in some manner, the composition of the sample has been found. The transmission of information about the mixture may be presented either on a gravimetric or volumetric basis. The gravimetric analysis of a gas in a mixture is the ratio of the mass of the particular gas to the mass of the entire mixture or

$$y = m/m_m \tag{13.1}$$

where y = gravimetric analysis of the particular gas, m = mass of the particular gas (lbm), and m_m = mass of the gas mixture (lbm).

Example 13.1. A gas sample consists of 5 lbm of hydrogen, 3 lbm of oxygen, and 7 lbm of carbon dioxide. Determine the gravimetric analysis of the mixture.

$$m_m = m_{H_2} + m_{O_2} + m_{CO_2}$$

$$= 5 + 3 + 7 = 15 \text{ (lbm)}$$

$$y_{H_2} = m_{H_2}/m_m = 5/15 = 0.333 \quad \text{or} \quad 33.3\% \ H_2$$

$$y_{O_2} = m_{O_2}/m_m = 3/15 = 0.200 \quad \text{or} \quad 20.0\% \ O_2$$

and

$$y_{CO_2} = m_{CO_2}/m_m = 7/15 = 0.467 \quad \text{or} \quad 46.7\% \ CO_2$$

VOLUMETRIC ANALYSIS

Assume that a mixture of gases in a container (Fig. 13.1) could be separated so that each gas occupied a portion of the container. If the pressure and temperature of the component gases remain the same as that of the original mixture, the total of the volume of each component gas will equal the original volume of the mixture, or

$$V_m = V_a + V_b + V_c + \cdots \tag{13.2}$$

where V_a, V_b, V_c, $\cdots$ = volumes of the components (ft^3) and V_m = volume of the mixture (ft^3). Let

$$x = V/V_m \tag{13.3}$$

where x = volumetric analysis, V = volume of the component gas (ft^3), and V_m = volume of the gas mixture (ft^3). From Eq. (5.5), the perfect gas equation,

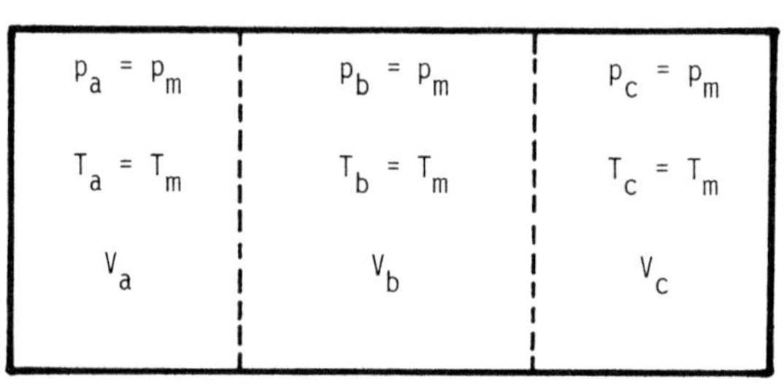

Figure 13.1

$$pV = mRT$$

and

$$R = \bar{R}/M$$

so that

$$pV = m\bar{R}T/M$$

where M is the molecular weight of the gas.

A mole of a gas is defined by

$$n = m/M \tag{13.4}$$

so that

$$pV = n\bar{R}T$$

where $\bar{R}$ is the universal gas constant and

$$\bar{R} = 1545 \text{ (ft-lbf/lb}_{\text{mole}}\text{-R)}$$

By solving for V,

$$V = \frac{n\bar{R}T}{p}$$

By substituting into Eq. (13.3),

$$x = V/V_m = (n\bar{R}T/p)/(n\bar{R}T/p)_m$$

since

$$p = p_m \quad \text{and} \quad T = T_m$$

$$x = n/n_m$$

where n = moles of the particular gas, n_m = moles of the gas mixture, and

$$n_m = n_a + n_b + n_c + \cdots$$

where n_a, n_b, n_c, = moles of the component gases.

Example 13.2. A gas mixture consists of 2 moles of helium, 3 moles of nitrogen, 4 moles of carbon monoxide, and 7 moles of air. Determine the volumetric analysis of the sample.

$$n_m = n_{He} + n_{N_2} + n_{CO} + n_{air}$$

$$= 2 + 3 + 4 + 7 = 16 \text{ (moles)}$$

$$x_{He} = n_{He}/n_m = 2/16 = 0.125 \quad \text{or} \quad 12.5\%$$

$$x_{N_2} = n_{N_2}/n_m = 3/16 = 0.188 \quad \text{or} \quad 18.8\%$$

$$x_{CO} = n_{CO}/n_m = 4/16 = 0.250 \quad \text{or} \quad 25.0\%$$

and

$$x_{air} = n_{air}/n_m = 7/16 = 0.437 \quad \text{or} \quad 43.7\%$$

The volumetric analysis may be obtained from various types of instruments. Experiments on gas samples generally result in volumetric rather than gravimetric data, and a technique is required to convert from one system to the other.

CONVERSION OF GAS ANALYSIS

Assume that the volumetric analysis of a gas mixture is known and that we wish to obtain the gravimetric analysis of the mixture. The following steps are necessary:

1. Express x in decimal form ($\text{mole}/\text{mole}_m$).
2. Obtain the molecular weight of each gas, M (lb/mole), from Appendix B.
3. Multiply x by M (lb/mole_m).
4. Sum the values of the products xM for each component gas ($\text{lb}_m/\text{mole}_m$). This is equal to M_m, the molecular weight of the mixture. (lb_m designates the mass of the mixture.)
5. Divide xM of each gas by M_m (lb/lb_m). This is the gravimetric analysis y in decimal form.

Example 13.3. Convert the previous volumetric analysis to a gravimetric analysis.

Gas	x	M	xM	y	Gravimetric Analysis, %
He	0.125	4	0.500	0.0197	2.0 He
N_2	0.188	28	5.264	0.2069	20.7 N_2
CO	0.250	28	7.000	0.2752	27.5 CO
Air	0.437	29	12.673	0.4982	49.8 Air
		M_m =	25.437	1.0000	100.0

where $y_{He} = 0.500/25.437 = 0.0197$, etc. If the gravimetric analysis is known, the volumetric analysis may be obtained by reversing the above procedure.

Gas	y	M	y/M	x
He	0.0197	4	0.00492	0.125
N_2	0.2069	28	0.00739	0.188
CO	0.2752	28	0.00983	0.250
Air	0.4982	29	0.01718	0.437
		$1/M_m$ =	0.03932	1.000

where $x_{He} = 0.00492/0.0393 = 0.125$, etc. Since air is approximately 79% N_2 and 21% O_2 by volume,

$$0.79(0.437) = 0.345 \text{ (mole of } N_2\text{) per mole mixture}$$

$$0.21(0.437) = 0.092 \text{ (mole of } O_2\text{) per mole mixture}$$

where the moles of air are 0.437, so that the volumetric and gravimetric analyses are

Gas	x	M	xM	y	Gravimetric, %
He	0.125	4	0.500	0.0197	2.0 He
N_2	0.533	28	14.924	0.5883	58.8 N_2
CO	0.250	28	7.000	0.2759	27.6 CO
O_2	0.092	32	2.944	0.1161	11.6 O_2
	1.000		25.368	1.0000	100.0

where $x_{N_2} = 0.188 + 0.345 = 0.533$ mole of N_2/mol_m. Note that $M_m = 25.4$ lbm/mole$_m$ is the same in both cases and that the gravimetric analyses of He and CO remain constant, neglecting round-off errors.

PROPERTIES OF GAS MIXTURES

From the previous calculations the molecular weight of the mixture may be obtained from either a gravimetric or volumetric gas analysis.

Since

$$R = \bar{R}/M$$

the gas constant for the mixture is

$$R_m = \bar{R}/M_m \tag{13.5}$$

<u>Example 13.4</u>. Given the gas mixture in Example 13.3, determine the gas constant for the mixture.

For a molecular weight M_m of 25.368, the value of the specific gas constant is

$$R_m = \bar{R}/M_m = 1545/25.368$$

$$= 60.90 \text{ (ft-lbf/lbm-R)}$$

<u>Example 13.5</u>. See Fig. 13.2. Determine the properties of the mixture after the valve is opened. The system is insulated so that Q = 0.

	O_2	H_2
p	20	30 psia
t	60	120 F
V	5	15 ft^3

Therefore

$$m_{O_2} = (pV/RT)_{O_2} = [20(144)(5)]/[48.3(520)]$$

$$= 0.573 \text{ (lbm)}$$

and

$$m_{H_2} = (pV/RT)_{H_2} = [30(144)(15)]/[767(580)] = 0.146 \text{ (lbm)}$$

When the valve is opened, the gas mixture is

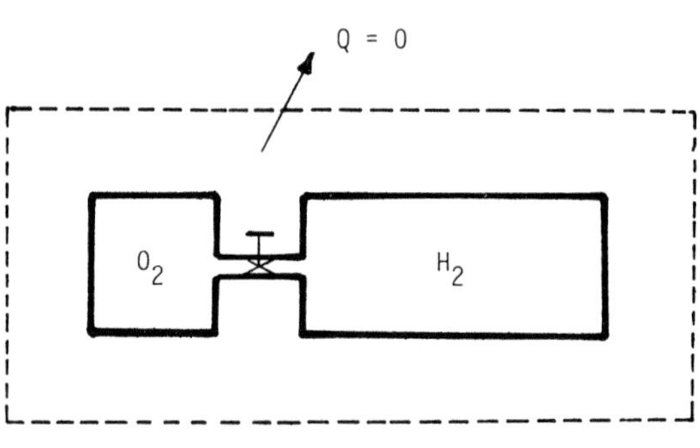

Figure 13.2

Gas	m	y	M	y/M
O_2	0.573	0.797	32	0.0249
H_2	0.146	0.203	2	0.1015
	0.719	1.000	$1/M_m$ =	0.1264

where $y_{O_2} = 0.573/0.719 = 0.797$ and $M_m = 1/0.1264 = 7.91$, so that

$$R_m = \bar{R}/M_m = 1545/7.91 = 195 \text{ (ft-lbf/lbm-R)}$$

The total volume of the mixture is

$$V_m = V_{O_2} + V_{H_2} = 5 + 15 = 20 \text{ (ft}^3\text{)}$$

Since Q and W_{nf} do not exist in this problem, the internal energy change of the mixture is

$$Q = \Delta U_{O_2} + \Delta U_{H_2} + W_{nf}$$

$$\Delta U_{O_2} + \Delta U_{H_2} = 0$$

so that

$$[mc_v(T_m - T)]_{O_2} + [mc_v(T_m - T)]_{H_2} = 0$$

and

$$0.573(0.157)(T_m - 520) + 0.146(2.465)(T_m - 580) = 0$$

or

$$T_m = 568 \text{ (R)}$$

and

$$p_m = (mRT/V)_m = [0.719(195)(568)]/[20(144)] = 27.65 \text{ (psia)}$$

SPECIFIC HEAT OF GAS MIXTURES

The specific heat of a gas mixture is given by

$$c_{pm} = y_a c_{pa} + y_b c_{pb} + y_c c_{pc} + \cdots \qquad (13.6)$$

$$c_{vm} = y_a c_{va} + y_b c_{vb} + y_c c_{vc} + \cdots \qquad (13.7)$$

where c_{pm} and c_{vm} are the specified heats at constant pressure and constant volume for the mixture, y_a, y_b, y_c, ··· = gravimetric analysis of component gases, and c_{pa}, c_{va}, ··· = specific heats of the component gases.

Since

$$k = c_p/c_v$$

then

$$k_m = c_{pm}/c_{vm} \tag{13.8}$$

Example 13.6. Determine c_{pm}, c_{vm}, and k_m for the gas in Example 13.3.

Gas	y	c_p	c_v	yc_p	yc_v
He	0.0197	1.234	0.741	0.024	0.015
N_2	0.5883	0.249	0.178	0.146	0.105
CO	0.2759	0.249	0.178	0.069	0.049
O_2	0.1161	0.219	0.157	0.025	0.018
	1.0000			c_{pm} = 0.264	c_{vm} = 0.187 B/lbm-R

Values of c_p and c_v are obtained from Appendix B. From Eq. (13.8)

$$k_m = c_{pm}/c_{vm} = 0.264/0.187 = 1.412.$$

DALTON'S LAW OF PARTIAL PRESSURE

Assume that the mixture of gases in Fig. 13.3 can be separated so that each component gas will occupy a container having the same volume as the mixture. The temperature of each component gas will equal that of the gaseous mixture. The component gas will exert a pressure, as given by the perfect gas law, of

$$p_a = (m_a R_a T_a)/V_a$$

Likewise,

$$p_b = (m_b R_b T_b)/V_b \quad \text{and} \quad p_c = (m_c R_c T_c)/V_c$$

The pressures p_a, p_b, and p_c are the partial pressures of the component gases in the mixture. Dalton's law states that

$$p_m = p_a + p_b + p_c \text{ (psia)} \tag{13.9}$$

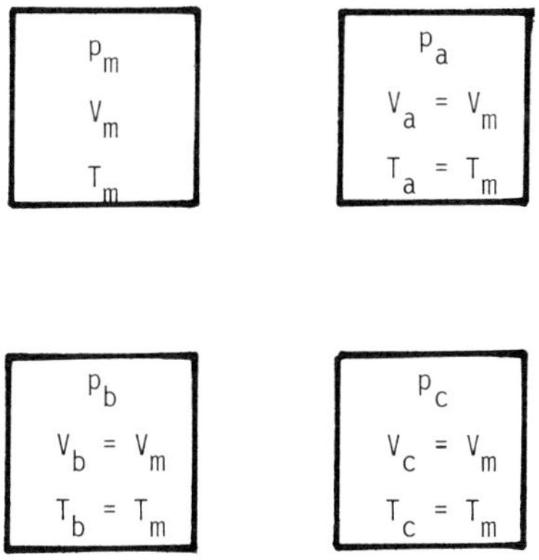

Figure 13.3

To determine the partial pressure for each component where

$$pV = n\bar{R}T$$

divide the partial pressure of each gas by the mixture pressure or

$$p_a/p_m = (n\bar{R}T/V)_a/(n\bar{R}T/V)_m$$

but

$$T_a = T_m \quad \text{and} \quad V_a = V_m$$

so that

$$p_a/p_m = n_a/n_m = x_a$$

and

$$p_a = x_a p_m$$

Likewise,

$$p_b = x_b p_m \quad \text{and} \quad p_c = x_c p_m$$

Example 13.7. Determine the partial pressures exerted by each gas in Example 13.3, where p_m = 20.0 psia.

Gas	x	p_m	p
He	0.125	20.00	2.50
N_2	0.533	20.00	10.66
CO	0.250	20.00	5.00
O_2	0.092	20.00	1.84
	1.000		20.00

Example 13.8. If the temperature of the mixture and each component gas is 40 F, determine the volume occupied by the mixture and each component at its partial pressure. The mass of mixture is 1 lb. $R_m = 60.90$ from Example 13.4.

$$V_m = (m_m R_m T_m)/p_m = [1(60.90)(500)]/[20(144)]$$

$$= 10.6\ (\text{ft}^3)$$

$$V_{He} = [0.0197(386)(500)]/[2.50(144)]$$

$$= 10.6\ (\text{ft}^3)$$

Likewise,

$$V_{N_2} = 10.6, \quad V_{CO} = 10.6, \quad \text{and} \quad V_{O_2} = 10.6\ (\text{ft}^3)$$

POLYTROPIC PROCESS FOR A GAS MIXTURE

Example 13.9. A perfect gas mixture consists of 50% He, 20% O_2, and 30% N_2 as measured by a volumetric analysis. The initial temperature and pressure of the mixture are 40 F and 15 psia. The final pressure and temperature are 340 F and 30 psia. For 4 lbm of the gas, determine the change in properties for the process.

Gas	x	M	xM	y	yc_p	yc_v
He	0.5	4	2.0	0.119	0.147	0.088
O_2	0.2	32	6.4	0.381	0.083	0.060
N_2	0.3	28	8.4	0.500	0.124	0.089
	1.0		M_m = 16.8	1.000	0.354	0.237

since

$$c_{pm} = \Sigma y c_p = 0.354\ (\text{B/lbm-R})$$

and

$$c_{vm} = \Sigma y c_v = 0.237\ (\text{B/lbm-R})$$

so that

$$k_m = c_{pm}/c_{vm} = 0.354/0.237 = 1.494$$

$$R_m = \bar{R}/M_m = 1545/16.8 = 91.96\ (\text{ft-lbf/lbm-R})$$

The subscript m for mixture will be omitted in the following calculations.

$$V_1 = mRT_1/p_1 = [4(91.96)(500)]/[15(144)]$$

$$= 85.1\ (ft^3)$$

$$V_2 = mRT_2/p_2 = [4(91.96)(800)]/[30(144)]$$

$$= 68.1\ (ft^3)$$

For a polytropic process

$$T_2/T_1 = (p_2/p_1)^{(n-1)/n}$$

$$800/500 = (30/15)^{(n-1)/n}$$

$$1.6 = (2)^{(n-1)/n}$$

$$(n - 1)/n = \ln(1.6)/\ln(2) = 0.6781$$

so that

$$n = 3.106$$

The specific heat for a polytropic process is given by

$$c_n = c_v[(k - 1)/(1 - n)]$$

$$= 0.237[(1.494 - 3.106)/(1 - 3.106)]$$

$$= 0.181\ (B/lbm\text{-}R)$$

The value of the property changes and heat and work are obtained from

$$U_2 - U_1 = mc_v(T_2 - T_1) = 4(0.237)(800 - 500) = 284.4\ (B)$$

$$H_2 - H_1 = mc_p(T_2 - T_1) = 4(0.354)(800 - 500) = 424.8\ (B)$$

$$Q = mc_n(T_2 - T_1) = 4(0.181)(800 - 500) = 217.2\ (B)$$

$$W_{nf} = Q - \Delta U = 217.2 - 284.4 = -67.2\ (B)$$

$$W_{sf} = Q - \Delta H = 217.2 - 424.8 = -207.6\ (B) \qquad (\Delta KE = \Delta PE = 0)$$

$$S_2 - S_1 = mc_n \ln(T_2/T_1) = 4(0.181) \ln(800/500)$$

$$= 0.340\ (B/R)$$

The above examples illustrate the principles involved in determining the changes of property values for an arbitrary mixture of gases which undergo a polytropic process.

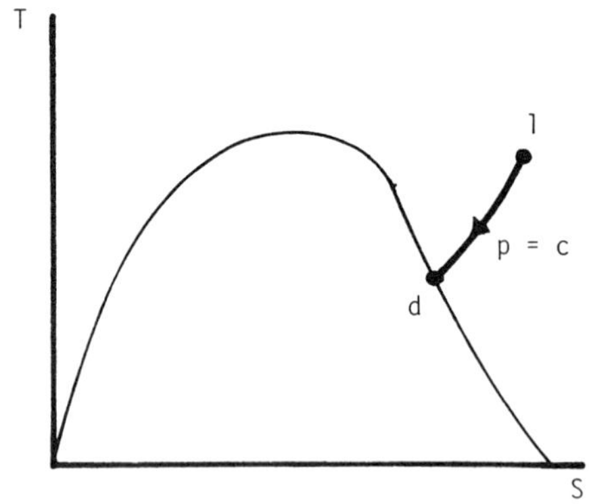

Figure 13.4

PERFECT GAS AND VAPOR MIXTURES

We shall now consider mixtures of perfect gases and vapors. The most common mixture of this type is air and water vapor. This is the working substance for all systems that use atmospheric air in the thermodynamic device. The water carried in the air is in the superheated state. Consider the mixture of air and water sketched in Fig. 13.4. If the mixture is cooled at constant pressure from state 1 to state d, moisture will condense when the saturated vapor line is reached. Experimentally this temperature may be obtained by placing an ice-water mixture in a polished metal beaker. The mixture may be stirred with a mercury in glass thermometer, and when condensation takes place on the outer surface, the value indicated by the thermometer is the dew point temperature.

RELATIVE HUMIDITY

The relative humidity is the ratio of the amount of water vapor in the air to the quantity of water vapor that the air could contain if it were completely saturated. This is also the ratio of the partial pressure of the water in the air-water vapor mixture to its saturation pressure at the temperature of the mixture. In equation form, the relative humidity is

$$\varphi_1 = p_{v1}/p_{vr1} \tag{13.10}$$

The partial pressure of the water in the air-water vapor mixture, p_{v1}, may be obtained by cooling the mixture at constant pressure to the saturation line, state d in Figs. 13.4 and 13.5. This is the dew point temperature, and the corresponding partial pressure may be obtained from the

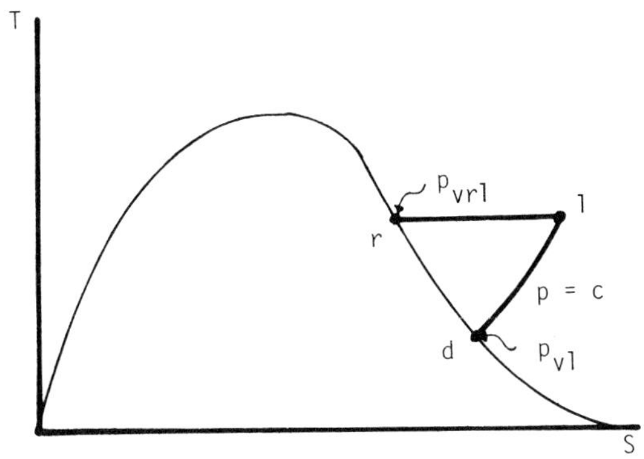

Figure 13.5

saturated steam tables. The saturation pressure of the water vapor at the mixture temperature is located at point r in Fig. 13.5. It is, therefore, possible to obtain the relative humidity of any perfect gas-vapor mixture if the temperature of the mixture, as read by an ordinary thermometer, and the dew point of the vapor are known. The temperature read by a thermometer for state 1 is referred to as the dry bulb temperature.

Example 13.10. Determine the relative humidity of an air-water vapor mixture having a dry bulb temperature of 80 F and a dew point of 60 F. From the saturated steam tables,

$$p_{vr1} = 0.5069 \text{ (psia)} \qquad \text{at } 80 \text{ F}$$

and

$$p_{v1} = 0.2563 \text{ (psia)} \qquad \text{at } 60 \text{ F}$$

so that

$$\varphi_1 = 0.2563/0.5069 = 0.506$$

$$= 50.6\%$$

ABSOLUTE HUMIDITY

The total mass of an air-water vapor mixture consists of the mass of the water vapor and the mass of the dry air. In a thermodynamic process the mass of the dry air remains constant during the process. The mass of the water vapor and, as a result, the mass of the mixture may be a variable.

For this reason calculations for air conditioning systems and other problems using gas-vapor mixtures are based on the quantity of dry air in the process.

The absolute humidity, also referred to as the humidity ratio or specific humidity, is the ratio of mass of the water vapor to the dry air in an air-water vapor mixture.

$$\omega = m(\text{water vapor})/m(\text{dry air})$$

By use of the perfect gas law where

$$m = pV/RT$$

then

$$\omega = (pV/RT)_v/(pV/RT)_{da}$$

The partial pressure of the water vapor at the temperatures normally encountered in air-water vapor problems places the water vapor well into the superheat region. As a vapor becomes more and more superheated, the use of the perfect gas law increases in validity.

Since the volume occupied by the vapor and the air are the same and since both components are at the same temperature,

$$\omega = (p_v R_{da})/(p_{da} R_v)$$

where p_v and p_{da} are the partial pressures of the water vapor and of the dry air. The values of the gas constants are $R = \bar{R}/M$ or

$$R_{da} = 1545/M_a = 1545/29 = 53.3$$

and

$$R_v = 1545/M_v = 1545/18 = 85.8 \text{ (ft-lbf)/(lbm-R)}$$

where M_a and M_v are the molecular weights of the two components. Substituting these values into the absolute humidity equation gives

$$\omega = 53.3p_v/85.8p_{da} = 0.621(p_v/p_{da})$$

From Dalton's law of partial pressures

$$p_m = p_{da} + p_v$$

or

$$p_{da} = p_m - p_v$$

where p_m is the pressure of the mixture, so that

$$\omega = 0.621p_v/(p_m - p_v) \quad (13.11)$$

Equation (13.11) applies only to mixtures of air and water vapor since the specific gas constants were for these particular gases.

Example 13.11. Determine the absolute humidity of the mixture in Example 13.10 when the mixture pressure is 14.696 psia.

From Example 13.10

$$p_{v1} = 0.2563 \text{ (psia)}$$

for the mixture at 80 F with a relative humidity of 50.6%. The absolute humidity is

$$\omega_1 = 0.621p_{v1}/(p_m - p_{v1}) = [0.621(0.256)]/(14.696 - 0.256)$$

$$= 0.0110 \text{ (lbm-v/lbm-da)}$$

The absolute humidity is often expressed in grains of water per pound of dry air where

$$1 \text{ lbm} = 7000 \text{ grains}$$

resulting in

$$\omega_1 = 0.0110(7000) = 77.1 \text{ (gr-v/lbm-da)}$$

Example 13.12. Determine the relative and absolute humidity for a mixture of ammonia and carbon dioxide having a dry bulb temperature of 80 F and a dew point of 40 F. The mixture pressure is 300 psia.

From the saturated ammonia tables (Appendix E),

$$p_{vr1} = 153.0 \text{ (psia)} \qquad \text{at } 80 \text{ F}$$

and

$$p_{v1} = 73.72 \text{ (psia)} \qquad \text{at } 40 \text{ F}$$

The relative humidity is

$$\varphi_1 = p_{v1}/p_{vr1} = 73.72/153.0 = 0.482$$

$$= 48.2\%$$

The absolute humidity is

$$\omega_1 = (p_v R_g)/(p_g R_v)$$

where, for CO_2,

$$R_g = 1545/M_g = 1545/44 = 35.1$$

and, for NH_3,

$$R_v = 1545/M_v = 1545/17 = 90.9 \text{ (ft-lbf/lbm-R)}$$

since the molecular weights of carbon dioxide and ammonia are 44 and 17, respectively.

From Dalton's law

$$p_g = p_m - p_v = 300 - 73.72$$

$$= 226.3 \text{ (psia)}$$

and

$$\omega_1 = (p_v R_g)/(p_g R_v) = [73.72(35.1)]/[226.3(90.9)]$$

$$= 0.126 \text{ (lbm-v/lbm-g)}$$

$$= 882 \text{ (gr/lbm-g)}$$

ADIABATIC SATURATION PROCESS

An air-water vapor process of importance is the adiabatic saturation process. In this steady-flow process a mixture of air and water vapor passes through a fine spray of water from a nozzle. The entering mixture gains moisture until it is saturated. No heat transfer or external work occurs during the process, and the changes in kinetic and potential energy are assumed to be negligible. The apparatus is sketched in Fig. 13.6.

An energy balance gives

$$\Delta H = 0$$

or

$$H_{\text{mixture in}} + H_{\text{water}} = H_{\text{mixture out}}$$

By using the specific enthalpies

$$m_{a1}h_{a1} + m_{v1}h_{v1} + m_w h_w = m_{a2}h_{a2} + m_{v2}h_{v2}$$

where the water m_w is assumed to be added at the temperature of the leaving

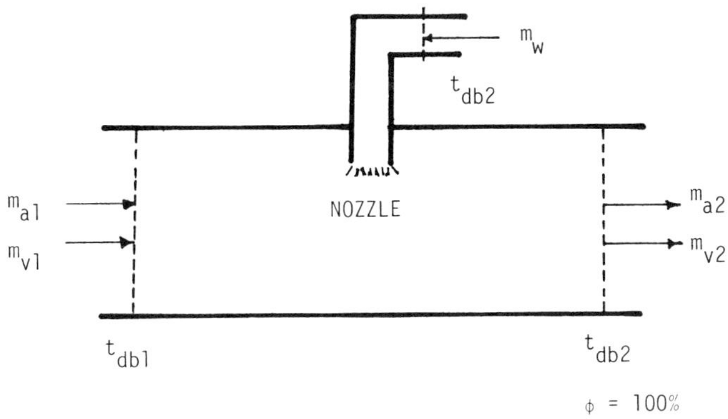

Figure 13.6

mixture. Subscript a1 is for the entering dry air, a2 for the leaving dry air, v1 for the entering water, and v2 for the leaving water.

Since the mass of dry air remains constant,

$$m_a = m_{a1} = m_{a2}$$

Dividing by the mass of the air gives

$$h_{a1} + (m_{v1}/m_a)h_{v1} + (m_w/m_a)h_w = h_{a2} + (m_{v2}/m_a)h_{v2}$$

and from the definition of the absolute humidity

$$\omega = m_v/m_a$$

so that

$$h_{a1} + \omega_1 h_{v1} + \omega_w h_w = h_{a2} + \omega_2 h_{v2}$$

A mass balance for the system gives

$$m_{a1} + m_{v1} + m_w = m_{a2} + m_{v2}$$

and by dividing by m_a

$$1 + \omega_1 + \omega_w = 1 + \omega_2$$

so that

$$\omega_w = \omega_2 - \omega_1$$

By substituting into the energy equation,

$$h_{a1} + \omega_1 h_{v1} + (\omega_2 - \omega_1)h_w = h_{a2} + \omega_2 h_{v2}$$

and

$$\omega_1(h_{v1} - h_w) = h_{a2} - h_{a1} - \omega_2 h_w + \omega_2 h_{v2}$$

so that

$$\omega_1 = [(h_{a2} - h_{a1}) + \omega_2(h_{v2} - h_w)]/(h_{v1} - h_w)$$

By assuming the air behaves as a perfect gas,

$$h_{a2} - h_{a1} = c_p(t_2 - t_1)$$

Even though the water vapor entering and leaving with the mixture is in the superheat region, no substantial error is caused by assuming

$$h_{v1} = h_{g1}$$

and

$$h_{v2} = h_{g2}$$

Since the water enters at the temperature of the leaving mixture,

$$h_w = h_{f2}$$

and

$$h_{v2} - h_w = h_{g2} - h_{f2} = h_{fg2}$$

The absolute humidity of the entering mixture is

$$\omega_1 = [\omega_2 h_{fg2} - c_p(t_1 - t_2)]/(h_{g1} - h_{f2}) \tag{13.12}$$

Example 13.13. Assume the mixture enters the adiabatic saturation device in Fig. 13.6 at 80 F and leaves at 66.8 F. Determine the absolute humidity of the mixture entering the device for a mixture pressure of 14.696 psia.

At a temperature of 66.8 F, by interpolation,

$$p_{v2} = \varphi_2 p_{vr2} = 0.3289 \text{ (psia)}$$

Since the mixture is saturated, the relative humidity at state 2 is 100%. From Eq. (13.11)

$$\omega_2 = 0.621p_{v2}/(p_m - p_{v2})$$
$$= [0.621(0.3289)]/(14.696 - 0.329)$$
$$= 0.0142 \text{ (lbm-v/lbm-da)}$$

The values of the enthalpies from the saturated steam tables are

$$h_{g1} = 1096.6 \text{ (B/lbm)} \qquad \text{at } 80 \text{ F}$$

$$h_{f2} = 34.8 \text{ (B/lbm)} \qquad \text{at } 66.8 \text{ F}$$

and

$$h_{fg2} = 1056.1 \text{ (B/lbm)} \qquad \text{at } 66.8 \text{ F}$$

Substitution of the data into Eq. (13.12) gives

$$\omega_1 = [0.0142(1056.1) - 0.24(80 - 66.8)]/(1096.6 - 34.8)$$
$$= (15.00 - 3.17)/1061.6$$
$$= 0.0111 \text{ (lbm-v/lbm-da)}$$

Example 13.14. From the results of Example 13.13, calculate the relative humidity and the dew point of the entering mixture.

From Eq. (13.11)

$$\omega_1 = 0.621p_{v1}/(p_m - p_{v1})$$

so that

$$\omega_1 p_m - \omega_1 p_{v1} = 0.621p_{v1}$$

or

$$p_{v1}(\omega_1 + 0.621) = \omega_1 p_m$$

and

$$p_{v1} = \omega_1 p_m/(\omega_1 + 0.621)$$

Therefore

$$p_{v1} = [0.0111(14.696)]/(0.0111 + 0.621)$$
$$= 0.258 \text{ (psia)}$$

and from the saturated steam tables, the dew point temperature is

$$t_{dp} = 60.2 \text{ (F)}$$

From Eq. (13.10)

$$\varphi_1 = p_{v1}/p_{vr1}$$

$$= 0.258/0.5069 = 0.507$$

$$= 50.7\%$$

where $p_{vr1} = 0.5069$ psia at the entering dry temperature of 80 F. This value is identical to that in Example 13.10 so that at state 1 the properties of the air-water vapor mixture are

$\varphi_1 = 50.7\%$, $\omega_1 = 0.0111$ (lbm-v/lbm-da)

$t_1 = 80$ (F), the dry bulb temperature

$t_2 = 66.8$ (F), the adiabatic saturation temperature

and

$t_{dp} = 60$ (F), the dew point temperature for a mixture pressure of 14.696 psia

Carrier's equation may be used to determine the partial pressure of the entering water vapor where

$$p_{v1} = p_{vr2} - \frac{(p_m - p_{vr2})(t_1 - t_2)}{2830 - 1.44t_2} \tag{13.13}$$

Example 13.15. Determine the partial pressure of the entering water vapor by use of Carrier's equation from the data in Examples 13.13 and 13.14 where $t_1 = 80$ F, $t_2 = 66.8$ F, $p_{vr2} = 0.3289$ psia, and $p_m = 14.696$ psia.

From Eq. (13.13)

$$p_{v1} = 0.3289 - [(14.696 - 0.3289)(80 - 66.8)]/[2830 - 1.44(66.8)]$$

$$= 0.3289 - 0.0694 = 0.259 \text{ (psia)}$$

This is the same value as determined in Example 13.13. The partial pressure of the entering water vapor may be used to calculate the relative and absolute humidity as in the previous examples.

For mixtures of gases and vapors it is necessary to know the mixture pressure and any two independent thermodynamic properties.

WET BULB TEMPERATURE

The properties of an air-water vapor mixture can be determined if the mixture pressure, the dry bulb temperature, and one other property are known. The remaining properties include the dew point, the relative humidity, the absolute humidity, and the adiabatic saturation temperature. It is easy to obtain the mixture pressure and the dry bulb temperature. The remaining properties can be evaluated by means of a hygrometer for relative humidity, actual separation of the components for absolute humidity, visual observation for the dew point, or the adiabatic saturation device described in the previous section. All of the above pose difficulties when applied to engineering problems. Some do not have a sufficient degree of accuracy, others are difficult to operate and require expensive or large-sized equipment, and finally most of the above take an appreciable amount of time to obtain a value for the property. The instrument in Fig. 13.7 is probably the one most often used for the determination of a second independent property. Air is blown across a thermometer equipped with a wetted cloth. The water on the cloth evaporates, causing a reduction in the temperature at the thermometer due to evaporative cooling. This is similar in concept to the adiabatic saturation device except that heat is transferred by conduction along the thermometer, heat may be radiated from the thermometer, and heat may be gained from the air flowing across the thermometer. For air-water vapor mixtures the above factors tend to cancel one another so

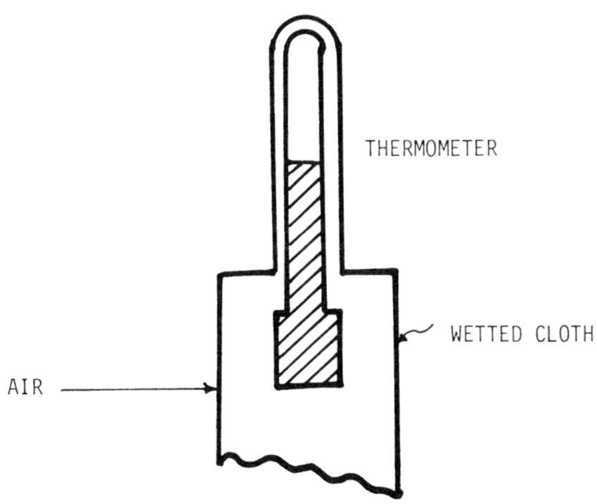

Figure 13.7

that the wet bulb temperature is nearly equivalent to the adiabatic saturation temperature.

PSYCHROMETRIC CHART

The psychrometric chart (Appendix I) consists of curves representing the various properties of the air-water vapor mixture. Lines of constant property values are sketched in Fig. 13.8. Vertical lines represent lines of constant dry bulb temperature, horizontal lines are constant absolute humidity values, and the properties plotted diagonally include the enthalpy of the mixture per pound of dry air, the wet bulb temperature, and the volume of the mixture per pound of dry air. The curved lines are of constant relative humidity. The only restriction placed on the use of the psychrometric chart for air-water vapor mixtures is that the mixture pressure must be 14.696 psia.

Example 13.16. Given dry bulb and wet bulb temperatures of 80 F and 60 F, determine property values by calculations and by use of the psychrometric chart for a mixture pressure of 14.696 psia.

Since the wet bulb temperature is a good approximation of the adiabatic saturation temperature, Eq. (13.12) gives

$$\omega_1 = \frac{\omega_2 h_{fg2} - c_p(t_1 - t_2)}{h_{g1} - h_{f2}}$$

where $h_{f2} = 28.1$ and $h_{fg2} = 1059.9$ B/lbm at 60 F and $h_{g1} = 1096.6$ B/lbm at 80 F. The partial pressure of the water vapor, where $\varphi_2 = 100\%$, is

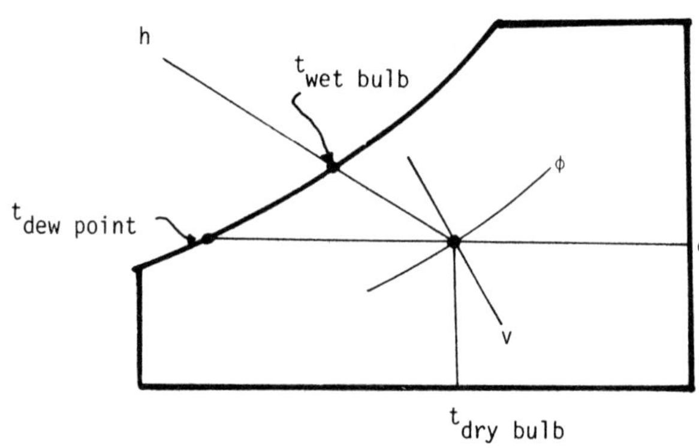

Figure 13.8

$$p_{v2} = 0.2563 \text{ (psia)} \qquad \text{at } 60 \text{ F}$$

and from Eq. (13.11)

$$\omega_2 = 0.621 p_{v2}/(p_m - p_{v2})$$

$$= [0.621(0.2563)]/(14.696 - 0.2563)$$

$$= 0.0110 \text{ (lbm-v/lbm-da)}$$

Substitution of the properties into the absolute humidity equation gives

$$\omega_1 = \frac{0.0110(1059.9) - 0.24(80 - 60)}{1096.6 - 28.1}$$

$$= 0.00642 \text{ (lbm-v/lbm-da)}$$

Again from Eq. (13.11)

$$\omega = 0.621 p_v/(p_m - p_v)$$

or

$$0.00642 = 0.621 p_{v1}/(14.696 - p_{v1})$$

so that

$$p_{v1} = 0.1504 \text{ (psia)}$$

For this pressure the dew point temperature is 45.5 F. At 80 F the partial pressure of the saturated water vapor is

$$p_{vr1} = 0.5069 \text{ (psia)}$$

and the relative humidity from Eq. (13.10) is

$$\varphi_1 = p_{v1}/p_{vr1} = 0.1504/0.5069$$

$$= 0.297 \quad \text{or} \quad 29.7\%$$

From the psychrometric chart (Appendix I),

$$\omega_1 = 0.00643 \text{ (lbm-v/lbm-da)}$$

$$t_{\text{dew point}} = 45.5 \text{ (F)}$$

and

$$\varphi = 30\%$$

Example 13.17. Repeat Example 13.16 for a mixture pressure of 30.0 psia. Values from the psychrometric chart will be the same as in Example 13.16 and are indicated by square brackets in this example.

From Eq. (13.11)

$$\omega_2 = [0.621(0.2563)]/(30.0 - 0.2563)$$

$$= 0.00535 \text{ (lbm-v/lbm-da) } [0.0110]$$

and

$$\omega_1 = \frac{0.00535(1059.9) - 0.24(80 - 60)}{1096.6 - 28.1}$$

$$= 0.00081 \text{ (lbm-v/lbm-da) } [0.00642]$$

so that

$$p_{v1} = 0.0391 \text{ (psia) } [0.1504]$$

and

$$\varphi = p_{v1}/p_{vr1} = 0.0391/0.5069$$

$$= 0.077 \quad \text{or} \quad 7.7\% \ [29.7\%]$$

which illustrates that the psychrometric chart is valid only at a mixture pressure of 14.696 psia.

SPECIFIC VOLUME AND ENTHALPY

The enthalpy of the mixture is

$$H_m = H_v + H_a$$

where m is the subscript for the mixture, v for the vapor, and a for the dry air. Dividing each term by the mass of air m_a in the mixture yields

$$H_m/m_a = (H_v/m_a) + (H_a/m_a)$$

$$= (m_v h_v/m_a) + (m_a h_a/m_a)$$

$$= \omega h_v + h_a \text{ (B/lbm-da)} \tag{13.14}$$

The mass balance equation for the mixture is

$$m_m = m_v + m_a$$

Since all three components occupy the same volume,

$$\frac{m_m}{V_m} = \frac{m_v}{V_v} + \frac{m_a}{V_a} \text{ (lbm-m/ft}^3\text{)}$$

$$\rho_m = \rho_v + \rho_a$$

$$= (p_v/R_vT) + (p_a/R_aT)$$

and

$$v_m' = 1/\rho_m \text{ (ft}^3\text{/lbm-m)}$$

Multiplying the above by the ratio m_m/m_{da} gives

$$v_m = v_m'(m_m/m_{da}) = v_m'(1 + \omega) \text{ (ft}^3\text{/lbm-da)} \quad (13.15)$$

Example 13.18. Determine the enthalpy and volume of the mixture per pound of dry air for the mixture in Example 13.16.

From Example 13.16, t_{db} = 80 F, ω_1 = 0.00642 lbm-v/lbm-da, and p_v = 0.1504 psia. $h_v = h_g$ = 1096.6 B/lbm at 80 F.

From Eq. (13.14)

$$H_m/m_a = \omega h_v + h_a = \omega h_g + c_p t$$

$$= 0.00642(1096.6) + 0.24(80)$$

$$= 7.0 + 19.2 = 26.2 \text{ (B/lbm-da)}$$

Note that t is used in F.

From the psychrometric chart

$$H_m/m_a = 26.4 \text{ (B/lbm-da)}$$

The density of the vapor and of the air is

$$\rho_v = \frac{p_v}{R_vT} = \frac{0.1504(144)}{1545(540)}(18) = 0.0005 \text{ (lbm-v/ft}^3\text{)}$$

$$\rho_a = \frac{(14.696 - 0.1504)144}{53.3(540)} = 0.0728 \text{ (lbm-da/ft}^3\text{)}$$

so that

$$\rho_m = \rho_v + \rho_a = 0.0005 + 0.0728$$

$$= 0.0733 \text{ (lbm-m/ft}^3\text{)}$$

and the specific volume of the mixture is

$$v_m' = 1/\rho_m = 1/0.0733 = 13.64 \text{ (ft}^3\text{/lbm-m)}$$

The mass of the mixture is for 1 lb of dry air

$$m_m = 1 + \omega = 1.00642 \text{ (lbm/lbm-da)}$$

so that from Eq. (13.15)

$$v_m = v_m'(1 + \omega) = 13.64(1.00642) = 13.73 \text{ (ft}^3\text{/lbm-da)}$$

From the psychrometric chart

$$V_m/m_{da} = 13.73 \text{ (ft}^3\text{/lbm-da)}$$

Since the absolute humidity is required for the enthalpy calculation and since the volume calculations require the use of both the partial pressure and absolute humidity terms, the values obtained from the psychrometric chart are valid only for a mixture pressure of 14.696 psia.

PROBLEMS

13.1 Given a gas sample consisting of 5 lbm of H_2, 8 lbm of CO, 2 lbm of CO_2, and 3 lbm of N_2, determine the gravimetric and volumetric analyses of the sample.

13.2 Given a gas sample consisting of 10 moles of He, 5 moles of H_2O, 2 moles of O_2, and 3 moles of air, determine the gravimetric and volumetric analyses of the sample.

13.3 Determine M and R for the gas sample in Prob. 13.1. If the gas has a pressure and temperature of 20 psia and 40 F, determine the volume occupied by the sample.

13.4 Determine the value of c_p, c_v, and k for the gas sample in Prob. 13.1.

13.5 The gas in Prob. 13.1 has a mixture pressure of 50 psia. Determine the partial pressure of each component gas.

13.6 The gas in Prob. 13.1 undergoes a polytropic process for which $n = 1.2$. The initial pressure and temperature are 50 psia and 40 F. The final temperature is 540 F. Determine ΔU, ΔH, c_n, ΔS, Q, W_{nf}, and W_{sf} (ΔKE, $\Delta PE = 0$).

13.7 Determine the relative humidity of an air-water vapor mixture having a dry bulb temperature of 100 F and a dew point temperature of 50 F.

13.8 An air-water vapor mixture has a dry bulb temperature of 80 F and a relative humidity of 40%. Determine the dew point temperature.

13.9 Repeat Prob. 13.7 for air-ammonia and air-Freon-12 mixtures.

13.10 Determine the absolute humidity for the air-water vapor mixture in Prob. 13.7. The mixture pressure is 14.696 psia.

13.11 Determine the absolute humidity in Prob. 13.9 for an air-ammonia mixture. The mixture pressure is 200 psia.

13.12 An air-water vapor mixture enters an adiabatic saturation device with a temperature of 100 F and leaves at a temperature of 70 F. The mixture pressure is 14.696 psia. Determine φ, ω, the dew point temperature, and the partial pressure of the entering mixture.

13.13 Determine the partial pressure of the air-water mixture in Prob. 13.12 by means of Carrier's equation.

13.14 Determine the dry bulb temperature, the wet bulb temperature, the relative humidity, and the absolute humidity for the mixture in Prob. 13.12 by use of the psychrometric chart.

13.15 Determine the enthalpy and the volume of the mixture per pound of dry air for the mixture in Prob. 13.12 by (a) calculations and (b) use of the psychrometric chart.

14

APPLICATION OF MIXTURES

In this chapter we shall consider the use of air-water vapor mixtures as the working substance for a cooling tower and various air conditioning processes.

COOLING TOWER

In a cooling tower hot water enters at the top of the tower and flows downward over a series of baffles (Fig. 14.1). Air is drawn through the tower by means of a fan. The water is cooled by evaporation of part of the

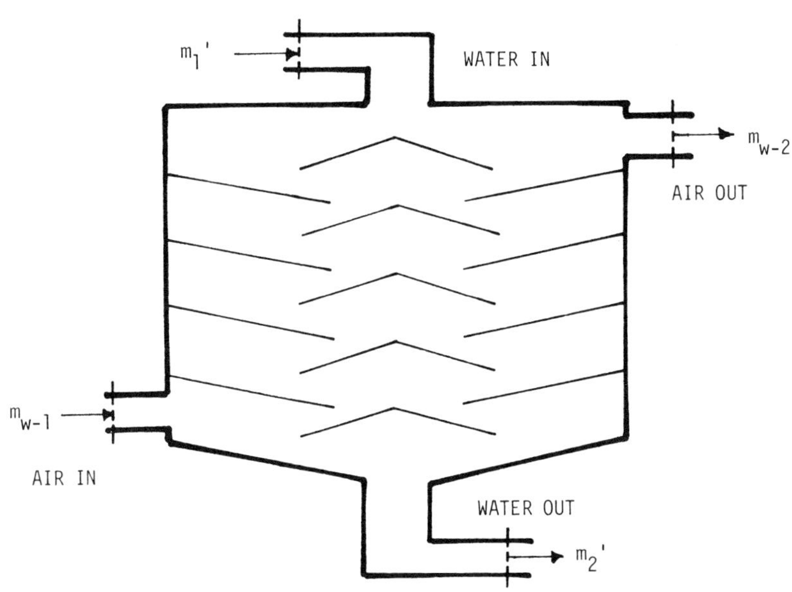

Figure 14.1

water, which represents a loss that must be made up by adding new subcooled liquid to the system. The mass balance for the water is

$$m_1' + m_{w-1} = m_2' + m_{w-2}$$

where m_1' = mass of water entering the tower, m_{w-1} = mass of water entering with the air, m_2' = mass of water leaving the tower, and m_{w-2} = mass of water leaving with the air.

The mass of air remains constant during the process, and dividing by the air flow gives

$$m_1 + \omega_1 = m_2 + \omega_2 \tag{14.1}$$

where $m_1 = m_1'/m_a$ is the mass of water entering per pound of air, $m_2 = m_2'/m_a$ is the mass of water leaving per pound of air, and ω_1, ω_2 are the absolute humidities of the entering and leaving air.

The energy balance, assuming no heat transfer and negligible changes in kinetic and potential energy and omitting the work imparted to the air by the fan, is

$$m_1'h_1 + m_{w-1}h_{g1} + m_a h_{a1} = m_2'h_2 + m_{w-2}h_{g2} + m_a h_{a2}$$

where h_1, h_2 = enthalpy of the entering and leaving water, h_{g1}, h_{g2} = enthalpy of the entering and leaving water in the air, neglecting superheat; and h_{a1}, h_{a2} = enthalpy of entering and leaving dry air.

Again dividing by the mass of the air gives

$$m_1h_1 + \omega_1h_{g1} + h_{a1} = m_2h_2 + \omega_2h_{g2} + h_{a2} \tag{14.2}$$

Substituting for the quantity of water leaving the tower gives, from Eq. (14.1),

$$m_2 = m_1 + \omega_1 - \omega_2$$

so that Eq. (14.2) becomes

$$m_1h_1 + \omega_1h_{g1} + h_{a1} = (m_1 + \omega_1 - \omega_2)h_2 + \omega_2h_{g2} + h_{a2} \tag{14.3}$$

The enthalpy of the subcooled water may be obtained from the steam tables, or it may be approximated by

$$h = c(t - 32) = (t - 32)$$

This equation is quite accurate, being in error by only 0.2 B/lbm at a temperature of 212 F. The enthalpy change for the air is

$$h_{a2} - h_{a1} = c_p(t_{a2} - t_{a1})$$

Solving for the quantity of water entering the tower per pound of dry air gives

$$m_1(h_1 - h_2) = (\omega_1 - \omega_2)h_2 + \omega_2 h_{g2} + c_p(t_{a2} - t_{a1}) - \omega_1 h_{g1}$$

or

$$m_1 = [(\omega_1 - \omega_2)h_2 + \omega_2 h_{g2} + c_p(t_{a2} - t_{a1}) - \omega_1 h_{g1}]/(h_1 - h_2)$$

$$= [(\omega_1 - \omega_2)(t_2 - 32) + \omega_2 h_{g2} + c_p(t_{a2} - t_{a1}) - \omega_1 h_{g1}]/(t_1 - t_2) \quad (14.4)$$

If the condition of the entering air and water is known together with the condition of the water and air leaving, the quantity of water cooled per pound of dry air can be calculated.

Example 14.1. Water enters a cooling tower at 180 F and leaves at 80 F. The air enters at 70 F with a relative humidity of 20%. The air leaves the tower at a temperature of 90 F and a relative humidity of 90%. Determine the air flow to cool 1000 gal/hr of water. Barometric pressure is 14.70 psia. From the psychrometric chart

$$\omega_1 = 0.0031 \quad \text{and} \quad \omega_2 = 0.0277 \qquad (\omega_2 \text{ by calculation})$$

so that

$$\omega_1 - \omega_2 = -0.0246 \text{ (lbm-w/lbm-da)}$$

From the steam tables

$$h_{g1} = 1092.3 \qquad \text{at 70 F}$$

and

$$h_{g2} = 1100.9 \text{ (B/lbm)} \qquad \text{at 90 F}$$

Substituting into Eq. (14.4) gives

$$m_1 = \frac{-0.0246(80 - 32) + 0.0277(1100.9) + 0.24(90 - 70) - 0.0031(1092.3)}{100}$$

$$= (-1.18 + 30.49 + 4.80 - 3.39)/100$$

$$= 0.307 \text{ (lbm-w/lbm-da)}$$

and the required air flow is

$$m_a = 1/0.307 = 3.26 \text{ (lbm-da/lbm-w)}$$

The total water flow, $\rho_\omega = 60.57$ lbm/ft^3 at 180 F, is

$$m_w = (1000/7.481)60.57 = 8097 \text{ (lbm-w/hr)}$$

since there are 7.481 gal in 1 ft^3. The total mass of air required is

$$m_a = 3.26(8097) = 26{,}400 \text{ (lbm-da/hr)}$$

From the definition of the relative humidity

$$\varphi_1 = p_{v1}/p_{vr1}$$

so that

$$p_{v1} = \varphi p_{vr1}$$

$$= 0.2(0.3631) = 0.07 \text{ (psia)}$$

The specific volume of the entering dry air is

$$v_1 = RT/p = [53.3(530)]/[(14.70 - 0.07)144]$$

$$= 13.41 \text{ (ft}^3\text{/lbm-da)}$$

The total volume of the entering air is

$$V = m_a v_1 = [(26{,}400)(13.41)]/60 = 5900 \text{ (cfm)}$$

The quantity of makeup water required is

$$m_1 - m_2 = \omega_1 - \omega_2 = 0.003 - 0.028$$

$$= -0.025 \text{ (lbm-w/lbm-da)}$$

$$\Delta m_w = -0.025(26{,}400) = -660.0 \text{ (lbm/hr)}$$

$$= (-660.0/62.4)7.481 = -79.1 \text{ (gal/hr)}$$

The makeup water requirements are 7.9% of the quantity of water circulated.

If the velocity of the air passing through the tower were reduced, the leaving temperature and moixture content of the air-water vapor mixture would be increased. Lower air velocities also reduce the possibility of liquid particles being carried out of the tower by the moving air stream. If the leaving air conditions were increased, the absolute humidity would again have to be calculated since the state point does not lie on the psychrometric chart. Two other psychrometric charts are available; one extends the low-temperature range for air-water vapor mixtures, and the other is for high-temperature applications.

Example 14.2. Rework Example 14.1 for an exit air temperature of 120 F and a relative humidity of 100%.

By calculation, the leaving absolute humidity is

$$\omega_2 = 0.621[p_{v2}/(p_m - p_{v2})] = 0.621[1.6924/(14.70 - 1.69)]$$

$$= 0.081\ [0.028]\ (\text{lbm-da})$$

and from the steam tables

$$h_{g2} = 1113.7\ [1100.8]\ (\text{B/lbm})$$

By use of Eq. (14.4), the mass of water required per pound mass of dry air is

$$m_1 = 0.951\ [0.308]\ (\text{lbm-w/lbm-da})$$

and the required air flow is

$$m_a = 1.0/0.951 = 1.052\ [3.25]\ (\text{lbm-da/lbm-w})$$

The total quantity of water supplied to the tower and the specific volume of the entering air are the same as in Example 14.1. Other values from Example 14.1 are in the square brackets.

The total air flow is

$$m_a = 1.052(8097) = 8520\ [26{,}400]\ (\text{lbm-da})$$

or by volume

$$V = [8520(13.41)]/60 = 1904\ [5740]\ (\text{cfm})$$

Compared to the original air flow, the higher temperature and relative humidity reduce the required quantity of air by 67%. This reduces the dimensions of the fan as well as the power required to drive the fan. The physical dimensions of the tower will also be reduced.

The makeup water requirement is

$$m_2 - m_1 = \omega_1 - \omega_2 = 0.003 - 0.081$$

$$= -0.078\ [-0.025]\ (\text{lbm-w/lbm-da})$$

or

$$\Delta m_w = -0.078(8520) = -664.6\ [-660.0]\ (\text{lbm-w/min})$$

$$= -79.7\ [-79.1]\ (\text{gal/hr})$$

This increase in makeup water represents an increase in the cost of operation but is slight in comparison to the savings resulting from the decreased air flow.

In the above calculations, we have specified the condition of the leaving air. In practice, this is dependent to a large extent on the physical design of the water tower. For example, the number of baffles and their size and location within the tower control the contact of the air with the liquid water surface and thereby the quantity of water evaporated. In the design of a cooling tower, prior knowledge of the probable leaving condition of the air is based on empirical data obtained from years of experience in constructing such devices.

Low air velocities are advantageous as noted in the examples. Low entering air temperature and relative humidity also increase the performance of the tower. Unfortunately, we have no control over the latter variables.

AIR CONDITIONING PROCESSES

Our second application of the air-water vapor mixture as the working substance in a process will be for the conditioning of air. While heating or cooling alone can be used to adjust the temperature of the air to acceptable values, it is generally necessary to vary the absolute humidity for optimum comfort of people living or working in a controlled environment. Since various people require different air conditions depending on age, sex, activity involved, type of clothes, etc., the correct temperature and absolute humidity may vary over a considerable range. Calculations may be performed by the use of the equations presented in the previous sections, or for a mixture pressure of 14.7 psia the psychrometric chart may be used. In the following discussion, we shall restrict ourselves to the use of the chart.

COOLING PROCESSES

The air is first cooled from its dry bulb temperature to the dew point temperature by the removal of sensible heat (Fig. 14.2). Moisture then condenses on the cooling coil as the temperature is reduced to the desired value. The heat removed serves to lower the temperature of the air, sensible heat, as well as the condensation of part of the water in the air, latent heat.

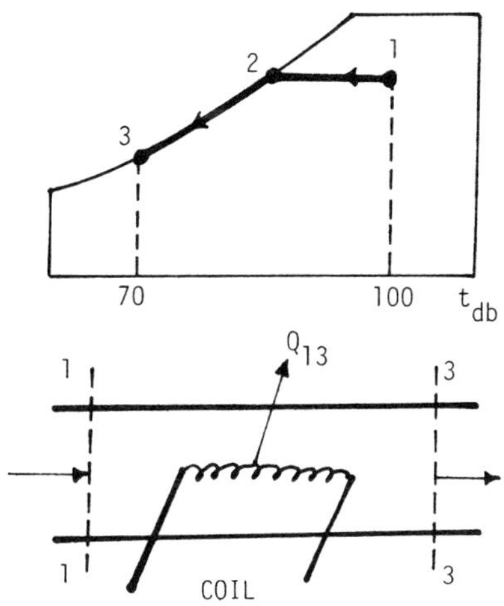

Figure 14.2

The principal disadvantage of the cooling process is the high relative humidity (100%) of the cooled air. When this air is added to the system being treated, it may result in a moisture content which lies outside the comfort zone for humans.

Example 14.3. Assume that the air-water vapor mixture entering the system in Fig. 14.2 has a dry bulb temperature of 100 F and a relative humidity of 50%. The air is to be cooled to a temperature of 70 F. Mixture pressure is 14.70 psia. Determine the heat and water removed per pound of dry air.

From the psychrometric chart at state 1, for t_{db-1} = 100 F and φ_1 = 50%, the enthalpy and absolute humidity are

$$h_1 = 47.3 \text{ (B/lbm-da)} \quad \text{and} \quad \omega_1 = 145 \text{ (gr-w/lbm-da)}$$

At state 2

$$h_2 = 41.3 \text{ (B/lbm-da)}$$

with a dew point temperature of 77.9 F. For state 3

$$h_3 = 34.1 \text{ (B/lbm-da)}$$

$$\varphi_3 = 100\% \quad \text{and} \quad \omega_3 = 111 \text{ (gr/lbm-da)}$$

and the dry bulb temperature is 70 F. The moisture removed is

$$\Delta\omega = \omega_3 - \omega_1 = 111 - 145 = -34 \text{ (gr/lbm-da)}$$

$$= -0.00485 \text{ (lbm-w/lbm-da)}$$

The heat removed from the air-vapor mixture is

$$Q_{13} = Q_{12} + Q_{23} = (h_2 - h_1) + (h_3 - h_2)$$

$$= h_3 - h_1 = 34.1 - 47.3$$

$$= -13.2 \text{ (B/lbm-da)}$$

The air leaving the system has the required dry bulb temperature of 70 F, and the relative humidity is 100%.

COOLING AND DEHUMIDIFYING

In Fig. 14.3 the air is cooled until the moisture content reaches the value required for the final mixture. The air is then reheated to the desired temperature.

Example 14.4. Given is an air-water vapor mixture with a dry bulb temperature of 100 F and a relative humidity of 50%. The final condition of the mixture is to be 70 F and 40%. The entering conditions are the same as in Example 14.3, and the leaving conditions for

$$t_{db-4} = 70 \text{ F} \quad \text{and} \quad \varphi_4 = 40\%$$

are

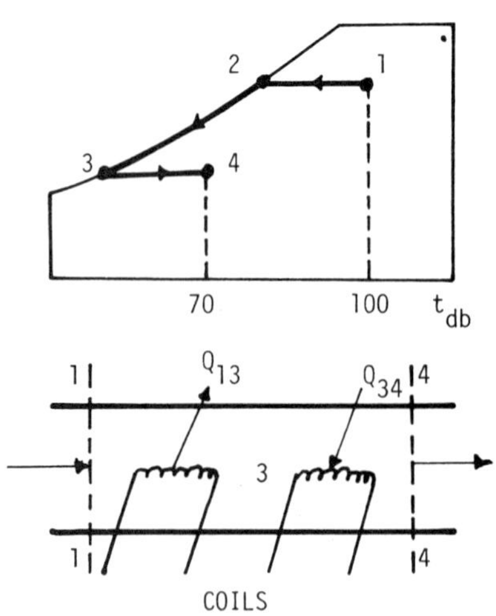

Figure 14.3

$$h_4 = 23.7 \text{ (B/lbm-da)} \quad \text{and} \quad \omega_4 = 44.0 \text{ (gr/lbm-da)}$$

Since only sensible heat is added between states 3 and 4, the moisture content at state 3 is

$$\omega_3 = \omega_4 = 44.0 \text{ (gr/lbm-da)}$$

and the relative humidity at state 3 is 100%, with a dry bulb temperature of 45 F. The enthalpy at state 3 for these conditions is

$$h_3 = 17.6 \text{ (B/lbm-da)}$$

The heat removed by the cooling coil is

$$Q_{13} = h_3 - h_1 = 17.6 - 47.3$$

$$= -29.7 \text{ (B/lbm-da)}$$

and the required amount of heat to be added is

$$Q_{34} = h_4 - h_3 = 23.7 - 17.6$$

$$= 6.1 \text{ (B/lbm-da)}$$

The moisture removed from the system is

$$\Delta\omega = \omega_3 - \omega_1 = 44.0 - 145 = -101.0 \text{ (gr-w/lbm-da)}$$

$$= -0.0144 \text{ (lbm-w/lbm-da)}$$

The temperature leaving the cooling coil, state 3, is 45 F as compared to 70 F for the simple cooling process. For the cooling and dehumidifying process, this requires a heat removal of -29.7 B/lbm-da in the cooling coil compared to -13.2 B/lbm-da for the simple cooling process. The water removed is -0.00485 lbm-w/lbm-da for the cooling process and -0.0144 lbm-w/lbm-da for the cooling-dehumidifying process.

HEATING

In Fig. 14.4, sensible heat is added to the system to increase the dry bulb temperature. Pure heating of the air may lower the air humidity in a system to an unacceptable level.

Example 14.5. The air enters the system with a dry bulb temperature of 30 F and a relative humidity of 40%. The air is to leave the system with a dry bulb temperature of 70 F. Determine the heat added in the heating process. For the given inlet conditions,

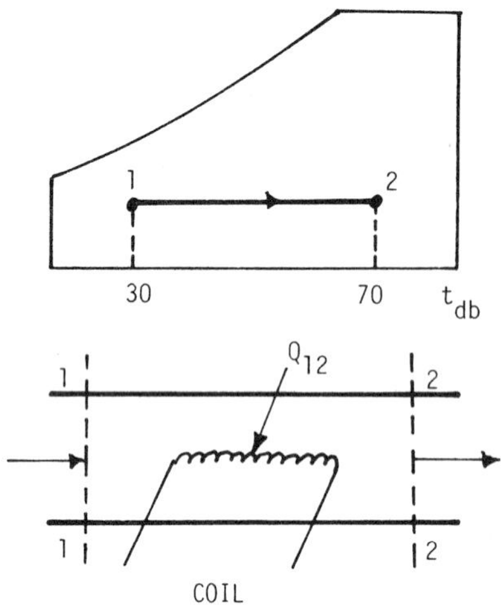

Figure 14.4

$$h_1 = 8.5 \text{ (B/lbm-da)} \quad \text{and} \quad \omega_1 = 10.0 \text{ (gr-w/lbm-da)}$$

Since only sensible heat is added to the system,

$$\omega_2 = \omega_1 = 10 \text{ (gr-w/lbm-da)} \quad \text{and } t_{db-2} = 70 \text{ (F)}$$

for which

$$h_2 = 18.4 \text{ (B/lbm-da)} \quad \text{and} \quad \varphi \approx 9\%$$

The heat added is

$$Q_{12} = h_2 - h_1 = 18.4 - 8.5 = 9.9 \text{ (B/lbm-da)}$$

If air with a relative humidity of only 9% is added to a closed system, the relative humidity within the system will probably drop to an unacceptable level.

HEATING AND HUMIDIFYING

Given is an entering air-water vapor mixture with a specified dry bulb temperature and relative humidity. A sensible heat coil may be used to raise the dry bulb temperature to such a value that an adiabatic saturation device will supply the required moisture content in the leaving air (Fig. 5). A second, sensible heat coil may then be employed to obtain the required final temperature.

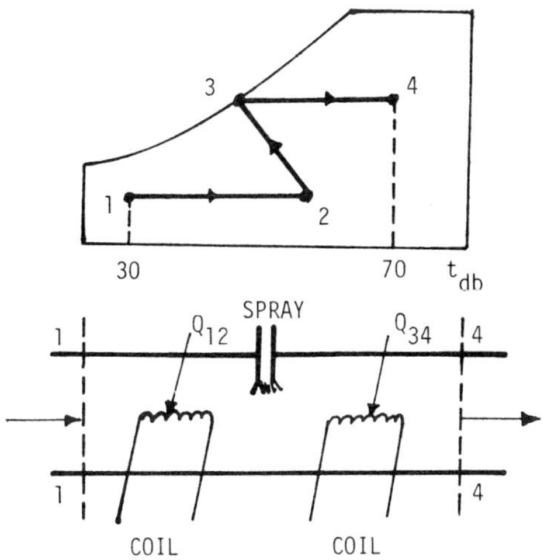

Figure 14.5

Example 14.6. The air enters the system with a dry bulb temperature of 30 F and a relative humidity of 40%. The air is to be conditioned to 70 F and 40% relative humidity. Determine the heat added in the coils and the required temperature of the water spray in the adiabatic saturation device.

For t_1 = 30 F and φ_1 = 40%,

$$h_1 = 8.5 \text{ (B/lbm-da)} \quad \text{and} \quad \omega_1 = 10.0 \text{ (gr-w/lbm-da)}$$

For t_4 = 70 F and φ_4 = 40%,

$$h_4 = 23.7 \text{ (B/lbm-da)} \quad \text{and} \quad \omega_4 = 44.0 \text{ (gr-w/lbm-da)}$$

For φ_3 = 100% and ω_3 = 44.0 gr-w/lbm-da,

$$h_3 = 17.7 \text{ (B/lbm-da)} \quad \text{and} \quad t_3 = 45 \text{ (F)}$$

Finally, for $\omega_2 = \omega_1$ = 10.0 gr-w/lbm-da and $h_2 = h_3$ = 17.7 B/lbm-da,

$$t_{db-2} = 66.5 \text{ (F)}$$

The heat added in the first coil is

$$Q_{12} = h_2 - h_1 = 17.7 - 8.5 = 9.2 \text{ (B/lbm-da)}$$

and in the second coil

$$Q_{34} = h_4 - h_3 = 23.7 - 17.7 = 6.0 \text{ (B/lbm-da)}$$

The water temperature in the water spray is $t_3 = 45$ F, and the increase of moisture content in the air is

$$\Delta\omega = \omega_3 - \omega_2 = 44.0 - 10.0 = 34.0 \text{ (gr-w/lbm-da)}$$

While this system will result in the required condition for exit air, the use of two heaters and the availability of spray water at 45 F may make the system uneconomical.

If warmer water is employed, it may be possible to obtain the desired results with a simple spray device (Fig. 14.6).

Example 14.7. Let the inlet conditions be those of Example 14.6, and the required leaving conditions are a dry bulb temperature of 70 F and a relative humidity of 40%.

For 70 F and 40% relative humidity,

$$\omega_2 = 44.0 \text{ (gr/lbm-da)} \quad \text{and} \quad h_2 = 23.7 \text{ (B/lbm-da)}$$

Since the initial conditions are the same as those of Example 14.6,

$$\omega_1 = 10.0 \text{ (gr/lbm-da)} \quad \text{and} \quad h_1 = 8.5 \text{ (B/lbm-da)}$$

The total gain in moisture of the system is

$$\Delta\omega = \omega_2 - \omega_1 = 44.0 - 10.0 = 34.0 \text{ (gr/lbm-da)} = 0.00486 \text{ (lbm-v/lbm-da)}$$

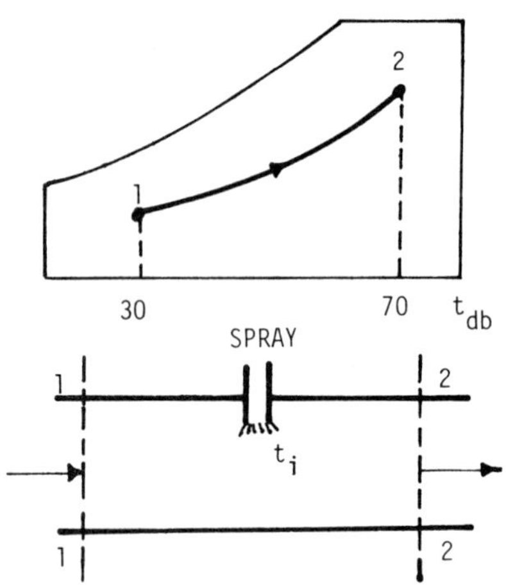

Figure 14.6

and the heat added is

$$Q_{12} = h_2 - h_1 = 23.7 - 8.5 = 15.2 \text{ (B/lbm-da)}$$

To obtain the conditions at state 2 after the mixing process, the mass balance equation is

$$m_{a1} + m_{v1} + m_{wi} = m_{a2} + m_{v2} + m_{wf}$$

where m_{wi} and m_{wf} are the quantity of liquid water sprayed into and drained from the system. Since $m_{a1} = m_{a2} = m_a$, dividing by the mass of air gives

$$1 + \omega_1 + (m_{wi}/m_a) = 1 + \omega_2 + (m_{wf}/m_a)$$

so that

$$(m_{wi} - m_{wf})/m_a = \omega_2 - \omega_1$$

or

$$(m_{wi} - m_{wf})/m_a = 0.00486 \text{ (lbm-w/lbm-da)}$$

For calculation purposes we may assume $m_{wi} = m_{wf} = m_w$.

The energy balance, assuming $Q = W_{sf} = 0$ and $\Delta KE = \Delta PE = 0$, is

$$h_1 + (m_w/m_a)t_i = h_2 + (m_w/m_a)t_f$$

or

$$(m_w/m_a)(t_i - t_f) = h_2 - h_1$$

$$= 15.2 \text{ (B/lbm-da)}$$

Many combinations of properties that satisfy the above equations are possible.

If the ratio of water to air flow were 10, the water temperature would drop:

$$t_i - t_f = 15.2/10 = 1.52 \text{ (F)}$$

Whether this can occur depends on the temperature and size of the water droplets leaving the spray or the velocity of the air through the system. In general, there must be sufficient time and contact area for the air-vapor mixture to gain the required quantity of moisture and heat. In practical applications, the variables can be adjusted until the desired air-water vapor properties are obtained.

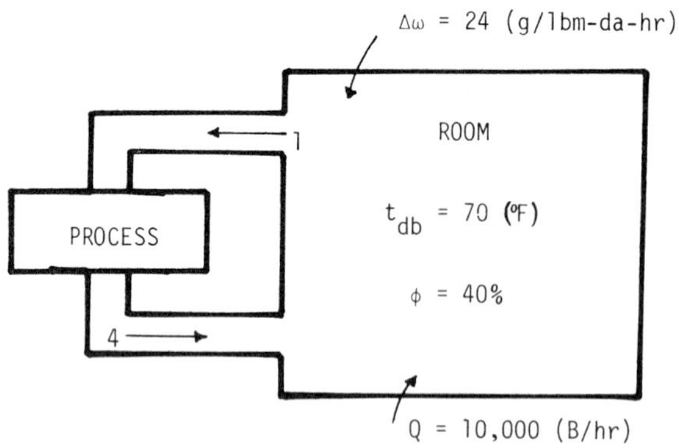

Figure 14.7

CLOSED SYSTEM

Given the closed system in Fig. 14.7, the conditions in the room are to be maintained at 70 F dry bulb temperature and 40% relative humidity. Due to people and the process the moisture generated within the room amounts to 24 gr/lbm-da/hr, and heat gains are 10,000 B/hr from the process and heat transfer through the structure. The air conditioning device completely recirculates the air in the room four times per hour. We wish to design the air conditioning system.

Since we have both a heat and moisture gain, we shall need a cooling and dehumidifying system.

For a room having dimensions of

50 x 30 x 10 (ft)

the volume of air in the room is

$$V_t = 15{,}000 \text{ (ft}^3\text{)}$$

For a mixture pressure of 14.70 psia and a relative humidity of 40% at a dry bulb temperature of 70 F,

$$\varphi = p_{v1}/p_{vr1}$$

so that

$$p_{v1} = 0.40(0.3631) = 0.15 \text{ (psia)}$$

The weight of dry air in the room is

$$m_t = pV/RT = [(14.70 - 0.15)144(15{,}000)]/[53.3(530)]$$

$$= 1113 \text{ (lbm)}$$

The quantity of air passing through the air conditioner is

$$m = 1113(4) = 4450 \text{ (lbm/hr)}$$

for four changes per hour. The heat to be removed from the air is

$$h_4 - h_1 = -10{,}000/4450 = -2.25 \text{ (B/lbm-da)}$$

See Fig. 14.3 for state point location. The total moisture gain is

$$\Delta\omega_t = 24(1113) = 26{,}700 \text{ (gr/hr)}$$

The moisture to be removed from the air is

$$\Delta\omega = -26{,}700/4450 = -6.0 \text{ (gr/lbm-da)}$$

Since the air enters the conditioner at 70 F dry bulb temperature and 40% relative humidity,

$$h_1 = 23.7 \text{ (B/lbm-da)} \quad \text{and} \quad \omega_1 = 44.0 \text{ (gr/lbm-da)}$$

it must leave with

$$h_4 = 23.7 - 2.2 = 21.5 \text{ (B/lbm-da)}$$

and

$$\omega_4 = 44.0 - 6.0 = 38.0 \text{ (gr/lbm-da)}$$

To accomplish the moisture is removed first so that

$$\omega_2 = 38.0 \text{ (gr/lbm-da)} \qquad \text{at } \varphi_2 = 100\%$$

where

$$h_2 = 15.8 \text{ (B/lbm-da)} \quad \text{and} \quad t_{db\text{-}3} = 41.2 \text{ (F)}$$

The air would then be heated to

$$h_4 = 21.5 \text{ (B/lbm-da)} \quad \text{and} \quad \omega_4 = 38.0 \text{ (gr/lbm-da)}$$

which requires a dry bulb temperature of

$$t_{db\text{-}4} = 64.8 \text{ (F)}$$

Since a closed system would be undesirable due to buildup of contaminates, assume that 20% of the air supplied to the air conditioner is drawn from the outside. Outside conditions are 90 F dry bulb temperature and 60% relative humidity. For the outside air

$$\omega = 127 \text{ (gr/lbm-da)} \quad \text{and} \quad h = 42 \text{ (B/lbm-da)}$$

The condition entering the conditioner is

$$\omega_1 = [127 + 4(44.0)]/5 = 60.6 \text{ (gr/lbm-da)}$$

and

$$h_1 = [42 + 4(23.7)]/5 = 27.4 \text{ (B/lbm-da)}$$

To balance the conditions in the room the conditions at states 2 and 3 remain the same as in the previous problem. Therefore

$$\Delta\omega = \omega_4 - \omega_1 = 38.0 - 60.6 = -22.6 \text{ (gr/lb-da)}$$

of moisture must be removed, and

$$\Delta h = h_4 - h_1 = 21.5 - 27.4 = -5.9 \text{ (B/lbm-da)}$$

of heat is removed. Of course, 4450(0.2) = 89.0 lbm-da/hr must be removed from the room to accommodate the fresh air.

PROBLEMS

14.1 Water is cooled from 160 to 70 F by means of a cooling tower. The air enters the tower at a temperature of 50 F and a relative humidity of 10%. The air leaves at a temperature of 100 F and a relative humidity of 80%. Barometric pressure is 14.7 psia. Determine the air flow and the makeup water required to cool 2500 gal/hr.

14.2 Air at a temperature of 100 F dry bulb and 75 wet bulb is to be cooled to 50 F. Determine the heat and water removed from the air vapor mixture. See Fig. 14.2.

14.3 The air in Prob. 14.2 is to be cooled to 50 F with a relative humidity of 60%. Determine the heat transfer in the two coils and the moisture removed from the mixture. See Fig. 14.3.

14.4 Air at a temperature of 30 F and a relative humidity of 20% is heated to 80 F. Determine the heat added in the coil. See Fig. 14.4.

14.5 The air in Prob. 14.4 is heated to 80 F with a relative humidity of 60%. Determine the heat added in the two coils, the temperature of the spray water, and the moisture added to the air. See Fig. 14.5.

15

COMBUSTION EQUATIONS

Combustion is a chemical reaction between various elements or compounds and oxygen. It represents the oxidation of these materials and is associated with the release of energy during the reaction process. The basic reactive elements involved in engineering combustion processes are carbon, hydrogen, and occasionally sulfur.

The combustion process is described by means of a chemical equation; for the above three elements we may write

$$C + O_2 \rightarrow CO_2$$

$$S + O_2 \rightarrow SO_2$$

and

$$H_2 + 1/2\ O_2 \rightarrow H_2O$$

The last equation could also be written as

$$2\ H_2 + O_2 \rightarrow 2\ H_2O$$

The arrow is used to designate the direction of the process.

Each of the above equations must have the same number of atoms of each element on the opposite sides of the equation. The mass of the reactants, which are the terms on the left side of the equation, must equal the mass of the products, the terms on the right side of the equation.

The atomic weights of the elements involved in the three equations are

$$C = 12, \qquad O = 16, \qquad H = 1, \qquad S = 32$$

The above values are the integer values for the various elements; real values may differ by a small amount due to isotopes.

The mass combustion equations are

$$12\ (\text{lbm})\ C + 32\ (\text{lbm})\ O_2 \rightarrow 44\ (\text{lbm})\ CO_2$$

$$32\ (\text{lbm})\ S + 32\ (\text{lbm})\ O_2 \rightarrow 64\ (\text{lbm})\ SO_2$$

$$2\ (\text{lbm})\ H_2 + 16\ (\text{lbm})\ O_2 \rightarrow 18\ (\text{lbm})\ H_2O$$

The mole of a substance is equal to

$$n = m/M$$

where M is the molecular weight of the element and m is the mass of the element. The three equations could also be written as

$$1\ \text{mole}\ C + 1\ \text{mole}\ O_2 \rightarrow 1\ \text{mole}\ CO_2$$

$$1\ \text{mole}\ S + 1\ \text{mole}\ O_2 \rightarrow 1\ \text{mole}\ SO_2$$

$$1\ \text{mole}\ H_2 + 0.5\ \text{mole}\ O_2 \rightarrow 1\ \text{mole}\ H_2O$$

For perfect gases under equal conditions of pressure and temperature, the volume occupied by a mole of gas is the same for all gases. For the hydrogen equation, assuming that the pressure and temperature of the reactants and products are the same, we could have, for a particular mass, pressure, and temperature,

$$1\ (\text{ft}^3)\ H_2 + 0.5\ (\text{ft}^3)\ O_2 \rightarrow 1\ (\text{ft}^3)\ H_2O$$

The volume of the reactants is 1.5 ft^3 and of the products, 1 ft^3.

The combustion equation represents a mass balance but not a volume balance.

COMBUSTION OF A FUEL IN OXYGEN

Example 15.1. Consider the combustion of normal octane, a fuel which approximates gasoline. Gasoline is actually a mixture of various hydrocarbon fuels. The equation for this hydrocarbon is C_8H_{18}, and its combustion equation in terms of oxygen requirements is

$$8\ C + 8\ O_2 \rightarrow 8\ CO_2$$

$$9\ H_2 + 9(1/2)\ O_2 \rightarrow 9\ H_2O$$

$$8\ C + 9\ H_2 + 12.5\ O_2 \rightarrow 8\ CO_2 + 9\ H_2O$$

$$C_8H_{18} + 12.5\ O_2 \rightarrow 8\ CO_2 + 9\ H_2O$$

By checking the combustion equation by balancing the atoms of each element

	Left Side	Right Side
C	8	8
H	18	18
O	25	16 + 9 = 25

The mass balance equation is

$$[8(12) + 18(1)]\ (\text{lbm})\ C_8H_{18} + 12.5(32)\ (\text{lbm})\ O_2 \rightarrow 8(12 + 32)\ (\text{lbm})\ CO_2 + 9(2 + 16)\ (\text{lbm})\ H_2O$$

$$114\ (\text{lbm})\ C_8H_{18} + 400\ (\text{lbm})\ O_2 \rightarrow 352\ (\text{lbm})\ CO_2 + 162\ (\text{lbm})\ H_2O$$

$$514\ (\text{lbm})\ \text{reactants} = 514\ (\text{lbm})\ \text{products}$$

COMBUSTION OF A FUEL IN DRY AIR

Combustion of a fuel does not generally take place in oxygen; a notable exception is in testing for the energy contained in a specific fuel. Most combustion processes occur in atmospheric air.

In Chapter 13 dry air was defined to be 79% N_2 by volume and 21% O_2 by volume. Since under equal conditions of pressure and temperature, $x = n/n_m$, where x is the volumetric fraction and n is the moles, the ratio of nitrogen to oxygen in air is

$$x_{N_2}/x_{O_2} = n_{N_2}/n_{O_2} = 0.79/0.21 = 3.76$$

For each mole of oxygen entering with the air, 3.76 moles of nitrogen must also be introduced. Since nitrogen is an inert gas, it does not take part in the reaction but appears on both sides of the combustion equation.

Example 15.2. In Example 15.1, 12.5 moles of O_2 were required for the combustion process. When n-octane is burned in air there must be

$$12.5(3.76) = 47 \text{ moles of } N_2$$

in the products and reactants, or

$$47(28) = 1316\ (\text{lbm})\ N_2$$

where the molecular weight of nitrogen is 28. The mass equation of the combustion of normal octane in air is

$$114\ (\text{lbm})\ C_8H_{18} + 400\ (\text{lbm})\ O_2 + 1316\ (\text{lbm})\ N_2 \rightarrow 352\ (\text{lbm})\ CO_2 + 162\ (\text{lbm})\ H_2O + 1316\ (\text{lbm})\ N_2$$

$$1830\ (\text{lbm})\ \text{reactants} = 1830\ (\text{lbm})\ \text{products}$$

AIR-FUEL RATIO, FUEL-AIR RATIO, STOICHIOMETRIC AIR

The stoichiometric air is the amount of air required in the combustion process so that there is just sufficient oxygen to completely react with each element in the fuel.

Example 15.3. Determine the air-fuel ratio AF for the combustion of octane in dry air.

In Example 15.2 there were

$$400\ (\text{lbm})\ O_2 + 1316\ (\text{lbm})\ N_2 = 1716\ (\text{lbm})\ \text{air}$$

or in general

$$1\ (\text{lbm})\ O_2 + 3.29\ (\text{lbm})\ N_2 \rightarrow 4.29\ (\text{lbm})\ \text{air}$$

The stoichiometric air for the combustion of normal octane is 1716 lbm. Stoichiometric air processes are also described as 100% theoretical air processes. The air-fuel ratio is

$$AF = m_{air}/m_{fuel} = 1716/114 = 15.1\ (\text{lbm-air/lbm-fuel})$$

and the fuel-air ratio is

$$FA = 1/AF = m_{fuel}/m_{air} = 1/15.1$$

$$= 0.0664\ (\text{lbm-fuel/lbm-air})$$

DEW POINT OF THE PRODUCTS

Example 15.4. Assume the products of combustion in Examples 15.1 and 15.2 have a mixture pressure of 14.7 psia. Determine the dew point of the products of combustion. For combustion in oxygen the products are

8 moles of CO_2 and 9 moles of H_2O

The volume fraction of water in the mixture is

$$x_{H_2O} = n_{H_2O}/n_{mixture} = 9/17 = 0.529$$

From Dalton's law of partial pressures

$$p_{H_2O} = x_{H_2O}p_m = 0.529(14.7) = 7.776\ (\text{psia})$$

giving a dew point, from the steam tables, of

$$t_{dp} = 181.5\ (\text{F})$$

For combustion in 100% theoretical air, the products are 8 moles of CO_2, 9 moles of H_2O, and 47 moles of N_2, so that

$$x_{H_2O} = n_{H_2O}/n_m = 9/(8 + 9 + 47) = 0.1406$$

and

$$p_{H_2O} = 2.067 \text{ (psia)}$$

so that

$$t_{dp} = 127.1 \text{ (F)}$$

The above calculation is important because if the dew point is reached in the exhaust system, rusting or corrosion may occur due to the condensation of the water vapor.

EXCESS AIR

Since most combustion reactions occur rather quickly, use of 100% theoretical air seldom results in complete combustion of the fuel. The reason for this incomplete combustion is that there is insufficient time for the oxygen in the air to locate and combine with the reactive components of the fuel. One solution is to provide excess air, which increases the chances of the oxygen finding a fuel component.

We would like to ensure complete combustion for two reasons: If a component of fuel is not burned or not burned completely, energy is lost. Second, complete combustion of a hydrocarbon produces only water and carbon dioxide, strictly speaking, neither of which is a pollutant.

In the steady-flow devices that we have previously considered, as the gas turbine or steam power plant, the use of excess air is feasible and results in a more complete combustion of the fuel. However, there is a balance point since increasing the excess air decreases the temperature of the combustion products, which in turn reduces the net work of the cycle.

In the spark-ignited internal combustion engine, an increase in the excess air may result in such a "lean" mixture of fuel and air that combustion is not possible.

The amount of air supplied may be stated in two forms:

$$\text{Excess air} = (m_a - m_{sa})/m_{sa} \tag{15.1}$$

where m_a = mass of air supplied, m_{sa} = mass of stoichiometric air for the given fuel, when $m_a > m_{sa}$.

The second form is to state the air actually supplied as a percentage of the theoretical requirement where stoichiometric air is 100% theoretical air.

Example 15.5. Assume the quantity of dry air supplied to burn normal octane is 3432 lbm. Determine the percentage of excess air.

From Eq. (15.1)

$$\text{Excess air} = (m_a - m_{sa})/m_{sa}$$

$$= (3432 - 1716)/1716 = 1.0$$

where 1716 lbm is the stoichiometric air requirement. This represents 100% excess air or 200% of the theoretical air requirements.

Example 15.6. Determine the mass equations for the combustion of normal octane in 200% theoretical dry air. Also determine the air-fuel ratio and the dew points of the products.

The combustion equations for 100% theoretical dry air were

$$1 \text{ mole } C_8H_{18} + 12.5 \text{ moles } O_2 + 47 \text{ moles } N_2 \rightarrow 8 \text{ moles } CO_2 + 9 \text{ moles } H_2O + 47 \text{ moles } N_2$$

and

$$114 \text{ (lbm) } C_8H_{18} + 400 \text{ (lbm) } O_2 + 1316 \text{ (lbm) } N_2 \rightarrow 352 \text{ (lbm) } CO_2 + 162 \text{ (lbm) } H_2O + 1316 \text{ (lbm) } N_2$$

For 200% theoretical dry air the moles of O_2 and N_2 supplied are

$$\text{Moles } O_2 = \text{stoichiometric } O_2\text{(theoretical air)}$$

$$= 12.5(2) = 25$$

and

$$\text{Moles } N_2 = \text{stoichiometric } N_2\text{(theoretical air)}$$

$$= 47(2) = 94$$

Since only 12.5 moles of O_2 can combine with the carbon and hydrogen in the fuel, the remaining 12.5 moles must appear in the products of combustion.

The combustion equations for 200% theoretical dry air are

$$1 \text{ mole } C_8H_{18} + 25 \text{ moles } O_2 + 94 \text{ moles } N_2 \rightarrow 8 \text{ moles } CO_2 + 9 \text{ moles } H_2O + 12.5 \text{ moles } O_2 + 94 \text{ moles } N_2$$

and

$$114\ (\text{lbm})\ C_8H_{18} + 800\ (\text{lbm})\ O_2 + 2632\ (\text{lbm})\ N_2 \rightarrow 352\ (\text{lbm})\ CO_2 + 162\ (\text{lbm})\ H_2O + 400\ (\text{lbm})\ O_2 + 2632\ (\text{lbm})\ N_2$$

$$3546\ (\text{lbm})\ \text{reactants} = 3546\ (\text{lbm})\ \text{products}$$

The mass of dry air supplied is

$$m_a = 800 + 2632 = 3432\ (\text{lbm})$$

The air-fuel ratio is

$$AF = m_a/m_f = 3432/114$$

$$= 30.1\ (\text{lbm-a/lbm-fuel})$$

which is too lean for combustion in a spark-ignited engine. The dew point, for

$$n_{H_2O} = 9\ \text{moles}\ H_2O$$

in the products and

$$n_{\text{products}} = 8 + 9 + 12.5 + 94 = 123.5$$

so that

$$x_{H_2O} = n_{H_2O}/n_{\text{products}} = 9/123.5 = 0.073$$

and for a mixture pressure of 14.7 psia

$$p_{H_2O} = x_{H_2O}p_m = 0.073(14.7) = 1.073\ (\text{psia})$$

giving a dew point of

$$t_{dp} = 103.8\ (\text{F})$$

from the steam tables.

COMBUSTION OF FUEL IN ATMOSPHERIC AIR

Combustion calculations are usually based on the quantity of dry air used in the reaction process. Actually, combustion occurs in atmospheric air which contains water vapor. Since the water vapor in the air is an uncontrolled variable, the reason for using dry air as a basis for calculations is obvious.

Example 15.7. Determine the variation in the theoretical air, the air-fuel ratio, and the dew point for the combustion process in Example 15.6 using atmospheric air. For the atmospheric air the dry bulb temperature is 80 F, the relative humidity is 60%, and the mixture pressure is 14.7 psia.

In Example 15.6 the dry air supplied was

$$m_{da} = 3432 \text{ (lbm)}$$

From the psychrometric chart (Appendix I),

$$\omega = 92 \text{ (gr } H_2O/\text{lbm-da)} = 0.01314 \text{ (lbm-}H_2O/\text{lbm-da)}$$

The quantity of water vapor entering with the dry air is

$$m_{H_2O} = 3432(0.01314) = 45.1 \text{ (lbm-}H_2O)$$

The amount of atmospheric air supplied is

$$m_a = 3432 + 45 = 3477 \text{ (lbm)}$$

The stoichiometric requirements using atmospheric air would be

$$m_{sa} = m_{da} + m_{da}\omega$$

$$= 1716 + 1716(0.01314) = 1738 \text{ (lbm)}$$

From Eq. (15.1)

$$\text{Excess air} = (m_a - m_{sa})/m_{sa}$$

$$= (3477 - 1738)/1738 = 1.0$$

or 200% theoretical air.

The moles of water vapor entering with the atmospheric air are

$$n_{H_2O} = m_{H_2O}/M_{H_2O} = 45.1/18 = 2.51 \text{ (moles)}$$

and the combustion equations for 200% theoretical air are

1 mole C_8H_{18} + 25 moles O_2
+ 94 moles N_2 + 2.51 moles H_2O → 8 moles CO_2 + 11.51 moles H_2O
+ 12.5 moles O_2 + 94 moles N_2

or

114 (lbm) C_8H_{18} + 800 (lbm) O_2
+ 2632 (lbm) N_2 + 45 (lbm) H_2O → 352 (lbm) CO_2 + 207 (lbm) H_2O
+ 400 (lbm) O_2 + 2632 (lbm) N_2

3591 (lbm) reactants = 3591 (lbm) products

In the future, to simplify equation writing, we shall designate the combustion equation as being on a molar or mass basis and drop the unit term for each element.

To determine the air-fuel ratio

$$m_a = 800 + 2632 + 45 = 3477 \text{ (lbm)}$$

and

$$m_f = 114 \text{ (lbm)}$$

so that

$$AF = m_a/m_f = 3477/114 = 30.5$$

as compared to 30.1 for the dry air solution.

The dew point, for

$$n_{H_2O} = 11.51 \text{ moles}$$

in the products and

$$n_{products} = 8 + 11.51 + 12.5 + 94 = 126.01 \text{ (moles)}$$

so that

$$x_{H_2O} = n_{H_2O}/n_{products} = 11.51/126.01 = 0.0913$$

and

$$p_{H_2O} = x_{H_2O}p_m = 0.0913(14.7) = 1.342 \text{ (psia)}$$

is

$$t_{dp} = 111.6 \text{ (F)}$$

compared to 103.8 for the dry air analysis.

DEFICIENT AIR SUPPLY

If the air supply is less than 100% theoretical, complete combustion cannot occur. This is an undesirable situation since incomplete combustion represents a loss of fuel energy. Because the reaction between hydrogen and oxygen is more active than that between carbon and oxygen, we shall assume that the formation of water vapor occurs first. Oxygen remaining after this process will be used to form carbon dioxide and carbon monoxide. The formation of CO indicates that complete combustion has not taken place.

Example 15.8. Assume that n-octane is burned in 90% of the theoretical air requirements. Determine the combustion equations and the air-fuel ratio using dry air.

The stoichiometric dry air requirements (Example 15.2) were 12.5 moles of O_2 and 47.0 moles of N_2. For 90% theoretical air, the moles of O_2 and N_2 are

$$n_{O_2} = 0.9(12.5) = 11.25$$

$$n_{N_2} = 0.9(47) = 42.30$$

Since 9 moles of H_2O are formed in the combustion of C_8H_{18}, the O_2 remaining for the combustion of the carbon is

$$11.25 - (9/2) = 11.25 - 4.5 = 6.75 \text{ (moles)}$$

The carbon reaction equation is

$$8\ C + 6.75\ O_2 \rightarrow xCO_2 + yCO$$

where x and y are the moles of CO_2 and CO in the products. Since the number of atoms of the elements must balance on each side of the equation, for the carbon

$$8 = x + y$$

and for the oxygen

$$6.75(2) = 2x + y$$

Solving for x and y gives

$$x = 5.5 \quad \text{and} \quad y = 2.5$$

The combustion equations are as follows: molar,

$$C_8H_{18} + 11.25\ O_2 + 42.3\ N_2 \rightarrow 9\ H_2O + 5.5\ CO_2 + 2.5\ CO + 42.3\ N_2$$

and mass,

$$114\ C_8H_{18} + 360\ O_2 + 1184.4\ N_2 \rightarrow 162\ H_2O + 242\ CO_2 + 70\ CO + 1184.4\ N_2$$

$$1658.4 \text{ (lbm) reactants} = 1658.4 \text{ (lbm) products}$$

The air-fuel ratio, where

$$m_a = 360 + 1184.4 = 1544.4 \text{ (lbm)}$$

and

$$m_f = 114 \text{ (lbm)}$$

is

$$AF = m_a/m_f = 1544.4/114 = 13.55$$

compared to 15.1 for stoichiometric air.

ORSAT ANALYSIS

To determine the theoretical air or the air-fuel ratio for a combustion process, an analysis of the products of combustion must be made. The orsat analyzer (Fig. 15.1) is capable of determining the percentage by volume of CO_2, O_2, and CO in the products. The system consists of a measuring column calibrated to contain 100 cc of gas, a water bottle used to pump the gas sample to the various chemical solutions, and three chemical solutions capable of selectively removing CO_2, O_2, and CO from the gas sample. Water is first pumped to the top of the measuring column by opening the three-way valve to the atmosphere. The valve is then opened to the gas source, and the water bottle is lowered, drawing the gas sample into the measuring column. This action may be repeated several times to purge the lines and equipment. The water bottle is then held level with the water height in the measuring column. This is necessary in all measurement steps to ensure that the compressible gas sample is under atmospheric pressure. The three-way valve and the first chemical solution valve are opened, and the gas sample is pumped to the solution capable of removing CO_2. If sulfur is present in the fuel, SO_2 will also be removed by this chemical. The gas is then returned to the measuring column, and a reading is made. It is advisable to repeat this procedure several times to ensure that all of the CO_2 has been absorbed. The same technique is used for the second and third

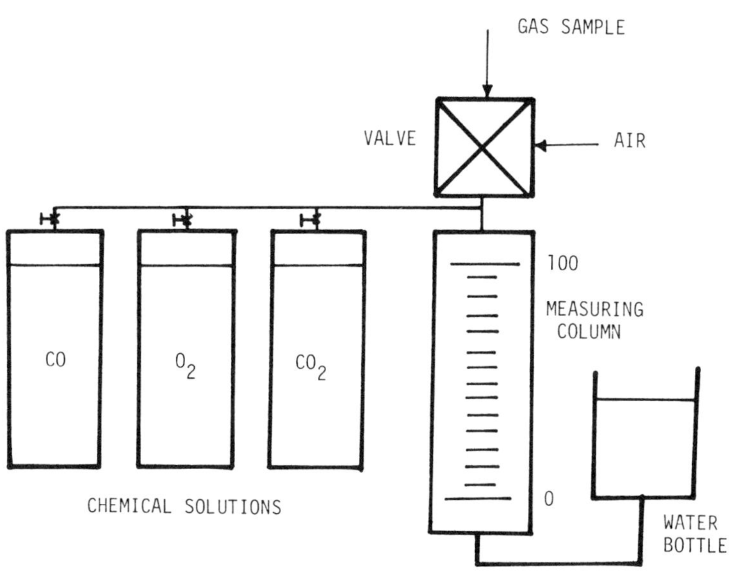

Figure 15.1

solutions that remove O_2 and CO. Removal of gas components must always be made in the order of CO_2, O_2, and CO.

The amount of water vapor in the gas sample cannot be obtained from the simple orsat since the sample is saturated by the water used to pump the gas.

The gas remaining after the separating processes is inert, at least to the chemicals used in the orsat, and is assumed to be nitrogen.

Example 15.9. The stable readings obtained from an orsat analysis of the products of combustion are 13.7%, 18.0%, and 18.5%. Determine the products of combustion.

Since the first reading is the result of the gas sample being placed in contact with the CO_2 absorbing chemical,

$$\%\ CO_2 = 13.7$$

The second reading consists of the CO_2 removed by the first chemical, potassium hydroxide, and the O_2 removed by the second chemical, pyrogallic acid; the percentage by volume of O_2 is

$$\%\ O_2 = 18.0 - 13.7 = 4.3$$

Likewise, for the third chemical solution, cuprous chloride,

$$\%\ CO = 18.5 - 18.0 = 0.5$$

By difference,

$$N_2 = 100 - 18.5 = 81.5\%$$

The presence of O_2 in the products of combustion indicates that air was available to complete the combustion process, but the presence of CO in the products means that part of the fuel was not completely burned. In actual combustion processes, both O_2 and CO can be present for either excess or deficient air, since even with excess air all of the fuel may not be burned, while with an air deficiency the oxygen may not find a combustible component in the fuel.

Example 15.10. An Oklahoma natural gas has a volumetric analysis of 84.1% CH_4, 6.7% C_2H_6, 0.8% CO_2, and 8.4% N_2. When the gas is burned in dry air, the resulting orsat analysis is 6.02% CO_2, 3.88% CO, and 5.04% O_2. Determine the combustion equation, the air-fuel ratio, and the percentage of theoretical air for the combustion process.

The molar combustion equation is

$$x(0.841\ CH_4 + 0.067\ C_2H_6 + 0.008\ CO_2 + 0.084\ N_2) + yO_2 + 3.76\ yN_2 \rightarrow 6.02\ CO_2 + 3.88\ CO + 5.04\ O_2 + (0.084x + 3.76y)\ N_2 + zH_2O$$

The moles of CO_2, CO, and O_2 in the products are obtained from the orsat analysis.

The unknown quantities in the combustion equation are

x = moles of fuel required to produce the orsat values

y = moles of O_2 entering with the dry air

z = moles of H_2O formed in the combustion process

The number of atoms of each element in the reactants and products must be equal.

The carbon (C) balance yields

$$0.841x + 0.067(2)x + 0.008x = 6.02 + 3.88$$

$$x = 10.07$$

The hydrogen (H) balance is

$$0.841(4)x + 0.067(6)x = 2z$$

$$z = 18.96$$

The oxygen (O) balance gives

$$0.008(2)x + 2y = 2(6.02) + 3.88 + 2(5.04) + z$$

$$2y = 44.80$$

$$y = 22.40$$

The nitrogen (N) balance is

$$0.084(2)x + 3.76(2)y = 2(0.084x + 3.76y)$$

The nitrogen balance equation must balance for all values of x and y and is of no use in determining the quantities x, y, and z.

The nitrogen in the products is

$$2[0.084(10.07) + 3.76(22.40)] = 170.14$$

so that the moles of N_2 in the products are 85.07.

A check for experimental accuracy is obtained by determining the moles of inert gas in the orsat analysis.

Assuming that the inert gas is N_2,

$$\text{Moles } N_2 = 100 - \text{moles } CO_2 - \text{moles } CO - \text{moles } O_2$$

$$= 100 - 6.02 - 3.88 - 5.04 = 85.06$$

The two separate determinations of the moles of N_2 in the products are in close agreement. Actually, the agreement is considerably closer than can be expected from a typical orsat analysis.

The combustion equation is

$$10.07(0.841\ CH_4 + 0.067\ C_2H_6 + 0.008\ CO_2 + 0.084\ N_2) + 22.4\ O_2 + 84.22\ N_2 \rightarrow 6.02\ CO_2 + 3.88\ CO + 5.04\ O_2 + 85.07\ N_2 + 18.96\ H_2O$$

The air-fuel ratio for

$$m_f = 10.07[0.841(16) + 0.067(30) + 0.008(44) + 0.084(28)]$$

$$= 10.07\ [18.17] = 183.0\ (\text{lbm})$$

and

$$m_a = 22.4(32) + 84.22(28) = 3075.0\ (\text{lbm})$$

is

$$AF = m_a/m_f = 3075.0/183.0 = 16.8$$

The stoichiometric air for the above quantity of fuel can be obtained from

$$\text{C in fuel} = 10.07(0.841) + 10.07(0.067)(2) + 10.07(0.008)$$

$$= 9.899\ (\text{moles})$$

$$\text{H in fuel} = 10.07(0.841)(4) + 10.07(0.067)(6)$$

$$= 37.923\ (\text{moles})$$

$$\text{O required to form } CO_2 = 9.899(2) = 19.798\ (\text{moles})$$

$$\text{O required to form } H_2O = 37.923/2 = 18.962\ (\text{moles})$$

$$\text{Gross O required} = 19.798 + 18.962 = 38.760\ (\text{moles})$$

$$\text{O in fuel} = 10.07(0.008)(2) = 0.161\ (\text{moles})$$

$$\text{O from air} = 38.760 - 0.161 = 38.599\ (\text{moles})$$

$$O_2 \text{ from air} = 19.300\ (\text{moles}) = 617.6\ (\text{lbm})$$

$$N_2 \text{ in air} = 19.300(3.76) = 72.568\ (\text{moles}) = 2031.9\ (\text{lbm})$$

$$\text{Stoichiometric air} = 617.6 + 2031.9 = 2649.5\ (\text{lbm})$$

From Eq. (15.1), where $m_a = 3075.0$ and $m_{sa} = 2649.5$ lbm,

$$\text{Excess air} = (m_a - m_{sa})/m_{sa} = (3075.0 - 2649.5)/2649.5 = 0.16$$

The excess air is 16% or 116% theoretical air.

Example 15.11. Assume the same data as in Example 15.10 except the air has a dry bulb temperature of 80 F and a relative humidity of 60%.

From Example 15.7

$$\omega = 0.01314(29) = 0.381 \text{ (lbm-}H_2O\text{/mole dry air)}$$

$$= 0.381(1/4.76) = 0.080 \text{ (lbm-}H_2O\text{/mole } O_2\text{)}$$

where the molecular weight of air is 29.

As in Example 15.10

$$x(0.841\ CH_4 + 0.067\ C_2H_6 + 0.008\ CO_2 + 0.084\ N_2) + yO_2 + 0.080\ yH_2O + 3.76\ yN_2 \rightarrow 6.02\ CO_2 + 3.88\ CO + 5.04\ O_2 + (0.084x + 3.76y)N_2 + (z + 0.080y)H_2O$$

The carbon balance is unchanged, so that

$$x = 10.07$$

The hydrogen balance gives

$$0.841(4)x + 0.067(6)x + 0.080(2)y = 2z + 0.080(2)y$$

Since the term in y is the same on both sides of the equation,

$$z = 18.96$$

as before. The oxygen balance yields

$$0.008(2)x + (2 + 0.080)y = 2(6.02) + 3.88 + 2.504 + (z + 0.080)y$$

and

$$y = 22.40$$

as before. As in Example 15.10, the mass of fuel is

$$m_f = 183.0 \text{ (lbm)}$$

and

$$m_a = 22.4(32) + 0.080(22.4)(18) + 3.76(22.4)(28) = 3107.3$$

so that

$$AF = m_a/m_f = 3107.3/183 = 17.0$$

compared to 16.8 for the dry air analysis. The stoichiometric air (wet) is

$$m_{sa} = 2645.9 + 2645.9(0.01314) = 2680.7 \text{ (lbm)}$$

and

$$\text{Excess air} = (3107.3 - 2680.7)/2680.7 = 0.16$$

or 120% theoretical air compared to 116% for dry air combustion.

COMBUSTION OF A HYDROCARBON

When a hydrocarbon is burned in air it is possible to determine its equivalent equation by use of the orsat analysis. The only assumption necessary is that the inert gas remaining after the orsat analysis is nitrogen. The equation of the fuel is expressed in the form C_xH_y. The unknowns x and y may have fractional values since the hydrocarbon may be a mixture of any number of basic fuels.

Example 15.12. The orsat analysis resulting from the combustion of a hydrocarbon is CO_2 = 5.18%, CO = 3.46%, and O_2 = 6.91%. Determine, in the form C_xH_y, the equation of the hydrocarbon.

The amount of nitrogen is

$$n_{N_2} = 100 - 5.18 - 3.46 - 6.91 = 84.45 \text{ (moles)}$$

The combustion equation is

$$C_xH_y + aO_2 + 3.76\ aN_2 \rightarrow 5.18\ CO_2 + 3.46\ CO + 6.91\ O_2 + bH_2O + 84.45\ N_2$$

From the carbon balance

$$x = 5.18 + 3.46 = 8.64$$

From the nitrogen balance

$$3.76(2)a = 84.45(2)$$

$$a = 22.46$$

From the oxygen balance

$$2a = 2(5.18) + 3.46 + 6.91(2) + b$$

$$b = 17.28$$

From the hydrogen balance

$$y = 2b = 34.56$$

The equation of the fuel is

$$C_{8.64}H_{34.56} = 8.64\ CH_4$$

and the fuel could be pure methane.

COMBUSTION OF SOLID FUELS

The principal solid fuel available for combustion purposes is coal. The United States is fortunate in that it possesses a large reserve of this important source of energy. Factors that have limited its use in the past are cost of production, cost of transportation, difficulty in handling in the power plant, and air pollution when used.

As the price and scarcity of alternate fuels increase, the first two factors are reduced in importance, even though these factors have also shown a marked increase. When burning coal air pollution is caused by solid particles and gaseous particles, primarily fly ash and sulfur dioxide, leaving the smoke stacks. The removal of these materials, while expensive, is feasible. It is somewhat easier to remove the solid particles, and since they contain unburned fuel, they may be recycled in the combustion process. Removal of the gaseous particles by washing the gaseous products is more difficult and expensive.

Coal may be roughly classified as anthracite, bituminous, subbituminous, and lignite. There are also subclassifications within these major groupings. Anthracitic fuels have a carbon content from 86% to 98% and this carbon content distinguishes them from bituminous coal. Bituminous coal has a heating value from 11,000 to 14,000 B/lbm, and this together with a low moisture content, less than 10%, separates bituminous from subbituminous or lignite. Subbituminous coal has a moisture content from 10% to 30% and a heating value from 8500 to 11,000 B/lbm. Lignite has a moisture content in excess of 30% and a heating value from 6000 to 8500 B/lbm.

Coal is analyzed by two different techniques, a proximate analysis and an ultimate analysis. The proximate analysis will include the moisture content, the volatile matter, the fixed carbon, and the ash present in the fuel. These quantities may be obtained by grinding a representative coal sample, and the difference between the mass of the sample before and after drying at a modest temperature determines the moisture content. This drying is usually performed in two steps: air drying at 50 to 60 F above room temperature until mass equilibrium is reached and 1 hr of drying at 220 to 230 F.

The volatile matter, which is a measure of the original carbohydrates in the fuel that are convertible to hydrocarbons, is then removed from the sample by heating at a higher temperature. To determine the carbon content the fuel is burned, and the component that is not combustible is designated as the ash.

An ultimate analysis is required for combustion calculations and includes the moisture content and the percentages by weight of carbon, hydrogen, sulfur, nitrogen, oxygen, and ash.

The heating value of the fuel is usually determined by the combustion of the fuel in a fixed volume bomb calorimeter. The fuel is burned in pure oxygen, under pressure, and the resulting temperature rise of the water surrounding the bomb is a measure of the heat contained in a fuel sample of known mass. ASTM standards describe the exact method to be employed in testing for the heating value of a solid fuel.

ULTIMATE ANALYSIS (WET)

All coals vary in their composition, so that there is no uniform fuel. The ultimate analysis differs depending on the geographic location of the fuel and, in the case of the moisture content, the conditions to which the mined coal has been exposed. To remove the latter variable, the ultimate analysis on a wet basis is converted to a dry analysis. Typical values of the wet ultimate analysis for a West Virginia bituminous coal* are

Component	% by Weight, Wet	% by Weight, Dry
Moisture	2.5	—
Carbon	75.0	76.9
Hydrogen	5.0	5.1
Sulfur	2.3	2.4
Nitrogen	1.5	1.5
Oxygen	6.7	6.9
Ash	7.0	7.2
	100.0	100.0

The analysis was converted to a dry basis by dividing each term, except the moisture content, in the wet analysis by

$$(100 - \%\ \text{moisture})/100 = (100 - 2.5)/100 = 0.975$$

*"Steam/Its Generation and Use," Babcock & Wilcox House Publication, New York, N. Y., 1972.

Example 15.13. Determine the stoichiometric quantity of dry air required to burn 1 lb of a West Virginia coal sample. The equations for the combustion of the reactants C, H_2, and S are

$$C + O_2 \rightarrow CO_2 \text{ (molar)}$$

$$12\ C + 32\ O_2 \rightarrow 44\ CO_2 \text{ (lbm)}$$

$$C + 32/12\ O_2 \rightarrow 44/12\ CO_2 \text{ (1 lbm-C)}$$

$$H_2 + 1/2\ O_2 \rightarrow H_2O \text{ (molar)}$$

$$2\ H_2 + 16\ O_2 \rightarrow 18\ H_2O \text{ (lbm)}$$

$$H_2 + 8\ O_2 \rightarrow 9\ H_2O \text{ (1 lbm-}H_2\text{)}$$

$$S + O_2 \rightarrow SO_2 \text{ (molar)}$$

$$32\ S + 32\ O_2 \rightarrow 64\ SO_2 \text{ (lbm)}$$

$$S + O_2 \rightarrow 2\ SO_2 \text{ (1 lbm-S)}$$

For the West Virginia coal, for 1 lbm of coal, the oxygen requirement for combustion is

Component	lbm/lbm-f	Multi.	Pounds of O_2 Required (+) or in Fuel (-)
C	0.769	32/12	+2.051
H_2	0.051	8	+0.408
S	0.024	1	+0.024
N_2	0.015	–	–
O_2	0.069	–	-0.069
Ash	0.072	–	–
	1.000 lbm		2.414 lbm of O_2 for dry air

By assuming air is 79% N_2 and 21% O_2 by volume,

$$0.21(32)\ O_2 + 0.79(28)\ N_2 = 28.84 \text{ air (lbm)}$$

$$O_2 + 3.29\ N_2 \rightarrow 4.29 \text{ air (1 lbm } O_2\text{)} \qquad (15.2)$$

and

$$0.233\ O_2 + 0.767\ N_2 \rightarrow \text{air (1 lbm air)} \qquad (15.3)$$

The stoichiometric air requirements are

$$\text{Air} = 2.414(4.29) = 10.36 \text{ (lbm-da)}$$

$$AF = m_a/m_f = 10.36 \text{ (lbm-da/lbm-fuel)}$$

Example 15.14. Determine the results of the combustion of the coal in Example 15.13 where the air-fuel ratio is 15.0. For an air-fuel ratio of

15, the dry air supplied per pound of fuel is $m_a = 15$. The excess air [Eq. (15.1)] is

$$\text{Excess air} = (m_a - m_{sa})/m_{sa} = (15 - 10.36)/10.36$$

$$= 0.448$$

or 144.8% theoretical air.

The quantity of oxygen and nitrogen supplied for an air-fuel ratio of 15 [from Eq. (15.3)] is

$$O_2 = 15(0.233) = 3.495 \text{ (lbm)}$$

and

$$N_2 = 15(0.767) = 11.505 \text{ (lbm)}$$

The total quantity of oxygen and nitrogen in the fuel and dry air is

$$O_2 = 3.495 + 0.069 = 3.564 \text{ (lbm)}$$

and

$$N_2 = 11.505 + 0.015 = 11.520 \text{ (lbm)}$$

The products of combustion include 11.520 lbm of N_2 and 0.072 lbm of ash, which are inert and do not react in the process. For 0.769 lbm of C,

$$0.769(44/12) = 2.820 \text{ (lbm) } CO_2$$

are formed. For 0.051 lbm of H_2,

$$0.051(9) = 0.459 \text{ (lbm) } H_2O$$

is formed. For 0.024 lbm of S,

$$0.024(2) = 0.048 \text{ (lbm) } SO_2$$

is formed. Excess oxygen is

$$3.495 - 2.414 = 1.081 \text{ (lbm) } O_2$$

The mass balance for the combustion process is

Component	Initial (lbm/lbm-f)	Final (lbm/lbm-f)
C	0.769	–
H_2	0.051	–
S	0.024	–
N_2	11.520	11.520
O_2	3.564	1.081

Component	Initial (lbm/lbm-f)	Final (lbm/lbm-f)
Ash	0.072	0.072
CO_2	–	2.820
H_2O	–	0.459
SO_2	–	0.048
	16.000	16.000

Example 15.15. Determine the dew point and the volume of the products for a mixture pressure of 14.7 psia and a temperature of 540 F. Use the products of combustion from Example 15.14.

To calculate the dew point and the gas constant for the product mixture, it is necessary to determine the number of moles of each component:

Gaseous Component	lbm/lbm-f	M	moles/lbm-f	x	xM
N_2	11.520	28	0.4114	0.768	21.50
O_2	1.081	32	0.0338	0.063	2.02
CO_2	2.820	44	0.0641	0.120	5.28
H_2O	0.459	18	0.0255	0.048	0.86
SO_2	0.048	64	0.0008	0.001	0.06
	15.928		0.5356	1.000	29.72

where, for N_2, moles/lbm-f = 11.520/28 = 0.4114 and the same procedure is used for the remaining gases.

The percentage by volume, mole fraction, for the water vapor in the products is

$$x_{H_2O} = n_{H_2O}/n_{products} = 0.0255/0.5356$$

$$= 0.0476$$

and the partial pressure of the water vapor, from Dalton's law of partial pressure, is

$$p_{H_2O} = x_{H_2O}p_m = 0.0476(14.7) = 0.6997 \text{ (psia)}$$

and from the steam tables

$$t_{dp} = 90.1 \text{ (F)}$$

The molecular weight of the products is

$$M_m = \Sigma x M = 29.72$$

giving a gas constant of

$$R_m = \bar{R}/M_m = 1545/29.72 = 51.99 \text{ (ft-lbf/lbm-R)}$$

so that

$$V = \frac{m_m R_m T_m}{p_m} = \frac{15.928(51.99)(1000)}{14.7(144)}$$

$$= 391.2 \text{ (ft}^3\text{/lbm-fuel)}$$

for the given mixture conditions.

Example 15.16. The West Virginia coal produces an orsat analysis of 12.72% CO_2, 2.60% CO, and 4.72% O_2. Determine the theoretical air for the combustion process. Since the orsat analysis is a volume analysis, it is necessary to express the fuel components in moles:

Component	lbm	Moles
C	0.769	0.0641
H_2	0.051	0.0255
S	0.024	0.0008
N_2	0.015	0.0005
O_2	0.069	0.0022

The ash, being inert and solid, is omitted from the calculations. Since the orsat includes both the CO_2 and SO_2 in a single reading,

$$CO_2 + SO_2 = 12.72$$

Let d be equal to the moles of SO_2; then the moles of CO_2 are 12.72 - d.

The combustion equation, molar, is

$$a(0.0641\ C + 0.0255\ H_2 + 0.0008\ S + 0.0005\ N_2 + 0.0022\ O_2) + bO_2 + 3.76\ bN_2 \rightarrow (12.72 - d)CO_2 + dSO_2 + 2.60\ CO + 4.72\ O_2 + (0.0005a + 3.76b)N_2 + eH_2O$$

For the sulfur balance

$$0.0008a = d$$

For the carbon balance

$$0.0641a = 12.72 - d + 2.60$$

$$0.0649a = 15.32$$

$$a = 236.1$$

and

$$d = 0.0008a = 0.0008(236) = 0.19$$

For the hydrogen (H_2) balance

$$0.0255a = e$$

$$e = 6.02$$

For the oxygen (O_2) balance

$$0.0022a + b = (12.72 - d) + d + (2.60/2) + 4.72 + (e/2)$$

$$0.52 + b = 12.72 + 1.30 + 4.72 + 3.01$$

$$b = 21.23$$

The nitrogen in the products, from the combustion equation, is

$$N_2 = 0.0005a + 3.76b = 0.12 + 79.82$$

$$= 79.94 \text{ (moles)}$$

From the orsat analysis

$$N_2 = 100 - 12.72 - 2.60 - 4.72 = 79.96 \text{ (moles)}$$

The amount of oxygen supplied, from the dry air, is

$$O_2 = b = 21.23 \text{ (moles)}$$

$$= 21.23(32) = 679.4 \text{ (lbm)}$$

From Eq. (15.2), the mass of air supplied is

$$m_a = 679.4(4.29) = 2915 \text{ (lbm)}$$

The mass of carbon burned is

$$m_c = 0.0641a(M_c) = 0.0641(236.1)(12) = 181.6 \text{ (lbm)}$$

and the weight of fuel, where $m_f = a$, is

$$m_f = m_c(m_f/m_c) = 181.6/0.769 = 236.1 \text{ (lbm)}$$

From Example 15.13, the stoichiometric air required to burn this quantity is

$$m_{sa} = 236.1(10.36) = 2446 \text{ (lbm)}$$

From Eq. (15.1)

$$\text{Excess air} = (m_a - m_{sa})/m_{sa}$$

$$= (2915 - 2446)/2446$$

$$= 0.192$$

and the theoretical air is 119%.

The sulfur balance can be eliminated in all problems by increasing the mass analysis for the carbon by, where m_S is the lbm-S in the fuel,

$$\Delta\ (\text{lbm-C}) = M_C/M_S\ (m_S) \qquad (15.4)$$

$$= 12/32\ (m_S)$$

Example 15.17. Rework Example 15.16 by use of Eq. (15.4).

From Eq. (15.4), for 2.4% S in the fuel,

$$\Delta\ (\text{lbm-C}) = (12/32)0.024 = 0.009\ (\text{lbm})$$

and

$$m_c = 0.769 + 0.009 = 0.778\ (\text{lbm})$$

The fuel sample now consists of

Component	lbm	Moles
C	0.778	0.0648
H_2	0.051	0.0255
N_2	0.015	0.0005
O_2	0.069	0.0022
Ash	0.072	—
	0.985	

Note that the total of the lbm column does not equal 1 lbm but is the equivalent analysis as if 1 lbm of fuel were being burned.

The molar combustion equation is

$$a(0.0648\ C + 0.0255\ H_2 + 0.0005\ N_2 + 0.0022\ O_2) + bO_2 + 3.76\ bN_2 \rightarrow 12.72\ CO_2 + 2.60\ CO + 4.72\ O_2 + (0.0005a + 3.76b)N_2 + eH_2O$$

For the carbon balance

$$0.0648a = 12.72 + 2.60$$

$$a = 236.4$$

For the hydrogen (H_2) balance

$$0.0255a = e$$

$$e = 6.03$$

For the oxygen (O_2) balance

$$0.0022a + b = 12.72 + (2.60/2) + 4.72 + (e/2)$$

$$b = 21.24$$

The nitrogen in the products is

$$N_2 = 0.0005a + 3.76b = 79.98 \text{ (moles)}$$

From Eq. (15.2)

$$m_a = 21.24(32)(4.29) = 2916 \text{ (lbm)}$$

and

$$m_c = 0.0648aM_c = 0.0648(236.4)(12) = 183.8 \text{ (lbm)}$$

so that

$$m_f = M_c(m_f/m_c) = 183.8/0.778 = 236.2$$

Also

$$m_{sa} = 236.2(10.36) = 2447 \text{ (lbm)}$$

giving

$$\text{Excess air} = (m_a - m_{sa})/m_{sa} = (2916 - 2447)/2447 = 0.192$$

as in Example 15.16.

CARBON IN REFUSE

In previous discussions it was assumed that all of the carbon in the fuel sample reacted with the oxygen. In burning solid fuels an appreciable part of the carbon may remain unburned. For large coal particles there will be insufficient time to burn the sample completely. For powdered coal combustion a portion of the unburned carbon may be found in the stack gases.

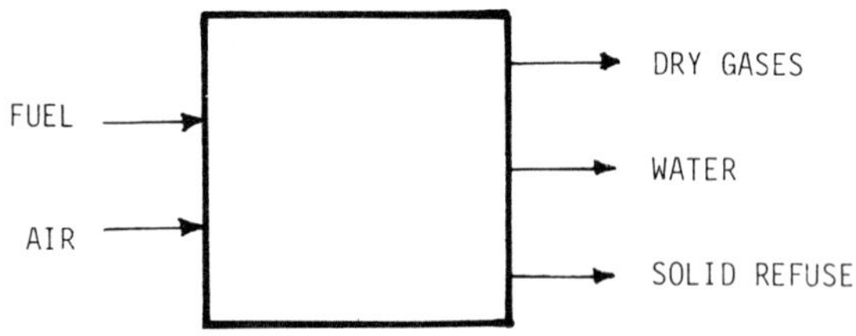

Figure 15.2

The amount of unburned carbon is determined by measuring the carbon in the refuse. A refuse sample of known mass is burned at high temperatures. The loss of weight of the sample represents the carbon in the refuse sample.

The mass balance for a combustion process is sketched in Fig. 15.2. In equation form this is

$$m_f + m_a = m_{dg} + m_{H_2O} + m_r \tag{15.5}$$

It is possible to determine the mass of fuel and refuse used in a particular combustion process. In large-sized power plants this would represent large quantities of material to be handled. The mass of air actually supplied, the dry gaseous products, and the water vapor are difficult to evaluate.

To simplify the calculations a carbon balance for each of the components in Fig. 15.2 will be used, where

$$m_{C-f} + m_{C-a} = m_{C-dg} + m_{C-H_2O} + m_{C-r}$$

Since there is no carbon in the dry air or water vapor,

$$m_{C-a} = m_{C-H_2O} = 0$$

and

$$m_{C-dg} = m_{C-f} - m_{C-r}$$

Dividing by the mass of fuel burned yields

$$m_{C-dg}/m_f = (m_{C-f}/m_f) - (m_{C-r}/m_f) \tag{15.6}$$

which permits calculations based on a unit mass of fuel. The ratio m_{C-f}/m_f is known from the ultimate analysis. To determine the ratio m_{C-r}/m_f, let

$$m_{C-r}/m_f = (m_{C-r}/m_{ash})(m_{ash}/m_f) \tag{15.7}$$

The ratio m_{ash}/m_f is also known from the ultimate analysis. We still must determine the ratio m_{C-r}/m_{ash}. A mass balance for the refuse gives

$$m_r = m_{C-r} + m_{ash}$$

and by dividing by the mass of refuse

$$m_r/m_r = 1 = (m_{C-r}/m_r) + (m_{ash}/m_r)$$

$$m_{ash}/m_r = 1 - (m_{C-r}/m_r) \tag{15.8}$$

The burning of a refuse sample gives the value of m_{C-r}/m_r, so the ratio m_{ash}/m_r may be calculated.

Multiplying the ratio m_r/m_{ash} by the ratio m_{C-r}/m_r, both of which are known, gives

$$m_{C-r}/m_{ash} = (m_r/m_{ash})(m_{C-r}/m_r) \tag{15.9}$$

which is the ratio to be determined for use in Eq. (15.7).

Example 15.18. Given is the West Virginia coal ultimate dry mass analysis of

C = 76.9%	N_2 = 1.5%
S = 2.4%	O_2 = 6.9%
H_2 = 5.1%	ash = 7.2%

The refuse is found to contain 25% unburned carbon. Determine m_{C-dg}/m_f. Correcting for the sulfur content as in Example 15.17 gives

$$m_{C-f}/m_f = 0.769 + (12/32)(0.024)$$

$$= 0.778$$

From the refuse data

$$m_{C-r}/m_r = 0.25$$

From Eq. (15.8)

$$m_{ash}/m_r = 1 - (m_{C-r}/m_r) = 1 - 0.25 = 0.75$$

From Eq. (15.9)

$$m_{C-r}/m_{ash} = (m_r/m_{ash})(m_{C-r}/m_r)$$

$$= (1/0.75)0.25 = 0.333$$

From the ultimate fuel analysis

$$m_{ash}/m_f = 0.072$$

and from Eq. (15.7)

$$m_{C-r}/m_f = (m_{C-r}/m_{ash})(m_{ash}/m_f)$$

$$= 0.333(0.072) = 0.024$$

From Eq. (15.6)

$$m_{C-dg}/m_f = (m_{C-f}/m_f) - (m_{C-r}/m_f)$$

$$= 0.778 - 0.024 = 0.754$$

compared to 0.778 if all of the carbon and sulfur had been burned.

Example 15.19. Given are the results of Example 15.18 and an orsat analysis of CO_2 = 11.03%, O_2 = 6.79%, and CO = 2.29% when the coal is burned in dry air. Determine the theoretical air and air-fuel ratio.

The carbon content in the gaseous products represents the carbon actually burned in the combustion process. Therefore, the fuel burned was

Component	lbm/lbm-f	Moles/lbm-f
C	0.754	0.0628
H_2	0.051	0.0255
N_2	0.015	0.0005
O_2	0.069	0.0022

The combustion equation (see Example 15.17) is

$$a(0.0628\ C + 0.0255\ H_2 + 0.0005\ N_2 + 0.0022\ O_2) + bO_2 + 3.76\ bN_2 \rightarrow 11.03\ CO_2 + 6.79\ O_2 + 2.29\ CO + (0.0005a + 3.76b)N_2 + eH_2O$$

For the carbon balance

$$0.0628a = 11.03 + 2.29$$

$$a = 212.1$$

For the hydrogen (H_2) balance

$$0.0255a = e$$

$$e = 5.41$$

For the oxygen (O_2) balance

$$0.0022a + b = 11.03 + 6.79 + (2.29/2) + (e/2)$$

$$b = 21.20$$

The nitrogen in the products is

$$N_2 = 0.0005a + 3.76b = 79.82 \text{ (moles)}$$

or from the orsat analysis

$$N_2 = 100 - 11.03 - 6.79 - 2.29 = 79.89 \text{ (moles)}$$

The mass of carbon burned is

$$m_C = 0.0628aM_C = 0.0628(212.1)(12)$$

$$= 159.8 \text{ (lbm)}$$

and the mass of fuel burned is

$$m_f = m_C(m_f/m_{C\text{-}dg}) = 159.8/0.754 = 212.0 \text{ (lbm)}$$

The mass of air supplied is

$$m_a = bM_{O_2}(4.29) = 21.20(32)(4.29) = 2910 \text{ (lbm)}$$

The air-fuel ratio is

$$AF = m_a/m_f = 2910/212 = 13.73$$

for the combustion process which results in 25% unburned carbon in the refuse. The mass of fuel actually supplied is

$$m_f' = m_f(m'_{c\text{-}f}/m_f)(m_f/m_{c\text{-}f})$$

$$= [212.0(0.778)]/0.754 = 218.7$$

where $m'_{c\text{-}f}/m_f = 0.778$ from Example 15.17, so that

$$AF' = m_a/m_f' = 2910/218.7 = 13.31$$

The excess air is

$$\text{Excess air} = (m_a - m_{sa})/m_{sa} = (13.31 - 10.36)/10.36 = 0.285$$

based on the actual air and fuel supplied.

A second technique may be employed for the solution of Example 15.19. Assuming the nitrogen content in the products can be accurately determined from the orsat analysis, we can calculate the ratio

$$m_{dg}/m_{C-dg}$$

The ratio m_{dg}/m_f can then be determined from

$$m_{dg}/m_f = (m_{C-dg}/m_f)(m_{dg}/m_{C-dg}) \tag{15.10}$$

The quantity of water vapor in the gaseous products can come only from the fuel since the air supply to the combustion process is assumed to be dry; therefore,

$$m_{H_2O}/m_f = (m_{H_2}/m_f)(m_{H_2O}/m_{H_2}) \tag{15.11}$$

where m_{H_2}/m_f is obtained from the ultimate analysis and $m_{H_2O}/m_{H_2} = 9$ is obtained from Example 15.13. The ratio m_r/m_f can be obtained from

$$m_r/m_f = (m_{ash}/m_f)(m_r/m_{ash}) \tag{15.12}$$

Dividing the mass balance equation [Eq. (15.5)] by the mass of fuel gives

$$1 + (m_a/m_f) = (m_{dg}/m_f) + (m_{H_2O}/m_f) + (m_r/m_f)$$

$$m_a/m_f = (m_{dg}/m_f) + (m_{H_2O}/m_f) + (m_r/m_f) - 1 \tag{15.13}$$

which is the air-fuel ratio.

Example 15.20. Rework Example 15.19 by the second method.

For the dry gaseous products

Component	Moles/100 Moles of Gas	M = lb/Mole	lbm/100 Moles of Gas	lbm-C/100 Moles of Gas
CO_2	11.03	44	485	132.4
O_2	6.79	32	217	–
CO	2.29	28	64	27.5
N_2	79.89	28	2237	–
			3003	159.9

where lbm-C/100 moles of gas for CO_2 is

$$11.03(12) = 132.4$$

To calculate the ratio m_{dg}/m_{C-dg},

$$m_{dg}/m_{C-dg} = (\text{lbm-dg}/100 \text{ moles gas})(100 \text{ moles gas}/\text{lbm-C-dg})$$

$$= 3003/159.9 = 18.78$$

From Eq. (15.10)

$$m_{dg}/m_f = (m_{C\text{-}dg}/m_f)(m_{dg}/m_{C\text{-}dg})$$

$$= 0.754(18.78) = 14.16$$

where the ratio $m_{C\text{-}dg}/m_f$ is the result of Example 15.18. From Eq. (15.11)

$$m_{H_2O}/m_f = (m_{H_2}/m_f)(m_{H_2O}/m_{H_2})$$

$$= 0.051(9) = 0.46$$

From Eq. (15.12)

$$m_r/m_f = (m_{ash}/m_f)(m_r/m_{ash})$$

$$= 0.072/0.75 = 0.10$$

where m_{ash}/m_f is obtained from the ultimate analysis and m_r/m_{ash} is obtained from Eq. (15.8). From Eq. (15.13)

$$m_a/m_f = (m_{dg}/m_f) + (m_{H_2O}/m_f) + (m_r/m_f) - 1$$

$$= 14.16 + 0.46 + 0.10 - 1$$

$$= 13.72$$

which is identical to Example 15.19. Since the stoichiometric quantity of air required to completely burn 1 lb of fuel in dry air is

$$m_{sa}/m_f = 10.36$$

from Example 15.13, the excess air is

$$\text{Excess air} = [13.72(0.754/0.778) - 10.36]/10.36 = 0.283$$

where

$$m'_{c\text{-}f}/m_f = 0.778 \quad \text{and} \quad m_{c\text{-}f}/m_f = 0.754$$

as in Example 15.19.

COMBUSTION WITH WATER IN AIR AND FUEL

Example 15.21. Calculate the air-fuel ratio and excess air if the West Virginia coal is burned in air having a dry bulb temperature of 60 F and a relative humidity of 60%. Due to outside storage, the moisture content of the fuel as fired is 10%.

For the West Virginia coal the ultimate analysis for the dry fuel is

Component	% by Weight, Dry	% by Weight, Wet	Moles/lbm-fuel
Carbon	0.769	0.692	0.0577
Hydrogen	0.051	0.046	0.0230
Sulfur	0.024	0.022	0.0007
Nitrogen	0.015	0.013	0.0005
Oxygen	0.069	0.062	0.0019
Ash	0.072	0.065	–
Moisture	–	0.100	0.0056
	1.000	1.000	

where

$$\% \ C_{wet} = (1 - \text{moisture})(\% \ C_{dry}) = 0.9(0.769) = 0.692$$

Correcting for the sulfur content gives

$$m_{C\text{-}f}/m_f = 0.692 + (12/32)(0.022) = 0.700$$

or

$$0.700/12 = 0.0583 \ (\text{mole } C/m_f)$$

The specific humidity of the air is 46 gr/lbm-dry air from the psychrometric chart (Appendix I), so that

$$\omega = 46/7000 = 0.00657 \ (\text{lbm-}H_2O/\text{lbm-dry air})$$

$$= 0.00657(4.29) = 0.02818 \ (\text{lbm-}H_2O/\text{lbm-}O_2)$$

$$= (0.02818)(32/18) = 0.0501 \ (\text{mole } H_2O/\text{mole } O_2)$$

The orsat analysis is 10.7% CO_2, 2.7% CO, and 6.9% O_2. The combustion equation is

$$a(0.0583 \ C + 0.0230 \ H_2 + 0.0005 \ N_2 + 0.00190 \ O_2 + 0.0056 \ H_2O) + 0.0501 \ bH_2O + bO_2 + 3.76 \ bN_2 \rightarrow 10.7 \ CO_2 + 2.7 \ CO + 6.9 \ O_2 + (0.0005a + 3.76b)N_2 + eH_2O$$

For the carbon balance

$$0.0583a = 10.7 + 2.7 = 13.4$$

$$a = 229.8$$

For the hydrogen (H_2) balance

$$0.023a + 0.0056a + 0.0501b = e$$

$$e = 6.5723 + 0.0501b$$

For the oxygen (O_2) balance

$$0.0019a + (0.0056/2)a + (0.0501/2)b + b = 10.7 + (2.7/2) + 6.9 + (e/2)$$

$$b = 17.433 + 0.4878e$$

By substituting for b in the hydrogen equation

$$e = 6.5723 + 0.8733 + 0.0244e$$

so that

$$e = 7.63$$

and

$$b = 21.16$$

The nitrogen in the products is

$$N_2 = 0.0005a + 3.76b = 79.7 \text{ (moles)}$$

or by difference, from the orsat analysis,

$$N_2 = 100 - CO_2 - CO - N_2 = 100 - 10.7 - 2.7 - 6.9$$

$$= 79.7 \text{ (moles)}$$

The mass of carbon burned is

$$m_C = 0.0583aM_C = 0.0583(229.8)(12)$$

$$= 160.8 \text{ (lbm)}$$

and the mass of fuel burned is

$$m_f = m_C(m_f/m_C) = 160.8/0.700 = 229.7 \text{ (lbm)}$$

The mass of air supplied is

$$m_a = bM_{O_2}(4.29) = 21.16(32)(4.29) = 2905 \text{ (lbm)}$$

giving an air-fuel ratio of

$$AF = m_a/m_f = 2905/229.7 = 12.6$$

The stoichiometric air is determined from the ratio of reactants where

Fuel Component	Amount (moles)	Equation	Moles of O_2 Required
C	0.0577	$C + O_2 \rightarrow CO_2$	0.0577
H_2	0.0230	$H_2 + 1/2\ O_2 \rightarrow H_2O$	0.0115
S	0.0007	$S + O_2 \rightarrow SO_2$	0.0007
			0.0699
Less oxygen in the fuel			0.0019
Moles of oxygen/m_f, 100% theoretical air			0.0680

$$\text{Stoichiometric air} = 0.0680(m_f)(M_{O_2})(4.29)$$

$$= 0.0680(229.7)(32)(4.29) = 2144 \text{ (lbm)}$$

The excess air is

$$\text{Excess air} = (m_a - m_{sa})/m_{sa} = (2905 - 2144)/2144 = 0.35$$

or 35.0% excess air, which is 135.0% theoretical air.

PROBLEMS

15.1 n-Pentane, C_5H_{12}, is completely burned in oxygen. Determine the moles and pound mass of each reactant and product.

15.2 Determine the air-fuel ratio when n-pentane is burned in 100% theoretical dry air.

15.3 Determine the air-fuel ratio when ethyl alcohol, C_2H_5OH, is burned in stoichiometric air.

15.4 Determine the dew point of the products of combustion in Prob. 15.2 for a product mixture pressure of 14.7 psia. Combustion is in dry air.

15.5 Determine the dew point of the products of combustion in Prob. 15.2 for a product mixture pressure of 14.7 psia. Combustion is in air with a dry bulb temperature of 80 F and a relative humidity of 60%.

15.6 Determine the air-fuel ratio and the volumetric analysis of the products of combustion if n-pentane is burned in 200% theoretical dry air.

15.7 Determine the dew point in Prob. 15.6 if the fuel is burned in air with a dry bulb temperature of 80 F and a relative humidity of 60%. $p_m = 14.7$ psia.

15.8 Determine the moles of the products and the air-fuel ratio when n-pentane is burned in 80% theoretical air.

15.9 Repeat Prob. 15.8 for 40% theoretical air.

15.10 n-Pentane is burned in dry air, yielding an orsat analysis of 4.2% CO_2, 7.4% CO, and 6.9% O_2. Determine the combustion equation, the air-fuel ratio, and the percentage of theoretical air for the process.

15.11 Determine the equation of a hydrocarbon fuel that produces orsat values of 12.6% CO_2, 1.9% CO, and 3.6% O_2.

15.12 Given is a coal with the gravimetric analysis of C = 73.0%, H_2 = 6.0%, S = 3.1%, N_2 = 1.3%, O_2 = 6.7%, and H_2O = 3.0%. The balance is ash. Determine the quantity of dry air to burn 1 lbm of dry coal.

15.13 Determine the excess air, the theoretical air, and the mass combustion equation when the dry coal in Prob. 15.12 is burned with an air-fuel ratio of 12.

15.14 Determine the dew point and the volume of the products for a mixture pressure of 14.7 psia and a temperature of 1040 F using the fuel and the air-fuel ratio in Prob. 15.13.

15.15 The fuel in Prob. 15.12 is burned dry and produces orsat readings of 11.38% CO_2 and SO_2, 3.63% CO, and 4.71% O_2. Determine the air-fuel ratio, the excess air, the theoretical air, and the percentage of N_2 in the products from the orsat analysis and from the combustion equation.

15.16 Determine the air-fuel ratio and the theoretical air for Prob. 15.15 by use of Eq. (15.4).

15.17 The dry fuel in Prob. 15.12 is burned in dry air. The refuse is found to contain 15% unburned carbon. Determine the carbon in the dry gases per pound of fuel burned. Convert the sulfur content in the fuel to an equivalent carbon content.

15.18 The coal as burned in Prob. 15.17 produces an orsat analysis of 10.3% CO_2, 3.3% CO, and 6.5% O_2. Determine the actual air-fuel ratio, the air-fuel ratio based on the coal actually supplied, and the actual amount of theoretical air supplied.

15.19 Determine the actual air-fuel ratio by means of the second method, Example 15.20.

16

COMBUSTION REACTIONS

In Chap. 15 combustion equations were used to determine the quantity of reactants and products that are present in a given combustion process. Combustion reactions are necessary to calculate either the heat transfer during the combustion process or the temperature of the products after the process. To determine either the heat transfer or the temperature of the products three additional sets of tabular values are required:

1. Enthalpy of reaction (Appendix J)
2. Heat of formation (Appendix K)
3. Internal energy and enthalpy of various gases (Appendix L)

ENTHALPY OF REACTION

The difference between the enthalpy of the reactants H_{r0} and the enthalpy of the products H_{p0} at a specified temperature, normally 77 F, is the enthalpy of reaction. If the reactants enter the combustion process at 77 F and combustion is allowed to occur, cooling of the products to 77 F will require the removal of a quantity of heat from the system. For a constant pressure combustion process in which

$$\Delta KE = \Delta PE = W_{sf} = 0$$

from the energy equation,

$$Q = H_{p0} - H_{r0} = \Delta H_R \text{ (B)} \tag{16.1}$$

where ΔH_R is the enthalpy of reaction at the reference temperature t_0.

Tabular values for the enthalpy of reaction for 1 lbm of fuel Δh_R are given in Appendix J.

It is necessary to designate two different values for the enthalpy of reaction if hydrogen is present in the fuel. If the water formed in the combustion process is condensed to liquid, the value in the $H_2O(l)$ column is used. If the water remains in the gaseous state, the $H_2O(g)$ value is used.

Example 16.1. Given n-octane as a gaseous fuel, determine the enthalpy of reaction if the water remains in the gaseous phase. From Appendix J, for the water as a liquid,

$$\Delta h_R[H_2O(l)] = -20{,}750 \text{ (B/lbm)}$$

From the molar combustion equation

$$C_8H_{18}(g) + 12.5\ O_2 \rightarrow 8\ CO_2 + 9\ H_2O$$

The enthalpy required to condense 1 lbm of water at 77 F, from the steam tables (Appendix D), is

$$h_{fg} = 1050.3 \text{ (B/lbm)}$$

For 9 moles,

$$h_{fg} = 9(18)(1050.3) = 170{,}150 \text{ (B)}$$

which is the heat necessary to condense the water formed by the combustion of 114 lbm of fuel. Therefore,

$$\Delta h_R[H_2O(g)] = -20{,}750 + (170{,}150/114)$$

$$= -20{,}750 + 1492 = -19{,}260 \text{ (B/lbm)}$$

which is reasonably close to the tabular value of -19,200.

HEAT FOR CONSTANT PRESSURE COMBUSTION PROCESS

For any process in which the fuel is completely burned, the enthalpy of reaction in Appendix J is the negative of the heat of combustion. These values are often described as the heating value of the fuel where the higher heating value is

$$\text{HHV} = -\Delta h_R[H_2O(l)]$$

and the lower heating value is

$$\text{LHV} = -\Delta h_R[H_2O(g)]$$

The HHV assumes that the water formed in the combustion process is condensed.

To determine the heat transfer during a constant pressure process for which

$$\Delta KE = \Delta PE = W_{sf} = 0$$

the energy equation is

$$Q = H_p - H_r$$

where the products and reactants may leave or enter the process at any temperature. The energy equation may be rewritten as

$$Q = (H_p - H_{p0}) - (H_r - H_{r0}) + H_{p0} - H_{r0}$$

$$= (H_p - H_{p0}) - (H_r - H_{r0}) + \Delta H_R$$

and

$$Q_{in} + Q_{out} = (H_p - H_{p0}) - (H_r - H_{r0}) + \Delta H_R$$

For most processes, $Q_{in} = 0$, and Q_{out} is assigned a positive sign, rather than its conventional negative connotation:

$$Q_{out} = (H_r - H_{r0}) - (H_p - H_{p0}) - \Delta H_R \qquad (16.2)$$

Example 16.2. Determine the heat transfer when 1 lbm of octane, as a gas, is burned in an open system, constant pressure process. Reactants enter at 60 F, and products leave at 800 F. Fuel is burned in 100% theoretical dry air. The molar combustion equation, for complete combustion, is

$$C_8H_{18} + 12.5\ O_2 + 47.0\ N_2 \rightarrow 8\ CO_2 + 9\ H_2O + 47\ N_2$$

For hydrocarbon fuels, no appreciable error results if we assume

$$c_p = 0.5\ (B/lbm\text{-}R)$$

For the reactants,

$$H_r - H_{r0} = [mc_p(t_1 - t_0)]_{C_8H_{18}} + [mc_p(t_1 - t_0)]_{O_2} + [mc_p(t_1 - t_0)]_{N_2}$$

where t_1 is the reactant temperature and to = 77 F. Therefore,

$$H_r - H_{r0} = [114(0.5) + 12.5(32)(0.219) + 47(28)(0.249)](60 - 77)$$

$$= -8030\ (B)$$

From Appendix J, Δh_R = -19,200 B/lbm since the water will be a vapor at a product temperature of 800 F. For the products,

$$H_p - H_{p0} = [mc_p(t_2 - t_0)]_{CO_2} + [mc_p(t_2 - t_0)]_{H_2O} + [mc_p(t_2 - t_0)]_{N_2}$$

where t_2 is the product temperature. Therefore,

$$H_p - H_{p0} = [8(44)(0.202) + 9(18)(0.445) + 47(28)(0.249)](800 - 77)$$

$$= 340,440 \text{ (B)}$$

From Eq. (16.2), where the mass of 1 mole of fuel is 114 lbm,

$$Q_{out} = (H_r - H_{r0}) - (H_p - H_{p0}) - \Delta H_R$$

$$= -8030 - 340,440 + 114(19,200)$$

$$= 1,840,350 \text{ (B/mole)}$$

and

$$Q_{out} = 1,840,350/114 = 16,140 \text{ (B/lbm)}$$

Example 16.3. If the octane in Example 16.2 is burned in 50% excess air, determine the product temperature for the same heat transfer.

The molar combustion equation is

$$C_8H_{18} + 18.75\ O_2 + 70.50\ N_2 \rightarrow 8\ CO_2 + 9\ H_2O + 6.25\ O_2 + 70.50\ N_2$$

For the reactants,

$$H_r - H_{r0} = [(mc_p)_{C_8H_{18}} + (mc_p)_{O_2} + (mc_p)_{N_2}](t_1 - t_0)$$

$$= [114(0.5) + 18.75(32)(0.219) + 70.50(28)(0.249)](60 - 77)$$

$$= -11,560 \text{ (B)}$$

From Eq. (16.2)

$$H_p - H_{p0} = (H_r - H_{r0}) - \Delta H_R - Q_{out}$$

$$= -11,560 - 114(-19,200) - 1,840,350$$

$$= 336,890 \text{ (B)}$$

For the products,

$$[(mc_p)_{CO_2} + (mc_p)_{H_2O} + (mc_p)_{O_2} + (mc_p)_{N_2}](t_2 - t_0) = 336,890$$

and

$$[8(44)(0.202) + 9(18)(0.445) + 6.25(32)(0.219)$$

$$+ 70.50(28)(0.249)](t_2 - 77) = 336,890$$

$$678.5(t_2 - 77) = 336,890$$

so that

$$t_2 = 574 \text{ (F)}$$

The effect of the excess air is to reduce the temperature, for the same heat transfer, by 226 F.

HEAT OF FORMATION

If the fuel is not completely burned, the enthalpy of reaction given in Appendix J cannot be used. Instead the enthalpy of reaction for the incomplete combustion process must be calculated. For this calculation, the heats of formation (Appendix K) are required. The heat of formation for elements such as carbon, nitrogen, oxygen, etc. is zero.

The enthalpy of reaction is calculated from the equation

$$\Delta H_R = (\Delta H_f)_{products} - (\Delta H_f)_{reactants} \tag{16.3}$$

Example 16.4. Determine the enthalpy of reaction for gaseous octane when burned in stoichiometric oxygen at 77 F.

The molar combustion equation is

$$C_8H_{18} + 12.5\ O_2 \rightarrow 8\ CO_2 + 9\ H_2O$$

For the products from Appendix K, water as a gas,

$$(\Delta H_f)_{products} = 8\ \Delta h_{f\text{-}CO_2} + 9\ \Delta h_{f\text{-}H_2O}$$

$$= 8(-169{,}290) + 9(-104{,}040)$$

$$= -2{,}290{,}680 \text{ (B)}$$

For the reactants,

$$(\Delta H_f)_{reactants} = \Delta h_{f\text{-}C_8H_{18}} + 12.5\Delta h_{f\text{-}O_2}$$

$$= -89{,}640 + 12.5(0) = -89{,}640 \text{ (B)}$$

From Eq. (16.3)

$$\Delta H_R = (\Delta H_f)_{products} - (\Delta H_f)_{reactants}$$

$$= -2{,}290{,}680 + 89{,}640 = -2{,}201{,}140$$

and for 1 lbm

$$\Delta h_R = -2{,}201{,}140/114 = -19{,}310 \text{ (B/lbm)}$$

compared to -19,200 B/lbm in Appendix J.

Example 16.5. Determine the enthalpy of reaction for the combustion of gaseous octane in 90% theoretical air at 77 F. Water in the products is gaseous. See Example 15.8.

The molar combustion equation is

$$C_8H_{18} + 0.9(12.5)\ O_2 + 42.3\ N_2 \rightarrow aCO_2 + bCO + 9\ H_2O + 42.3\ N_2$$

From the carbon balance

$$8 = a + b$$

and from the oxygen balance

$$0.9(12.5) = a + (b/2) + (9/2)$$

so that a = 5.5 and b = 2.5.

For the products from Appendix K

$$(\Delta H_f)_{products} = 5.5\ \Delta h_{f\text{-}CO_2} + 2.5\ \Delta h_{f\text{-}CO} + 9\ \Delta h_{f\text{-}H_2O}$$

$$= 5.5(-169{,}290) + 2.5(-47{,}560) + 9(-104{,}040)$$

$$= -1{,}986{,}360\ (B)$$

For the reactants

$$(\Delta H_f)_{reactants} = \Delta h_{f\text{-}C_8H_{18}} = -89{,}640$$

From Eq. (16.3)

$$\Delta H_R = -1{,}986{,}360 + 89{,}640 = -1{,}896{,}720\ (B)$$

and

$$\Delta h_R = -1{,}896{,}720/114 = -16{,}640\ (B/lbm)$$

The magnitude of the enthalpy of reaction has changed by 2560 B/lbm due to incomplete combustion.

Example 16.6. Gaseous octane is burned in dry air. The orsat analysis is 8.5% CO_2, 3.9% CO, and 4.6% O_2. Determine the enthalpy of reaction, water as a gas.

Since the combustion is incomplete, as indicated by the presence of carbon monoxide in the orsat analysis, the heats of formation must be used to calculate the enthalpy of reaction.

The molar combustion equation is

$$aC_8H_{18} + bO_2 + 3.76\ bN_2 \rightarrow 8.5\ CO_2 + 3.9\ CO + 4.6\ O_2 + dH_2O + 3.76\ bN_2$$

From the carbon balance

$$8a = 8.5 + 3.9 \quad \text{or} \quad a = 1.55$$

From the hydrogen H balance

$$18a = 2d \quad \text{or} \quad d = 9(1.55) = 13.95$$

From the oxygen O_2 balance

$$b = 8.5 + (3.9/2) + 4.6 + (d/2)$$

and

$$b = 22.02$$

For the nitrogen

$$n_{N_2} = 3.76b = 82.8 \text{ (moles)}$$

or from the orsat analysis

$$n_{N_2} = 100 - 8.5 - 3.9 - 4.6 = 83.0 \text{ (moles)}$$

The heat of formation for the products is

$$(\Delta H_f)_p = 8.5\ \Delta h_{f\text{-}CO_2} + 3.9\ \Delta h_{f\text{-}CO} + 13.95\ \Delta h_{f\text{-}H_2O}$$

where the elements oxygen and nitrogen are omitted. Therefore,

$$(\Delta H_f)_p = 8.5(-169,290) + 3.9(-47,560) + 13.95(-104,040)$$

$$= -3,075,810 \text{ (B)}$$

For the reactants,

$$(\Delta H_f)_r = 1.55\Delta h_{f\text{-}C_8H_{18}} = 1.55(-89,640)$$

$$= -138,940$$

For the products,

$$\Delta H_R = -3,075,810 + 138,940 = -2,936,870 \text{ (B)}$$

and

$$\Delta h_R = -2,936,870/[114(1.55)] = -16,620 \text{ (B/lbm)}$$

CONSTANT VOLUME COMBUSTION

For a closed system, constant volume process, $Q = \Delta U$. For a process where the reactants and products enter at t_0, 77 F,

$$Q = U_{p0} - U_{r0} = \Delta U_R$$

From the definition of enthalpy

$$H = U + pV$$

or

$$\Delta H = \Delta U + \Delta pV$$

and

$$\Delta U_R = \Delta H_R - \Delta pV \qquad (16.4)$$

where ΔU_R is the internal energy of the reaction.

To evaluate ΔpV

$$\Delta pV = (pV)_{products} - (pV)_{reactants}$$

$$= \bar{R}T(n_p - n_r)$$

where

$$n_p = \text{moles of gaseous products}$$

and

$$n_r = \text{moles of gaseous reactants}$$

and

$$\bar{R} = 1545/778 = 1.986 \text{ (B/mole-R)}$$

Example 16.7. Determine the internal energy of the reaction for gaseous octane at 77 F.

The molar combustion equation is

$$C_8H_{18}(g) + 12.5\ O_2 \rightarrow 8\ CO_2 + 9\ H_2O(g)$$

for water as a gas. Therefore, $n_p = 17$ and $n_r = 13.5$, so that

$$\Delta pV = 1.986(537)(17 - 13.5) = 3732 \text{ (B)}$$

and for 1 lbm of octane

$$\Delta pV = 3732/114 = 33 \text{ (B/lbm)}$$

and

$$\Delta u_R = \Delta h_R - \Delta pV = -19{,}200 - 33 = -19{,}230 \text{ (B/lbm)}$$

If the water is condensed, $n_p = 8$ and $n_r = 13.5$, so that

$$\Delta pV = 1.986(537)(8 - 13.5) = -5865 \text{ (B)}$$

$$= -5865/114 = -51 \text{ (B/lbm)}$$

and

$$\Delta u_R = \Delta h_R - \Delta pV = -20{,}750 + 51 = -20{,}700 \text{ (B/lbm)}$$

HEAT FOR CONSTANT VOLUME COMBUSTION PROCESS

To determine the heat transfer in a closed system, constant volume process

$$\begin{aligned} Q &= U_p - U_r \\ &= (U_p - U_{p0}) - (U_r - U_{r0}) + (U_{p0} - U_{r0}) \end{aligned}$$

and

$$Q_{in} + Q_{out} = (U_p - U_{p0}) - (U_r - U_{r0}) + \Delta U_R$$

so that, for $Q_{out} > 0$,

$$Q_{out} = (U_r - U_{r0}) - \Delta U_R - (U_p - U_{p0}) \qquad (16.5)$$

Example 16.8. Repeat Example 16.2 for a closed system, constant volume process. From Example 16.2, t_r = 60 F and t_p = 800 F with combustion in 100% theoretical dry air.

The specific heat at constant volume, for the gaseous fuel, is

$$\begin{aligned} c_v &= c_p - (R/J) = 0.50 - \{1545/[114(778)]\} \\ &= 0.48 \text{ (B/lbm-R)} \end{aligned}$$

For the molar combustion equation

$$C_8H_{18} + 12.5\ O_2 + 47.0\ N_2 \rightarrow 8\ CO_2 + 9\ H_2O + 47\ N_2$$

the internal energy of the products is

$$\begin{aligned} U_p - U_{p0} &= [mc_v(t_2 - t_0)]_{CO_2} + [mc_v(t_2 - t_0)]_{H_2O} + [mc_v(t_2 - t_0)]_{N_2} \\ &= [8(44)(0.157) + 9(18)(0.335) + 47(28)(0.178)](800 - 77) \\ &= 248{,}550 \text{ (B)} \end{aligned}$$

and for the reactants

$$\begin{aligned} U_r - U_{r0} &= [mc_v(t_1 - t_0)]_{C_8H_{18}} + [mc_v(t_1 - t_0)]_{O_2} + [mc_v(t_1 - t_0)]_{N_2} \\ &= [114(1)(0.48) + 12.5(32)(0.157) + 47(28)(0.178)](60 - 77) \\ &= -5980 \text{ (B)} \end{aligned}$$

From Example 16.7

$$\Delta u_R = -19{,}230 \text{ (B/lbm)}$$

when the water in the products is a gas. From Eq. (16.5)

$$Q_{out} = (U_r - U_{r0}) - \Delta U_R - (U_p - U_{p0})$$

$$= -5980 + 19{,}230(114) - 248{,}550$$

$$= 1{,}937{,}690$$

or

$$Q_{out} = 1{,}937{,}690/114 = 17{,}000 \text{ (B/lbm)}$$

compared to

$$Q_{out} = 16{,}140 \text{ (B/lbm)}$$

for the open system, constant pressure process.

ADIABATIC FLAME TEMPERATURE

If no heat is transferred in a combustion process, Eq. (16.2),

$$Q_{out} = (H_r - H_{r0}) - \Delta H_R - (H_p - H_{p0})$$

may be rewritten as

$$H_p - H_{p0} = (H_r - H_{r0}) - \Delta H_R \tag{16.6}$$

for the open system, constant pressure process, where

$$\Delta KE = \Delta PE = W_{sf} = 0$$

For the closed system, constant volume process, Eq. (16.5),

$$Q_{out} = (U_r - U_{r0}) - \Delta U_R - (U_p - U_{p0})$$

for an adiabatic process, results in

$$U_p - U_{p0} = (U_r - U_{r0}) - \Delta U_R \tag{16.7}$$

Even though heat may be transferred from the system in a real combustion device, the quantity is usually small in comparison to the total internal energy or enthalpy changes found in the process.

The product temperature calculated from Eq. (16.6) or (16.7) will not be reached if the reactants and products reach equilibrium before the combustion process is completed. If disassociation of the products occurs, the temperature will be further reduced. The calculation of the flame temperature in the following sections will not include either of the above factors, and the resulting temperature will exceed that which can actually be attained. For the present we shall assume that the products of combustion act as perfect gases.

Example 16.9. Determine the adiabatic flame temperature when gaseous octane is burned in the theoretical correct amount of oxygen for complete combustion. The reactants enter at 60 F.

The molar combustion equation is

$$C_8H_{18}(g) + 12.5\ O_2 \rightarrow 8\ CO_2 + 9\ H_2O(g)$$

By use of Eq. (16.6) for constant pressure combustion

$$H_p - H_{p0} = (H_r - H_{r0}) - \Delta H_R$$

$$[mc_p(t_2 - t_0)]_{CO_2} + [mc_p(t_2 - t_0)]_{H_2O} = [mc_p(t_1 - t_0)]_{C_8H_{18}} + [mc_p(t_1 - t_0)]_{O_2} - m_{C_8H_{18}}\,\Delta h_R$$

where $t_0 = 77$ F and $\Delta h_R = -19{,}200$ B/lbm from Appendix J.

For the octane

$$mc_p(t_1 - t_0) = 114(0.5)(60 - 77) = -970\ (B)$$

For the oxygen

$$mc_p(t_1 - t_0) = 12.5(32)(0.219)(60 - 77) = -1490\ (B)$$

Therefore

$$H_p - H_{p0} = -970 - 1490 - (-19{,}200)(114) = 2{,}186{,}000\ (B)$$

and

$$8(44)(0.202)(t_2 - 77) + 9(18)(0.445)(t_2 - 77) = 2{,}186{,}000$$

$$143.2(t_2 - 77) = 2{,}186{,}000$$

$$t_2 - 77 = 15{,}265$$

$$t_2 = 15{,}342\ (F)$$

The high temperature of the products is due to combustion in oxygen.

Example 16.10. Repeat Example 16.9 for combustion in dry stoichiometric air, constant pressure combustion in an open system.

The molar combustion equation must include

$$n_{N_2} = 12.5(3.76) = 47.0\ N_2\ \text{(moles)}$$

For the nitrogen as a reactant

$$mc_p(t_1 - t_0) = 47.0(28)(0.249)(60 - 77) = -5570\ \text{(B)}$$

and

$$H_p - H_{p0} = -970 - 1490 - 5570 + 19{,}200(114) = 2{,}181{,}000$$

so that

$$143.2(t_2 - 77) + 47(28)(0.249)(t_2 - 77) = 2{,}181{,}000$$

$$470.9(t_2 - 77) = 2{,}181{,}000$$

$$t_2 - 77 = 4632$$

$$t_2 = 4709\ \text{(F)}$$

The use of air lowers the adiabatic flame temperature by 10,633 F.

Example 16.11. Repeat Example 16.10 for constant volume combustion in dry air.

From Example 16.7

$$\Delta u_R = -19{,}230\ \text{(B/lbm)}$$

and from Example 16.8

$$c_{v\text{-}C_8H_{18}} = 0.48\ \text{(B/lbm-R)}$$

For the octane

$$mc_v(t_1 - t_0) = 114(0.48)(60 - 77) = -930\ \text{(B)}$$

For the oxygen in the reactants

$$mc_v(t_1 - t_0) = 12.5(32)(0.157)(60 - 77) = -1070\ \text{(B)}$$

For the nitrogen in the reactants

$$mc_v(t_1 - t_0) = 47.0(28)(0.178)(60 - 77) = -3980$$

From Eq. (16.7)

$$U_p - U_{p0} = -930 - 1070 - 3980 + 114(19{,}230) = 2{,}186{,}000 \text{ (B)}$$

and

$$[8(44)(0.157) + 9(18)(0.335) + 47.0(28)(0.178)](t_2 - 77) = 2{,}186{,}000$$

$$343.8(t_2 - 77) = 2{,}186{,}000$$

$$t_2 - 77 = 6358$$

$$t_2 = 6435 \text{ (F)}$$

The temperature for the constant volume process exceeds that of the constant pressure process by 1726 F.

Example 16.12. Repeat Example 16.10 for constant pressure combustion in 200% theoretical air.

The molar combustion equation is

$$C_8H_{18}(g) + 25\ O_2 + 25(3.76)\ N_2 \rightarrow 8\ CO_2 + 9\ H_2O(g) + 12.5\ O_2 + 94\ N_2$$

For the reactants, $H_r - H_{r0}$,

$$C_8H_{18}: \quad mc_p(t_1 - t_0) = 114(0.5)(60 - 77) = -970 \text{ (B)}$$

$$O_2: \quad mc_p(t_1 - t_0) = 25(32)(0.219)(60 - 77) = -2980 \text{ (B)}$$

$$N_2: \quad mc_p(t_1 - t_0) = 94(28)(0.249)(60 - 77) = -11{,}140 \text{ (B)}$$

so that

$$H_p - H_{p0} = -970 - 2980 - 11{,}140 + 19{,}200(114) = 2{,}174{,}000$$

For the products, $H_p - H_{p0}$,

$$[mc_p(t_2 - t_0)]_{CO_2} + [mc_p(t_2 - t_0)]_{H_2O}$$

$$+ [mc_p(t_2 - t_0)]_{O_2} + [mc_p(t_2 - t_0)]_{N_2} = [8(44)(0.202) + 9(18)(0.445)$$

$$+ 12.5(32)(0.219)$$

$$+ 94(28)(0.249)](t_2 - t_0)$$

$$= 886.2(t_2 - t_0)$$

and

$$886.2(t_2 - t_0) = 2{,}174{,}000$$

$$t_2 - 77 = 2453$$

$$t_2 = 2530 \text{ (F)}$$

The addition of the excess air has reduced the constant pressure adiabatic flame temperature by 2179 F, compared to 100% theoretical air.

Example 16.13. Determine the adiabatic flame temperature for combustion of gaseous octane in 90% theoretical air. Constant pressure process. From Example 16.5 $\Delta h_R = -16{,}640$ (B/lbm)

and

$$C_8H_{18}(g) + 11.25\ O_2 + 42.3\ N_2 \rightarrow 5.5\ CO_2 + 2.5\ CO + 9\ H_2O + 42.3\ N_2$$

For the reactant, $H_r - H_{r0}$,

$$C_8H_{18}: \quad mc_p(t_1 - t_0) = 114(0.5)(60 - 77) = -970 \text{ (B)}$$

$$O_2: \quad mc_p(t_1 - t_0) = 11.25(32)(0.219)(60 - 77) = -1340 \text{ (B)}$$

$$N_2: \quad mc_p(t_1 - t_0) = 42.3(28)(0.249)(60 - 77) = -5013 \text{ (B)}$$

and

$$H_p - H_{p0} = -970 - 1340 - 5013 + 114(16{,}640) = 1{,}889{,}640 \text{ (B)}$$

For the product, $H_p - H_{p0}$,

$$[5.5(44)(0.202) + 2.5(28)(0.249) + 9(18)(0.445) + 42.3(28)(0.249)](t_2 - t_0) = 433.32(t_2 - t_0)$$

and

$$433.32(t_2 - 77) = 1{,}889{,}640$$

$$t_2 = 4438 \text{ (F)}$$

The temperature has decreased by 271 F when gaseous octane is burned in 90% theoretical air compared to burning in stoichiometric air.

Example 16.14. Determine the adiabatic flame temperature for Example 16.6. For this problem the enthalpy of reaction was

$$\Delta H_R = -2,936,870 \text{ (B)}$$

for a molar combustion given by

$$1.55\ C_8H_{18}(g) + 22.02\ O_2 + 82.8\ N_2 \rightarrow 8.5\ CO_2 + 3.9\ CO + 4.6\ O_2 + 82.8\ N_2 + 13.95\ H_2O(g)$$

For the reactants, where the temperature of the reactants is 60 F,

$$C_8H_{18}: \quad mc_p(t_1 - t_0) = 1.55(114)(0.5)(60 - 77) = -1500 \text{ (B)}$$

$$O_2: \quad mc_p(t_1 - t_0) = 22.02(32)(0.219)(60 - 77) = -2620 \text{ (B)}$$

$$N_2: \quad mc_p(t_1 - t_0) = 82.8(28)(0.249)(60 - 77) = -9810 \text{ (B)}$$

and

$$H_p - H_{p0} = -1500 - 2620 - 9810 + 2,936,870 = 2,922,940 \text{ (B)}$$

For the products, $H_p - H_{p0}$,

$$[8.5(44)(0.202) + 3.9(28)(0.249) + 4.6(32)(0.219) + 82.8(28)(0.249) + 13.95(18)(0.445)](t_2 - t_0) = 824.0(t_2 - t_0)$$

and

$$824.0(t_2 - 77) = 2,922,940$$

$$t_2 = 3624 \text{ (F)}$$

for a process with both incomplete combustion and oxygen in the products.

Again due to equilibrium and disassociation, the above temperatures will exceed those that would be achieved in a real process.

OTTO CYCLE

The adiabatic flame temperature for a constant volume closed system may be used for the combustion process in the Otto cycle. This temperature will be the maximum temperature for the cycle (Fig. 16.1).

Example 16.15. Determine the theoretical horsepower for a four-stroke cycle Otto engine operating at 3600 rpm. The engine has six cylinders with a bore of 3.439 in., a stroke of 4.25 in., and a compression ratio of 9. The mixture of gaseous octane and dry air enters the engine at 60 F and 14.7 psia. The octane is burned in the stoichiometric quantity of air.

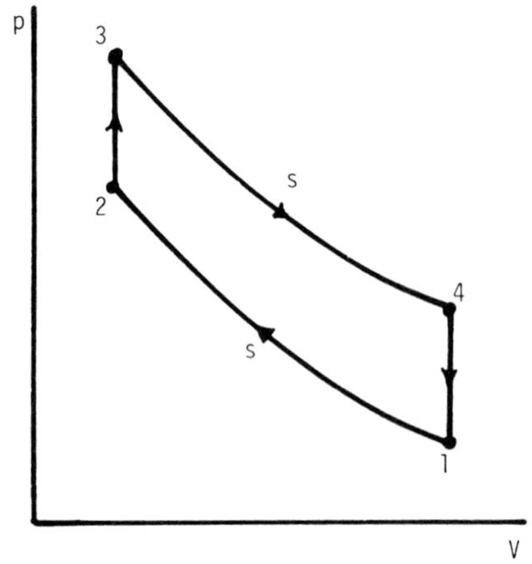

Figure 16.1

For the combustion process

$$C_8H_{18}(g) + 12.5\ O_2 + 47.0\ N_2 \rightarrow 8\ CO_2 + 9\ H_2O(g) + 47.0\ N_2$$

The mass of fuel is $m_f = 114$ lbm, and the mass of air is

$$m_a = 12.5(32) + 47.0(28) = 1716\ \text{(lbm)}$$

and the air-fuel ratio is

$$AF = m_a/m_f = 1716/114 = 15.05$$

For the air-fuel mixture entering the engine (see Chap. 13),

Component	m	y	c_p	c_v	yc_p	yc_v
C_8H_{18}	1.0	0.0623	0.5	0.48	0.0312	0.0299
Air	15.05	0.9377	0.24	0.1714	0.2250	0.1607
	16.05	1.0000			$c_{pm} = 0.256$	$c_{vm} = 0.191$ B/lbm-R

and

$$k = c_{pm}/c_{vm} = 0.256/0.191 = 1.340$$

The temperature at the end of the isentropic compression process is

$$T_2 = T_1 r_k^{\,k-1} = 520(9)^{0.340} = 1098\ \text{(R)}$$

For the reactants, $U_r - U_{r0}$,

C_8H_{18}: $mc_v(t_2 - t_0) = 114(0.48)(1098 - 537) = 30{,}700$ (B)

O_2: $mc_v(t_2 - t_0) = 12.5(32)(0.157)(1098 - 537) = 35{,}200$ (B)

N_2: $mc_v(t_2 - t_0) = 47.0(28)(0.178)(1098 - 537) = 131{,}400$ (B)

and

$$U_p - U_{p0} = 30{,}700 + 35{,}200 + 131{,}400 + 19{,}230(114)$$

where $\Delta U_R = -19{,}230$ B/lbm, from Example 16.7, so that

$$U_p - U_{p0} = 2{,}390{,}000 \text{ (B)}$$

For the products, $U_p - U_{p0}$,

$$[8(44)(0.157) + 9(18)(0.335) + 47(28)(0.178)](T_3 - 537) = 2{,}390{,}000$$

$$343.78(T_3 - 537) = 2{,}390{,}000$$

$$T_3 = 7489 \text{ (R)}$$

Even though 114 lbm of octane is not involved in the engine combustion process, the theoretical combustion temperature has been correctly calculated since the relative masses are fixed by the combustion equation.

For the combustion products

Component	n	m	y	c_p	c_v	yc_p	yc_v
CO_2	8	352	0.192	0.202	0.157	0.039	0.030
H_2O	9	162	0.089	0.445	0.335	0.040	0.030
N_2	47	1316	0.719	0.249	0.178	0.179	0.128
		1830	1.000			$c_{pm} = 0.258$	$c_{vm} = 0.188$ B/lbm-R

and

$$k = c_{pm}/c_{vm} = 0.258/0.188 = 1.372$$

The temperature after the isentropic expansion is

$$T_4/T_3 = (V_3/V_4)^{k-1} = 1/r_k^{\,k-1} = 1/(9)^{0.372} = 0.4416$$

$$T_4 = 0.4416(7489) = 3307 \text{ (R)}$$

For the expansion process

$$W_{34} = \frac{p_4 v_4 - p_3 v_3}{J(1 - k)} = \frac{R(T_4 - T_3)}{J(1 - k)}$$

where

$$R/J = c_{pm} - c_{vm} = 0.258 - 0.188 = 0.070 \text{ (B/lbm-R)}$$

for the products, so that

$$W_{34} = \frac{0.070(3307 - 7489)}{1 - 1.372} = 786.9 \text{ (B/lbm)}$$

For the compression process

$$W_{12} = \frac{p_2 v_2 - p_1 v_1}{J(1 - k)} = \frac{R(T_2 - T_1)}{J(1 - k)}$$

where

$$R/J = c_{pm} - c_{vm} = 0.256 - 0.191 = 0.065 \text{ (B/lbm-R)}$$

for the reactants, so that

$$W_{12} = \frac{0.065(1098 - 520)}{1 - 1.340} = -110.5 \text{ (B/lbm)}$$

The net work, for the cycle, is

$$\Sigma W = W_{12} + W_{34} = -110.5 + 786.9 = 676.4 \text{ (B/lbm)}$$

The displacement volume is

$$V_D = (\Pi/4)(\text{bore})^2(\text{stroke})$$

$$= (\Pi/4)(3.439)^2(4.25) = 39.5 \text{ (in.}^3\text{)}$$

$$= 0.0228 \text{ (ft}^3\text{)}$$

Since

$$r_k = \frac{V_C + V_D}{V_C} = \frac{V_C + 0.0228}{V_C} = 9$$

then

$$V_C = 0.0028 \text{ (ft}^3\text{)}$$

and

$$V_1 = V_C + V_D = 0.0028 + 0.0228 = 0.0256 \ (\text{ft}^3)$$

From the specific heat data for the reactants

$$R/J = 0.065 \ (\text{B/lbm-R})$$

and

$$R = 50.57 \ (\text{ft-lbf/lbm-R})$$

For the given pressure and temperature

$$v_1 = RT_1/p_1 = [50.57(520)]/[14.7(144)] = 12.42 \ (\text{ft}^3/\text{lbm})$$

so that the mass of reactants or products is

$$m = V_1/v_1 = 0.0256/12.42$$

$$= 0.00206 \ (\text{lbm})$$

The work per cycle is

$$\Sigma W = 676.4(0.00206) = 1.393 \ (\text{B})$$

and for a four-stroke cycle at 3600 rpm

$$\Sigma W = 1.393(3600/2) = 2507 \ (\text{B/min})$$

For six cylinders

$$\Sigma W = 2507(6) = 15{,}042 \ (\text{B/min})$$

and the theoretical horsepower is

$$HP = [15{,}042(778)]/33{,}000 = 355$$

To determine the cycle efficiency

$$Q_{41} = c_v(T_1 - T_4) = 0.188(520 - 3307)$$

$$= -524.0 \ (\text{B/lbm})$$

and

$$Q_{23} = \Sigma W - Q_{41} = 676.4 + 524.0 = 1200.4 \ (\text{B/lbm})$$

which yields

$$e = \Sigma W/Q_{in} = 676.4/1200.4 = 0.563$$

For the cycle using air

$$e = 1 - (1/r_k^{\ k-1}) = 1 - [1/(9)^{0.4}] = 0.585$$

To determine the value of k that would yield an efficiency of 56.3%,

$$1/r_k^{\,k-1} = 1/(9)^{k-1} = 1 - 0.563 = 0.437$$

and k = 1.376, which would be the "hot air" value for this theoretical cycle. Since the maximum temperature of 7489 R cannot be attained in a "real engine," hot air values of k are normally from 1.32 to 1.36.

COMBUSTION IN THE GAS TURBINE CYCLE

The combustion process for the gas turbine cycle may be assumed to be an open system, constant pressure process. In Fig. 16.2, the combustion occurs between states 2 and 3.

Example 16.16. A gas turbine receives air at 60 F and has a pressure ratio $r_p = 5$. The fuel, gaseous octane, is burned in 200% theoretical air. For $e_t = e_c = 100\%$, determine the cycle efficiency and the temperature at the end of the combustion process.

Since only air is present in the compressor, k = 1.4, and

$$T_2 = T_1 r_p^{(k-1)/k} = 520(5)^{0.286} = 824 \text{ (R)}$$

The molar combustion equation, for 200% theoretical air, is

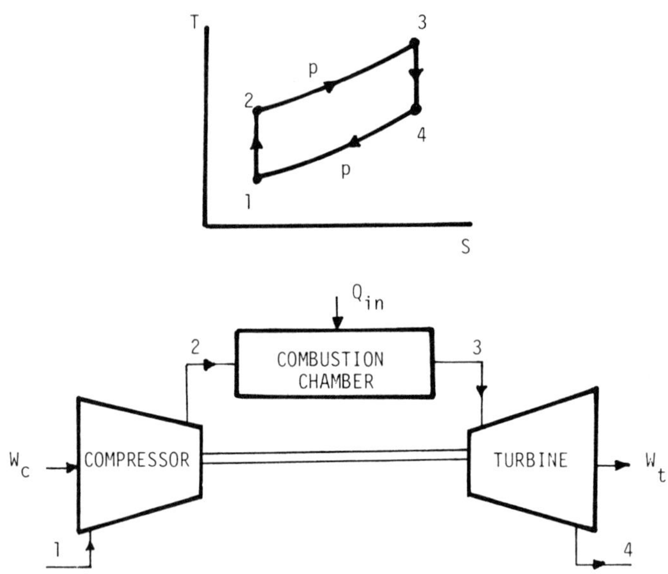

Figure 16.2

$$C_8H_{18} + 25\ O_2 + 94\ N_2 \rightarrow 8\ CO_2 + 9\ H_2O + 12.5\ O_2 + 94\ N_2$$

For the reactants

$$C_8H_{18}: \quad mc_p(T_2 - T_0) = 114(0.5)(824 - 537) = 16,360\ (B)$$

$$O_2: \quad mc_p(T_2 - T_0) = 25(32)(0.219)(824 - 537) = 50,280\ (B)$$

$$N_2: \quad mc_p(T_2 - T_0) = 94(28)(0.249)(824 - 537) = \underline{188,090}\ (B)$$

$$H_r - H_{r0} = 254,730\ (B)$$

For the products

$$CO_2: \quad mc_p(T_3 - T_0) = 8(44)(0.202)(T_3 - 537) = 71.10(T_3 - 537)\ (B)$$

$$H_2O: \quad mc_p(T_3 - T_0) = 9(18)(0.445)(T_3 - 537) = 72.09(T_3 - 537)\ (B)$$

$$O_2: \quad mc_p(T_3 - T_0) = 12.5(32)(0.219)(T_3 - 537) = 87.60(T_3 - 537)\ (B)$$

$$N_2: \quad mc_p(T_3 - T_0) = 94(28)(0.249)(T_3 - 537) = \underline{655.37}(T_3 - 537)\ (B)$$

$$H_p - H_{p0} = 886.16(T_3 - 537)\ (B)$$

From Appendix J

$$\Delta h_R = -19,200\ (B/lbm)$$

and

$$\Delta H_R = -19,200(114) = -2,188,800\ (B)$$

From Eq. (16.1)

$$H_p - H_{p0} = (H_r - H_{r0}) - \Delta H_R$$

$$886.16(T_3 - 537) = 254,730 + 2,188,800$$

and

$$T_3 = 3294\ (R)$$

To determine the properties of the products,

Component	n	m	y	c_p	c_v	yc_p	yc_v
CO_2	8	352	0.099	0.202	0.157	0.020	0.016
H_2O	9	162	0.046	0.445	0.335	0.020	0.015
O_2	12.5	400	0.113	0.219	0.157	0.025	0.018
N_2	94	2632	0.742	0.249	0.178	0.185	0.132
		3546	1.000			0.250	0.181

so that

$$c_{pm} = 0.250 \quad \text{and} \quad c_{vm} = 0.181 \ (\text{B/lbm-R})$$

and

$$k_m = c_{pm}/c_{vm} = 0.250/0.181 = 1.381$$

The temperature leaving the turbine is

$$T_4 = T_3/r_p^{(k-1)/k} = 3294/(5)^{0.276}$$

$$= 2113 \ (\text{R})$$

From the combustion equation, $m_f = 114$ lbm, and

$$m_a = 25(32) + 94(28) = 3432 \ (\text{lbm})$$

so that the air-fuel ratio is

$$AF = m_a/m_f = 3432/114 = 30.11$$

The compressor work, for 1 lbm of air, is

$$W_{12} = c_p(T_1 - T_2) = 0.24(520 - 824) = -73.0 \ (\text{B/lbm})$$

For the combustion process, the mass of products per pound of air is

$$m_p/m_a = 1 + (1/30.11) = 1.033 \ (\text{lbm})$$

and the turbine work, per pound of air, is

$$W_{34} = (m_p/m_a)c_{pm}(T_3 - T_4)$$

$$= 1.033(0.250)(3294 - 2113) = 305.0 \ (\text{B/lbm})$$

The net cycle work, per pound of air, is

$$\Sigma W = W_{12} + W_{34} = -73.0 + 305.0 = 232.0 \ (\text{B/lbm})$$

The heat added, per pound of air, is

$$Q_{23} = (1/AF)(-\Delta h_R) = (1/30.11)19{,}200$$

$$= 637.7 \ (\text{B/lbm})$$

The cycle efficiency is

$$e = \Sigma W/Q_{in} = 232.0/637.7 = 0.363$$

The cycle efficiency, using air, is

$$e = 1 - 1/r_p^{(k-1)/k} = 1 - [1/(5)^{0.286}] = 0.369$$

Example 16.17. Determine the cycle efficiency when the maximum temperature in the cycle is limited to 1540 F.

From Example 16.16, 200% theoretical air results in a maximum cycle temperature of 3294 R. To reduce the maximum cycle temperature the air-fuel ratio must be increased.

The molar combustion equation is

$$C_8H_{18}(g) + xO_2 + 3.76\ xN_2 \rightarrow 8\ CO_2 + 9\ H_2O + (x - 12.5)\ O_2 + 3.76\ xN_2$$

where

$$x > 12.5$$

From Example 16.16, $T_2 = 824$ R.

For the reactants

$$C_8H_{18}: \quad mc_p(T_2 - T_0) = 114(0.5)(824 - 537) = 16,360 \text{ (B)}$$

$$O_2: \quad mc_p(T_2 - T_0) = x(32)(0.219)(824 - 537) = 2011x \text{ (B)}$$

$$N_2: \quad mc_p(T_2 - T_0) = 3.76x(28)(0.249)(824 - 537) = 7524x \text{ (B)}$$

$$H_r - H_{r0} = 16,360 + 9535x \text{ (B)}$$

For the products

$$CO_2: \quad mc_p(T_3 - T_0) = 8(44)(0.202)(2000 - 537) = 104,030 \text{ (B)}$$

$$H_2O: \quad mc_p(T_3 - T_0) = 9(18)(0.445)(2000 - 537) = 105,470 \text{ (B)}$$

$$O_2: \quad mc_p(T_3 - T_0) = (x - 12.5)(32)(0.219)(2000 - 537)$$

$$= 10,253x - 128,160 \text{ (B)}$$

$$N_2: \quad mc_p(T_3 - T_0) = 3.76x(28)(0.249)(2000 - 537) = 38,352x \text{ (B)}$$

$$H_p - H_{p0} = 81,340 + 48,605x \text{ (B)}$$

From Example 16.16

$$\Delta H_R = -2,188,800 \text{ (B)}$$

and from Eq. (16.6)

$$H_p - H_{p0} = (H_r - H_{r0}) - \Delta H_R$$

$$81,340 + 48,605x = 16,360 + 9535x + 2,188,800$$

$$39{,}070x = 2{,}123{,}820$$

$$x = 54.4$$

To determine the properties of the products,

Component	n	m	y	yc_p	yc_v
CO_2	8	352	0.047	0.009	0.007
H_2O	9	162	0.021	0.009	0.007
O_2	41.9	1341	0.177	0.039	0.028
N_2	204.5	5726	0.755	0.188	0.134
		7581	1.000	0.245	0.176

and

$$k = c_{pm}/c_{vm} = 0.245/0.176 = 1.392$$

Since the air content of the products is increasing, k approaches 1.4. The temperature leaving the turbine is

$$T_4 = T_3/r_p^{(k-1)/k} = 2000/(5)^{0.282} = 1271 \text{ (R)}$$

The air-fuel ratio, m_f = 114 lbm, is

$$m_a = 54.4(32) + 54.4(3.76)(28) = 7468 \text{ (lbm)}$$

and

$$AF = m_a/m_f = 7468/114 = 65.5$$

as compared to 30.10 in Example 16.16. The compressor work is

$$W_{12} = c_p(T_1 - T_2) = 0.24(520 - 824) = -73.0 \text{ (B/lbm)}$$

as in Example 16.16. The turbine work is

$$W_{34} = [1 + (1/AF)]c_{pm}(T_3 - T_4)$$

$$= [1 + (1/65.5)](0.245)(2000 - 1271)$$

$$= 181.6 \text{ (B/lbm)}$$

and

$$\Sigma W = W_{12} + W_{34} = -73.0 + 181.6 = 108.6 \text{ (B/lbm)}$$

The heat added is

$$Q_{23} = (1/AF)(-\Delta h_R) = (1/65.5)19{,}200 = 293.1 \text{ (B/lbm)}$$

so that

$$e = \Sigma W/Q_{in} = 108.6/293.1 = 0.371$$

The cycle efficiency remains essentially constant even though the maximum cycle temperature is reduced 1294 F since $r_p = 5$ in both examples.

SEMIPERFECT GASES

Semiperfect gases, as discussed in Chap. 7, obey the perfect gas equation but have specific heats that vary with temperature. Since the variation of the specific heat values from those of the perfect gas increases with temperature, combustion calculations should be based on the tabular values of the enthalpy or of the internal energy. Appendix L, from the gas tables by Keenan and Kaye, lists the values of the enthalpy and internal energy in B/mole for the various gases that are used in the combustion process.

Example 16.18. Rework Example 16.2 assuming the reactants and products are semiperfect gases.

From Example 16.2, $t_1 = 60$ F, $t_2 = 800$ F, and for 100% theoretical air

$$C_8H_{18}(g) + 12.5\ O_2 + 47.0\ N_2 \rightarrow 8\ CO_2 + 9\ H_2O(g) + 47\ N_2$$

From Appendix L at 520 R, $h_{O_2} = 3606.1$ and $h_{N_2} = 3611.3$ B/mole; at 537 R, $h_{O_2} = 3725.1$, $h_{N_2} = 3729.5$, $h_{CO_2} = 4030.2$, and $h_{H_2O} = 4258.3$ B/mole; and at 1260 R, $h_{CO_2} = 11661.0$, $h_{H_2O} = 10354.9$, and $h_{N_2} = 8859.3$ B/mole. For the reactants, note that for octane the specific heat value is used due to a lack of tabular data:

$$C_8H_{18}:\quad mc_p(t_1 - t_0) = 114(0.5)(60 - 77) = -969 \text{ (B)}$$

$$O_2:\quad n(h_1 - h_0) = 12.5(3606.1 - 3725.1) = -1488 \text{ (B)}$$

$$N_2:\quad n(h_1 - h_0) = 47(3611.3 - 3729.5) = \underline{-5555} \text{ (B)}$$

$$H_r - H_{r0} = -8012 \text{ (B)}$$

Using the perfect gas relationship, the enthalpy change for the reactants was -8030 B. This very small error indicates the perfect gas calculations are satisfactory for low gas temperatures.

For the products

CO_2: $n(h_2 - h_0) = 8(11,661.0 - 4030.2) = 61,046$ (B)

H_2O: $n(h_2 - h_0) = 9(10,354.9 - 4258.3) = 54,869$ (B)

N_2: $n(h_2 - h_0) = 47(8859.3 - 3729.5) = \underline{241,101}$ (B)

$H_p - H_{p0} = 357,016$ (B)

which differs from the perfect gas solution of 340,440 B by +4.6%. For the combustion of gaseous octane, yielding water as a gas,

$$\Delta h_R = -19,200 \text{ (B/lbm)}$$

and from Eq. (16.2)

$$\begin{aligned} Q_{out} &= (H_r - H_{r0}) - (H_p - H_{p0}) - \Delta H_R \\ &= -8012 - 357,016 + 114(19,200) \\ &= 1,823,770 \text{ (B)} \\ &= 1,823,770/114 = 16,000 \text{ (B)} \end{aligned}$$

The error of +0.88% compared to the perfect gas value of 16,140 B/lbm is due to the variation of the specific heats for the gaseous products.

Example 16.19. Repeat Example 16.8 using the gas tables. The solution is identical to Example 16.18, except internal energy values replace the enthalpy values.

From Appendix L, the internal energies in B/mole are

	520 (R)	537 (R)	1260 (R)
u_{O_2}	2573.4	2658.7	—
u_{N_2}	2578.6	2663.1	6357.2
u_{CO_2}	—	2963.8	9158.8
u_{H_2O}	—	3191.9	7852.7

For the reactants

C_8H_{18}: $mc_v(t_1 - t_0) = 114(0.48)(60 - 77) = -930$ (B)

O_2: $m(u_1 - u_0) = 12.5(2573.4 - 2658.7) = -1066$ (B)

N_2 $m(u_1 - u_0) = 47(2578.6 - 2663.1) = \underline{-3972}$ (B)

$U_r - U_{r0} = -5968$ (B)

For the products

$$CO_2: \quad m(u_2 - u_0) = 8(9158.8 - 2963.8) = 49{,}560 \text{ (B)}$$

$$H_2O: \quad m(u_2 - u_0) = 9(7852.7 - 3191.9) = 41{,}947 \text{ (B)}$$

$$N_2: \quad m(u_2 - u_0) = 47(6357.2 - 2663.1) = \underline{173{,}623} \text{ (B)}$$

$$U_p - U_{p0} = 265{,}130 \text{ (B)}$$

From Eq. 16.5, where $\Delta u_R = -19{,}230$ B/lbm,

$$Q_{out} = (U_r - U_{r0}) - \Delta U_R - (U_p - U_{p0})$$

$$= -5968 - 114(-19{,}230) - 265{,}130$$

$$= 1{,}921{,}120 \text{ (B)}$$

or

$$= 16{,}850 \text{ (B/lbm)}$$

compared to 16,140 B/lbm for the perfect gas solution.

Again the error is caused by the variation in the specific heat values. For a product temperature of 800 F the error in both the constant pressure and constant volume calculations is small enough to permit reasonable calculations with the perfect gas equations.

Example 16.20. Determine the adiabatic flame temperature for the constant pressure process used in Example 16.12.

For 200% theoretical air, the molar combustion equation is

$$C_8H_{18}(g) + 25\ O_2 + 94\ N_2 \rightarrow 8\ CO_2 + 9\ H_2O + 12.5\ O_2 + 94\ N_2$$

For the reactants at 60 F

$$C_8H_{18}: \quad mc_p(t_1 - t_0) = 114(0.5)(60 - 77) = -970 \text{ (B)}$$

$$O_2: \quad m(h_1 - h_0) = 25(3606.1 - 3725.1) = -2{,}980 \text{ (B)}$$

$$N_2: \quad m(h_1 - h_0) = 94(3611.3 - 3729.5) = \underline{-11{,}110} \text{ (B)}$$

$$H_r - H_{r0} = -15{,}060 \text{ (B)}$$

For the products

$$CO_2: \quad m(h_2 - h_0) = 8(h_{CO_2} - 4030.2) = 8h_{CO_2} - 32{,}240 \text{ (B)}$$

$$H_2O: \quad m(h_2 - h_0) = 9(h_{H_2O} - 4258.3) = 9h_{H_2O} - 38{,}320 \text{ (B)}$$

$$O_2: \quad m(h_2 - h_0) = 12.5(h_{O_2} - 3725.1) = 12.5h_{O_2} - 46{,}560 \text{ (B)}$$

$$N_2: \quad m(h_2 - h_0) = 94(h_{N_2} - 3729.5) = 94h_{N_2} - 350{,}570 \text{ (B)}$$

$$8h_{CO_2} + 9h_{H_2O} + 12.5h_{O_2} + 94h_{N_2} - 467{,}690 \text{ (B)}$$

From Eq. (16.6), where $\Delta h_R = -19{,}200$ B/lbm,

$$H_p - H_{p0} = (H_r - H_{r0}) - \Delta H_R$$

or

$$8\ h_{CO_2} + 9\ h_{H_2O} + 12.5\ h_{O_2} + 94\ h_{N_2} - 467{,}690 = 15{,}060 - 114(-19{,}200)$$

and

$$8\ h_{CO_2} + 9\ h_{H_2O} + 12.5\ h_{O_2} + 94\ h_{N_2} = 2{,}641{,}430\ (B)$$

To determine the adiabatic flame temperature we must assume a value of the temperature, determine the enthalpies from Appendix L, and check the summation against the right-hand value of the equation. A logical starting point would be the value obtained from the perfect gas solution.

From Example 16.12, $t_2 = 2530$ F = 2990 R, so assume the temperature is 3000 R:

Gas	3000 R, mh			2800 R, mh	2700 R, mh	2740 R, mh
$8\ h_{CO_2}$	8(34806.6)	=	278,450	255,860	244,650	249,120
$9\ h_{H_2O}$	9(28386.3)	=	255,480	234,790	224,610	228,670
$12.5\ h_{O_2}$	12.5(23817.7)	=	297,720	275,270	264,790	269,160
$94\ h_{N_2}$	94(22761.5)	=	2,139,580	1,981,620	1,903,160	1,934,500
			2,971,230	2,747,540	2,637,210	2,681,450

By linear interpolation, the adiabatic flame temperature is

$$t_2 = 2703\ (R) = 2243\ (F)$$

compared to 2530 F for the perfect gas solution.

The 200% theoretical air data presented in Appendix L were derived for hydrocarbon fuels having the equation $(CH_2)_n$. These data would be correct for fuels such as ethylene (C_2H_4), propylene (C_3H_6), butene (C_4H_8), and others.

The tables are usable, without serious error, for fuels from $(CH)_n$ to $(CH_3)_n$. For an 8-carbon fuel the coverage would be from C_8H_8 to C_8H_{24}, so that octane, C_8H_{18}, may be used. This removes the necessity of using a trial and error solution for the 200% theoretical air solution.

Example 16.21. Determine the adiabatic flame temperature for Example 16.20 by use of the 200% theoretical air data.

For the reactants, from Example 16.20,

$$H_r - H_{r0} = -15{,}060 \text{ (B)}$$

and

$$\Delta h_R = -19{,}200 \text{ (B/lbm)}$$

$$H_p - H_{p0} = H_r - H_{r0} - \Delta h_R$$

$$= -15{,}060 - 114(-19{,}200) = 2{,}173{,}740 \text{ (B)}$$

For the products

$$H_p - H_{p0} = 8(h_2 - h_0) + 9(h_2 - h_0) + 12.5(h_2 - h_0) + 94(h_2 - h_0)$$

$$= 123.5(h_2 - h_0)$$

and

$$123.5(h_2 - 3774.9) = 2{,}173{,}740$$

so that

$$h_2 = 21{,}376.0 \text{ (B)}$$

or

$$h_2 = 21{,}376.0 \text{ (B/mole)}$$

since 1 mole of fuel was burned.

From Appendix L,

$$h_2 = 21{,}293.8 \text{ (B/mole)} \qquad \text{at } 2700 \text{ (R)}$$

and

$$h_2 = 21{,}659.9 \text{ (B/mole)} \qquad \text{at } 2740 \text{ (R)}$$

By interpolation

$$t_2 = 2709 \text{ (R)} = 2249 \text{ (F)}$$

which differs from Example 16.20 by 6 F.

Example 16.22. Repeat Example 16.14 by use of the gas tables.

From Example 16.14, $\Delta H_R = -2{,}936{,}870$ for a molar combustion process given by

$$1.55\ C_8H_{18}(g) + 22.02\ O_2 + 82.8\ N_2 \rightarrow 8.5\ CO_2 + 3.9\ CO + 4.6\ O_2$$
$$+ 82.8\ N_2 + 13.95\ H_2O$$

For the reactants the reactant temperature is 60 F.

C_8H_{18}: $mc_p(t_1 - t_0) = 1.55(114)(0.5)(60 - 77) = -1,500$ (B)

O_2: $m(h_1 - h_0) = 22.02(3606.1 - 3725.1) = -2,620$ (B)

N_2: $m(h_1 - h_0) = 82.8(3611.3 - 3729.5) = \underline{-9,790}$ (B)

$H_r - H_{r0} = -13,910$ (B)

From Eq. (16.6)

$$H_p - H_{p0} = (H_r - H_{r0}) - \Delta H_R$$

$$= -13,910 + 2,936,870 = 2,922,960 \text{ (B)}$$

For the products

CO_2: $m(h_2 - h_0) = 8.5(h_{CO_2} - 4030.2) = 8.5h_{CO_2} - 34,260$

CO: $m(h_2 - h_0) = 3.9(h_{CO} - 3729.5) = 3.9h_{CO} - 14,550$

O_2: $m(h_2 - h_0) = 4.6(h_{O_2} - 3725.1) = 4.6h_{O_2} - 17,140$

N_2: $m(h_2 - h_0) = 82.8(h_{N_2} - 3729.5) = 82.8h_{N_2} - 308,800$

H_2O: $m(h_2 - h_0) = 13.95(h_{H_2O} - 4258.3) = 13.9h_{H_2O} - 59,400$

$$H_p - H_{p0} = 8.5h_{CO_2} + 3.9h_{CO} + 4.6h_{O_2} + 82.8h_{N_2} + 13.95h_{H_2O} - 434,150$$

and

$$8.5h_{CO_2} + 3.9h_{CO} + 4.6h_{O_2} + 82.8h_{N_2} + 13.95h_{H_2O} = 3,357,110$$

Again, by selecting various values for the temperature, the above equation may be balanced:

	3500 R	4000 R
$8.5h_{CO_2}$	356,700	418,470
$3.9h_{CO}$	106,320	123,260
$4.6h_{O_2}$	130,060	150,910
$82.8h_{N_2}$	2,236,920	2,594,070
$13.95h_{H_2O}$	478,810	564,820
	3,308,810	3,851,530

By interpolation, t_2 = 3509 R = 3049 F, compared to 3624 F for the perfect gas solution. The error, when assuming the products act as perfect gases, is 18.9%. Since the evaluation of $H_r - H_{r0}$ is essentially the same for both solutions, the error results from the variation in specific heats at the high product temperatures.

PROBLEMS

Perfect gases are used except where noted.

16.1 Determine the difference between $\Delta h_R[H_2O(l)]$ and $\Delta h_R(H_2O(g)]$ for xylene from Appendix J and by calculation.

16.2 Determine the heat transfer when 1 lbm of xylene, as a gas, is burned in an open system, constant pressure process. The reactants enter at 100 F; the products leave at 1000 F. Fuel is burned in 100% theoretical dry air.

16.3 Given the same heat transfer and properties as in Prob. 16.2 except the combustion takes place in 200% theoretical air, determine the temperature of the leaving products.

16.4 Calculate the enthalpy of reaction from the heats of formation when hexane is burned in stoichiometric oxygen at 77 F. The water in products is gaseous.

16.5 Determine the enthalpy of reaction when hexane is burned in 80% theoretical air at 77 F. The water in products is gaseous.

16.6 Hexane is burned in dry air which produces an orsat analysis of 5.0% CO_2, 8.7% CO, and 4.4% O_2. Determine the enthalpy of reaction when the product water is a gas.

16.7 Determine Δu_R for the combustion of xylene, as a gas, for combustion in stoichiometric air at 77 F. The product water is a gas.

16.8 Repeat Prob. 16.2 for constant volume combustion of the fuel.

16.9 Determine the adiabatic flame temperature when hexane is burned in the theoretical correct amount of oxygen for complete combustion. Reactants enter at 100 F; constant pressure combustion.

16.10 Repeat Prob. 16.9 for combustion in dry stoichiometric air.

16.11 Repeat Prob. 16.9 for combustion in 200% theoretical dry air.

16.12 Repeat Prob. 16.9 for combustion in 80% theoretical dry air.

16.13 Repeat Prob. 16.9 for combustion in dry stoichiometric air during a constant volume combustion process.

16.14 Determine the adiabatic flame temperature when hexane is burned in dry air during a constant pressure process. The orsat analysis is 5.0% CO_2, 8.7% CO, and 4.4% O_2. The reactant temperature is 77 F.

16.15 Determine the theoretical horsepower for a four-stroke cycle Diesel operating at 3000 rpm. The engine has four cylinders with a bore of 3.5 in. and a stroke of 4.0 in. The compression ratio is 16. The pressure and temperature at the start of the compression process are 14.7 psia and 40 F. The engine is supplied with liquid benzene at 77 F. The theoretical air for the combustion process is 200%.

16.16 A gas turbine receives air at 40 F and has a pressure ratio of $r_p = 5.5$. The fuel, liquid benzene, is burned in 200% theoretical air. For $e_t = e_c = 100\%$, determine the cycle efficiency and the temperature at the end of the combustion process.

16.17 Determine the cycle efficiency in Prob. 16.16 when the maximum cycle temperature is limited to 2000 F.

16.18 Repeat Prob. 16.2 assuming the gases are semiperfect.

16.19 Repeat Prob. 16.10 assuming the gases are semiperfect.

16.20 Repeat Prob. 16.11 assuming the gases are semiperfect.

16.21 Repeat Prob. 16.12 assuming the gases are semiperfect.

16.22 Repeat Prob. 16.14 assuming the gases are semiperfect.

16.23 Repeat Prob. 16.11 by use of the 200% theoretical air data for semiperfect gases.

17

SECOND LAW OF THERMODYNAMICS

The First Law of Thermodynamics is concerned with the conservation of energy for a cycle and is expressed by

$$\Sigma Q = \Sigma W$$

so that the net heat of a cycle is equivalent to the net work of the cycle.

Even though a cycle may be proposed that conforms with the First Law of Thermodynamics, it may not be an operable cycle unless it also satisfies the requirements of the Second Law of Thermodynamics.

To determine if a cycle or process can occur we shall propose a new cycle, the Carnot cycle. The inequality of Clausius and the concept of entropy production will also be used to determine if a cycle or process is possible or impossible or if they are reversible or irreversible.

Irreversible processes will be discussed in terms of entropy production by use of an isolated thermodynamic system. Unavailable and available heat will be examined in terms of processes, cycles, and the heat exchanger.

CARNOT CYCLE

The Carnot cycle (Fig. 17.1) consists of two reversible isothermal processes and two reversible isentropic processes. The cycle, when operated as a heat engine, receives heat from a source having a constant temperature T_s and rejects heat to a sink having a constant temperature T_0. From the TS diagram, $Q_{12} = T_1(S_2 - S_1)$, $Q_{23} = 0$, $Q_{34} = T_3(S_4 - S_3)$, and $Q_{41} = 0$. Therefore, from the First Law of Thermodynamics, for a cycle

$$\begin{aligned}\Sigma Q = \Sigma W &= Q_{12} + Q_{23} + Q_{34} + Q_{41} \\ &= T_1(S_2 - S_1) + T_3(S_4 - S_3)\end{aligned}$$

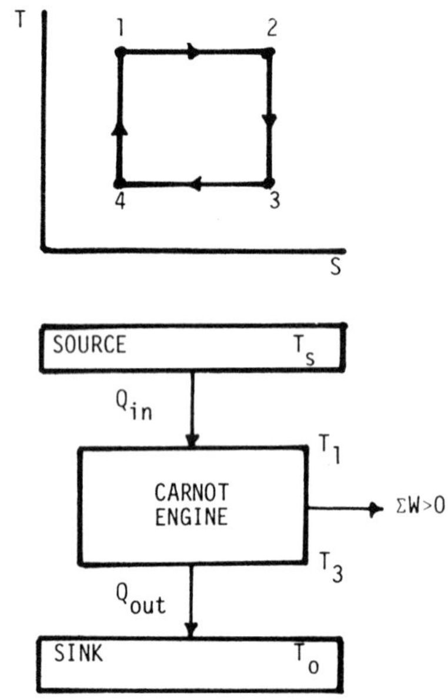

Figure 17.1

and since

$$S_4 = S_1 \quad \text{and} \quad S_3 = S_2$$

$$\Sigma Q = \Sigma W = T_1(S_2 - S_1) + T_3(S_1 - S_2) = (T_1 - T_3)(S_2 - S_1)$$

Note that $Q_{12} = Q_{in}$ since $Q_{12} > 0$ and that $Q_{34} = Q_{out}$ since $Q_{34} < 0$. For the Carnot cycle, the efficiency is

$$e = \Sigma W/Q_{in} = [(T_1 - T_3)(S_2 - S_1)]/[T_1(S_2 - S_1)]$$

$$= (T_1 - T_3)/T_1 \qquad \text{(Carnot engine)} \tag{17.1}$$

The upper limit for the value of T_1 is the source temperature T_s, and the lower limit for T_3 is T_0, where T_0 is the sink temperature. An increase in T_1 or a decrease in T_3 would result in an increase in the efficiency of the Carnot cycle.

Example 17.1. Determine the Carnot cycle efficiency for $t_s = t_1 =$ 1540 F and $t_0 = t_3 = 40$ F.

From Eq. (17.1)

$$e = (T_1 - T_3)/T_1 = (2000 - 500)/2000 = 0.75$$

$$= 75\%$$

REVERSED CARNOT CYCLE

If the cycle in Fig. 17.1 is reversed, heat is removed from the source at T_s and added to the sink at T_0; in this case $T_0 > T_s$, and the reversed cycle is pumping the heat from a low-temperature source to a high-temperature sink.

For the unit as a heat pump

$$\gamma_P = |Q_{out}|/|\Sigma W|$$

and for a refrigerator

$$\gamma_R = |Q_{in}|/|\Sigma W|$$

where γ designates the coefficient of performance.

For the cycle in Fig. 17.2, $Q_{12} = T_1(S_2 - S_1) = Q_{in}$, $Q_{34} = T_3(S_4 - S_3) = T_3(S_1 - S_2) = Q_{out}$, and $Q_{14} = Q_{23} = 0$. From the First Law of Thermodynamics,

$$\Sigma W = \Sigma Q = T_1(S_2 - S_1) + T_3(S_1 - S_2) = (T_1 - T_3)(S_2 - S_1)$$

for the cycle. Since $T_3 > T_1$, the net work is negative, indicating that the work is done on the system.

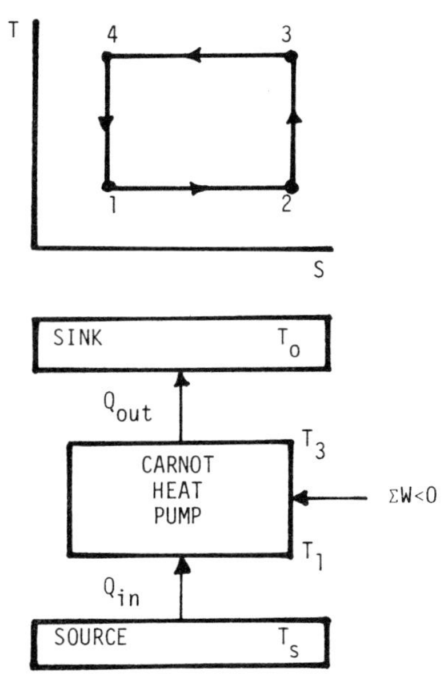

Figure 17.2

Substitution of the heat and work terms gives

$$\gamma_P = |Q_{out}|/|\Sigma W| = |T_3(S_1 - S_2)|/|(T_1 - T_3)(S_2 - S_1)|$$

$$= T_3/(T_3 - T_1) \qquad \text{(Carnot heat pump)} \qquad (17.2)$$

for the pump. For the refrigerator

$$\gamma_R = |Q_{in}|/|\Sigma W| = |T_1(S_2 - S_1)|/|(T_1 - T_3)(S_2 - S_1)|$$

$$= T_1/(T_3 - T_1) \qquad \text{(Carnot refrigerator)} \qquad (17.3)$$

The limiting temperature values are $T_3 = T_0$ and $T_1 = T_s$ for the reversed Carnot cycle.

Example 17.2. Determine the coefficient of performance for the temperature limits given in Example 17.1 for the reversed cycle.

From Example 17.1, $t_0 = t_3 = 1540$ F, and $t_s = t_1 = 40$ F. By use of Eq. (17.2)

$$\gamma_P = T_3/(T_3 - T_1) = 2000/1500 = 1.33$$

and from Eq. (17.3)

$$\gamma_R = T_1/(T_3 - T_1) = 500/1500 = 0.33$$

for the reversed Carnot cycle. For any reversed cycle,

$$\gamma_P - \gamma_R = 1.0$$

SECOND LAW OF THERMODYNAMICS

The First Law of Thermodynamics for a cycle is, in equation form,

$$\Sigma Q = \Sigma W$$

and for a process

$$Q - W_{nf} = \Delta U \qquad \text{(closed system)}$$

$$Q - W_{sf} = \Delta H + \Delta PE + \Delta KE \qquad \text{(open system)}$$

The first law for a cycle tells us that the net heat and work of a cycle must be equal. For the process the difference between the heat and work terms is equal to the changes in the energy terms.

The second law places more restrictions on both the thermodynamic process and the thermodynamic cycle.

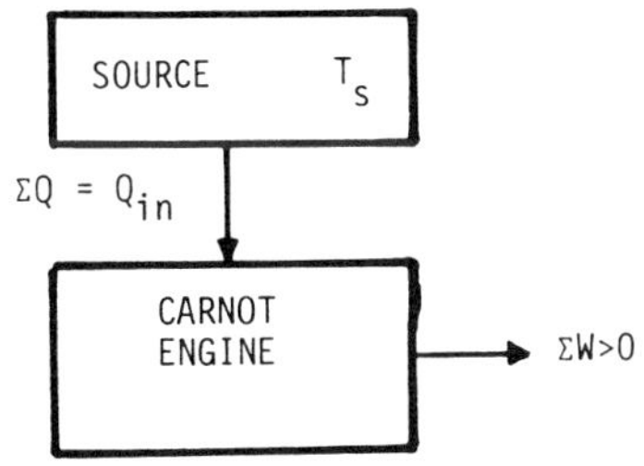

Figure 17.3

Consider the Carnot cycle, acting as an engine and receiving heat from a source at a single temperature, T_s (Fig. 17.3).

The first law is satisfied since

$$\Sigma Q = \Sigma W$$

but it violates the Kelvin-Planck statement of the second law, which says, "It is impossible to construct a device that will operate in a cycle that produces no other effect than the conversion of heat from a source at a fixed temperature to work."

In Fig. 17.3, the source could be the surrounding air or the ocean, and we could draw heat from this source and produce work. This would result in a perpetual motion machine of the second kind. A perpetual motion machine of the first kind violates the First Law of Thermodynamics. Implied by the Kelvin-Planck statement is the existence of a sink, so that some of the heat supplied to the cycle must be rejected to the sink at a lower temperature.

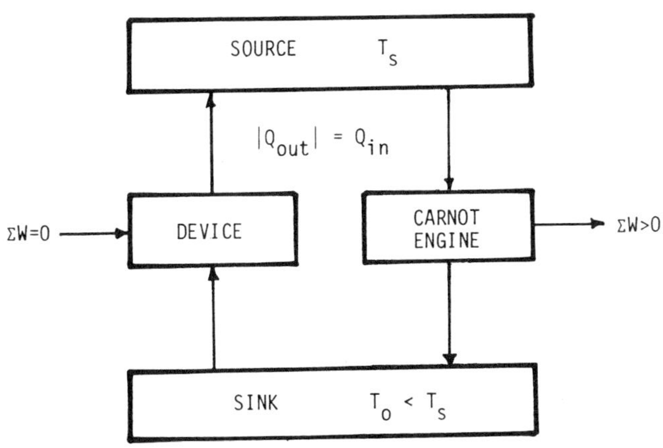

Figure 17.4

Another statement of the second law is that of Clausius: "It is impossible to construct a device that produces no other effect than the transfer of heat from a body at a low temperature to a body at a higher temperature." If this were not true, the Carnot cycle in Fig. 17.1 could be operated perpetually by use of a heat pump device to supply the necessary heat to run the engine (Fig. 17.4).

REVERSIBLE CYCLE EFFICIENCY

Let us assume that a reversible or irreversible cycle can be operated between the source and sink having an efficiency greater than that of the reversible Carnot cycle (Fig. 17.5). By designating this new cycle by a prime ($'$), from our assumption that $e' > e$ and by assuming that each cycle is supplied with the same quantity of heat,

$$Q_{in}' = Q_{in}$$

From the efficiency equation, $e = \Sigma W/Q_{in}$, so that

$$e' = \Sigma W'/Q_{in}' > \Sigma W/Q_{in} = e$$

and

$$\Sigma W' > \Sigma W$$

Since the Carnot cycle in Fig. 17.5 is reversible, the two cycles may now be operated as in Fig. 17.6. Since the work required to drive the

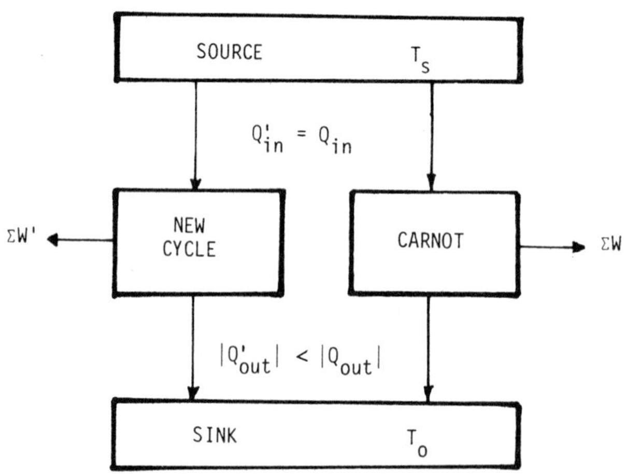

Figure 17.5

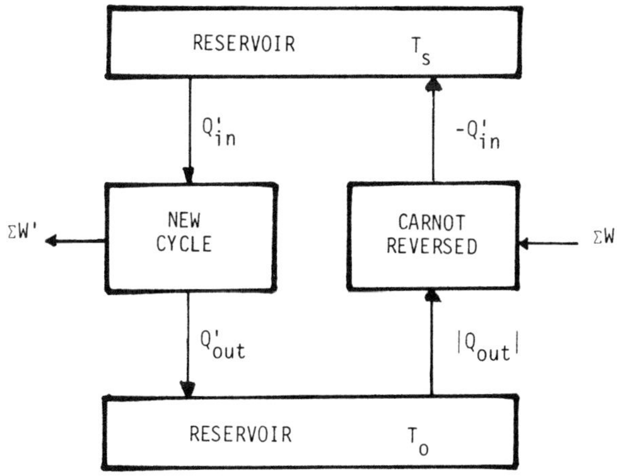

Figure 17.6

Carnot cycle is ΣW, we must obtain a quantity of work

$$\Sigma W_{net} = \Sigma W' - \Sigma W$$

every time the two cycles are completed. This quantity of work is obtained from the sink. Since the heat rejected by the reversed Carnot cycle is at the source temperature and the quantity of heat added to the engine is of the same amount and also at the source temperature, the source may be eliminated (Fig. 17.7). But this cycle violates the Kelvin-Planck statement of the second law because if we consider both cycles to be the system, heat

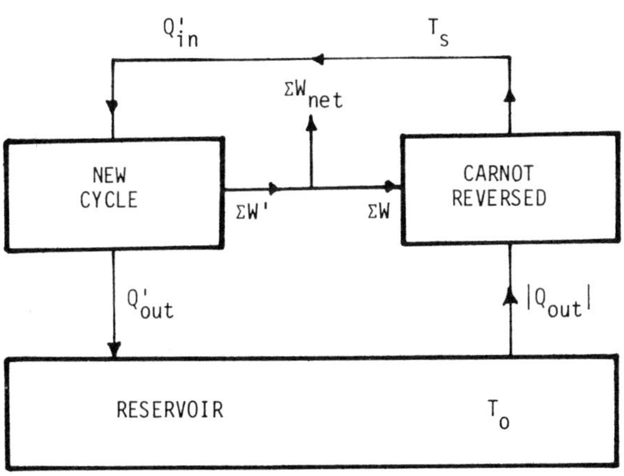

Figure 17.7

is being transferred from only one reservoir and work ΣW_{net} is being produced.

The following statement is obtained from the above discussion: "No engine operating between a constant temperature source and sink can have an efficiency greater than that of a Carnot cycle that receives and transfers heat at the source and sink temperatures."

Example 17.3. An inventor claims his engine will develop 100 HP when burning 0.25 lbm/min of fuel having a heating value of 20,000 B/lbm. The maximum and minimum cycle temperatures are 60 F and 2500 F. Is the above claim justified?

The efficiency of a Carnot cycle having these temperature values is

$$e = (T_{max} - T_{min})/T_{max} = (2960 - 520)/2960 = 0.824$$

For the inventor's engine

$$\Sigma W = 100(33,000)/778 = 4242 \text{ (B/min)}$$

and

$$Q_{in} = 0.25(20,000) = 5000 \text{ (B/min)}$$

so that

$$e = \Sigma W/Q_{in} = 4242/5000 = 0.848$$

Since the efficiency of the inventor's engine exceeds that of a Carnot cycle engine operating between the same temperature limits, it is impossible. Either the power produced or the fuel consumed has been incorrectly calculated.

Example 17.4. An Otto cycle has a compression of 9 and a maximum cycle temperature of 2540 F. At the start of the compression process the air temperature is 40 F. Compare the efficiency of this reversible cycle with that of the Carnot cycle.

For the Otto cycle engine

$$e = 1 - (1/r_k{}^{k-1})$$

$$e = 1 - [1/(9)^{0.4}] = 0.585$$

and for the Carnot cycle

$$e = (T_{max} - T_{min})/T_{max} = (3000 - 500)/3000 = 0.833$$

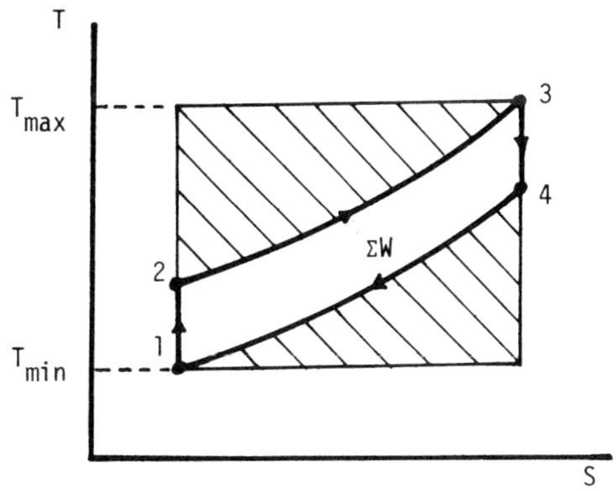

Figure 17.8

Even though both cycles are reversible, the efficiency of the Carnot cycle exceeds that of the Otto cycle by 24.8%. The reason for this is indicated in Fig. 17.8. The dashed areas represent net work obtained from the Carnot cycle that is not produced by the Otto cycle. The heat source and sink for the Otto cycle do not operate at a single temperature but at an infinite number of temperatures between T_2 and T_3 as well as between T_4 and T_1.

Example 17.5. Determine the values of Q_{in}, Q_{out}, ΣW, and e for the Otto and Carnot cycles in Fig. 17.8.

From Example 17.4

$$e_{Otto} = 0.585 \quad \text{and} \quad e_{Carnot} = 0.833$$

For the Otto cycle

$$T_2 = T_1 r_k^{k-1} = 500(9)^{0.4} = 1204 \text{ (R)}$$

and

$$s_3 - s_2 = c_v \ln(T_3/T_2) = 0.1714 \ln(3000/1204)$$

$$= 0.1565 \text{ (B/lbm-R)}$$

For the Otto cycle

$$Q_{in} = c_v(T_3 - T_2) = 0.1714(3000 - 1204) = 307.8 \text{ (B/lbm)}$$

$$\Sigma W = eQ_{in} = 0.585(307.8) = 180.1 \text{ (B/lbm)}$$

and

$$Q_{out} = \Sigma W - Q_{in} = 180.1 - 307.8 = -127.7 \text{ (B/lbm)}$$

For the Carnot cycle

$$Q_{in} = T_3(S_3 - S_2) = 3000(0.1565) = 469.5 \text{ (B/lbm)}$$

$$\Sigma W = eQ_{in} = 0.833(469.5) = 391.1 \text{ (B/lbm)}$$

$$Q_{out} = \Sigma W - Q_{in} = 391.1 - 469.5 = -78.4 \text{ (B/lbm)}$$

Even though the Carnot cycle requires a greater addition of heat, more of the heat is converted to work, and less heat is rejected by the cycle.

STIRLING CYCLE

The Stirling cycle (Fig. 17.9) consists of two reversible constant volume processes and two reversible isothermal processes. For the cycle

$$Q_{12} = T_1(S_2 - S_1), \qquad Q_{12} > 0$$

$$Q_{23} = mc_v(T_3 - T_2), \qquad Q_{23} < 0$$

$$Q_{34} = T_3(S_4 - S_3), \qquad Q_{34} < 0$$

$$Q_{41} = mc_v(T_1 - T_4), \qquad Q_{14} > 0$$

$$= mc_v(T_2 - T_3)$$

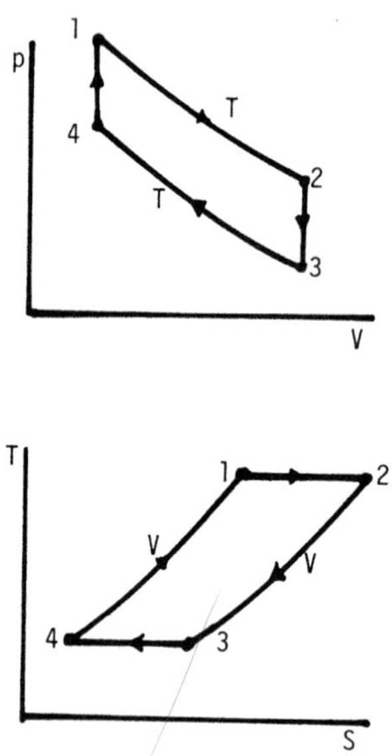

Figure 17.9

since

$$T_1 = T_2 \quad \text{and} \quad T_4 = T_3$$

The general equation for the entropy change in terms of temperatures and volumes is

$$S_2 - S_1 = mc_v \ln(T_2/T_1) + (mR/J) \ln(V_2/V_1)$$

and for the isothermal process

$$S_2 - S_1 = (mR/J) \ln(V_2/V_1)$$

Likewise for the isothermal process from state 3 to state 4

$$S_4 - S_3 = (mR/J) \ln(V_4/V_3)$$

$$= (mR/J) \ln(V_1/V_2) = -(mR/J) \ln(V_2/V_1)$$

since

$$V_4 = V_1 \quad \text{and} \quad V_3 = V_2$$

Therefore

$$S_2 - S_1 = -(S_4 - S_3)$$

so that

$$Q_{12} = T_1 \, \Delta S \quad \text{and} \quad Q_{34} = -T_3 \, \Delta S$$

where

$$\Delta S = S_2 - S_1$$

Example 17.6. Determine the efficiency of a Stirling cycle that operates between temperature limits of 1540 F and 40 F. $s_2 - s_1 = 0.1$ B/lbm-R.

$$Q_{12} = T_1 \, \Delta s = 2000(0.1) = 200.0 \text{ (B/lbm)}$$

$$Q_{23} = c_v(T_3 - T_2) = 0.1714(500 - 2000) = -257.1 \text{ (B/lbm)}$$

$$Q_{31} = -T_3 \, \Delta s = -500(0.1) = 50.0 \text{ (B/lbm)}$$

$$Q_{41} = c_v(T_2 - T_3) = 0.1714(2000 - 500) = 257.1 \text{ (B/lbm)}$$

For the cycle

$$\Sigma Q = \Sigma W = Q_{12} + Q_{23} + Q_{31} + Q_{41} = 200.0 - 257.1 - 50.0 + 257.1$$

$$= 150.0 \text{ (B/lbm)}$$

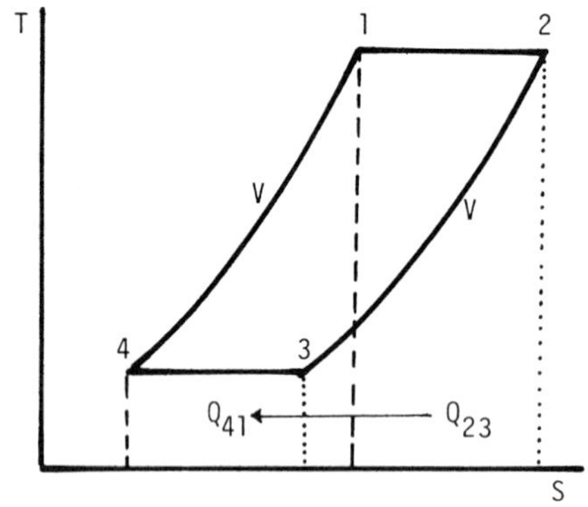

Figure 17.10

and

$$Q_{in} = Q_{12} + Q_{41} = 200 + 257.1 = 457.1$$

so that

$$e = \Sigma W/Q_{in} = 150/457.1 = 0.328$$

Since the heat transfers during the constant volume processes are of the same magnitude and at the same temperature levels, the heat rejected during process 23 may be supplied to the engine during process 41 (Fig. 17.10). This is an internal regenerative process, and for this cycle

$$Q_{in} = Q_{12} = T_1\ \Delta s \quad \text{and} \quad Q_{out} = Q_{34} = -T_3\ \Delta s$$

so that the heat terms Q_{23} and Q_{41} do not represent heat that crosses the boundaries of the system.

Example 17.7. Determine the efficiency of the Stirling cycle from Example 17.6 for a regenerative cycle. Compare the efficiency of this cycle to that of a Carnot cycle operating between the same temperature limits.

For the Stirling cycle

$$Q_{12} = T_1\ \Delta S$$

$$Q_{34} = -T_3\ \Delta S$$

so that

$$\Sigma Q = \Sigma W = T_1\ \Delta S - T_3\ \Delta S = (T_1 - T_3)\ \Delta S$$

and

$$e = \Sigma W/Q_{in} = \Sigma W/Q_{12} = [(T_1 - T_3)\ \Delta S]/(T_1\ \Delta S)$$

$$= (T_1 - T_3)/T_1 = (2000 - 500)/2000 = 0.75$$

From Example 17.1 and Eq. (17.1), the efficiency of a Carnot cycle is

$$e = (T_1 - T_3)/T_1 = 0.75$$

The Stirling cycle with regeneration has the same efficiency as the Carnot cycle when operating between the same temperature limits.

In Example 17.4, the Otto cycle operating between 500 and 3000 R had an efficiency of 58.5%. The comparable Carnot cycle efficiency was 83.3%. A regenerative Stirling cycle operating between these limits would also have an efficiency of 83.3%. For the Carnot and Stirling cycles it is difficult to construct a high-speed engine that can transfer heat isothermally. Large temperature differences between the source and the engine are required for the transfer of large quantities of heat in a short time period. For the Stirling cycle, the effectiveness of the regenerator also limits the efficiency that can be obtained in a real device.

IRREVERSIBLE ADIABATIC PROCESS

We have previously considered the irreversible adiabatic process in connection with the gas turbine cycle, the Rankine cycle, and other devices in which there was a compressor or turbine efficiency. In these discussions the possible end points of the process were based on the assumption that the compressor or turbine efficiency of an actual process must be less than 100%. For the compressor this meant that more work was required for actual adiabatic compression, between two stated pressures, than was required for the isentropic compression. For the turbine, less work was obtained for the actual adiabatic expansion, between two stated pressures, than that obtained during a reversible adiabatic or isentropic process. It was also noted that for the irreversible adiabatic process, for either the compressor or turbine, the entropy must increase.

Consider the Carnot cycle sketched in Fig. 17.11. Assume that the process from state 2 to state 3 is replaced by an irreversible adiabatic process from state 2 to state 3′ or state 3″.

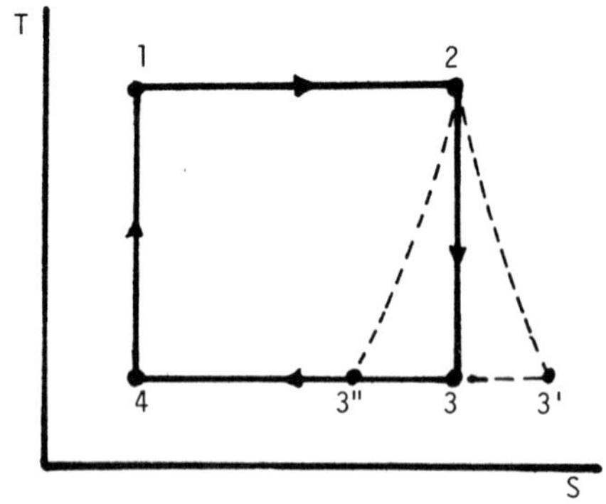

Figure 17.11

For the Carnot cycle

$$Q_{in} = Q_{12} = T_1(S_2 - S_1)$$

and

$$Q_{out} = Q_{34} = T_3(S_4 - S_3) \qquad |Q_{out}| = T_3(S_3 - S_4)$$

The cycle efficiency is

$$e = \frac{\Sigma W}{Q_{in}} = \frac{Q_{in} + Q_{out}}{Q_{in}} = \frac{Q_{in} - |Q_{out}|}{Q_{in}}$$

For the irreversible adiabatic process, terminating at 3′,

$$Q_{in} = Q_{12} = T_1(S_2 - S_1)$$

and

$$Q_{out}' = T_3(S_4 - S_3') = T_3(S_1 - S_3') = -T_3(S_3' - S_1)$$

Since

$$S_3' > S_3 = S_2$$

then

$$|Q_{out}'| > |Q_{out}|$$

and

$$\frac{Q_{in} - |Q_{out}'|}{Q_{in}} < \frac{Q_{in} - |Q_{out}|}{Q_{in}}$$

so that

$$e' < e$$

Therefore the efficiency of the irreversible cycle (a single irreversible process within a cycle makes the cycle irreversible) is less than that of the reversible cycle where both cycles are operating between a source and sink having fixed temperatures. This is in accordance with the premise that stated no cycle could have an efficiency greater than that of a reversible cycle operating between a source and sink having fixed temperatures.

For the irreversible process, terminating at 3″,

$$Q_{in} = Q_{12} = T_1(S_2 - S_1)$$

as before, and

$$Q_{out}'' = T_3(S_4 - S_3'') = -T_3(S_3'' - S_1)$$

Since

$$S_3'' < S_3 = S_2$$

then

$$|Q_{out}''| > |Q_{out}|$$

and

$$\frac{Q_{in} - |Q_{out}''|}{Q_{in}} > \frac{Q_{in} - |Q_{out}|}{Q_{in}}$$

so that

$$e'' > e$$

This violates our previous premise, and by coupling this cycle with a reversed Carnot cycle operating between the same source and sink, we could theoretically construct a perpetual motion machine of the second kind.

In Fig. 17.12, the reversible compression process from 4 to 1 has been replaced by an irreversible process. For the irreversible process, from state 4 to state 1′,

$$Q_{in} = T_1(S_2 - S_1')$$

and

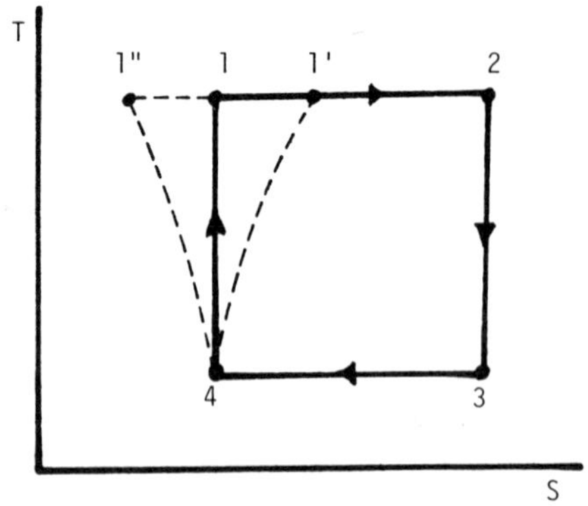

Figure 17.12

$$Q_{out} = T_3(S_4 - S_3) = -T_3(S_2 - S_1)$$

Since

$$S_1' > S_1$$

then

$$Q_{in}' < Q_{in}$$

For the irreversible cycle, where $Q_{41}' = 0$,

$$e' = \frac{Q_{in}' - |Q_{out}|}{Q_{in}'} = 1 - \frac{|Q_{out}|}{Q_{in}'}$$

and for the reversible cycle

$$e = 1 - \frac{|Q_{out}|}{Q_{in}}$$

Since

$$|Q_{out}|/Q_{in}'| > |Q_{out}|/Q_{in}$$

then

$$e' < e$$

This is in accordance with our premise.

For the irreversible adiabatic process from state 4 to state 1'',

$$Q_{in} = T_1(S_2 - S_1'')$$

and

$$Q_{out} = -T_3(S_2 - S_1)$$

where

$$S_1'' < S_1$$

so that

$$Q_{in}'' > Q_{in}$$

This leads to

$$Q_{out}/Q_{in}'' < Q_{out}/Q_{in}$$

or

$$e'' > e$$

which violates our premise.

Since $T_2 > T_3$, the cycle in Fig. 17.11 represents an expansion process, and since $T_1 > T_4$, the cycle in Fig. 17.12 is for a compression process. A notable difference between this argument and that used in the Brayton and Rankine processes is that we have substituted fixed initial and final temperatures for stated values of the initial and final pressures. This is necessary to accommodate a single source and sink temperature. If the arguments were carried out for a vapor where all of the end points were in the wet region (Fig. 17.13), the reasoning would apply to both situations.

Note that the permissible end states of the irreversible process result in an entropy increase for the irreversible adiabatic process.

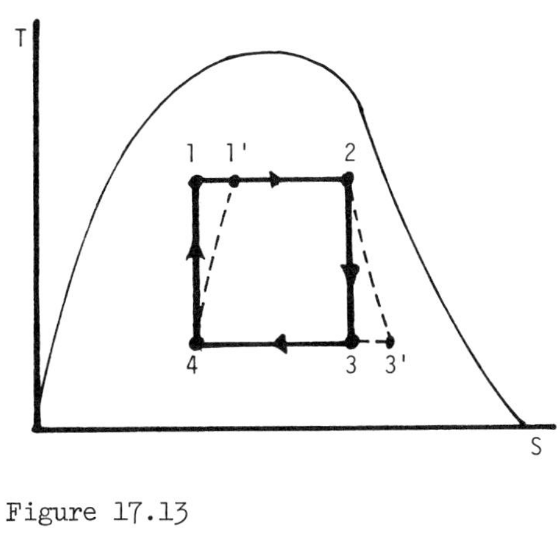

Figure 17.13

Example 17.8. Given is a Carnot cycle operating between a source and sink at 1540 F and 40 F (Fig. 17.11). The heat added to the cycle is 1000 B, and $S_3' - S_3 = 0.1$; also $S_3'' - S_3 = -0.1$ B/R. Determine Q_{out}, ΣW, and e for the reversible and irreversible cycles.

For the reversible cycle

$$S_2 - S_1 = Q_{12}/T_1 = 1000/2000 = 0.5 \text{ (B/R)}$$

and

$$Q_{out} = -T_3(S_2 - S_1) = -500(0.5) = -250 \text{ (B)}$$

so that

$$e = (Q_{in} + Q_{out})/Q_{in} = (1000 - 250)/1000$$

$$= 0.75$$

or

$$e = (T_1 - T_3)/T_1 = (2000 - 500)/2000 = 0.75$$

For the irreversible cycle which includes the process from state 2 to state 3′ where $S_4 - S_3 = -(S_2 - S_1)$,

$$S_4 - S_3' = (S_4 - S_3) + (S_3 - S_3')$$

$$= -(S_2 - S_1) - (S_3' - S_3)$$

$$= -0.5 - 0.1 = -0.6 \text{ (B/R)}$$

and

$$Q_{out}' = -T_3(S_3' - S_4)$$

$$= -500(0.6) = -300 \text{ (B)}$$

so that

$$e' = (Q_{in} + Q_{out}')/Q_{in} = (1000 - 300)/1000$$

$$= 0.70$$

This is less than the efficiency of the Carnot cycle and is theoretically possible.

For the irreversible cycle which includes the process from state 2 to state 3″,

$$S_4 - S_3'' = -(S_2 - S_1) - (S_3'' - S_3)$$

$$= -0.5 + 0.1 = -0.4 \text{ (B/R)}$$

and

$$Q_{out}'' = -T_3(S_3'' - S_4)$$

$$= -500(0.4) = -200 \text{ (B)}$$

so that

$$e'' = (Q_{in} + Q_{out}'')/Q_{in} = (1000 - 200)/1000$$

$$= 0.80$$

which exceeds that of the comparable Carnot cycle and is therefore impossible.

For the reversible cycle

$$\Sigma W = Q_{in} + Q_{out} = 1000 - 250 = 750 \text{ (B)}$$

for the irreversible cycle

$$\Sigma W' = Q_{in} + Q_{out} = 1000 - 300 = 700 \text{ (B)}$$

and for the impossible cycle

$$\Sigma W'' = Q_{in} + Q_{out}'' = 1000 - 200 = 800 \text{ (B)}$$

Example 17.9. Given is a Carnot cycle operating between a source and sink at 1540 F and 40 F (Fig. 17.12). Heat rejected by the cycle is -250 B, and $S_1' - S_1 = 0.1$; also $S_1'' - S_1 = -0.1$ B/R. Determine Q_{in}, ΣW, and e for the three cycles.

For the reversible cycle

$$e = (T_1 - T_3)/T_1 = 0.75$$

and

$$e = (Q_{in} + Q_{out})/Q_{in} = (Q_{in} - 250)/Q_{in} = 0.75$$

yielding

$$Q_{in} = 1000 \text{ (B)}$$

so that

$$W = eQ_{in} = 0.75(1000) = 750 \text{ (B)}$$

The entropy change for the heat addition process is

$$S_2 - S_1 = Q/T = 1000/2000 = 0.5\ (\text{B/R})$$

For the irreversible cycle including state 1′,

$$S_2 - S_1' = (S_2 - S_1) + (S_1 - S_1') = (S_2 - S_1) - (S_1' - S_1)$$

$$= 0.5 - 0.1 = 0.4\ (\text{B/R})$$

and

$$Q_{in}' = T_1(S_2 - S_1') = 2000(0.4) = 800\ (\text{B})$$

Also

$$\Sigma W' = Q_{in} + Q_{out} = 800 - 250 = 550\ (\text{B})$$

and

$$e' = \Sigma W/Q_{in} = 550/800 = 0.688$$

which is theoretically possible since $e = 0.750$.

For the irreversible cycle where

$$S_1'' - S_1 = -0.1$$

as before

$$S_2 - S_1'' = (S_2 - S_1) - (S_1'' - S_1) = 0.5 + 0.1 = 0.6\ (\text{B/hr})$$

so that

$$Q_{in}'' = T_1(S_2 - S_1'')$$

$$= 2000(0.6) = 1200\ (\text{B})$$

Also

$$\Sigma W'' = Q_{in} + Q_{out} = 1200 - 250 = 950\ (\text{B})$$

and

$$e'' = \Sigma W/Q_{in} = 950/1200 = 0.792$$

which is impossible since $e' > e$.

Example 17.10. Given is the reversed Carnot cycle in Fig. 17.14 operating between 1540 F and 40 F. Heat added to the cycle is 250 B, and $S_3' - S_3 = 0.1$; also $S_3'' - S_3 = -0.1$ B/R. Determine Q_{out}, ΣW, γ_p, and γ_R for the three cycles.

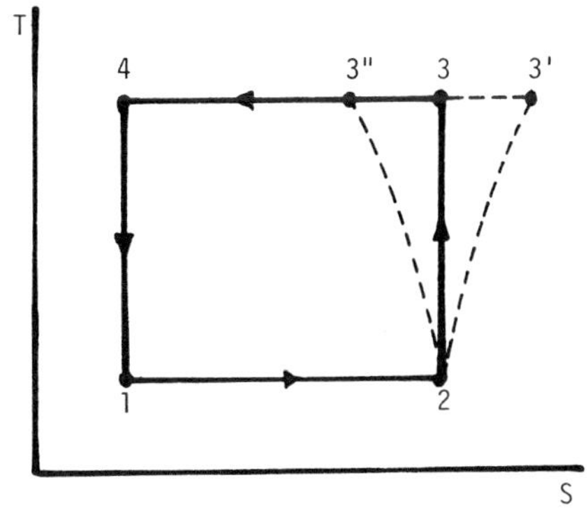

Figure 17.14

For all three cycles

$$Q_{in} = 250 = T_1(S_2 - S_1) = 500(S_2 - S_1)$$

and

$$S_2 - S_1 = 0.5 \text{ (B/R)}$$

For the Carnot cycle

$$Q_{out} = T_3(S_4 - S_3) = T_3(S_1 - S_2) = 2000(-0.5) = -1000 \text{ (B)}$$

and

$$\Sigma W = Q_{in} + Q_{out} = 250 - 100 = -750 \text{ (B)}$$

From Eq. (17.2)

$$\gamma_p = T_3/(T_3 - T_1) = 2000/1500 = 1.33$$

and from Eq. (17.3)

$$\gamma_R = T_1/(T_3 - T_1) = 500/1500 = 0.33$$

For the irreversible cycle including state point 3′,

$$S_4 - S_3' = (S_4 - S_3) + (S_3 - S_3') = +0.5 + 0.1 = 0.6 \text{ (B/R)}$$

and

$$Q_{out}' = T_3(S_3' - S_4) = 2000(-0.6) = -1200 \text{ (B)}$$

Also

$$\Sigma W' = Q_{in} + Q_{out} = 250 - 1200 = -950 \text{ (B)}$$

The coefficient of performance, as a heat pump, is

$$\gamma_P = |Q_{out}|/|\Sigma W| = 1200/950 = 1.26$$

and for the refrigerator, from Eq. (17.3),

$$\gamma_R = |Q_{in}|/|\Sigma W| = 250/950 = 0.26$$

Since $\gamma_P' < \gamma_P$ and $\gamma_R' < \gamma_R$, the cycle is irreversible and may exist.

For the irreversible cycle including state point 3″, using the same temperature as in the previous study,

$$S_4 - S_3'' = (S_4 - S_3) + (S_3 - S_3'') = +0.5 - 0.1 = +0.4 \text{ (B/R)}$$

and

$$Q_{out}'' = T_3(S_3'' - S_4) = 2000(-0.4) = -800 \text{ (B)}$$

Also

$$\Sigma W'' = Q_{in} + Q_{out} = 250 - 800 = -550 \text{ (B)}$$

From the coefficient of performance equation

$$\gamma_P'' = |Q_{out}|/|\Sigma W| = 800/550 = 1.45$$

and

$$\gamma_R'' = \gamma_P'' - 1 = 0.45$$

Since $\gamma_P'' > \gamma_P$ and $\gamma_R'' > \gamma_R$, the cycle is impossible.

Again note that for the irreversible adiabatic process the entropy must increase.

INEQUALITY OF CLAUSIUS

The inequality of Clausius states that for a reversible cycle the sum of the terms $\int dQ/T$ must equal zero. For an irreversible cycle the sum of the $\int dQ/T$ process terms is less than zero. For an impossible cycle the sum exceeds zero. In equation form,

$$\Sigma \int dQ/T \leq 0 \quad \text{or} \quad \oint dQ/T \leq 0$$

for the cycle. The notation $\oint$ represents the sum of the integrals for the cycle.

The inequality of Clausius may be used instead of the efficiency or coefficient of performance to establish if a cycle is reversible, irreversible, or impossible. Since the inequality of Clausius may be applied to cycles that receive and reject heat for all processes and is not limited to isothermal transfers of heat, it is a more usable criterion than the efficiency or coefficient of performance evaluation.

Example 17.11. Determine if the cycles in Examples 17.8, 17.9, and 17.10 are reversible, irreversible, or impossible by means of the inequality of Clausius. From Example 17.8 and Fig. 17.15, for the cycle including state point 3, $t_1 = 1540$ F, $t_3 = 40$ F, $Q_{in} = 1000$ B, and $Q_{out} = -250$ B. For the isothermal process from state 1 to state 2,

$$\int dQ/T = Q_{12}/T_1 = 1000/2000 = 0.5 \text{ (B/R)}$$

For the isothermal process from state 3 to state 4,

$$\int dQ/T = Q_{34}/T_3 = -250/500 = -0.5 \text{ (B/R)}$$

For the two isentropic processes, $Q = 0$, so that

$$\int dQ/T = 0.$$

By summing the integral terms,

$$\oint dQ/T = 0.5 + 0.0 - 0.5 + 0.0 = 0.0$$

and the cycle is reversible.

From Example 17.8 and Fig. 17.15, for the cycle including state point 3′, $t_1 = 1540$ F, $t_3 = 40$ F, $Q_{in} = 1000$ B, and $Q_{out}' = -300$ B. The only variation from the reversible cycle is for the isothermal process from state 3′ to 4, where

$$\int dQ/T = Q_{3'4}/T_3 = -300/500 = -0.6 \text{ (B/R)}$$

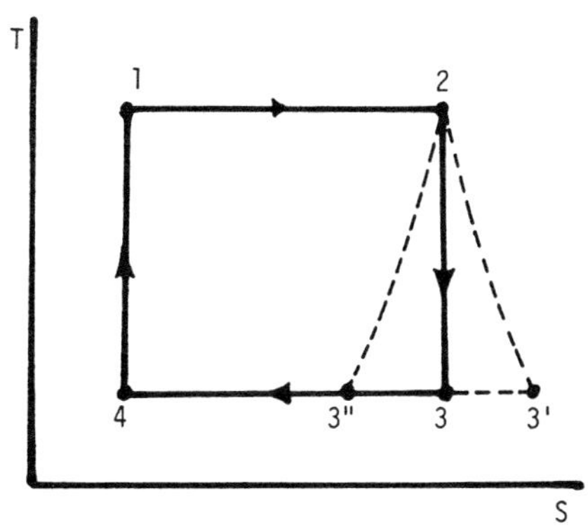

Figure 17.15

For the isentropic process and for the irreversible adiabatic process, Q is again zero. Therefore,

$$\oint dQ/T = 0.5 + 0.0 - 0.6 + 0.0 = -0.1 \text{ (B/R)}$$

and the cycle is irreversible.

From Example 17.8 and Fig. 17.15, $t_1 = 1540$ F, $t_3 = 40$ F, $Q_{in} = 1000$ B, and $Q_{out}'' = -200$ B. For the isothermal process from state 3″ to 4,

$$\int dQ/T = Q_{3''4}/T_3 = -200/500 = -0.4 \text{ (B/R)}$$

and

$$\oint dQ/T = 0.5 + 0.0 - 0.4 + 0.0 = 0.1 \text{ (B/R)}$$

and since the result of the summation is positive, the cycle is impossible.

For Example 17.9 and Fig. 17.16, $t_1 = 1540$ F, $t_3 = 40$ F, $Q_{in} = 1000$ B, $Q_{in}' = 800$ B, $Q_{in}'' = 1200$ B, and $Q_{out} = -250$ B. For the cycle with state point 1,

$$\oint dQ/T = (1000/2000) + 0 - (250/500) + 0$$
$$= 0.5 - 0.5 = 0$$

and the cycle 12341 is reversible; also

$$\oint dQ/T = (800/2000) + 0 - (250/500) + 0 = 0.4 - 0.5$$
$$= -0.1 \text{ (B/R)}$$

for the irreversible cycle 1′2341′. Also,

$$\oint dQ/T = (1200/200) + 0 - (250/500) + 0 = 0.6 - 0.5 = 0.1 \text{ (B/R)}$$

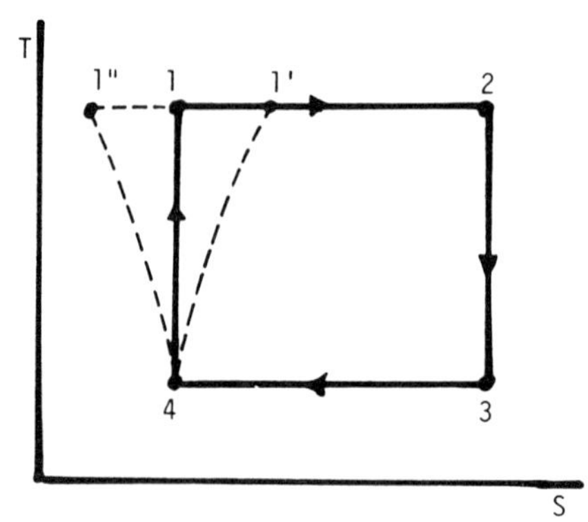

Figure 17.16

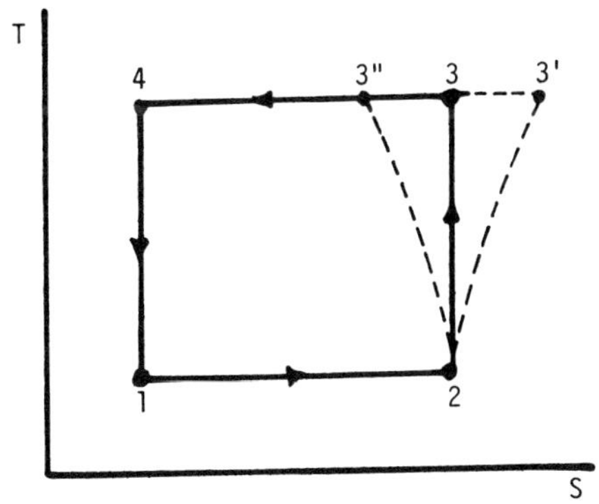

Figure 17.17

for the impossible cycle 1″2341″.

From Example 17.10 and Fig. 17.17, $t_1 = 40$ F, $t_3 = 1540$ F, $Q_{in} = 250$ B, $Q_{out} = -1000$ B, $Q_{out}{}' = -1200$ B, and $Q_{out}{}'' = -800$ B. For the cycle with state point 3,

$$\oint dQ/T = (250/500) - (1000/2000) = 0.5 - 0.5 = 0$$

and the cycle 12341 is reversible, and

$$\oint dQ/T = (250/500) - (1200/2000) = 0.5 - 0.6 = -0.1$$

for the irreversible cycle 123′41. Also,

$$\oint dQ/T = (250/500) - (800/2000) = 0.5 - 0.4 = 0.1$$

for the impossible cycle 123″41.

Example 17.12. Given is a Brayton cycle operating between temperatures of 40 F and 1540 F using air as a perfect gas (Fig. 17.18); $r_p = 5$, and $e_t = e_c = 100\%$. Determine the value of $\oint dQ/T$.

$$T_2/T_1 = (p_2/p_1)^{(k-1)/k} = r_p^{(k-1)/k}$$

$$= (5)^{0.286} = 1.585$$

and

$$T_2 = 500(1.585) = 792 \text{ (R)}$$

For a simple Brayton cycle

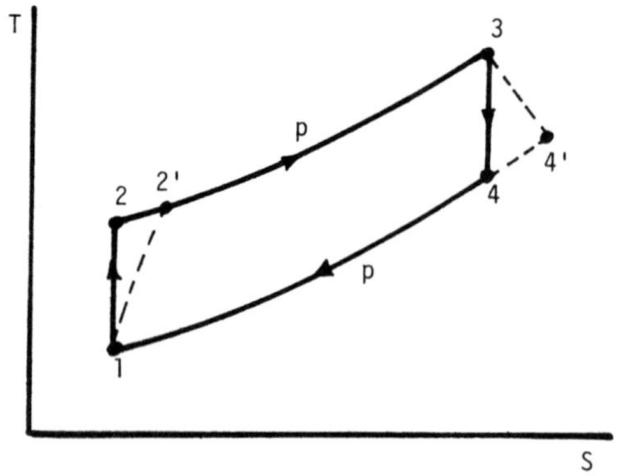

Figure 17.18

$$T_4 = T_1T_3/T_2 = [500(2000)]/792 = 1262.6 \text{ (R)}$$

For the isentropic process from state 1 to state 2, $\int dQ/T = 0$, since $Q = 0$. For the constant pressure process from state 2 to state 3,

$$\int dQ/T = \int_{T_2}^{T_3} (c_p \, dT) = c_p \ln(T_3/T_2)$$

$$= 0.24 \ln(2000/792) = 0.2223 \text{ (B/lbm-R)}$$

For the isentropic process from state 3 to state 4, $\int dQ/T = 0$, since $Q = 0$. For the constant pressure process from state 4 to state 1,

$$\int dQ/T = \int_{T_4}^{T_1} (c_p \, dT)/T = c_p \ln(T_1/T_4)$$

$$= 0.24 \ln(500/1262.6) = -0.2223 \text{ (B/lbm-R)}$$

and

$$\oint dQ/T = 0 + 0.2223 + 0 - 0.2223 = 0$$

for the reversible cycle.

Example 17.13. Repeat Example 17.12 for $e_c = 80\%$ and $e_t = 75\%$. For turbine and compressor efficiency less than 100%, the isentropic process in the reversible cycle is replaced by an irreversible adiabatic process between the same pressure limits.

For the compressor,

$$e_c = (T_2 - T_1)/(T_2' - T_1) = (792 - 500)/(T_2' - 500) = 0.8$$

and

$$T_2' = 865 \text{ (R)}$$

For the turbine,

$$e_t = (T_3 - T_4')/(T_3 - T_4) = (2000 - T_4')/(2000 - 1262.6) = 0.75$$

and

$$T_4' = 1447 \text{ (R)}$$

For the two irreversible adiabatic processes, $\oint dQ/T = 0$, since $Q = 0$. For the constant pressure, heat addition process,

$$\int dQ/T = \int_{T_2'}^{T_3} (c_p\ dT)/T = c_p \ln(T_3/T_2')$$

$$= 0.24 \ln(2000/865) = 0.2012 \text{ (B/lbm-R)}$$

and for the constant pressure, heat rejection process,

$$\int dQ/T = \int_{T_4'}^{T_1} (c_p\ dT)/T = c_p \ln(T_1/T_4')$$

$$= 0.24 \ln(500/1447) = -0.2550 \text{ (B/lbm-R)}$$

such that

$$\oint dQ/T = 0 + 0.2012 + 0 - 0.2550 = -0.0538 \text{ (B/lbm-R)}$$

The results confirm the prior knowledge that the cycle was irreversible.

ENTROPY PRODUCTION

The entropy production for a process is defined by

$$\Delta S_p = \Delta S - \int dQ/T \tag{17.4}$$

and for a reversible process

$$\Delta S_p = 0$$

For a cycle

$$\Sigma\Delta S_p = \Sigma\Delta S - \Sigma \int dQ/T$$

The change in any thermodynamic property for a cycle is zero so that

$$\Sigma\Delta S = 0$$

By using the notation for the cyclic integral,

$$\oint \Delta S_p = -\oint dQ/T \tag{17.5}$$

and $\oint \Delta S_p = 0$ for a reversible cycle. For an irreversible cycle

$$\int \Delta S_p > 0$$

and is equal to the magnitude of the quantity obtained by use of the inequality of Clausius.

Example 17.14. Repeat Example 17.13 to determine the entropy production of each process and of the cycle. See Fig. 17.18.

For the process from state 1 to state 2′, $T_1 = 500$ R, $T_2' = 865$ R, $p_2/p_1 = r_p = 5$.

The equations that relate properties for the polytropic process may be used for irreversible processes. The justification for the use of these equations is that properties are point functions and are therefore independent of the process, reversible or irreversible, used to attain these end points.

The general entropy equation that relates temperatures and pressure is

$$s_2' - s_1 = c_p \ln(T_2'/T_1) - (R/J) \ln(p_2/p_1)$$

$$= 0.24 \ln(865/500) - (53.3/778) \ln(5)$$

$$= 0.1315 - 0.1103 = 0.0212 \text{ (B/lbm-R)}$$

For the polytropic process

$$T_2'/T_1 = (p_2/p_1)^{(n-1)/n}$$

or

$$865/500 = (5)^{(n-1)/n}$$

and

$$(n - 1)/n = (\ln 1.730)/\ln(5) = 0.3406$$

so that

$$n = 1.5165$$

The polytropic specific heat, which was derived assuming reversible processes for the heat and work terms in the nonflow general energy equation, is

$$c_n = c_v(k - n)/(1 - n)$$

$$= 0.1714(1.4 - 1.5165)/(1 - 1.5165)$$

$$= 0.03866 \text{ (B/lbm-R)}$$

The entropy change, using the polytropic specific heat, is

$$s_2' - s_1 = c_n \ln(T_2'/T_1) = 0.03866 \ln(865/500)$$

$$= 0.0212 \text{ (B/lbm-R)}$$

Even though reversible processes were assumed in the derivation of the polytropic specific heat equation, the evaluation of the property entropy must be the same for any process connecting these state points, including the irreversible process.

Consider the path functions, heat and work. For the irreversible adiabatic process, by definition,

$$Q = 0$$

If we attempt to use the heat equation as derived for the reversible polytropic process,

$$Q = c_n(T_2' - T_1) = 0.03866(865 - 500)$$

$$= 14.1 \text{ (B/lbm)}$$

which is incorrect.

For the steady-flow work for the irreversible process,

$$W_{sf} = Q - \Delta H \qquad (\Delta KE = \Delta PE = 0)$$

and

$$W_{sf} = -\Delta H \qquad (\text{for } Q = 0)$$

$$= -c_p(T_2' - T_1) = -0.24(865 - 500)$$

$$= -87.6 \text{ (B/lbm)}$$

If we assume the process is reversible, where $W_{sf} = nW_{nf}$,

$$W_{sf} = \frac{n(p_2v_2' - p_1v_1)}{J(1 - n)} = \frac{nR(T_2' - T_1)}{J(1 - n)}$$

$$= [1.5165(53.3)(865 - 500)]/[778(1 - 1.5165)]$$

$$= -73.4 \text{ (B/lbm)}$$

which is incorrect.

The polytropic expressions for the heat and work can be used only for the reversible processes. This is logical since it was the restriction placed on their derivation. The value of the path n and the polytropic specific heat c_n may be used in conjunction with the determination of property changes.

For the process from 1-2′ from Eq. (17.4),

$$\Delta s_p = \Delta s - \int dQ/T$$

and

$$\Delta s_{p2'1} = \Delta s_{2'1} = 0.0212 \text{ (B/lbm-R)}$$

since Q = 0 for the irreversible adiabatic process.

For the reversible process from state 2′ to state 3 where T_3 = 2000 R,

$$\Delta s_{32'} = s_3 - s_{2'} = c_p \ln(T_3/T_{2'})$$

$$= 0.24 \ln(2000/865) = 0.2012 \text{ (B/lbm-R)}$$

and

$$\int dQ/T = \int_{T_2'}^{T_3} (c_p\, dT/T) = c_p \ln(T_3/T_2') = 0.2012 \text{ (B/lbm-R)}$$

From Eq. (17.4)

$$\Delta s_{p32'} = \Delta s_{32'} - \int dQ/T = 0.2012 - 0.2012 = 0$$

or

$$\Delta s_p = 0$$

for a reversible process.

For the irreversible adiabatic process from state 3 to state 4′ where T_4' = 1447 R and r_p = 5,

$$s_4' - s_3 = c_p \ln(T_4'/T_3) - (R/J) \ln(p_4/p_3)$$

Since

$$p_4/p_3 = p_1/p_2 = 1/r_p$$

then

$$s_4' - s_3 = c_p \ln(T_4'/T_3) + (R/J) \ln r_p$$

$$= 0.24 \ln(1447/2000) + (53.3/778) \ln(5)$$

$$= -0.0777 + 0.1103 = 0.0326 \text{ (B/lbm-R)}$$

Since $\int dQ/T = 0$ for the irreversible adiabatic process,

$$\Delta s_{p4'3} = 0.0326 \text{ (B/lbm-R)}$$

For the reversible process from state 4′ to state 1,

$$\Delta s_{p14'} = 0$$

For the cycle,

$$\oint \Delta s_p = \Delta s_{p2'1} + \Delta s_{p32'} + \Delta s_{p4'3} + \Delta s_{p14'}$$

$$= 0.0212 + 0 + 0.0326 + 0$$

$$= 0.0538 \text{ (B/lbm-R)}$$

From Example 17.13

$$\oint dQ/T = -0.0538 \text{ (B/lbm-R)}$$

which agrees with Eq. (17.5),

$$\oint \Delta s_p = -\oint dQ/T$$

for the cycle.

Example 17.15. Determine if the constant pressure, adiabatic combustion process in Example 16.10 (Chapter 16) is irreversible (Fig. 17.19). From Example 16.10, $t_1 = 60$ F, $t_2 = 4709$ F, and

$$C_8H_{18}(g) + 12.5\ O_2 + 47.0\ N_2 \rightarrow 8\ CO_2 + 9\ H_2O + 47.0\ N_2$$

Assume $p_1 = p_2 = p_0 = 14.69$ psia and $t_0 = 77$ F.

Construct the following chart to determine the entropy of the reactants:

Gas	n	x	m	y	$c_p \ln(T_1/T_0)$	$-(R/J) \ln x$	$s_1 - s_0$	$y(s_1 - s_0)$
C_8H_{18}	1	0.0165	114	0.0623	-0.0161	0.0715	0.0554	0.00345
O_2	12.5	0.2066	400	0.2186	-0.0070	0.0979	0.0909	0.01987
N_2	47	0.7769	1316	0.7191	-0.0080	0.0179	0.0099	0.00712
	60.5	1.0000	1830	1.0000				0.03044

For the octane,

$$x = n/\Sigma n = 1/60.5 = 0.0165$$

$$y = m/\Sigma m = 114/1830 = 0.0623$$

$$c_p \ln(T_1/T_0) = 0.5 \ln(520/537) = -0.0161 \text{ (B/lbm-R)}$$

$$-(R/J) \ln x = -[1545/114(778)] \ln(0.0165) = 0.0715 \text{ (B/lbm-R)}$$

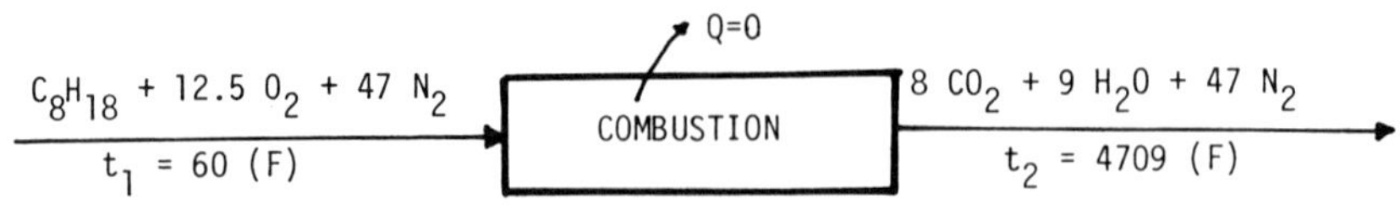

Figure 17.19

From Dalton's law of partial pressures

$$p = xp_m \quad \text{or} \quad x = p/p_m = p/p_0$$

where p is the partial pressure of the component gas.

The entropy change between states 1 and 0 is

$$s_1 - s_0 = c_p \ln(T_1/T_0) - (R/J) \ln(p/p_0)$$

$$= c_p \ln(T_1/T_0) - (R/J) \ln(x)$$

$$= -0.0161 + 0.0715 = 0.0554 \text{ (B/lbm-R)}$$

$$y(s_1 - s_0) = 0.0623(0.0554) = 0.00345 \text{ (B/lbm-R)}$$

for each pound of reactant. For the reactants

$$s_1 - s_0 = \Sigma y(\Delta s) = 0.03044 \text{ (B/lbm-R)}$$

For the products, where $T_2 = 5169$ R,

Gas	n	x	m	y	$c_p \ln(T_2/T_0)$	$-(R/J) \ln x$	$s_2 - s_0$	$y(s_2 - s_0)$
CO_2	8	0.1250	352	0.1923	0.4574	0.0938	0.5512	0.1060
H_2O	9	0.1406	162	0.0886	1.0077	0.2164	1.2241	0.1085
N_2	47	0.7344	1316	0.7191	0.5638	0.0219	0.5857	0.4212
	64	1.0000	1830	1.0000				0.6357

so that

$$s_2 - s_0 = 0.6357 \text{ (B/lbm-R)}$$

From Eq. (17.4)

$$\Delta s_p = \Delta s - \int dQ/T = (s_2 - s_0) - (s_1 - s_0) - \int dQ/T$$

$$= 0.6357 - 0.0304 - 0 = 0.6053 \text{ (B/lbm-R)}$$

and the process is irreversible. $\int dQ/T$ is zero for the adiabatic flame temperature process. The above procedure, where the entropy of the reactants of products is calculated with reference to an arbitrary datum plane, defined by the pressure and temperature, is necessary for processes where the compositions of the gases at the initial and final states differ.

ISOLATED SYSTEM

The use of an isolated system is a convenient technique to determine if a process is irreversible. The isolated system (Fig. 17.20) permits neither energy nor mass to cross the system boundaries. Therefore, it represents an adiabatic system, and processes occurring within the system are either isentropic or irreversible adiabatic.

FREE EXPANSION OF A GAS

Example 17.16. Given the isolated system in Fig. 17.21 where the air is allowed to occupy the evacuated spaces, let p_1 = 100 psia, t_1 = 60 F, and V_1 = 10 ft^3 for the properties before expansion. Assume the volume of the evacuated space is 5 ft^3. Since mass does not cross the system boundaries, it is a closed system, and $Q = \Delta U + W_{nf}$.

For an isolated system,

$$Q = W_{nf} = 0$$

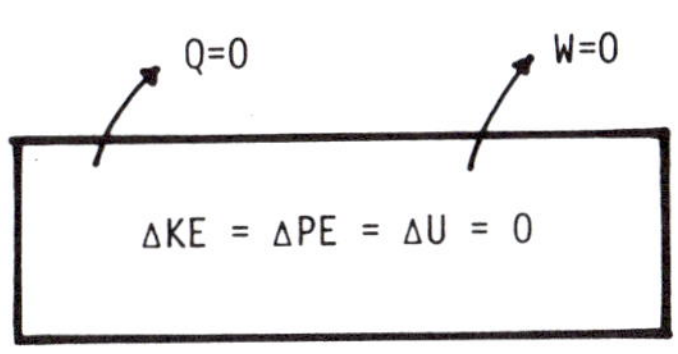

Figure 17.20

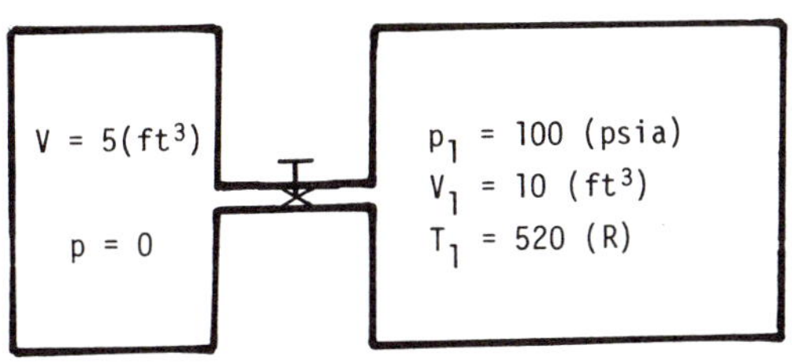

Figure 17.21

so that

$$\Delta U = 0$$

Therefore the temperature does not change, and $t_2 = t_1 = 60$ F.

The mass of air is

$$m_1 = \frac{p_1V_1}{RT_1} = \frac{100(144)(10)}{53.3(520)} = 5.20 \text{ (lbm)}$$

and

$$p_2 = \frac{m_2RT_2}{V_2} = \frac{5.20(53.3)(520)}{(10 + 5)144} = 66.7 \text{ (psia)}$$

where state 2 occurs after the valve is opened, and $m = m_1 = m_2$ for the isolated system.

The entropy change between the initial and final states is

$$S_2 - S_1 = mc_p \ln(T_2/T_1) - (mR/J) \ln(p_2/p_1)$$

$$= 5.2(0.240) \ln(1) - \{[5.2(53.3)]/778\} \ln(66.7/100)$$

$$= 0.1443 \text{ (B/R)}$$

The entropy production is

$$\Delta S_p = \Delta S - \int dQ/T = 0.1443 \text{ (B/R)}$$

and the free expansion process is irreversible.

Example 17.17. Assume the evacuated space in Example 17.16 contains air at a pressure of 20 psia and a temperature of 60 F. Determine the entropy production. For the high-pressure air, $m_1 = 5.20$, $p_1 = 100$ psia, $V_1 = 10$ ft^3, and $T_1 = 520$ R. For the low-pressure air,

$$m_2 = p_2V_2/RT_2 = [20(144)(5)]/[53.3(520)]$$

$$= 0.520 \text{ (lbm)}$$

The mass of air, after the valve is opened, is

$$m_3 = m_1 + m_2 = 5.20 + 0.52 = 5.72 \text{ (lbm)}$$

Again

$$\Delta U = 0$$

so that

$$T_1 = T_2 = T_3$$

since the internal energy of a perfect gas is only a function of temperature. The final pressure is

$$p_3 = m_3RT_3/V_3 = [5.72(53.3)(520)]/[15(144)]$$

$$= 73.4 \text{ (psia)}$$

The change in entropy using $p_0 = 14.7$ psia and $t_0 = 60$ F as the reference values is

$$\Delta S = (S_3 - S_0) - (S_2 - S_0) - (S_1 - S_0)$$

For each component,

$$mc_p \ln(T/T_0) = 0$$

since

$$T_1 = T_2 = T_3 = T_0$$

so that for $p_0 = 14.7$ psia,

$$\Delta S = (R/J)[-m_3 \ln(p_3/p_0) + m_1 \ln(p_1/p_0) + m_2 \ln(p_2/p_0)]$$

$$= 0.0685[-5.72(1.608) + 5.2(1.917) + 0.52(0.3079)]$$

$$= 0.0685(0.931) = 0.0638 \text{ (B/R)}$$

For the isothermal system with the free expansion of air to a lower pressure,

$$\Delta S_p = \Delta S = 0.0638 \text{ (B/R)}$$

and the process is irreversible, but less so than in Example 17.16 since the amount of free expansion has been reduced.

Example 17.18. Repeat Example 17.17 where $p_2 = 100$ psia and $t_2 = 140$ F. Let the reference temperature and pressure be $t_0 = 60$ F and $p_0 = 100$ psia. As before,

$$\Delta U = 0$$

so that

$$(U_3 - U_0) - (U_1 - U_0) - (U_2 - U_1) = 0$$

and

$$m_3c_v(t_3 - t_0) - m_1c_v(t_1 - t_0) - m_2c_v(t_2 - t_0) = 0$$

$$m_3t_3 - m_1t_1 - m_2t_2 - (m_3 - m_1 - m_2)t_0 = 0$$

But

$$m_3 = m_1 + m_2$$

giving

$$m_3t_3 - m_1t_1 - m_2t_2 = 0$$

or

$$m_3t_3 = m_1t_1 + m_2t_2$$

Therefore,

$$p_3V_3/R = (p_1V_1/R) + (p_2V_2/R)$$

Also

$$p_1 = p_2 = 100 \text{ (psia)}$$

and

$$p_3V_3 = p_1(V_1 + V_2)$$

Since

$$V_3 = V_1 + V_2$$

then

$$p_3 = p_1 = 100 \text{ (psia)}$$

From Example 17.17

$$m_1 = p_1V_1/RT_1 = [100(144)(10)]/[53.3(520)]$$

$$= 5.20 \text{ (lbm)}$$

and

$$m_2 = p_2V_2/RT_2 = [100(144)(5)]/[53.3(600)]$$

$$= 2.25 \text{ (lbm)}$$

so that

$$m_3 = m_1 + m_2 = (5.20 + 2.25) = 7.45 \text{ (lbm)}$$

The final temperature is

$$T_3 = p_3V_3/m_3R = [100(144)(15)]/[7.45(53.3)]$$

$$= 544\ (R) = 84\ (F)$$

or

$$m_3t_3 = m_1t_1 + m_2t_2$$

and

$$t_3 = (5.20/7.45)60 + (2.25/7.45)140$$

$$= 42 + 42 = 84\ (F)$$

The entropies at the initial and final states are

$$S_3 - S_0 = m_3c_p \ln(T_3/T_0) - (m_3R/J) \ln(p_3/p_0)$$

$$= 7.45(0.240) \ln(544/520) - \{[7.45(53.3)]/778\} \ln(100/100)$$

$$= 0.0807\ (B/R)$$

and

$$S_2 - S_0 = m_2c_p \ln(T_2/T_0) - (m_2R/J) \ln(p_2/p_0)$$

$$= 2.25(0.240) \ln(600/520) - \{[2.25(53.3)]/778\} \ln(100/100)$$

$$= 0.0773\ (B/R)$$

Also

$$S_1 - S_0 = m_1c_p \ln(T_1/T_0) - (m_1R/J) \ln(p_1/p_0)$$

$$= 5.20(0.240) \ln(520/520) - \{[5.20(53.3)]/778\} \ln(100/100)$$

$$= 0.0$$

so that

$$\Delta S_p = \Delta S - \int dQ/T = (S_3 - S_0) - (S_1 - S_0) - (S_2 - S_0) - \int dQ/T$$

$$= 0.0807 - 0.0773 - 0.0$$

$$= 0.0034\ (B/R)$$

The mixing of a common gas at a uniform pressure but at different temperatures is an irreversible process since the entropy production is positive for the isolated system.

Example 17.19. Repeat Example 17.18 for a semiperfect gas. The internal energies are, from the gas tables, $t_1 = t_0 = 60$ F and $t_2 = 140$ F; $u_1 = 88.62$, $u_2 = 102.34$, and $u_0 = 88.62$ B/lbm; and $m_1 = 5.20$, $m_2 = 2.25$, and $m_3 = 7.45$ lbm. Also, $\varphi_0 = \varphi_1 = 0.59173$ and $\varphi_2 = 0.62607$ B/lbm-R. From the internal energy balance

$$U_3 - U_0 = U_1 - U_0 + U_2 - U_0$$

$$7.45(u_3 - 88.62) = 5.20(88.62 - 88.62) + 2.25(102.34 - 88.62)$$

$$u_3 - 88.62 = 4.14$$

and

$$u_3 = 92.76 \text{ (B/lbm)}$$

From the gas tables

T	u	φ
540	92.04	0.60078
560	95.47	0.60950

where

$$f = (92.76 - 92.04)/(95.47 - 92.04) = 0.21$$

so that

$$T_3 = 540 + 0.21(20) = 544 \text{ (R)} = 84 \text{ (F)}$$

and

$$\varphi_3 = 0.60078 + 0.21(0.60950 - 0.60078)$$

$$= 0.60261 \text{ (B/lbm-R)}$$

The entropies at the initial and final states are

$$S_3 - S_0 = m_3(\varphi_3 - \varphi_0) - (mR/J) \ln(p_3/p_0)$$

$$= 7.45(0.60261 - 0.59173) - \{[7.45(53.3)]/778\} \ln(100/100)$$

$$= 0.08106 \text{ (B/R)}$$

and

$$S_2 - S_0 = 2.25(0.62607 - 0.59173) - \{[2.25(53.3)]/778\} \ln(100/100)$$

$$= 0.07726 \text{ (B/R)}$$

Also

$$S_1 - S_0 = 5.20(0.59173 - 0.59173) - \{[5.20(53.3)]/778\} \ln(100/100)$$

$$= 0.0$$

The entropy production for the isolated system is

$$\Delta S_p = (S_3 - S_0) - (S_2 - S_0) - (S_1 - S_0)$$

$$= 0.08106 - 0.07726 - 0 = 0.0038 \text{ (B/R)}$$

for the mixing of a common gas at the same pressure but different temperatures. The entropy production is positive for both the perfect and semi-perfect gas solutions and differs by only 0.0004 B/R.

Example 17.20. Given is the isolated system in Fig. 17.22 where $p_A = p_B = 20$ psia, $t_A = t_B = 60$ F, and $V_A = V_B = 10$ ft^3. The gas in A is air, and the gas in B is carbon dioxide. The mass of each gas is

$$m_A = p_A V_A / R_A T_A = [20(144)(10)]/[53.3(520)]$$

$$= 1.039 \text{ (lbm)}$$

and

$$m_B = p_B V_B / R_B T_B = [20(144)(10)]/[35.1(520)]$$

$$= 1.578 \text{ (lbm)}$$

From the internal energy balance

$$(U_m - U_{m0}) - (U_{A1} - U_{A0}) - (U_{B1} - U_{B0}) = 0$$

Let the reference temperature and pressure be 60 F and 20 psia; then

$$U_{A1} - U_{A0} = U_{B1} - U_{B0} = 0$$

and

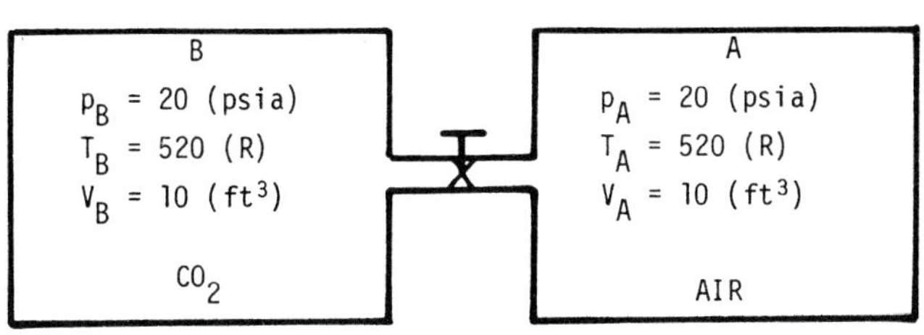

Figure 17.22

$$U_m - U_0 = m_m c_{vm}(t_m - t_0) = 0$$

so that

$$t_m = t_0 = 60 \text{ (F)}$$

and

$$t_{A2} = t_{B2} = 60 \text{ (F)}$$

The partial pressure of each component is

$$p_{A2} = p_{A1}(V_{A1}/V_{A2})(T_{A2}/T_{A1})$$

$$= 20(10/20)(520/520) = 10 \text{ (psia)}$$

In like manner, $p_{B2} = 10$ psia, and $p_m = p_{B2} + p_{A2} = 20$ psia. The entropy for each component at the initial and final states is

$$S - S_0 = mc_p \ln(T/T_0) - (mR/J) \ln(p/p_0)$$

Since $T = T_0$ in all cases, $mc_p \ln(T/T_0) = 0$, and

$$S_{A1} - S_{A0} = -(mR/J) \ln(p_{A1}/p_0)$$

$$= \{[1.039(53.3)]/778\} \ln(20/20) = 0$$

Also

$$S_{B1} - S_{B0} = 0$$

For the final states

$$S_{A2} - S_{A0} = -\{[1.039(53.3)]/778\} \ln(10/20) = 0.0493 \text{ (B/R)}$$

and

$$S_{B2} - S_{B0} = -\{[1.578(35.1)]/778\} \ln(10/20) = 0.0493 \text{ (B/R)}$$

so that

$$\Delta S_p = \Delta S = (S_{A2} - S_{A0}) - (S_{A1} - S_{A0}) + (S_{B2} - S_{B0}) - (S_{B1} - S_{B0})$$

$$= 0.0986 \text{ (B/R)}$$

The mixing of two dissimilar gases at the same pressure and temperature is an irreversible process.

Example 17.21. Repeat Example 17.20 where $p_B = 40$ psia, $V_B = 5$ ft^3, and $t_B = 140$ F. Assume the reference properties are $p_0 = 20$ psia and $t_0 = 60$ F. The mass of carbon dioxide is

$$m_B = [40(144)(5)]/[35.1(600)] = 1.368 \text{ (lbm)}$$

and

$$m_m = m_A + m_B = 1.039 + 1.368 = 2.407 \text{ (lbm)}$$

so that

$$c_{vm} = (m_A/m_m)c_{vA} + (m_B/m_m)c_{vB}$$

$$= (1.039/2.407)0.1714 + (1.368/2.407)0.157$$

$$= 0.163 \text{ (B/lbm-R)}$$

From the internal energy balance

$$(U_m - U_{m0}) - (U_{A1} - U_{A0}) - (U_{B1} - U_{B0}) = 0$$

$$2.407(0.163)(t_m - 60) - 1.039(0.1714)(60 - 60) - 1.368(0.157)(140 - 60) = 0$$

so that

$$t_m - 60 = 43.8$$

and

$$t_m = 103.8 \text{ (F)}$$

The partial pressure of each component is

$$p_{A2} = p_{A1}(V_{A1}/V_{A2})(T_{A2}/T_{A1}) = 20(10/15)(563.8/520)$$

$$= 14.5 \text{ (psia)}$$

and

$$p_{B2} = p_{B1}(V_{B1}/V_{B2})(T_{B2}/T_{B1}) = 40(5/15)(563.8/600)$$

$$= 12.5 \text{ (psia)}$$

The entropies at the initial and final states are

$$S_{A1} - S_{A0} = 0$$

$$S_{B1} - S_{B0} = 1.368(0.202) \ln(600/520) - \{[1.368(35.1)]/778\} \ln(40/20)$$

$$= 0.0395 - 0.0428 = -0.0033 \text{ (B/R)}$$

$$s_{A2} - s_{A0} = 1.039(0.24) \ln(563.8/520) - \{[1.039(53.3)]/778\} \ln(14.5/20)$$
$$= 0.0202 + 0.0229 = 0.0431 \text{ (B/R)}$$
$$s_{B2} - s_{B0} = 1.368(0.202) \ln(563.8/520) - \{[1.368(35.1)]/778\} \ln(12.5/20)$$
$$= 0.0223 + 0.0290 = 0.0513$$

and

$$S_p = S = (s_{A2} - s_{A0}) + (s_{B2} - s_{B0}) - (s_{A1} - s_{A0}) - (s_{B2} - s_{B0})$$
$$= 0.0431 + 0.0513 - 0 + 0.0033$$
$$= 0.0977 \text{ (B/R)}$$

which indicates the process is irreversible. The process is irreversible due to mixing of dissimilar gases, free expansion, and the temperature difference that exists between the gases before the mixing occurs.

HEAT TRANSFER ACROSS A FINITE TEMPERATURE DIFFERENCE

Example 17.22. In Fig. 17.23 the air in the two chambers is separated by a wall that is capable of transferring heat but does not permit mass flow. Assume air in both chambers and $p_{A1} = p_{B1} = 100$ psia, $V_{A1} = V_{B1} = 10$ ft^3, $t_{A1} = 60$ F, and $t_{B1} = 400$ F. Determine the entropy production after the system has reached thermal equilibrium. Reference conditions:, $p_0 = 100$ psia and $t_0 = 60$ F.

The mass of air in the two chambers is

$$m_A = p_A V_A / R T_A = [100(144)(10)]/[53.3(520)]$$
$$= 5.20 \text{ (lbm)}$$

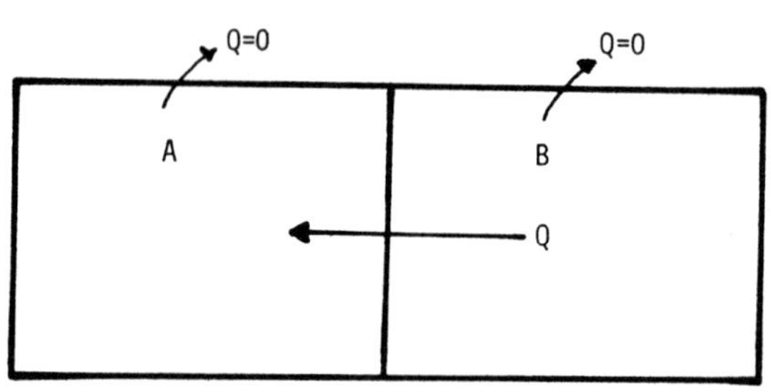

Figure 17.23

and

$$m_B = p_B V_B/RT_B = [100(144)(10)]/[53.3(860)]$$

$$= 3.14 \text{ (lbm)}$$

The final temperature may be obtained from the internal energy balance, where

$$(U_{A2} - U_{A0}) - (U_{A1} - U_{A0}) + (U_{B2} - U_{B0}) - (U_{B1} - U_{B0}) = 0$$

or

$$(U_{A2} - U_{A1}) + (U_{B2} - U_{B1}) = 0$$

and

$$m_A c_v(T_2 - T_{A1}) + m_B c_v(T_2 - T_{B1}) = 0$$

so that

$$5.20(T_2 - 520) + 3.14(T_2 - 860) = 0$$

yielding

$$T_2 = 648 \text{ (R)}$$

The final pressures in the two chambers are

$$p_{2A} = p_{1A}(T_2/T_{1A}) = 100(648/520) = 124.6 \text{ (psia)}$$

and

$$p_{2B} = p_{1B}(T_2/T_{1B}) = 100(648/860) = 75.3 \text{ (psia)}$$

For the air in chamber A,

$$S_{1A} - S_{A0} = m_A c_p \ln(T_{1A}/T_0) - (m_A R/J) \ln(p_{1A}/p_0)$$

$$= 5.20(0.24) \ln(520/520) - \{[5.20(53.3)]/778\} \ln(100/100)$$

$$= 0$$

and

$$S_{2A} - S_{A0} = m_A c_p \ln(T_2/T_0) - (m_A R/J) \ln(p_{2A}/p_0)$$

$$= 5.20(0.24) \ln(648/520) - \{[5.20(53.3)]/778\} \ln(124.6/100)$$

$$= 0.1963 \text{ (B/R)}$$

so that

$$\Delta S_A = (S_{2A} - S_O) - (S_{1A} - S_O) = 0.1963 \text{ (B/R)}$$

Likewise, for the air in chamber B,

$$S_{1B} - S_{BO} = 3.14(0.24) \ln(860/520) - \{[3.14(53.3)]/778\} \ln(100/100)$$
$$= 0.3791 \text{ (B/R)}$$

and

$$S_{2B} - S_{BO} = 3.14(0.24) \ln(648/520) - \{[3.14(53.3)]/778\} \ln(75.3/100)$$
$$= 0.2268 \text{ (B/R)}$$

so that

$$\Delta S_B = (S_{2B} - S_{AO}) - (S_{1B} - S_{BO}) = 0.2268 - 0.3791$$
$$= -0.1523 \text{ (B/R)}$$

and

$$\Delta S_p = \Delta S_A + \Delta S_B = 0.1963 - 0.1523 = 0.0440 \text{ (B/R)}$$

The transfer of heat across a finite temperature difference is an irreversible process.

IRREVERSIBLE PROCESSES

By use of the entropy production we have established that the following processes are irreversible:

1. Adiabatic expansion or compression between fixed pressures or temperatures where the work for the compression process exceeds that of the isentropic process and the work for the expansion process is less than that of the isentropic process. The entropy production is due to fluid or mechanical friction.
2. Combustion processes.
3. Free expansion processes.
4. Mixing of gases.
5. Heat transfer across a finite temperature difference.

EFFECT OF HEAT TRANSFER IN THE CARNOT CYCLE

Given is the Carnot cycle engine operating between a source temperature of 1540 F and a sink temperature of 40 F (Fig. 17.24). The heat transfer from the source to the engine is 1000 B. From Eq. (17.1)

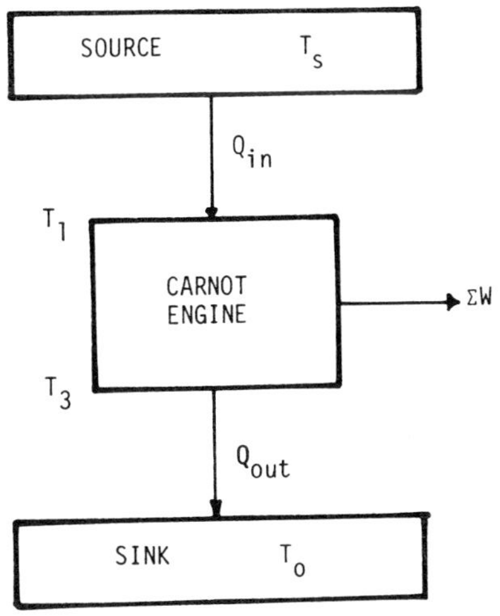

Figure 17.24

$$e = (T_1 - T_3)/T_1 = (2000 - 500)/2000$$

$$= 0.75$$

The net work is

$$\Sigma W = eQ_{in} = 0.75(1000) = 750 \text{ (B)}$$

and

$$\Sigma W = Q_{in} + Q_{out}$$

so that

$$Q_{out} = \Sigma W - Q_{in} = 750 - 1000 = -250 \text{ (B)}$$

If the engine receives the heat at a temperature less than that of the source or transfers heat to a receiver at a temperature greater than that of the sink, there will be a loss of cyclic work due to the irreversible transfer of heat across a finite temperature difference.

Example 17.23. Assume the engine in Fig. 17.24 receives heat at 1240 F and rejects heat at 140 F. Determine the net work and cycle efficiency. From Eq. (17.1)

$$e = (1700 - 600)/1700 = 0.647$$

and

$$\Sigma W = eQ_{in} = 0.647(1000) = 647 \text{ (B)}$$

so that

$$Q_{out} = \Sigma W - Q_{in} = 647 - 1000 = -353 \text{ (B)}$$

The net work has decreased by

$$\Delta W = 750 - 647 = 103 \text{ (B)}$$

and the heat rejected to the sink has increased by

$$|\Delta Q_{out}| = -250 + 353 = 103 \text{ (B)}$$

This is due to the transfer of heat across finite temperature differences from 1540 to 1240 F and from 140 to 40 F.

Example 17.24. Determine the entropy production for the system and surroundings in Example 17.23 (Fig. 17.25). For the Carnot engine, which is the system,

$$S_2 - S_1 = Q_{12}/T_1 = 1000/1700 = 0.588 \text{ (B/R)}$$

and

$$S_4 - S_1 = Q_{34}/T_3 = -353/600 = -0.588 \text{ (B/R)}$$

and

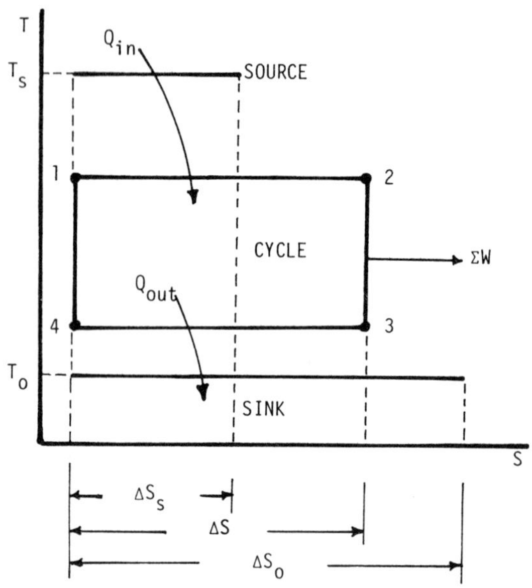

Figure 17.25

$$\Delta S_e = (S_2 - S_1) + (S_3 - S_2) + (S_4 - S_3) + (S_1 - S_4)$$

$$= 0.588 + 0 - 0.588 + 0 = 0$$

since the entropy change for a cycle must equal zero. For the heat source,

$$\Delta S_s = -Q_{12}/T_s = -1000/2000 = -0.500 \text{ (B/R)}$$

where the negative sign indicates the heat is transferred from the source, and for the sink,

$$\Delta S_0 = -Q_{34}/T_0 = 353/500 = +0.706 \text{ (B/R)}$$

where the positive sign indicates the heat is transferred into the sink. The entropy production for the surroundings is the sum of the entropy changes for the source and sink, so that

$$\Delta S_p = -0.500 + 0.706 = 0.206 \text{ (B/R)}$$

which is due to the irreversible heat transfer across finite temperature differences.

For the Carnot engine operating between 2000 and 500 R, which were also the source and sink temperatures,

$$\Sigma W_c = 750 \text{ (B)}$$

For the Carnot engine operating between 1700 and 600 R, with the same heat input,

$$\Sigma W_c = 647 \text{ (B)}$$

The loss in work production due to the irreversible heat transfer is

$$\Delta W = 750 - 647 = 103 \text{ (B)}$$

This is equivalent to the sink temperature multiplied by the entropy change of the surroundings,

$$\Delta W = T_0 \, \Delta S_p$$

$$= 500(0.206) = 103 \text{ (B)}$$

Example 17.25. Given is the engine cycle and surrounding conditions illustrated in Fig. 17.26, where t_s = 1540 F, t_0 = 40 F, t_1 = 1240 F, and t_3' = 140 F. $S_3' - S_2$ = 0.1 B/R and $S_1 - S_4'$ = 0.1 B/R. Determine the entropy production for the source, sink, and engine. The heat transfer from the source is 1000 B.

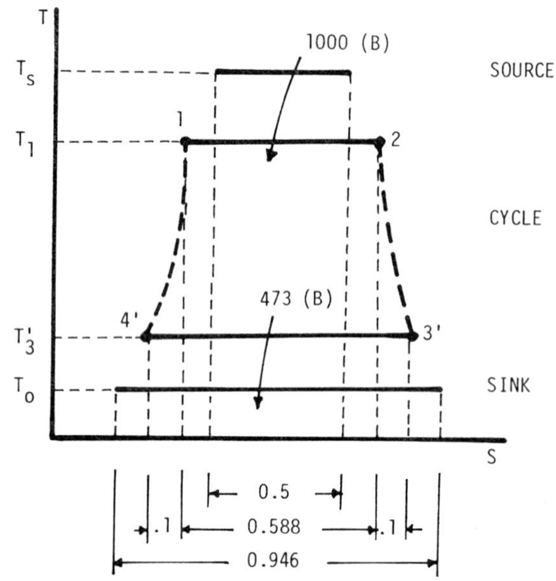

Figure 17.26

The entropy change for the source is

$$\Delta S_s = Q_{12}/T_3 = -1000/2000 = -0.5 \text{ (B/R)}$$

For the engine,

$$S_2 - S_1 = Q_{12}/T_1 = 1000/1700 = 0.588$$

and

$$S_4' - S_3' = (S_1 - S_2) + (S_2 - S_3') + (S_4 - S_1)$$

$$= -0.588 - 0.1 - 0.1 = -0.788 \text{ (B/R)}$$

For the engine, the entropy change is

$$\Delta S_e = (S_2 - S_1) + (S_3' - S_2) + (S_4' - S_3') + (S_1 - S_4')$$

$$= 0.588 + 0.1 - 0.788 + 0.1 = 0.0$$

The change of a property value for a cycle must be zero. The heat rejected by the cycle is

$$Q_{out}' = Q_{3'4'} = T_3'(S_4' - S_3') = 600(-0.788) = -473 \text{ (B)}$$

and the net work is

$$\Sigma W' = Q_{in} + Q_{out} = 1000 - 473 = 527 \text{ (B)}$$

which is

$$\Delta W = \Sigma W_c - \Sigma W'$$

$$= 750 - 527 = 223 \text{ (B)}$$

less than the Carnot cycle operating between the source and sink temperatures. The entropy change for the sink is

$$\Delta S_0 = -Q_{3'4'}/T_0 = 473/500 = 0.946 \text{ (B/R)}$$

For the surroundings the entropy production is

$$\Delta S_p = -0.5 + 0.946 = 0.446 \text{ (B/R)}$$

The net work loss is

$$\Delta W = T_0 \, \Delta S_p = 500(0.446) = 223 \text{ (B)}$$

Example 17.26. Given are the Otto cycle and the surroundings illustrated in Fig. 17.27. $t_0 = t_1 = 40$ F, $t_s = t_3$, $r_k = 9$, and $Q_{in} = Q_{23} = 300$ B/lbm. Determine the entropy production for the surroundings and the loss of work due to the irreversible heat transfers.

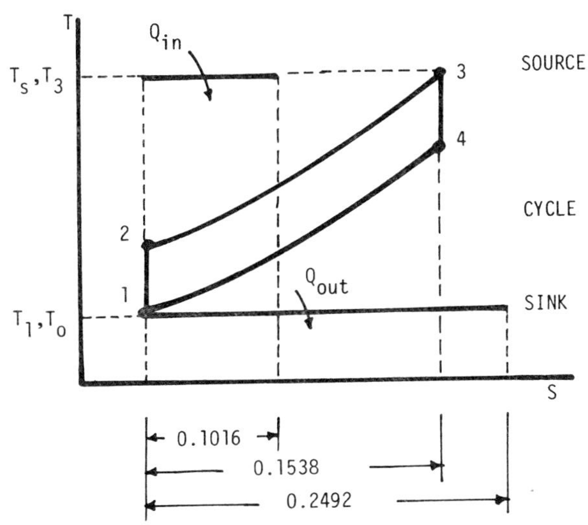

Figure 17.27

For the Otto cycle,

$$T_2 = T_1(V_1/V_2)^{k-1} = T_1 r_k^{k-1} = 500(9)^{0.4} = 1204 \text{ (R)}$$

and

$$Q_{23} = c_v(T_3 - T_2)$$

or

$$300 = 0.1714(T_3 - 1204)$$

so that

$$T_3 = 2954 \text{ (R)}$$

The temperature at state 4 is

$$T_4 = T_3(V_3/V_4)^{k-1} = T_3/r_k^{k-1} = 2954/(9)^{0.4} = 1227 \text{ (R)}$$

The entropy change for the source is

$$\Delta s_s = -Q_{23}/T_0 = Q/T_3 = -300/2954 = -0.1016 \text{ (B/lbm-R)}$$

For the heat transferred to the engine,

$$s_3 - s_2 = c_v \ln(T_3/T_2) = 0.1714 \ln(2954/1204) = 0.1538 \text{ (B/lbm-R)}$$

and from the engine,

$$s_1 - s_4 = c_v \ln(T_1/T_4) = 0.1714 \ln(500/1227) = -0.1538 \text{ (B/lbm-R)}$$

so that for the engine cycle

$$\Delta s_e = 0.0$$

The heat rejected by the cycle is

$$Q_{out} = Q_{41} = c_v(T_1 - T_4) = 0.1714(500 - 1227) = -124.6 \text{ (B/lbm)}$$

and the net cycle work is

$$\Sigma W = Q_{in} + Q_{out} = 300 - 124.6 = 175.4 \text{ (B/lbm)}$$

For the sink

$$\Delta s_0 = -Q_{41}/T_0 = 124.6/500 = 0.2492 \text{ (B/lbm-R)}$$

and $\Delta s_p = \Delta s_s + \Delta s_0$

$$= -0.1016 + 0.2492 = 0.1476 \text{ (B/lbm-R)}$$

For a Carnot cycle engine operating between T_s and T_0,

$$e = (T_s - T_0)T_3 = (2954 - 500)/2954 = 0.831$$

and

$$\Sigma W_c = eQ_{in} = 0.831(300) = 249.3 \text{ (B/lbm)}$$

The loss of work is

$$\Delta W = 249.3 - 175.4 = 73.9 \text{ (B/lbm)}$$

$$= T_0 \, \Delta S_p = 500(0.1476) = 73.8 \text{ (B/lbm)}$$

Example 17.27. Given the gas turbine cycle in Fig. 17.28, determine the entropy production for the universe and the loss of work due to the irreversible heat transfers. $t_0 = t_1 = 40$ F, $t_s = t_3 = 1540$ F for a maximum net work cycle. For the maximum net work cycle,

$$T_2 = T_4 = (T_1 T_3)^{1/2} = [500(2000)]^{1/2} = 1000 \text{ (R)}$$

The heat added to the cycle is

$$Q_{in} = Q_{23} = c_p(T_3 - T_2) = 0.24(2000 - 1000) = 240 \text{ (B/lbm)}$$

and

$$Q_{out} = Q_{41} = c_p(T_1 - T_4) = 0.24(500 - 1000) = -120 \text{ (B/lbm)}$$

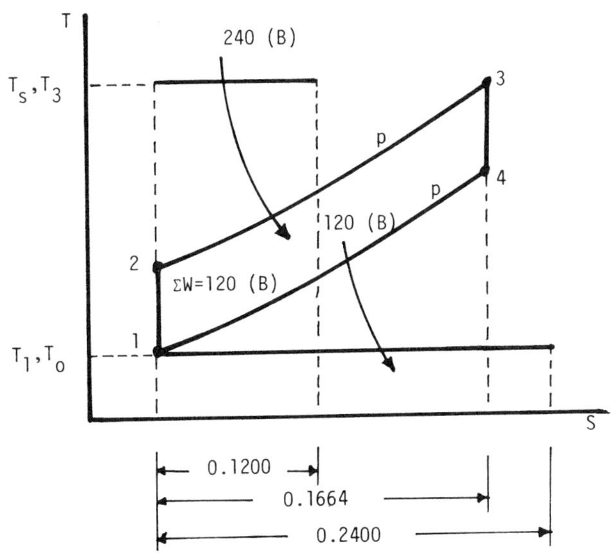

Figure 17.28

The entropy change for the cycle is

$$\Delta S_e = 0$$

For the source,

$$\Delta s_s = -Q_{23}/T_s = -240/2000 = -0.1200 \text{ (B/lbm-R)}$$

and for the sink,

$$\Delta s_0 = -Q_{41}/T_0 = 120/500 = 0.2400 \text{ (B/lbm-R)}$$

The entropy change of the surroundings is

$$\Delta s_p = \Delta s_s + \Delta s_0 = -0.1200 + 0.2400 = 0.1200 \text{ (B/lbm-R)}$$

For a Carnot cycle operating between T_s and T_0,

$$e = (T_s - T_0)/T_s = (2000 - 500)/2000 = 0.75$$

and

$$\Sigma W_c = eQ_{in} = 0.75(240) = 180 \text{ (B/lbm)}$$

The loss of work is

$$\Delta W = \Sigma W_c - (Q_{in} + Q_{out}) = 180 - (240 - 120) = 60 \text{ (B/lbm)}$$

or

$$\Delta W = T_0 \, \Delta s_p = 500(0.1200) = 60 \text{ (B/lbm)}$$

Example 17.28. Assume the compressor efficiency in Example 17.27 is 85% and the turbine efficiency is 75%. Determine the entropy production

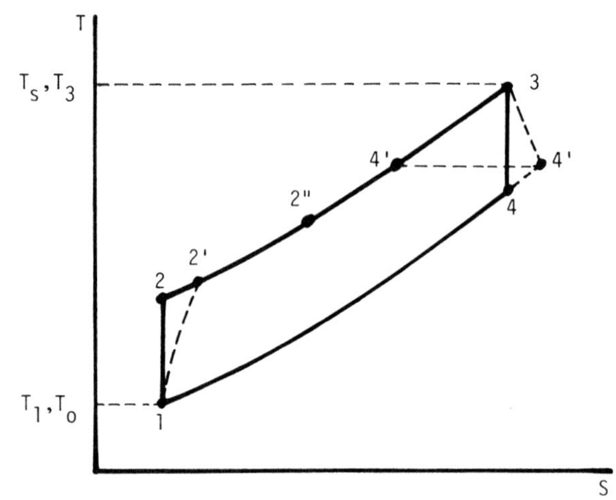

Figure 17.29

of the surroundings and the loss of work (Fig. 17.29). For the compressor,

$$e_c = (T_2 - T_1)/(T_2' - T_1)$$

or

$$0.85 = (1000 - 500)/(T_2' - 500)$$

and

$$T_2' = 1088 \text{ (R)}$$

For the turbine,

$$e_t = (T_3 - T_4')/(T_3 - T_4)$$

or

$$0.75 = (2000 - T_4')/(2000 - 1000)$$

and

$$T_4' = 1250 \text{ (R)}$$

The heat added to the cycle is

$$Q_{in}' = Q_{2'3} = c_p(T_3 - T_2') = 0.24(2000 - 1088) = 218.9 \text{ (B/lbm)}$$

and the heat rejected is

$$Q_{out}' = Q_{4'1} = c_p(T_1 - T_4') = 0.24(500 - 1250) = -180.0 \text{ (B/lbm)}$$

so that

$$\Sigma W' = Q_{in}' + Q_{out}' = 218.9 - 180.0 = 38.9 \text{ (B/lbm)}$$

For the Carnot cycle, e = 75%, and

$$\Sigma W_c' = 0.75(218.9) = 164.2 \text{ (B/lbm)}$$

and

$$\Delta W' = \Sigma W_c' - \Sigma W' = 164.2 - 38.9 = 125.3 \text{ (B/lbm)}$$

For the source,

$$\Delta s_s = -Q_{2'3}/T_s = -218.9/2000 = -0.1094 \text{ (B/lbm-R)}$$

and for the sink,

$$\Delta s_0 = -Q_{4'1}T_0 = 180.0/500 = 0.3600 \text{ (B/lbm-R)}$$

so that

$$\Delta s_p = \Delta s_s + \Delta s_0 = -0.1094 + 0.3600$$

$$= 0.2506 \text{ (B/lbm-R)}$$

and

$$\Delta W' = T_0 \Delta s_p = 500(0.2506) = 125.3 \text{ (B/lbm)}$$

Example 17.29. Repeat Example 17.27 for a cycle with an 85% regenerator (Fig. 17.29). For the regenerator,

$$e_g = (T_2'' - T_2')/(T_4' - T_2') = (T_2' - 1088)/(1250 - 1088) = 0.85$$

and

$$T_2'' = 1226 \text{ (R)}$$

The work is not changed by the use of a regenerator, so the cycle work is

$$\Sigma W'' = 38.9 \text{ (B/lbm)}$$

as in Example 17.27. The heat added to the cycle is

$$Q_{in}'' = c_p(T_3 - T_2'') = 0.24(2000 - 1226) = 185.8 \text{ (B/lbm)}$$

and

$$Q_{out}'' = \Sigma W'' - Q_{in}'' = 38.9 - 185.8 = -146.9 \text{ (B/lbm)}$$

For the source,

$$\Delta s_s = -Q_{in}''/T_s = -185.8/2000 = -0.0929 \text{ (B/lbm-R)}$$

and for the sink,

$$\Delta s_0 = -Q_{out}''/T_0 = 146.9/500 = 0.2938 \text{ (B/lbm-R)}$$

The entropy production is

$$\Delta s_p = \Delta s_0 + \Delta s_0 = -0.0929 + 0.2938 = 0.2009 \text{ (B/lbm)}$$

For the Carnot cycle, e = 75%, and

$$\Sigma W_c'' = eQ_{in}'' = 0.75(185.8) = 139.4 \text{ (B/lbm)}$$

so that

$$\Delta W'' = \Sigma W_c'' - \Sigma W'' = 139.4 - 38.9 = 100.5 \text{ (B/lbm)}$$

$$= T_0 \Delta s_p = 500(0.2009) = 100.4 \text{ (B/lbm)}$$

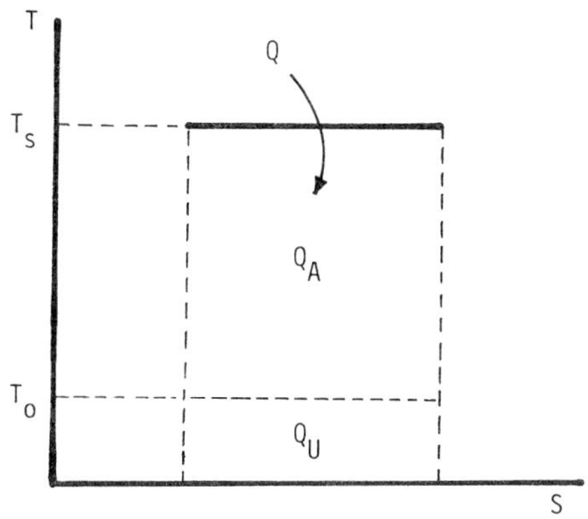

Figure 17.30

AVAILABLE HEAT

Consider a source at a temperature T_s and the surroundings or the sink having a temperature T_0 (Fig. 17.30). The quantity of heat available for conversion to work by means of a reversible cycle engine is the available heat. On the TS plane the area between the source temperature and the sink temperature represents the available heat. The area below the sink temperature represents the unavailable heat. The sum of the available and unavailable heat is equivalent to the heat transfer from the source, where

$$|Q| = Q_A + Q_U \tag{17.6}$$

LOSS OF AVAILABLE HEAT

Let the heat be transferred from the source to a receiver at a low temperature (Fig. 17.31). The heat transferred from the source is equal to that entering the receiver. The entropy change for the source is

$$|\Delta S_s| = |Q|/T_s$$

and the unavailable heat is

$$Q_{Us} = T_0|\Delta S_s|$$

From Eq. (17.6) the available heat is

$$Q_{As} = |Q| - Q_{Us}$$

For the heat in the receiver,

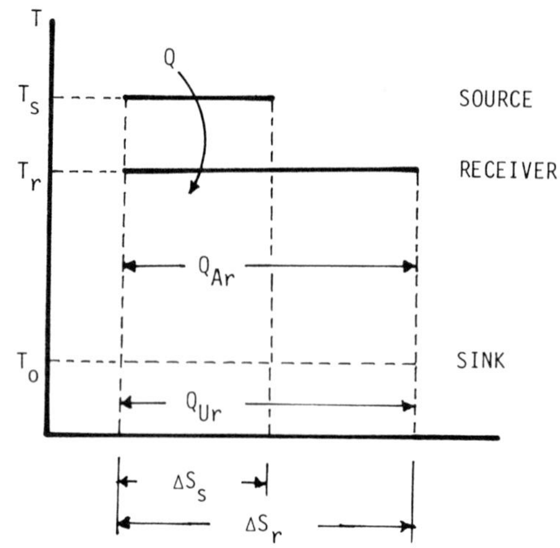

Figure 17.31

$$\Delta S_r = Q/T_r$$

and

$$Q_{Ur} = T_0 \; \Delta S_r$$

so that

$$Q_{Ar} = Q - Q_{Ur}$$

The loss in available heat due to the transfer of heat across a finite temperature difference is

$$\Delta Q_A = Q_{Ar} - Q_{As}$$

and the increase in the unavailable heat is

$$\Delta Q_U = Q_{Ur} - Q_{Us} = -\Delta Q_A$$

Example 17.30. In Fig. 17.31, $T_s = 2000$ R, $T_r = 1700$ R, $T_0 = 500$ R, and $Q = 1000$ B. Determine ΔQ_A and ΔQ_U.

The magnitude of the entropy change for the source is

$$|\Delta S_s| = |Q|/T_0 = 1000/2000 = 0.5 \text{ (B/R)}$$

and

$$Q_{Us} = T_0 |\Delta S_s| = 500(0.5) = 250 \text{ (B)}$$

From Eq. (17.6)

$$Q_{As} = |Q| - Q_{Us} = 1000 - 250 = 750 \text{ (B)}$$

For the receiver,

$$\Delta S_r = Q/T_r = 1000/1700 = 0.588 \text{ (B/R)}$$

and

$$Q_{Ur} = T_0 \, \Delta S_r = 500(0.588) = 294 \text{ (B)}$$

so that

$$Q_{Ar} = Q - Q_{Ur} = 1000 - 294 = 706 \text{ (B)}$$

The change in available and unavailable energy is

$$\Delta Q_A = Q_{Ar} - Q_{As} = 706 - 750 = -44 \text{ (B)}$$

and

$$\Delta Q_U = Q_{Ur} - Q_{Us} = 294 - 250 = 44 \text{ (B)}$$

or

$$\Delta Q_U = T_0(\Delta S_r - |\Delta S_s|) = 500(0.588 - 0.500) = 44 \text{ (B)}$$

Example 17.31. Assume that the heat in Example 17.30 is transferred to a constant volume receiver (Fig. 17.32). The receiver contains air as a perfect gas, and the air temperature increases from 140 to 540 F. Determine ΔQ_A and ΔQ_U.

From Example 17.30 for the source, Q = 1000 B, Q_{As} = 750 B, Q_{Us} = 250 B, and $|\Delta S_s|$ = 0.5 B/R. The mass of air in the constant volume receiver is

$$Q = mc_v(t_2 - t_1)$$

and

$$1000 = m(0.1714)(540 - 140)$$

so that

$$m = 14.59 \text{ (lbm)}$$

The entropy change in the receiver is

$$\Delta S_r = mc_v \ln(T_2/T_1) = 14.59(0.1714) \ln(1000/600)$$

$$= 1.277 \text{ (B/R)}$$

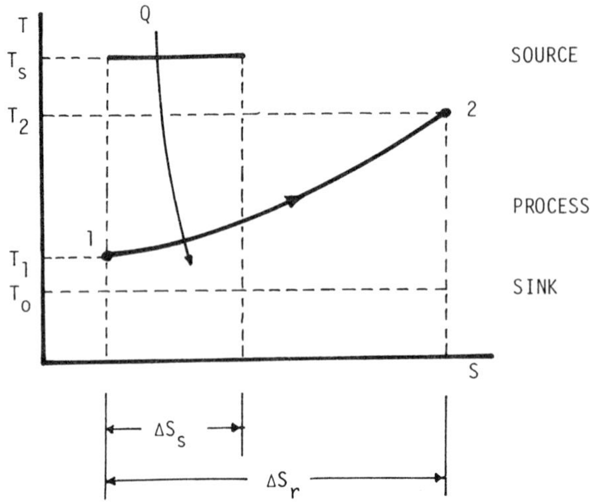

Figure 17.32

and the unavailable heat is

$$Q_{Ur} = T_0\ \Delta S_r = 500(1.277) = 638.5\ (B)$$

and the available heat is

$$Q_{Ar} = Q - Q_{Ur} = 1000 - 638.5 = 361.5\ (B)$$

The changes in the available and unavailable heat are

$$\Delta Q_A = Q_{Ar} - Q_{As} = 361.5 - 750 = -388.5\ (B)$$

and

$$\Delta Q_U = Q_{Ur} - Q_{As} = 638.5 - 250 = 388.5\ (B)$$

or

$$\Delta Q_U = T_0(\Delta S_r - |\Delta S_s|) = 500(1.277 - 0.5)$$

$$= 388.5\ (B)$$

Example 17.32. Given the source, sink, and Carnot cycle engine used in Example 17.23 (Fig. 17.33), determine ΔQ_U for the irreversible transfer of heat from the source to the engine and from the engine to the sink. $T_s = 2000$ R, $T_1 = 1700$ R, $T_3 = 600$ R, $T_0 = 500$ R, and $Q_{12} = 1000$ B.

From Example 17.30, where $T_1 = T_r$, $Q_{Us} = 250$, $Q_{U-12} = 294$, $\Delta Q_{Us} = 44$ B, and $S_2 - S_1 = 0.588$ B/R, the Carnot cycle efficiency is

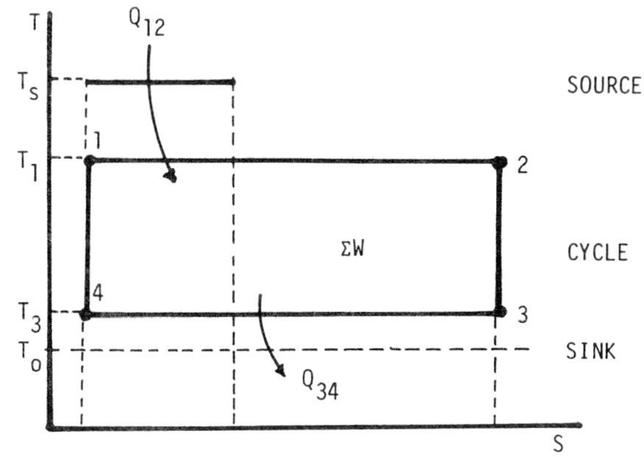

Figure 17.33

$$e = (T_1 - T_3)/T_1 = 1100/1700 = 0.647$$

and

$$\Sigma W = eQ_{12} = 0.647(1000) = 647 \text{ (B)}$$

so that

$$Q_{34} = \Sigma W - Q_{12} = 647 - 1000 = -353 \text{ (B)}$$

where

$$S_4 - S_3 = -(S_2 - S_1) = -0.588 \text{ (B/R)}$$

Of the heat rejected by the Carnot cycle, the unavailable heat portion is

$$Q_{U-34} = T_0|S_4 - S_3| = 500(0.588) = 294 \text{ (B)}$$

For the sink,

$$\Delta S_0 = Q_0/T_0 = |Q_{34}|/T_0 = 353/500 = 0.706 \text{ (B/R)}$$

and

$$Q_{U0} = T_0 \,\Delta S_0 = 500(0.706) = 353 \text{ (B)}$$

The available heat remaining at the sink temperature is

$$Q_{A0} = Q - Q_{U0} = 353 - 353 = 0$$

The available heat at the sink or surrounding temperature is zero. For the irreversible heat transfer from the engine to the sink the increase in unavailable heat is

$$\Delta Q_{U0} = Q_{U0} - Q_{U-34} = 353 - 294 = 59 \text{ (B)}$$

The increase of unavailable heat is

$$\Sigma \Delta Q_U = Q_{U0} - Q_{Us} = 353 - 250 = 103 \text{ (B)}$$

$$= \Delta Q_{Us} + \Delta Q_{U0} = 44 + 59 = 103 \text{ (B)}$$

If the Carnot engine had operated between the source and sink temperatures, the efficiency would have been

$$e = (T_s - T_0)/T_s = (2000 - 500)/2000 = 0.750$$

giving a net work of

$$\Sigma W = eQ = 750 \text{ (B)}$$

In this example the net work was 647 B. The loss of net work is

$$\Delta W = 750 - 647 = 103 \text{ (B)}$$

which is accounted for by the increase of unavailable heat. This is also equivalent to

$$\Delta W = T_0 \, \Delta S_p = 500(0.206) = 103 \text{ (B)}$$

where ΔS_p is the entropy production for the surroundings as found in Example 17.24.

Example 17.33. Given Example 17.25 (Fig. 17.34), determine $\Sigma \Delta Q_U$. From Example 17.25, $\Delta S_s = -0.5$, $T_s = 2000$ R, Q = 1000 B, $Q_{As} = 750$ B, and $Q_{Us} = 250$ B; also from Example 17.25,

$$S_2 - S_1 = 0.588 \text{ (B/R)}$$

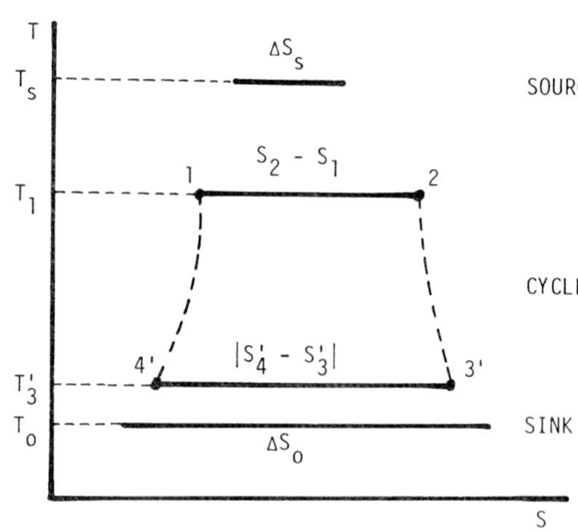

Figure 17.34

and

$$Q_{U-12} = T_0(S_2 - S_1) = 500(0.588)$$

$$= 294 \text{ (B)}$$

The increase in the unavailable heat due to the irreversible transfer from the source is

$$\Delta Q_{Us} = Q_{U-12} - Q_{Us} = 294 - 250 = 44 \text{ (B)}$$

From Example 17.25, $Q_{out}' = -473$ B, $\Sigma W' = 527$ B, and $S_{4'} - S_{3'} = -0.788$ B/R. The unavailable energy for the process from state 3′ to state 4′ is

$$Q_{U-3'4'} = T_0(S_{4'} - S_{3'}) = 500\ (0.788) = 394 \text{ (B)}$$

For the sink,

$$\Delta S_0 = Q/T_0 = 473/500 = 0.946 \text{ (B/R)}$$

and

$$Q_{U0} = T_0\ \Delta S_0 = 500(0.946) = 473 \text{ (B)}$$

so that

$$\Delta Q_{U0} = Q_{U0} - Q_{U-3'4'} = 473 - 394 = 79 \text{ (B)}$$

The increase in the unavailable energy from the source to the sink is

$$\Sigma\ \Delta Q_U = Q_{U0} - Q_{Us} = 473 - 250 = 223 \text{ (B)}$$

which is the loss of net work in Example 17.25; also

$$\Delta W = T_0\ \Delta S_p = 500(0.446) = 223 \text{ (B)}$$

The increase of unavailable heat due to the irreversible heat transfers is

$$\Delta Q_{Us} + \Delta Q_{U0} = 44 + 79 = 123 \text{ (B)}$$

The increase of unavailable heat due to the irreversible adiabatic processes is

$$\Delta Q_{U'} = T_0[(S_{3'} - S_2) + (S_1 - S_{4'})] = 500(0.2) = 100 \text{ (B)}$$

and

$$\Sigma\ \Delta Q_U = 123 + 100 = 223 \text{ (B)}$$

as before.

Example 17.34. Determine the increase in unavailable energy for the Otto cycle in Example 17.26 (Fig. 17.35). From Example 17.26, $T_0 = T_1 = 500$ R, $T_2 = 1204$ R, $T_3 = T_s = 2954$ R, $T_4 = 1227$ R, $Q_{in} = 300$ B/lbm, $Q_{out} = -124.6$ B/lbm, $\Sigma W = 175.4$ B/lbm, and $s_3 - s_2 = 0.1538$ B/lbm-R.

For the source,

$$\Delta s_s = -Q_{in}/T_s = -300/2954 = -0.1016 \text{ (B/lbm-R)}$$

and

$$Q_{Us} = T_0|\Delta s_s| = 500(0.1016) = 50.8 \text{ (B/lbm)}$$

For the constant volume process from state 2 to state 3,

$$s_3 - s_2 = 0.1538 \text{ (B/lbm-R)}$$

and

$$Q_{U-23} = T_0(s_3 - s_2) = 500(0.1538) = 76.9 \text{ (B/lbm)}$$

For the constant volume process from state 4 to state 1,

$$Q_{U-41} = T_0|s_1 - s_4| = T_0(s_3 - s_2) = 76.9 \text{ (B/lbm)}$$

For the sink,

$$\Delta s_0 = |Q_{out}|/T_0 = 124.6/500 = 0.2492 \text{ (B/lbm-R)}$$

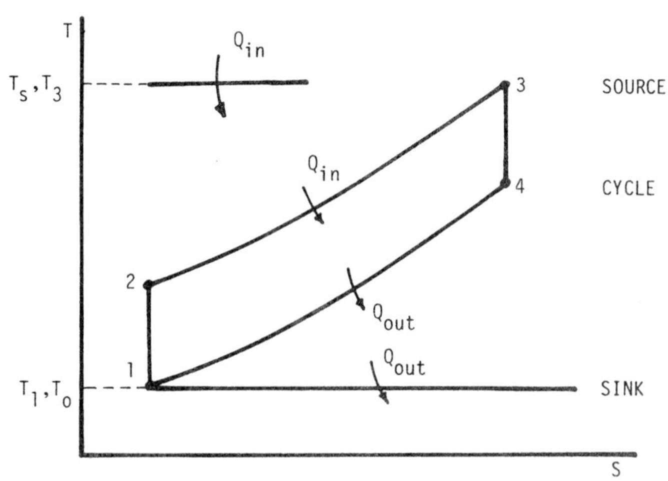

Figure 17.35

and

$$Q_{U0} = T_0 \, \Delta s_0 = 500(0.2492) = 124.6 \text{ (B/lbm)}$$

From the source to the sink,

$$\Sigma \, \Delta Q_U = Q_{U0} - Q_{Us} = 124.6 - 50.8 = 73.8 \text{ (B/lbm)}$$

$$= (Q_{U-12} - Q_{Us}) + (Q_{U0} - Q_{U-41}) = (76.9 - 50.8) + (124.6 - 76.9)$$

$$= 26.1 + 47.7 = 73.8 \text{ (B/lbm)}$$

The latter equation totals the increase of unavailable heat due to the irreversible transfer from the source to the Otto cycle and from the Otto cycle to the sink.

Example 17.35. Determine the increase in unavailable energy for the Brayton cycle in Example 17.28 (Fig. 17.36). From Example 17.28, $T_1 = 500$, $T_2 = T_4 = 1000$, $T_3 = 2000$, $T_{2'} = 1088$, and $T_{4'} = 1250$ R. Also, $Q_{in}' = 218.9$, $Q_{out}' = -180.0$, $\Sigma W' = 38.9$, and $\Delta W' = 125.3$ B/lbm.

For the source,

$$\Delta s_s = -Q_{in}/T_s = -218.9/2000 = -0.1094 \text{ (B/lbm-R)}$$

$$Q_{Us} = T_0 |\Delta s_s| = 500(0.1094) = 54.7 \text{ (B/lbm)}$$

For the process from state 2′ to state 3,

$$s_3 - s_{2'} = c_p \ln(T_3/T_{2'}) = 0.24 \ln(2000/1088) = 0.1461 \text{ (B/lbm-R)}$$

and

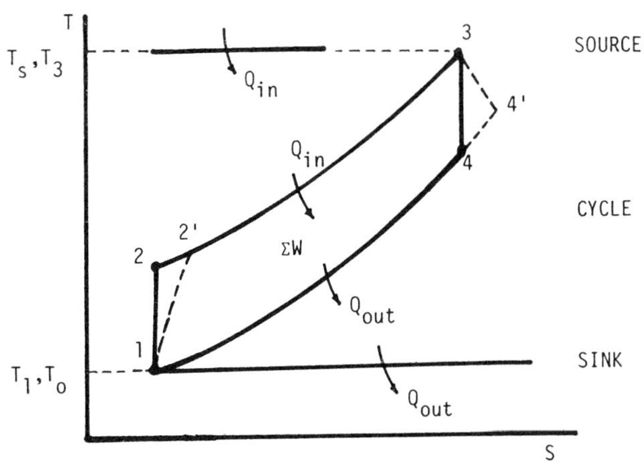

$$Q_{U-23'} = T_0(s_3 - s_2') = 500(0.1461) = 73.0 \text{ (B/lbm)}$$

For the process from state 4′ to state 1,

$$s_1 - s_4' = c_p \ln(T_1/T_{4'}) = 0.24 \ln(500/1250) = -0.2199 \text{ (B/lbm-R)}$$

and

$$Q_{U-4'1} = T_0|s_1 - s_4'| = 500(0.2199) = 110.0 \text{ (B/lbm)}$$

For the sink,

$$\Delta s_0 = -Q_{out}/T_0 = 180.0/500 = 0.3600 \text{ (B/lbm-R)}$$

and

$$Q_{U0} = T_0 \Delta s_0 = 500(0.3600) = 180.0 \text{ (B/lbm)}$$

The increase in unavailable heat between the source and sink is

$$\Sigma \Delta Q_U = Q_{U0} - Q_{Us} = 180 - 54.7 = 125.3 \text{ (B/lbm)}$$

and

$$\Sigma \Delta Q_U = \Delta W'$$

The increase in unavailable heat due to the irreversible transfer of heat from the source to the cycle and from the cycle to the sink is

$$\Delta Q_{Us} + \Delta Q_{U0} = (Q_{U-23'} - Q_{Us}) + (Q_{U0} - Q_{U-4'1})$$

$$= (73.0 - 54.7) + (180.0 - 110.0) = 18.3 + 70.0 = 88.3 \text{ (B/lbm)}$$

For the Brayton cycle consisting of only reversible processes,

$$s_3 - s_2 = c_p \ln(T_3/T_2) = 0.24 \ln(2000/1000)$$

$$= 0.1664 \text{ (B/lbm)}$$

so that

$$s_{2'} - s_2 = (s_3 - s_2) + (s_3 - s_{2'}) = 0.1664 - 0.1461 = 0.0203 \text{ (B/lbm-R)}$$

Likewise,

$$s_1 - s_4 = -0.1664 \text{ (B/lbm)}$$

and

$$s_{4'} - s_4 = (s_1 - s_4) - (s_1 - s_{4'}) = -0.1664 + 0.2199$$

$$= 0.0535 \text{ (B/lbm-R)}$$

The increase of the unavailable heat due to the irreversible adiabatic processes in the compressor and turbine is

$$\Delta Q_{U'} = T_0[(s_{2'} - s_2) + (s_{4'} - s_4)] = 500(0.0203 + 0.0538)$$

$$= 37.0 \text{ (B/lbm)}$$

and

$$\Sigma\, \Delta Q_U = 88.3 + 37.0 = 125.3 \text{ (B/lbm)}$$

which is the loss of net work compared to a Carnot cycle operating between the same source and sink temperatures.

HEAT EXCHANGERS

The purpose of a heat exchanger is to heat one substance by removing heat from another substance. In Fig. 17.37, hot air, perhaps from the flue gases in a combustion chamber, is used to heat water. By assuming no heat

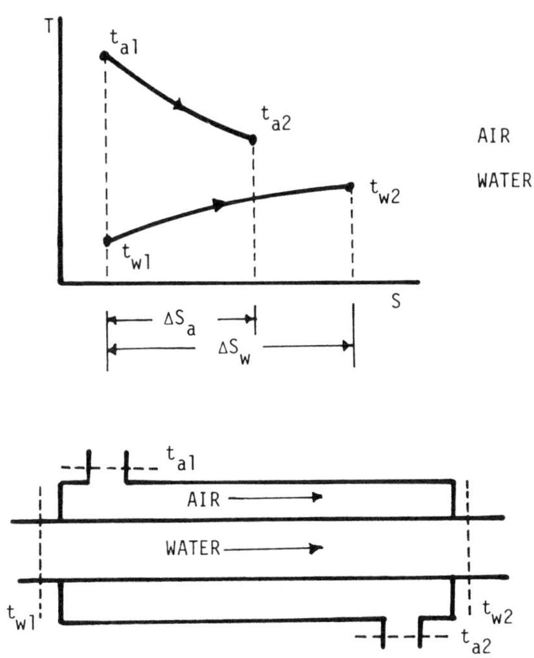

Figure 17.37

loss from the exchanger, the heat transfer from the air is equal in magnitude to the heat transfer to the water:

$$Q_{air} = \Delta H_a = m_a c_{pa}(t_{a2} - t_{a1})$$

and

$$Q_{water} = \Delta H_w = m_w c_w(t_{w2} - t_{w1})$$

where

$$W_{sf} = \Delta KE = \Delta PE = 0$$

for the steady-flow process. The change in entropy for the air and water is

$$\Delta S_a = m_a c_{pa} \ln(T_{a2}/T_{a1})$$

and

$$\Delta S_w = m_w c_w \ln(T_{w2}/T_{w1})$$

for p = c. The increase in unavailable heat for the irreversible transfer of heat is

$$\Delta Q_U = T_0(\Delta S_w - |\Delta S_a|) \tag{17.7}$$

Example 17.36. It is desired to heat 100 lbm/hr of water from 90 to 180 F by use of air which enters the heat exchanger at a temperature of 540 and leaves at a temperature of 240 F. The surrounding temperature is 60 F. The heat exchanger in Fig. 17.37 is classified as a parallel-flow exchanger since the fluids move in the same direction. Determine the increase in unavailable heat.

For the water,

$$Q_{water} = m_w c_w(t_{w2} - t_{w1}) = 100(1)(180 - 90)$$

$$= 9000 \text{ (B/hr)}$$

For the air,

$$Q_{air} = m_a c_{pa}(t_{a2} - t_{a1}) = m_a(0.24)(240 - 540) = -9000 \text{ (B/hr)}$$

so that

$$m_a = 125 \text{ (lbm/hr)}$$

The entropy changes for the air and water are

$$\Delta S_a = m_a c_{pa} \ln(T_{a2}/T_{a1}) = 125(0.24) \ln(700/1000)$$

$$= -10.700 \text{ (B/R-hr)}$$

and

$$\Delta S_w = m_w c_w \ln(T_{w2}/T_{w1}) = 100(1) \ln(640/550)$$

$$= 15.155 \text{ (B/R-hr)}$$

The unavailable heat in the water and air is

$$Q_{Ua} = T_0|\Delta S_a| = 500(10.700) = 5350 \text{ (B/hr)}$$

and

$$Q_{Uw} = T_0(\Delta S_a) = 500(15.155) = 7578 \text{ (B/hr)}$$

The available heat for the air and water is

$$Q_{Aa} = Q - Q_{Ua} = 9000 - 5350 = 3650 \text{ (B/hr)}$$

and

$$Q_{Aw} = Q - Q_{Uw} = 9000 - 7578 = 1422 \text{ (B/hr)}$$

The decrease in the available heat or the increase in unavailable heat is

$$\Delta Q_U = Q_{Uw} - Q_{Ua} = 7578 - 5350 = 2228 \text{ (B/hr)}$$

or from Eq. (17.7)

$$\Delta Q_U = T_0(\Delta S_w - |\Delta S_a|) = 500(15.155 - 10.700)$$

$$= 2228 \text{ (B/hr)}$$

and

$$\Delta Q_A = -2228 \text{ (B/hr)}$$

For this exchanger parallel-flow or counterflow operation would produce the same loss of available heat.

Example 17.37. It is desired to cool 100 lbm/hr of oil, c = 0.7 B/lbm-R, from 180 to 100 F by use of water available at the surrounding temperature of 60 F. Determine the loss of available heat for a parallel exchanger. For parallel-flow operation (Fig. 17.37) the theoretical temperature limit for the water leaving the exchanger is 100 F.

$$Q_{out} = m_o c_o(t_{o2} - t_{o1}) = 100(0.7)(-80) = -5600 \text{ (B/hr)}$$

and

$$Q_{water} = m_w c_w (t_{w2} - t_{w1}) = m_w(1)(100 - 60) = 5600$$

so that

$$m_w = 140.0 \text{ (lbm/hr)}$$

The entropy changes are

$$\Delta S_o = m_o c_o \ln(T_{o2}/T_{o1}) = 100(0.7) \ln(560/640) = -9.347 \text{ (B/R-hr)}$$

and

$$\Delta S_w = m_w c_w \ln(T_{w2}/T_{w1}) = 140.0(1) \ln(560/520) = 10.375 \text{ (B/R-hr)}$$

The increase in unavailable heat, from Eq. (17.7), is

$$\Delta Q_U = T_0(\Delta S_w - |\Delta S_o|) = 520(10.375 - 9.347)$$

$$= 534.6 \text{ (B/hr)}$$

and

$$\Delta Q_A = -534.6 \text{ (B/hr)}$$

which represents the loss of available heat due to the irreversible heat transfer.

Example 17.38. Repeat Example 17.37 for a counterflow heat exchanger (Fig. 17.38). The theoretical temperature limit of the water leaving the exchanger is 180 F.

$$Q_{oil} = -5600 \text{ (B/hr)}$$

and

$$Q_{water} = 5600 = m_w(1)(180 - 60)$$

so that

$$m_w = 46.7 \text{ (lbm/hr)}$$

The entropy changes are

$$\Delta S_o = -9.347 \text{ (B/R-hr)}$$

and

$$\Delta S_w = m_w c_w \ln(T_{w2}/T_{w1}) = 46.7(1) \ln(640/520)$$

$$= 9.697 \text{ (B/R-hr)}$$

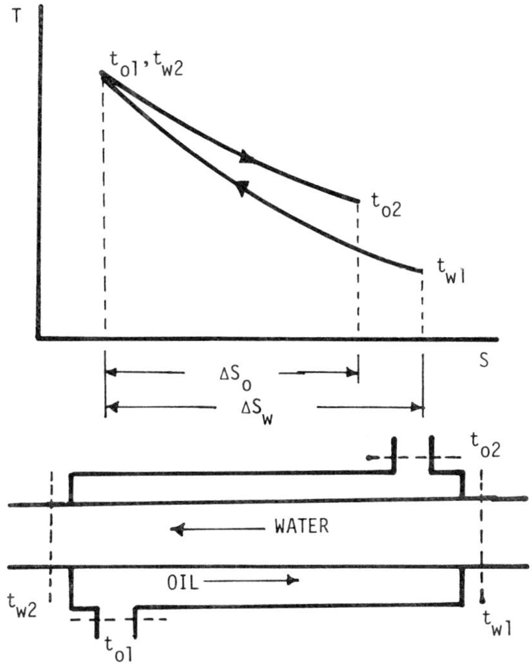

Figure 17.38

The increase in unavailable heat is

$$\Delta Q_U = T_0(\Delta S_w - |\Delta S_o|) = 520(9.697 - 9.347)$$

$$= 182.0 \text{ (B/hr)}$$

$$\Delta Q_A = -182.0 \text{ (B/hr)}$$

The loss of available heat has been reduced by 352.6 B/hr or 194%. The quantity of water required to cool the oil has been decreased by 93.3 lbm/hr or 200%. The counterflow heat exchanger performs considerably better than the parallel-flow exchanger.

Example 17.39. Assume that oil for the heat exchanger in Example 17.38 flows through the outer tube. The oil loses 600 B/hr to the surrounding air at 60 F. Determine the loss of available heat in the exchanger. Since the exchanger loses heat to the surrounding air, the water must remove

$$Q_w = 5600 - 600 = 5000 \text{ (B/hr)}$$

and

$$5000 = m_w(1)(120)$$

so that

$$m_w = 41.7 \text{ (lbm/hr)}$$

The entropy change for the oil remains at

$$\Delta S_o = -9.347 \text{ (B/R-hr)}$$

and

$$\Delta S_w = m_w c_w \ln(T_{w2}/T_{w1})$$

$$= 41.7(1) \ln(640/520)$$

$$= 8.659 \text{ (B/R-hr)}$$

The entropy change for the surrounding air is

$$\Delta S_a = Q/T_0 = 600/520 = 1.154 \text{ (B/R-hr)}$$

The total entropy change for the oil, water, and air is

$$\Delta S = \Delta S_o + \Delta S_w + \Delta S_a = -9.347 + 8.659 + 1.154$$

$$= 0.466 \text{ (B/R-hr)}$$

and

$$\Delta Q_U = T_0 \, \Delta S = 520(0.466) = 242.3 \text{ (B/hr)}$$

so that

$$\Delta Q_A = -242.3 \text{ (B/hr)}$$

For the oil,

$$Q_U = T_0 |S_o| = 520(9.347) = 4860 \text{ (B/hr)}$$

and

$$Q_A = Q - Q_U = 5600 - 4860 = 740 \text{ (B/hr)}$$

For the water,

$$Q_U = T_0 \, \Delta S_w = 520(8.659) = 4503 \text{ (B/hr)}$$

and

$$Q_A = Q - Q_U = 5000 - 4503 = 497 \text{ (B/hr)}$$

For the air,

$$Q_U = T_0 \Delta S_a = 520(1.154) = 600 \text{ (B/hr)}$$

and

$$Q_A = Q - Q_U = 600 - 600 = 0 \text{ (B/hr)}$$

The decrease in available heat is

$$\Delta Q_A = Q_{Aw} + Q_{Aa} - Q_{Ao}$$

$$= 497 + 0 - 740 = -243 \text{ (B/hr)}$$

Even though the loss of available heat has increased by 61 B/hr or 33.5%, the required water flow has decreased 5.0 lbm/hr or 12% compared to the lowest flow. Since conservation of water is probably the critical factor, placing the oil in the outer tube is desirable. If the transfer of heat from the oil to the surrounding air is objectionable, the oil may be transported through the inner tube.

Example 17.40. Determine the loss in available heat for the open heater in Fig. 17.39. The conditions of the steam entering the heater are $p_1 = 40$ psia and $t_1 = 300$ F. The water from the condenser enters at $t_2 = 80$ F and $p_2 = 40$ psia. The temperature of the surroundings is 60 F. For the steady-flow process where $W_{sf} = Q = \Delta KE = \Delta PE = 0$,

$$H_{in} = H_{out}$$

For 1 lbm of water leaving the exchanger at $p_3 = 40$ psia and $x_3 = 0.0$,

$$m_e h_1 + (1 - m_e)h_2 = h_3$$

From the steam tables, $h_1 = 1186.6$ B/lbm and $s_1 = 1.6994$ B/lbm-R for the

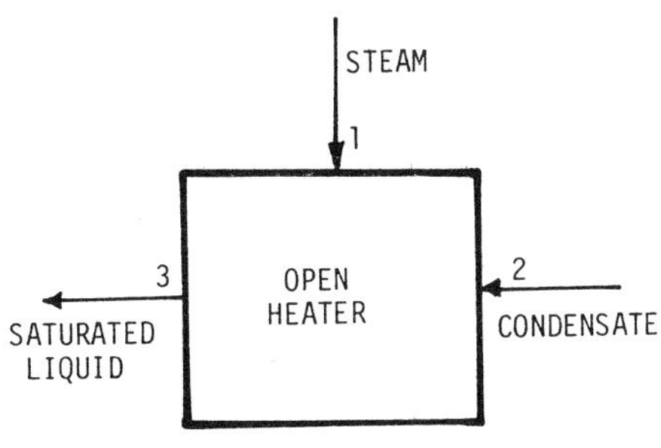

Figure 17.39

superheated steam, and $h_2 = 48.0$ B/lbm and $s_2 = 0.0932$ B/lbm-R for the subcooled liquid at 80 F; also $h_3 = 236.0$ B/lbm and $s_3 = 0.3919$ B/lbm-R for the saturated liquid. All three substances are at a pressure of 40 psia.

Solving for the quantity of extracted steam, one obtains

$$1186.6m_e + (1 - m_e)48.0 = 236.0$$

and

$$m_e = 0.1652 \text{ (lbm)}$$

The heat supplied by the extracted steam is

$$Q_e = m_e(h_3 - h_1) = 0.1652(236.0 - 1186.6)$$
$$= 157.0 \text{ (B)}$$

The heat gained by the condenser water is

$$Q_c = (1 - m_e)(h_3 - h_2) = 0.8348(236.0 - 48.0)$$
$$= 157.0 \text{ (B)}$$

The entropy change for the extracted steam is

$$\Delta s_e = m_e(s_3 - s_1) = 0.1652(0.3919 - 1.6994)$$
$$= -0.2159 \text{ (B/R)}$$

so that

$$Q_{Ue} = T_0|\Delta s_e| = 520(-0.2159) = 112.3 \text{ (B)}$$

and

$$Q_{Ae} = Q - Q_{Ue} = 157.0 - 112.3 = 44.7 \text{ (B)}$$

The entropy change for the condensate is

$$\Delta s_c = (1 - m_e)(s_3 - s_2) = 0.8348(0.3919 - 0.0932)$$
$$= 0.2494 \text{ (B/R)}$$

so that

$$Q_{Uc} = T_0 \,\Delta s_c = 520(0.2494) = 129.7 \text{ (B)}$$

and

$$Q_{Ac} = Q - Q_{Uc} = 157.0 - 129.7 = 27.3 \text{ (B)}$$

The loss of available heat is

$$\Delta Q_A = Q_{Ac} - Q_{Ae} = 27.3 - 44.7 = -17.4 \text{ (B)}$$

The loss of available heat may be also calculated by determining the available heat in each substance.

The heat transfer from the steam at 40 psia and 300 F to the sink at 60 F is

$$Q_e = m_e(h_1 - h_0) = 0.1652(1186.6 - 28.1)$$

$$= 191.4 \text{ (B)}$$

and

$$\Delta s_e = m_e(s_1 - s_0) = 0.1652(1.6994 - 0.0555)$$

$$= 0.2715 \text{ (B/R)}$$

The unavailable heat is

$$Q_{Ue} = T_0 \, \Delta s_e = 520(0.2715) = 141.2 \text{ (B/lbm)}$$

and

$$Q_{Ae} = Q - Q_U = 191.4 - 141.2 = 50.2 \text{ (B/lbm)}$$

For the water from the condenser,

$$Q_c = (1 - m_e)(h_2 - h_0) = 0.8348(48 - 28) = 16.7 \text{ (B/lbm)}$$

and

$$Q_{Uc} = (1 - m_e)T_0(s_2 - s_0) = 0.8348(520)(0.0932 - 0.0555)$$

$$= 16.4 \text{ (B/lbm)}$$

and

$$Q_{Ac} = Q - Q_U = 16.7 - 16.4 = 0.3 \text{ (B/lbm)}$$

For the water leaving the open heater,

$$Q_w = h_3 - h_0 = 236.0 - 28.0 = 208.0 \text{ (B/lbm)}$$

and

$$Q_{Uw} = T_0(s_3 - s_0) = 520(0.3919 - 0.0555)$$

$$= 175.0 \text{ (B/lbm)}$$

and

$$Q_{Aw} = Q - Q_U = 208.0 - 175.0 = 33.0 \text{ (B/lbm)}$$

The decrease in available heat is

$$\Delta Q_A = Q_{Aw} - Q_{Ae} - Q_{Ac}$$

$$= 33.0 - 50.2 - 0.3 = -17.5 \text{ (B/lbm)}$$

as before.

PROBLEMS

17.1 Determine the Carnot cycle efficiency and the coefficient of performance for an engine or reversed cycle operating between 1240 and 140 F.

17.2 The Carnot cycle engine in Prob. 17.1 has a power output of 30 HP. Determine Q_{in}, Q_{out}, and $|\Delta S|$.

17.3 A Stirling cycle engine operates between 1240 and 140 F. $s_2 - s_1 = 0.5$ B/lbm-R. Determine e, Q_{in}, ΣW, and Q_{out}.

17.4 Repeat Prob. 17.3 for a Stirling cycle with full regeneration.

17.5 A cycle consists of two isothermal processes acting at temperatures of 1240 and 140 F with Q_{in} = 90,000 B at the higher temperature and Q_{out} = -60,000 B. The remaining two processes are irreversible adiabatic. Determine the cycle efficiency and the net work for the cycle. Determine ΔS for the irreversible adiabatic processes.

17.6 Determine $\oint dQ/T$ and ΔW for the cycle in Prob. 17.5.

17.7 A reversed Brayton cycle using air as a perfect gas has a compressor inlet pressure and temperature of 20 psia and 40 F. The pressure ratio is 4, and Q_{in} = 20 B/lbm. For $e_t = e_c$ = 100%, determine γ_R and $\oint dQ/T$.

17.8 Using the reversible process temperatures for the reversible cycle in Prob. 17.7, determine γ_R and $\oint dQ/T$ when e_c = 80% and e_t = 70%.

17.9 A Diesel cycle using air as a perfect gas has a compression ratio of 16. $p_1 = 15$ psia, $t_1 = 40$ F, and $Q_{in} = 240$ B/lbm. The isentropic processes are replaced by an adiabatic compression process for which $n = 1.45$ and an adiabatic expansion process for which $n = 1.35$. Determine the efficiency of the reversible cycle and the irreversible cycle and the entropy production for each process and for the cycle.

17.10 Benzene is burned in a constant pressure adiabatic combustion process. Theoretical air is 200%. $t_1 = 40$ F and $p_1 = 14.7$ psia. Determine Δs_p for the process. Assume products and reactants are perfect gases.

17.11 Given is the isolated system in Fig. 17.21 containing air at $p_1 =$ 50 psia, $t_1 = 40$ F, and $V_1 = 10$ ft^3. The evacuated space has a volume of 10 ft^3. Determine ΔS_p for the process.

17.12 Given is the isolated system in Fig. 17.22. Container A is filled with air, and container B is filled with helium. $p_A = p_B = 100$ psia, and $t_A = t_B = 140$ F. Determine ΔS_p for the mixing process.

17.13 Repeat Prob. 17.12 with $p_B = 20$ psia and $t_B = 40$ F.

17.14 Given is the system in Fig. 17.23 where $p_A = 100$ and $p_B = 50$ psia. $V_A = 20$ ft^3, and $V_B = 10$ ft^3. $t_{A1} = 40$ F, and $t_{B1} = 540$ F. Determine ΔS_p for the process. Air in both chambers.

17.15 A gas turbine operates between temperatures of 40 and 1540 F in the maximum net work cycle. The source temperature is 2040 F, and the sink temperature is 40 F. Determine ΔS_p and ΔW.

17.16 Repeat Prob. 17.15 for $e_t = 80\%$ and $e_c = 70\%$.

17.17 Repeat Prob. 17.16 when $e_g = 50\%$.

17.18 Ten lbm of air, as a perfect gas, is heated from 40 to 240 F in a constant pressure receiver. The source temperature is 340 F, and $t_0 = 40$ F. Determine Q_{Ar}, ΔQ_A, and ΔQ_U.

17.19 Given a Diesel engine using air as a perfect gas and $t_0 = t_1 = 40$ F, $t_s = t_3$, $r_k = 16$, and $r_c = 1.5$, determine Δs_p, ΔW, and $\Sigma \Delta Q_U$.

17.20 It is desired to heat 150 lbm/hr of oil, $c = 0.8$ B/lbm-R, from 60 to 120 F by use of water which enters the heat exchanger at 180 F. The surrounding temperature is 40 F. The leaving water temperature exceeds the oil temperature by 20 F. Determine ΔQ_U for a parallel-flow and counterflow heat exchanger.

THERMODYNAMICS OF FLUID FLOW

The working substance present in the study of the flow of fluids in a steady state thermodynamic system consists of compressible and incompressible fluids. Incompressible fluids include all liquids and gases for which the pressure variation in the process is relatively small. Compressible fluids consist of gases that undergo a larger pressure drop as they flow through the thermodynamic system.

In this chapter we shall consider the flow of perfect gases, semi-perfect gases, vapors, and liquids. Stagnation properties will be defined, and an equation for the speed of sound will be derived. Flow in subsonic and supersonic nozzles and diffusers will be studied.

TOTAL OR STAGNATION PROPERTIES

Consider a fluid moving through a thermodynamic system that is brought to rest by an isentropic process in which no external work is transferred to or from the system.

The equation for a steady-flow process is

$$Q = \Delta PE + \Delta KE + \Delta U + \Delta FE + W_{sf}$$

and for $Q = 0$, since the process is isentropic, together with the assumptions that $\Delta PE = 0$ and $W_{sf} = 0$, the equation reduces to

$$\Delta U + \Delta FE + \Delta KE = 0$$

Let state 1 represent the moving fluid and state 0 the fluid at rest for the steady-flow system:

$$mu_1 + mp_1v_1/J + m\overline{V}_1{}^2/2g_0J = mu_0 + mp_0v_0/J + m\overline{V}_0{}^2/2g_0J$$

or since

$$h = u + pv/J$$

then

$$h_1 + \bar{V}_1^2/2g_0J = h_0 + \bar{V}_0^2/2g_0J$$

Since the system is at rest for state 0,

$$\bar{V}_0 = 0$$

and the stagnation or total enthalpy is

$$h_0 = h_1 + \bar{V}_1^2/2g_0J \tag{18.1}$$

For a perfect gas,

$$h_0 - h_1 = c_p(T_0 - T_1)$$

and Eq. (18.1) may be written as

$$c_p(T_0 - T_1) = \bar{V}_1^2/2g_0J$$

Solving for T_0 yields

$$T_0 = T_1 + \bar{V}_1^2/2g_0Jc_p \tag{18.2}$$

where T_0 is the stagnation temperature. Dividing Eq. (18.2) by T_1 yields

$$T_0/T_1 = 1 + (\bar{V}_1^2/2g_0Jc_pT_1) \tag{18.3}$$

Since the process from state 1 to state 0 is isentropic,

$$T_0/T_1 = (p_0/p_1)^{(k-1)/k} = (v_1/v_0)^{k-1} = (\rho_0/\rho_1)^{k-1}$$

where p_0 is the stagnation pressure, v_0 is the stagnation specific volume, and ρ_0 is the stagnation density.

Example 18.1. Air at a pressure of 20 psia and 80 F is moving at a velocity of 300 mph. Determine the stagnation properties for air as a (a) perfect gas and (b) semiperfect gas.

(a)

$$\bar{V}_1 = [300(5280)]/3600 = 440 \text{ (ft/sec)}$$

$$\bar{V}_1^2/2g_0Jc_p = (440)^2/[2(32.2)(778)(0.24)]$$

$$= 16 \text{ (R)}$$

From Eq. (18.2)

$$T_0 = T_1 + \bar{V}_1^2/2g_0Jc_p$$

$$= 540 + 16 = 556 \text{ (R)}$$

The stagnation pressure is obtained from

$$p_0/p_1 = (T_0/T_1)^{k/(k-1)} = (556/540)^{3.5} = 1.108$$

where k = 1.4 for air and

$$p_0 = 20(1.108) = 22.16 \text{ (psia)}$$

The specific volume for the moving air is

$$v_1 = RT_1/p_1$$

$$= [53.3(540)]/20(144)]$$

$$= 9.99 \text{ (ft}^3\text{/lbm)}$$

The stagnation specific volume is

$$v_0/v_1 = (T_1/T_0)^{1/(k-1)} = (540/556)^{2.5} = 0.9296$$

or

$$v_0 = 0.9296(9.99) = 9.29 \text{ (ft}^3\text{/lbm)}$$

The stagnation specific volume could also be calculated by

$$v_0 = RT_0/p_0 = [53.3(556)]/[22.16(144)]$$

$$= 9.29 \text{ (ft}^3\text{/lbm)}$$

The stagnation density is

$$\rho_0 = 1/v_0 = 1/9.29 = 0.108 \text{ (lbm/ft}^3\text{)}$$

(b) As a semiperfect gas, the values of c_p and k are not constant, and the property values must be obtained from the gas tables. For a temperature of 540 R, h_1 = 129.06 B/lbm, p_{r1} = 1.3860, and v_{r1} = 144.32. From Eq. (18.1)

$$h_0 = h_1 + \bar{V}_1^2/2g_0J$$

$$= 129.06 + (440)^2/[2(32.2)(778)]$$

$$= 129.06 + 3.86$$

$$= 132.92 \text{ (B/lbm)}$$

Interpolation for h_0 = 132.92 B/lbm in the gas tables gives T_0 = 556 R, p_{r0} = 1.5374, and v_{r0} = 134.24. For the isentropic process,

$$p_0/p_1 = p_{r0}/p_{r1} = 1.5374/1.3860 = 1.109$$

and

$$p_0 = 20(1.109) = 22.18 \text{ (psia)}$$

Also

$$v_0/v_1 = v_{r0}/v_{r1} = 134.24/144.32 = 0.9302$$

and

$$v_0 = 0.9302(9.99) = 9.29 \text{ (ft}^3\text{/lbm)}$$

As before

$$\rho_0 = 1/v_0 = 1/9.29 = 0.108 \text{ (lbm/ft}^3\text{)}$$

SPEED OF SOUND

If a small pressure wave is generated in a perfect gas by means of a piston (Fig. 18.1), a pressure wave moving at the speed of sound a will be generated in the gas. The properties in the region in front of the pressure wave are p, ρ, and h. In the region through which the wave has passed the properties are p + Δp, ρ + Δρ, and h + Δh. The velocity is decreased by an amount $\Delta\bar{V}$ across the wave. In Fig. 18.2, the system is drawn as if the pressure wave were stationary. The gas enters the system with a velocity a and leaves with a velocity of a - $\Delta\bar{V}$. The other properties remain as in Fig. 18.1. An energy balance for the system yields

$$h + a^2/2g_0J = (h + \Delta h) + (a - \Delta\bar{V})^2/2g_0J$$

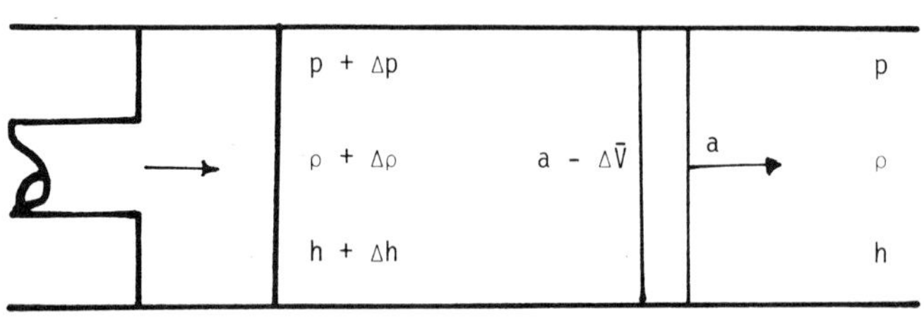

Figure 18.1

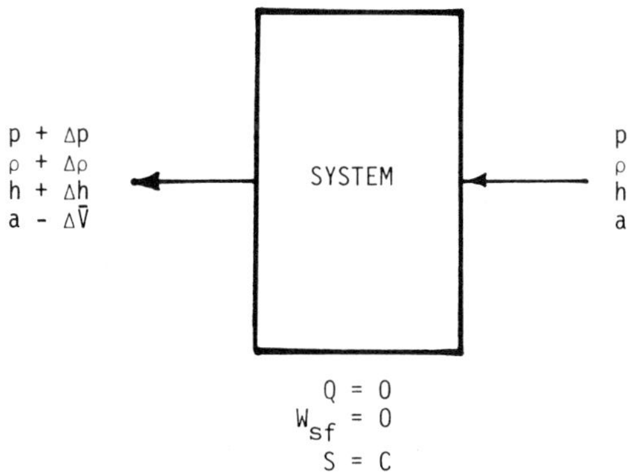

Figure 18.2

which assumes $Q = 0$, $W_{sf} = 0$, and $\Delta PE = 0$. Expanding the last term in the above equation gives

$$(a^2 - 2a\,\Delta\overline{V} + \Delta\overline{V}^2)/2g_0J$$

Since ΔV is assumed to be a small quantity, $\Delta\overline{V}^2$ may be omitted from the equation. By cancelling identical terms on the left and right sides of the original equation,

$$\Delta h - a\,\Delta\overline{V}/g_0J = 0$$

or

$$\Delta h = a\,\Delta\overline{V}/g_0J \tag{18.4}$$

From the equation of continuity

$$m = \rho A\overline{V}$$

or for the system in Fig. 18.2

$$\rho Aa = (\rho + \Delta\rho)A(a - \Delta\overline{V})$$

which assumes that the cross-sectional area A of the system is constant. Expanding the right-hand side of the equation, after the deletion of A, gives

$$\rho a = \rho a - \rho\,\Delta\overline{V} + a\,\Delta\rho - \Delta\overline{V}\,\Delta\rho$$

Again, by omitting the second-order term $\Delta\overline{V}\,\Delta\rho$,

$$\Delta\overline{V} = a\,\Delta\rho/\rho \tag{18.5}$$

Substitution of the velocity term into Eq. (18.4) gives

$$\Delta h = (a/g_0 J)(a\ \Delta\rho/\rho) = a^2\ \Delta\rho/\rho g_0 J \tag{18.6}$$

The enthalpy is defined by

$$h = u + pv/J$$

so that

$$\Delta h = \Delta u + \Delta(pv)/J$$

$$= \Delta u + p\ \Delta v/J + v\ \Delta p/J$$

or

$$\Delta u = \Delta h - p\ \Delta v/J - v\ \Delta p/J \tag{18.7}$$

For a reversible process

$$\Delta Q = T\ \Delta s$$

and for a nonflow process

$$\Delta Q = \Delta u + p\ \Delta v/J$$

or

$$T\ \Delta s = \Delta u + p\ \Delta v/J \tag{18.8}$$

Since Eq. (18.8) is an equation that relates properties, it is valid for both the nonflow and steady-flow systems. Substituting Eq. (18.7) into Eq. (18.8) gives

$$T\ \Delta s = \Delta h - p\ \Delta v/J - v\ \Delta p/J + p\ \Delta v/J$$

$$T\ \Delta s = \Delta h - v\ \Delta p/J = \Delta h - \Delta p/\rho J \tag{18.9}$$

If the pressure wave is small, it may be assumed to be an isentropic process so that

$$\Delta s = 0$$

and Eq. (18.9) may be written

$$\Delta h = \Delta p/\rho J$$

Equating Eq. (18.6) and the above equation gives

$$\Delta p/\rho J = a^2\ \Delta\rho/\rho g_0 J$$

or

$$\Delta p/\rho = a^2\ \Delta\rho/\rho g_0$$

and

$$\Delta p/\Delta \rho = a^2/g_0, \qquad s = c$$

Taking the limit of the above equation for an isentropic process gives

$$\lim_{\Delta\rho \to 0} (\Delta p/\Delta \rho) = dp/d\rho$$

and

$$dp/d\rho = a^2/g_0 \tag{18.10}$$

For an isentropic process

$$pv^k = c$$

or

$$p/\rho^k = c$$

Taking the derivative of the equation yields

$$\frac{dp}{d\rho} = \frac{\rho^k\ dp - pk\rho^{k-1}\ d\rho}{\rho^{2k}} = 0$$

and

$$\rho^k\ dp - pk\rho^{k-1}\ d\rho = 0$$

$$dp - (pk/\rho)\ d\rho = 0$$

or

$$dp/d\rho = kp/\rho \tag{18.11}$$

Equating Eqs. (18.10) and (18.11) yields

$$a^2/g_0 = kp/\rho = kpv$$

so that

$$a^2 = kpvg_0$$

From the perfect gas equation

$$pv = RT$$

and

$$a^2 = kg_0RT$$

so that

$$a = (kg_0RT)^{1/2} \tag{18.12}$$

which is the speed of sound in a perfect gas. The units of $(kg_0RT)^{1/2}$ are

$$[(\text{lbm-ft/lbf-sec}^2)(\text{ft-lbf/lbm-R})\ (R)]^{1/2} = (\text{ft}^2/\text{sec}^2)^{1/2}$$

so that the units of a are ft/sec.

MACH NUMBER

The Mach number of a moving fluid is the ratio of the velocity of the fluid to the speed of sound of the fluid:

$$M = \overline{V}/a \tag{18.13}$$

The symbol M has been previously used to designate the molecular weight of the working substance. Since both uses seldom appear in the same thermodynamic problem, we shall use the same symbol for both terms.

<u>Example 18.2</u>. Determine the Mach number and speed of sound for air, as a perfect gas, at a pressure and temperature of 20 psia and 80 F. The velocity of the air stream is 300 mph.

From Eq. (18.12)

$$a = (kg_0RT)^{1/2} = [1.4(32.2)(53.3)(540)]^{1/2}$$

$$= 1139\ (\text{ft/sec})$$

$$\overline{V} = 300(5280/3600) = 440\ (\text{ft/sec})$$

and from Eq. (18.13)

$$M = \overline{V}/a = 440/1139 = 0.386$$

<u>Example 18.3</u>. Assume the temperature of the air in Example 18.2 was 2000 R. Determine the acoustic velocity and Mach number for air as (a) a perfect gas and (b) a semiperfect gas.

(a) As a perfect gas

$$a = (kg_0RT)^{1/2} = [1.4(32.2)(53.3)(2000)]^{1/2}$$

$$= 2190\ (\text{ft/sec})$$

and

$$M = \overline{V}/a = 440/2190 = 0.201$$

(b) For air as a semiperfect gas when T_1 = 1980 and T_2 = 2020 R, h_1 = 499.17 B/lbm, h_2 = 510.26 B/lbm, u_1 = 363.43 B/lbm, and u_2 = 371.79 B/lbm, so that

$$\Delta h = h_2 - h_1 = 510.26 - 499.17 = 11.09 \text{ (B/lbm)}$$

and

$$\Delta u = u_2 - u_1 = 371.79 - 363.43 = 8.36 \text{ (B/lbm)}$$

The value of k may be obtained from

$$k = \Delta h/\Delta u = 11.09/8.36 = 1.33$$

so that

$$a = (kg_0RT)^{1/2} = [1.33(32.2)(53.3)(2000)]^{1/2}$$

$$= 2137 \text{ (ft/sec)}$$

and

$$M = \bar{V}/a = 440/2137 = 0.206$$

SUBSONIC FLOW THROUGH A NOZZLE

Subsonic flow occurs when the Mach number is less than 1, and supersonic flow occurs when the Mach number exceeds 1. In a subsonic nozzle the velocity increases as the fluid proceeds through the system. Figure 18.3 illustrates a subsonic nozzle, $p_2 < p_1$, for which the properties at states 1 and 2 are known. Assuming that the process from state 1 to state 2 is isentropic and that no external work is done on or by the system, the energy equation

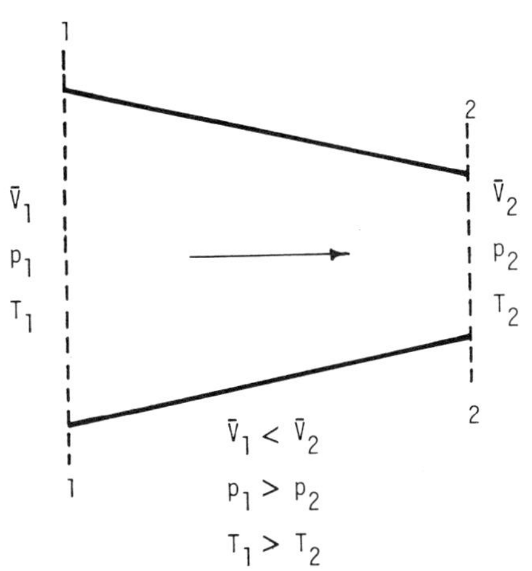

Figure 18.3

$$Q = \Delta H + \Delta KE + \Delta PE + W_{sf}$$

reduces to

$$\Delta H + \Delta KE = 0$$

when

$$\Delta PE = 0$$

Therefore, for steady flow,

$$H_1 + KE_1 = H_2 + KE_2$$

and

$$h_1 + \bar{V}_1^2/2g_0J = h_2 + \bar{V}_2^2/2g_0J \tag{18.14}$$

or

$$h_{01} = h_{02} = h_0$$

where h_0 is the stagnation enthalpy. For a perfect gas, from Eq. (18.14),

$$\bar{V}_2^2 - \bar{V}_1^2 = 2g_0Jc_p(T_1 - T_2) \tag{18.15}$$

and from the equation of continuity

$$m = \rho_1A_1\bar{V}_1 = \rho_2A_2\bar{V}_2$$

or

$$\bar{V}_1 = (\rho_2A_2/\rho_1A_1)\bar{V}_2$$

Substitution into Eq. (18.15) gives

$$\bar{V}_2^2[1 - (\rho_2A_2/\rho_1A_1)^2] = 2g_0Jc_p(T_1 - T_2)$$

or

$$\bar{V}_2 = [2g_0Jc_p(T_1 - T_2)]^{1/2}/F \tag{18.16}$$

where F is the velocity of approach factor and is equal to

$$F = [1 - (\rho_2A_2/\rho_1A_1)^2]^{1/2} \tag{18.17}$$

Example 18.4. Given is air as a perfect gas flowing through a subsonic nozzle where the inlet conditions are $p_1 = 100$ psia and $t_1 = 540$ F. The exit pressure is 80 psia. Inlet and outlet diameters are 4 and 1 in. Determine the mass flow, $\bar{V}_2$, a_2, M_2, and $\bar{V}_1$.

For an isentropic expansion process

$$T_2/T_1 = (p_2/p_1)^{(k-1)/k}$$

$$T_2 = 1000(80/100)^{0.286} = 938 \text{ (R)}$$

The densities are

$$\rho_1 = p_1/RT_1 = [100(144)]/[53.3(1000)]$$

$$= 0.270 \text{ (lbm/ft}^3\text{)}$$

and

$$\rho_2 = p_2/RT_2 = [80(144)]/[53.3(938)]$$

$$= 0.230 \text{ (lbm/ft}^3\text{)}$$

The ratio of the density and area terms is

$$\rho_2 A_2/\rho_1 A_1 = (0.230/0.270)(1/4)^2 = 0.0532$$

giving a velocity of approach factor, from Eq. (18.17), of

$$F = [1 - (0.0532)^2]^{1/2} = 0.9986$$

The velocity at the exit section is

$$\bar{V}_2 = [2g_0 Jc_p(T_1 - T_2)]^{1/2}/F$$

$$= (1/0.9986)[2(32.2)(778)(0.24)(1000 - 938)]^{1/2}$$

$$= 865 \text{ (ft/sec)}$$

The speed of sound, from Eq. (18.12), is

$$a_2 = (kg_0 RT_2)^{1/2}$$

$$= [1.4(32.2)(53.3)(938)]^{1/2}$$

$$= 1501 \text{ (ft/sec)}$$

From Eq. (18.13), the Mach number is

$$M_2 = \bar{V}_2/a_2 = 865/1501$$

$$= 0.576$$

The mass flow is

$$m = \rho_2 A_2 \bar{V}_2 = 0.230(\pi/4)(1/12)^2(865)$$

$$= 1.085 \text{ (lbm/sec)}$$

Table 18.1

p_2	T_2	ρ_2	F	$\overline{V}_2$	a_2	M_2	m
100	1000	0.270	0.9980	0	1550	0.000	0.00
90	970	0.251	0.9983	602	1527	0.394	0.82
80	938	0.230	0.9986	865	1501	0.576	1.08
70	903	0.209	0.9988	1081	1473	0.734	1.23
60	864	0.188	0.9991	1280	1441	0.888	1.31
52.8	833	0.171	0.9992	1418	1415	1.002	1.32
50	820	0.165	0.9993	1472	1404	1.048	1.32
40	769	0.141	0.9995	1667	1359	1.227	1.28
30	709	0.114	0.9997	1871	1305	1.434	1.16
20	631	0.086	0.9998	2107	1231	1.712	0.99
10	518	0.052	0.9999	2407	1116	2.157	0.68

From the equation of continuity the inlet velocity is

$$\overline{V}_1 = \rho_2 A_2 \overline{V}_2 / \rho_1 A_1$$

$$= [0.230(1)(865)]/[0.270(16)] = 46.1 \text{ (ft/sec)}$$

Example 18.5. Repeat Example 18.4 where p_2 varies from 100 to 10 psia. The inlet conditions remain constant.

The results of varying p_2 are indicated in Table 18.1, with the units as in Example 18.4.

In Fig. 18.4 the Mach number and mass flow are plotted with respect to the exit pressure for the nozzle. From Fig. 18.4 the mass flow increases as the exit pressure decreases until a maximum value is reached at 52.8 psia.

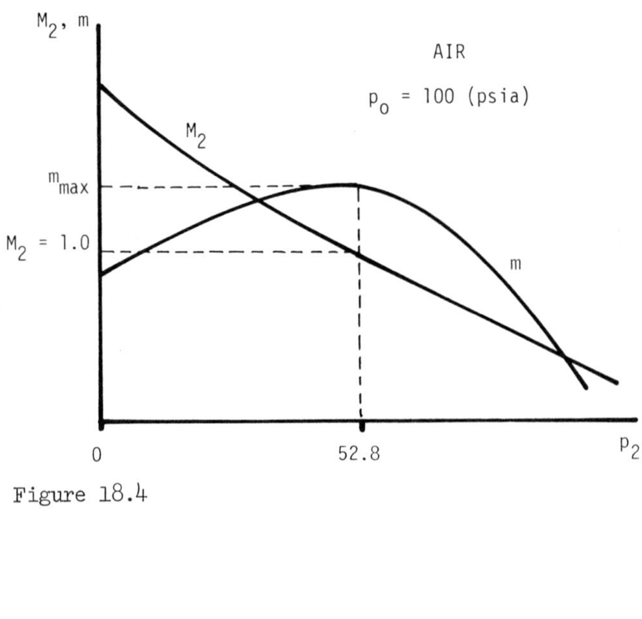

Figure 18.4

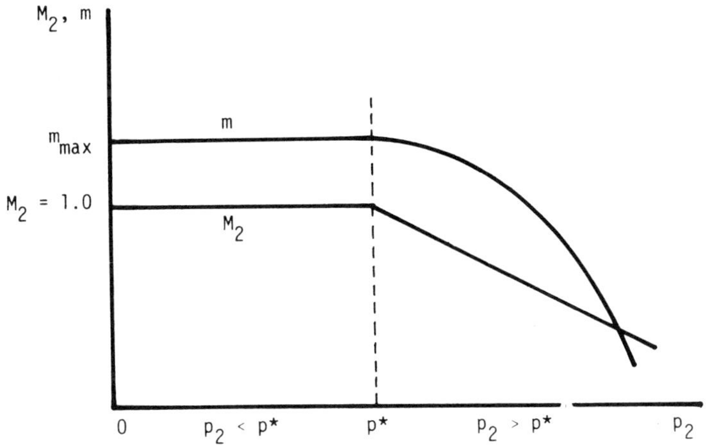

Figure 18.5

Further reduction in the pressure indicates that the flow will decrease. Logically, reduction in the exit pressure should always increase the flow. But in the subsonic nozzle there is a limiting value of the exit pressure beyond which the flow will not change. The actual mass flow and Mach number of the nozzle in Fig. 18.4 are sketched in Fig. 18.5. The properties at this limiting condition are referred to as the **critical** properties. In Table 18.1, $p^* = 52.8$ psia, $T^* = 833$ R, $\overline{V}^* = 1418$ ft/sec, $a^* = 1415$ ft/sec, and $M^* = 1.002$, where the superscript * is used to designate the critical value of the property. The critical condition actually occurs when the velocity at the exit is the speed of sound so that $\overline{V}^* = a^*$ and $M^* = 1.0$. By neglecting round-off errors Table 18.1 verified this fact.

STAGNATION AND CRITICAL PROPERTIES

From Eq. (18.14)

$$h_1 + \overline{V}_1^2/2g_0J = h_2 + \overline{V}_2^2/2g_0J$$

This may be rewritten, where state 2 is replaced by the critical state, as

$$h_0 = h^* + a^{*2}/2g_0J$$

and

$$h_0 = h_{0a}$$

by the substitution of the total enthalpy and critical properties into Eq. (18.14). By using Eq (18.12) for the speed of sound,

$$h_0 = h^* + (kg_0RT^*)/2g_0J$$

$$= h^* + kRT^*/2J$$

Since $c_p - c_v = R/J$, we may write

$$h_0 = h^* + (kT^*)(c_p - c_v)/2$$

and

$$h_0 = h^* + (kT^*)(c_p)[1 - (1/k)]/2$$

For the perfect gas

$$T_0 = T^* + (kT^*)[(k - 1)/2k]$$

$$= T^*[1 + (k - 1)/2]$$

$$= T^*[(k + 1)/2]$$

or

$$T^*/T_0 = 2/(k + 1) \tag{18.18}$$

For the isentropic process

$$p^* = p_0 = (T^*/T_0)^{k/(k-1)} = [2/(k + 1)]^{k/(k-1)} \tag{18.19}$$

For k = 1.4,

$$T^*/T_0 = 0.833 \tag{18.20}$$

and

$$p^*/p_0 = 0.528 \tag{18.21}$$

Example 18.6. Determine the critical values for the nozzle in Examples 18.4 and 18.5. First assume $\bar{V}_1 = 0$ so that $T_1 = T_0 = 1000$ R and $p_0 = 100$ psia. From Eqs. (18.20) and (18.21), $T^* = 833$ R and $p^* = 52.8$ psia. The critical density is

$$\rho^* = p^*/RT^* = [52.8(144)]/[53.3(833)]$$

$$= 0.1712 \text{ (lbm/ft}^3\text{)}$$

The critical velocity is

$$\bar{V}^* = a^* = (kg_0RT^*)^{1/2}$$

$$= [1.4(32.2)(53.3)(833)]^{1/2}$$

$$= 1415 \text{ (ft/sec)}$$

Therefore

$$m^* = \rho^* A^* a^* = 0.1712(\pi/4)(1/12)^2(1415)$$

$$= 1.321 \text{ (lbm/sec)}$$

Note the use of the area A_2 for the critical area. Solving for the velocity at the inlet position, one obtains

$$\overline{V}_1 = m/\rho_1 A_1$$

$$\overline{V}_1 = 1.321/[0.270(\pi/4)(4/12)^2]$$

$$= 56.1 \text{ (ft/sec)}$$

where $\rho_1 = 0.270$ lbm/ft^3 from Example 18.4. Solving for T_0, one obtains

$$T_0 = T_1 + V_1^2/2g_0 Jc_p$$

$$= 1000 + \{(56.1)^2/[2(32.2)(778)(0.24)]\} = 1000 + 0.3$$

$$= 1000.3 \text{ (R)}$$

For $T_0 = 1000.3$ R, the values of the critical properties will not vary by an appreciable amount from those obtained for $T_0 = 1000$ R.

SUPERSONIC FLOW THROUGH A NOZZLE

For supersonic flow to occur in the nozzle, the entering Mach number must be

$$M_1 \geq 1.0$$

The nozzle configuration is sketched in Fig. 18.6.

Example 18.7. Assume the initial conditions for the nozzle using air in Fig. 18.6 are $p_1 = 52.8$ psia, $T_1 = 833$ R, $M_1 = 1$, and $D_1 = 1$ in., which are the critical properties since $M_1 = 1$. Determine m_2, T_2, $\overline{V}_2$, a_2, and A_2 for $p_2 = 10$ psia. Since

$$m_2 = m_1 = \rho_1 A_1 \overline{V}_1$$

then

$$m_1 = 0.1712(\pi/4)(1/12)^2(1415)$$

$$= 1.321 \text{ (lbm/sec)}$$

where $\overline{V}_1 = 1415$ ft/sec and $\rho_1 = 0.1712$ lbm/ft^3 for $M_1 = 1$ (Example 18.6).

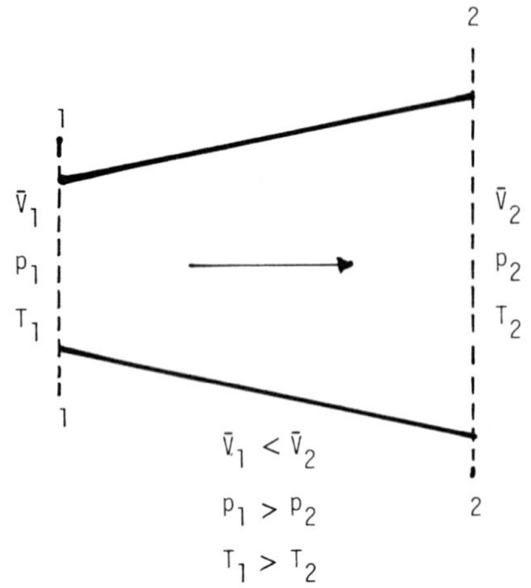

Figure 18.6

For the isentropic expansion process,

$$T_2/T_1 = (p_2/p_1)^{(k-1)/k}$$

so that

$$T_2 = 833(10/52.8)^{0.286}$$

$$= 518 \text{ (R)}$$

The density at the exit is

$$\rho_2 = p_2/RT_2 = [10(144)]/[53.3(518)]$$

$$= 0.0522 \text{ (lbm/ft}^3\text{)}$$

The velocity at the exit is obtained from

$$\bar{V}_2^2 = 2g_0Jc_p(T_1 - T_2) + \bar{V}_1^2$$

$$= 2(32.2)(778)(0.24)(833 - 518) + (1415)^2$$

$$= 5.790 \times 10^6$$

or

$$\bar{V}_2 = 2406 \text{ (ft/sec)}$$

The acoustic velocity and Mach number are

$$a_2 = (kg_0RT_2)^{1/2} = [1.4(32.2)(53.3)(518)]^{1/2} = 1116 \text{ (ft/sec)}$$

and

$$M_2 = \bar{V}_2/a_2 = 2406/1116 = 2.156$$

The area for the exit section is

$$A_2 = m_2/\rho_2\bar{V}_2$$
$$= 1.321/[0.0522(2406)]$$
$$= 0.0105 \text{ (ft}^2\text{)}$$

and

$$D_2^2 = (4/\pi)(0.0105)(144)$$
$$= 1.925 \text{ (in.}^2\text{)}$$
$$D_2 = 1.387 \text{ (in.)}$$

The supersonic nozzle increases from a diameter of 1 in. at the narrowest section to 1.387 in. at the exit.

SUBSONIC FLOW THROUGH A DIFFUSER

A diffuser (Fig. 18.7) is the opposite of a nozzle. In the diffuser the velocity decreases, and the pressure and temperature increase in the direction of the flow. We shall continue to assume that the process between the

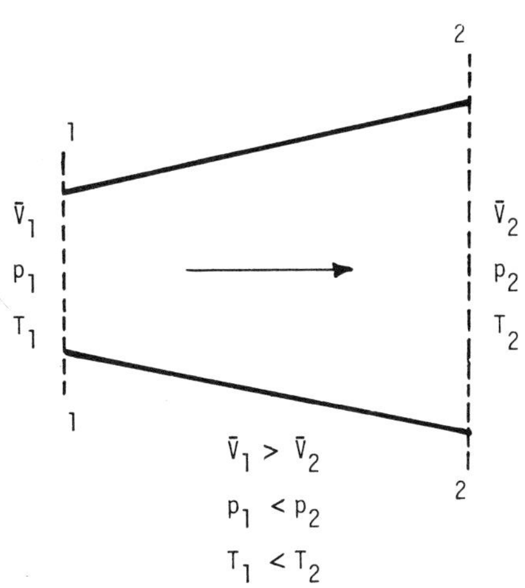

Figure 18.7

initial and final states is isentropic and that no external work is involved in the process. For a subsonic diffuser it is required that

$$M_1 \leq 1$$

Since we have assumed an isentropic, reversible process, the previous equations, as applied to the nozzle, may be used for the diffuser.

Example 18.8. Assume the initial conditions for the diffuser in Fig. 18.7 are $p_1 = 52.8$ psia, $T_1 = 833$ R, $M_1 = 1$, and $D_1 = 1$ in. Determine M_2, T_2, $\overline{V}_2$, a_2, and A_2 for $p_2 = 80$ psia when the working substance is air acting as a perfect gas.

From Eqs. (18.12) and (18.13), $M_1 = \overline{V}_1/a_1$ or $a_1 = \overline{V}_1$ since $M_1 = 1$.

$$a_1 = \overline{V}_1 = (kg_0RT_1)^{1/2} = [1.4(32.2)(53.3)(833)]^{1/2}$$

$$= 1415 \text{ (ft/sec)}$$

The density at the diffuser inlet is

$$\rho_1 = p_1/RT_1 = [52.8(144)]/[53.3(833)]$$

$$= 0.1712 \text{ (lbm/ft}^3\text{)}$$

For steady-flow conditions

$$m_2 = m_1 = \rho_1 A_1 \overline{V}_1$$

$$= 0.1712(\pi/4)(1/12)^2(1415)$$

$$= 1.321 \text{ (lbm/sec)}$$

From Eq. (18.18)

$$T^*/T_0 = 2/(k + 1)$$

where $T_1 = T^* = 833$ R since $M_1 = 1$. Solving for T_0, which is the stagnation temperature throughout the diffuser,

$$T_0 = T^*(k + 1)/2 = 833(1.4 + 1)/2$$

$$= 1000 \text{ (R)}$$

Also from Eq. (18.19), the stagnation pressure throughout the diffuser is

$$p_0 = p^*[(k + 1)/2]^{k/(k-1)}$$

$$= 52.8(1.2)^{3.5}$$

$$= 100 \text{ (psia)}$$

where $p_1 = p^* = 52.8$ psia for $M_1 = 1$. The temperature at the diffuser exit may be calculated from

$$T_2/T_0 = (p_2/p_0)^{(k-1)/k} = (80/100)^{0.286}$$

$$= 0.938$$

and

$$T_2 = 0.938(1000) = 938 \text{ (R)}$$

For the diffuser exit,

$$h_0 = h_2 + \bar{V}_2^2/(2g_0J)$$

or

$$V_2^2 = 2g_0J(h_0 - h_2)$$

and for a perfect gas,

$$\bar{V}_2^2 = 2g_0Jc_p(T_0 - T_2)$$

$$= [2(32.2)(778)(0.24)](1000 - 938)$$

$$= 7.455 \times 10^5 \text{ (ft}^2\text{/sec}^2\text{)}$$

$$\bar{V}_2 = 863 \text{ (ft/sec)}$$

From Eq. (18.12), the speed of sound at the diffuser exit is

$$a_2 = (kg_0RT_2)^{1/2} = [1.4(32.2)(53.3)(938)]^{1/2}$$

$$= 1501 \text{ (ft/sec)}$$

yielding an exit Mach number of

$$M_2 = \bar{V}_2/a_2 = 863/1501 = 0.575$$

The density at the exit is

$$\rho_2 = p_2/RT_2 = [80(144)]/[53.3(938)]$$

$$= 0.2304 \text{ (lbm/ft}^3\text{)}$$

The area of the exit section is

$$A_2 = m_2/\rho_2\bar{V}_2 = 1.321/[0.2304(863)]$$

$$= 0.00664 \text{ (ft}^2\text{)}$$

$$= 0.957 \text{ (in.}^2\text{)}$$

and the exit diameter is

$$D_2^2 = (4/\pi)A_2 = (4/\pi)0.957$$

$$= 1.218 \text{ (in.}^2\text{)}$$

and

$$D_2 = 1.104 \text{ (in.)}$$

The diameter at the diffuser inlet was given as

$$D_1 = 1.00 \text{ (in.)}$$

so that the area increases in the direction of the flow. To verify the solution, from Eq. (18.17), the velocity of approach factor is

$$F' = [(\rho_2 A_2/\rho_1 A_1)^2 - 1]^{1/2}$$

and Eq. (18.16) becomes

$$\overline{V}_2 = [2g_0 Jc_p(T_2 - T_1)]^{1/2}/F'$$

Since $T_2 > T_1$ for the diffuser, the order of the terms in the velocity of approach equation and velocity equation has been reversed. Since

$$\rho_2 A_2/\rho_1 A_1 = [0.2304(0.957)]/[0.1712(0.785)] = 1.641$$

where

$$A_1 = (\pi/4)(1)^2 = 0.785 \text{ (in.}^2\text{)}$$

then

$$F' = [(1.641)^2 - 1]^{1/2} = 1.301$$

Solving for $\overline{V}_2$ gives

$$\overline{V}_2 = [2(32.2)(778)(0.24)(938 - 833)]^{1/2}/1.301$$

$$= 864 \text{ (ft/sec)}$$

which checks with the previously calculated value of the exit velocity.

SUPERSONIC DIFFUSER

In a supersonic diffuser (Fig. 18.8) the velocity will decrease as the pressure and temperature increase in the direction of the flow. To ensure supersonic conditions throughout the nozzle, $M_1 > M_2$, where $M_2 \geq 1$. Again

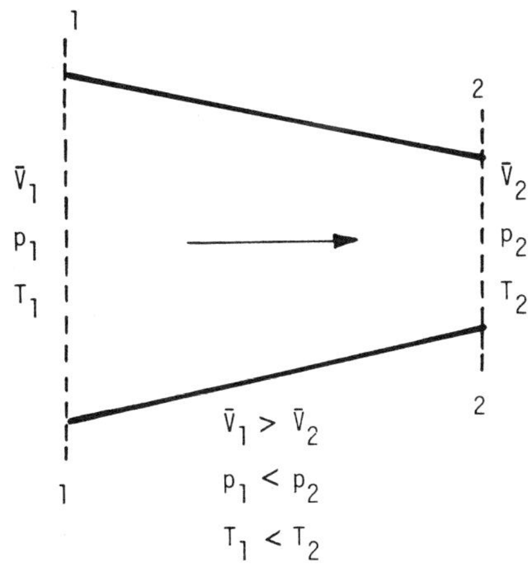

Figure 18.8

for an isentropic process with no external work, the previously derived equations may be used.

Example 18.9. Assume the initial conditions for the diffuser in Fig. 18.8 are $p_1 = 20$ psia, $t_1 = 40$ F, $M_1 = 2$, and $D_1 = 1$ in. The Mach number at the diffuser exit is 1. Determine T_2, $\bar{V}_2$, a_2, and A_2 using air as a perfect gas.

From Eq. (18.12) for the acoustic velocity

$$a_1 = (kg_0RT_1)^{1/2} = [1.4(32.2)(53.3)(500)]^{1/2}$$

$$= 1096 \text{ (ft/sec)}$$

The inlet velocity is

$$\bar{V}_1 = M_1a_1 = 2(1096)$$

$$= 2192 \text{ (ft/sec)}$$

and the density at the inlet is

$$\rho_1 = p_1/RT_1 = [20(144)]/[53.3(500)]$$

$$= 0.1081 \text{ (lbm/ft}^3\text{)}$$

The mass flow at all locations in the diffuser is

$$m = \rho_1A_1\bar{V}_1 = 0.1081(\pi/4)(1/12)^2(2192)$$

$$= 1.292 \text{ (lbm/sec)}$$

The stagnation temperature is obtained from Eq. (18.2), where

$$T_0 = T_1 + \bar{V}_1^2/2g_0Jc_p$$

$$= 500 + (2192)^2/[2(32.2)(778)(0.24)]$$

$$= 500 + 400 = 900 \text{ (R)}$$

Solving for the temperature where $M_2 = 1$ from Eq. (18.18) gives

$$T^* = 2T_0/(k + 1) = 2(900)/2.4 = 750 \text{ (R)}$$

The acoustic velocity for the preceding temperature is

$$a^* = (kg_0RT^*)^{1/2} = [1.4(32.2)(53.3)(750)]^{1/2}$$

$$= 1342 \text{ (ft/sec)}$$

Since $M_2 = 1$ at the diffuser exit,

$$T_2 = T^* = 750 \text{ (R)}$$

and

$$\bar{V}_2 = a_2 = a^* = 1342 \text{ (ft/sec)}$$

For an isentropic process between the diffuser inlet and exit

$$p_2/p_1 = (T_2/T_1)^{k/(k-1)}$$

$$p_2 = 20(750/500)^{3.5} = 82.7 \text{ (psia)}$$

The density at the diffuser exit is

$$\rho_2 = p_2/RT_2 = [82.7(144)]/[53.3(750)]$$

$$= 0.2979 \text{ (lbm/ft}^3\text{)}$$

The exit area is

$$A_2 = m/\rho_2\bar{V}_2 = 1.292/[0.2979(1342)]$$

$$= 0.00323 \text{ (ft}^2\text{)}$$

$$= 0.465 \text{ (in.}^2\text{)}$$

and

$$D_2^2 = (4/\pi)0.465 = 0.592$$

so that

$$D_2 = 0.769 \text{ (in.)}$$

The inlet diameter is

$$D_1 = 1.0 \text{ (in.)}$$

so that the area decreases in the direction of the flow.

CONVERGING-DIVERGING NOZZLE

In Fig. 18.9 a subsonic nozzle is joined to a supersonic nozzle. The conditions required for the converging and diverging sections to act as nozzles are $M_1 < M^*$, $M_2 > M^*$, where $M^* = 1$. In terms of pressure and temperature

$$p_1 > p^* > p_2$$

and

$$T_1 > T^* > T_2$$

The velocity will progressively increase in the direction of the flow.

Example 18.10. Given $p_1 = 100$ psia, $t_1 = 540$ F, $D_1 = 1.2$ in., $D^* = 1$ in., and $p_2 = 20$ psia, determine, using air as a perfect gas, m, $\overline{V}_2$, a_2, M_2, and D_2.

As in Example 18.6, assume

$$T_0 = T_1 = 1000 \text{ (R)}$$

and

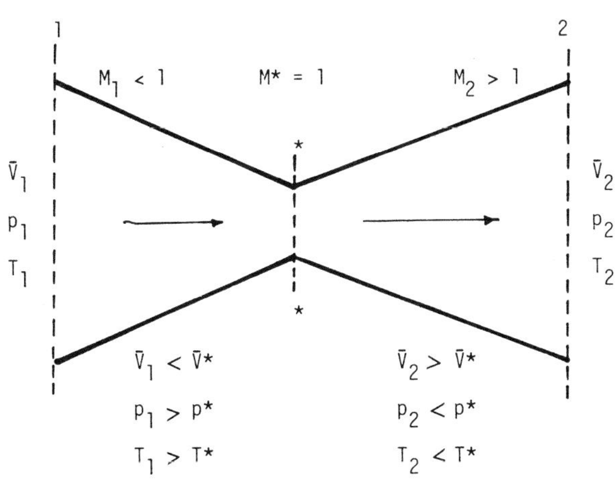

Figure 18.9

$$p_0 = p_1 = 100 \text{ (psia)}$$

which implies $\bar{V}_1 = 0$. From Eqs. (18.20) and (18.21) for $k = 1.4$,

$$T^* = 0.833T_0 = 833 \text{ (R)}$$

and

$$p^* = 0.528p_0 = 52.8 \text{ (psia)}$$

The critical properties, which occur at the narrowest section in the nozzle, commonly called the throat, are

$$\rho^* = p^*/RT^* = [52.8(144)]/[53.3(833)]$$

$$= 0.1712 \text{ (lbm/ft}^3\text{)}$$

and

$$\bar{V}^* = a^* = (kg_0RT^*)^{1/2} = [1.4(32.2)(53.3)(833)]^{1/2}$$

$$= 1415 \text{ (ft/sec)}$$

The mass flow is

$$m^* = \rho^*A^*a^* = 0.1712(\pi/4)(1/12)^2(1415)$$

$$= 1.321 \text{ (lbm/sec)}$$

The density at the inlet is

$$\rho_1 = p_1/RT_1 = [100(144)]/[53.3(1000)]$$

$$= 0.2702 \text{ (lbm/ft}^3\text{)}$$

and the inlet velocity is

$$\bar{V}_1 = m/\rho_1A_1 = 1.321/[0.2702(\pi/4)(1.2/12)^2]$$

$$= 622 \text{ (ft/sec)}$$

Solving for the stagnation temperature from Eq. (18.2) gives

$$T_0 = T_1 + \bar{V}_1{}^2/2g_0Jc_p$$

$$= 1000 + (622)^2/[2(32.2)(778)(0.24)]$$

$$= 1000 + 32 = 1032 \text{ (R)}$$

This differs from the assumption of $T_0 = 1000$ R by 32 R. Let $T_0 = 1032$ R, $T_1 = 1000$ R, and $p_1 = 100$ psia. Then

$$p_0 = p_1(T_0/T_1)^{k/(k-1)}$$

$$= 100(1032/1000)^{3.5} = 111.7 \text{ (psia)}$$

and

$$T^* = 0.833(1032) = 860 \text{ (R)}$$

Also

$$p^* = 0.528(111.7) = 59.0 \text{ (psia)}$$

The critical density is

$$\rho^* = p^*/RT^* = [59.0(144)]/[53.3(860)]$$

$$= 0.1853 \text{ (lbm/ft}^3\text{)}$$

and the velocity at the throat is

$$\bar{V}^* = a^* = (kg_0RT^*)^{1/2} = [1.4(32.2)(53.3)(860)]^{1/2}$$

$$= 1437 \text{ (ft/sec)}$$

The mass flow is

$$\dot{m}^* = \rho^*A^*a^* = 0.1853(\pi/4)(1/12)^2(1437)$$

$$= 1.452 \text{ (lbm/sec)}$$

The inlet velocity is

$$\bar{V}_1 = \dot{m}/\rho_1A_1 = 1.452/[0.2702(\pi/4)(1.2/12)^2]$$

$$= 684 \text{ (ft/sec)}$$

The stagnation temperature for these conditions is

$$T_0 = T_1 + \bar{V}_1^2/2g_0Jc_p$$

$$= 1000 + (684)^2/[2(32.2)(778)(0.24)]$$

$$= 1000 + 39 = 1039 \text{ (R)}$$

which differs from the previous value by 7 R.

Repeating the above procedure for T_0 = 1039 R gives p_0 = 114.3 psia, T^* = 865 R, p^* = 60.4 psia, ρ^* = 0.1886 lbm/ft^3, a^* = 1441 ft/sec, and $\dot{m}^*$ = 1.482 lbm/sec, which leads to $\bar{V}_1$ = 698 ft/sec and T_0 = 1041 R. Using T_0 = 1041 R gives p_0 = 115.1 psia, T^* = 867 R, p^* = 60.8 psia, ρ^* = 0.1894 lbm/ft^3, a^* = 1443 ft/sec, and $\dot{m}^*$ = 1.491 lbm/sec, and $\bar{V}_1$ = 702 ft/sec and T_0 = 1041 R as before. The mass flow through the nozzle is

$$m = 1.491 \text{ (lbm/sec)}$$

At the nozzle exit

$$T_2 = T_1(p_2/p_1)^{(k-1)/k} = 1000(20/100)^{0.286}$$

$$= 631 \text{ (R)}$$

and

$$\rho_2 = p_2/RT_2 = [20(144)]/[53.3(631)]$$

$$= 0.0856 \text{ (lbm/ft}^3\text{)}$$

The exit velocity is

$$\overline{V}_2^{\,2} = 2g_0Jc_p(T_0 - T_2) = 2(32.2)(778)(0.24)(1041 - 631)$$

$$= 4.930 \times 10^6 \text{ (ft}^2\text{/sec}^2\text{)}$$

or

$$\overline{V}_2 = 2220 \text{ (ft/sec)}$$

The acoustic velocity and Mach number, at the nozzle exit, are

$$a_2 = (kg_0RT_2)^{1/2} = [1.4(32.2)(53.3)(631)]^{1/2}$$

$$= 1231 \text{ (ft/sec)}$$

and

$$M_2 = \overline{V}_2/a_2 = 2220/1231 = 1.803$$

The exit area is

$$A_2 = m/\rho_2\overline{V}_2 = 1.491/[0.0856(2220)]$$

$$= 0.00785 \text{ (ft}^2\text{)}$$

$$= 1.1304 \text{ (in.}^2\text{)}$$

and

$$D_2^{\,2} = (4/\pi)A_2 = (4/\pi)1.1304$$

$$= 1.439 \text{ (in.}^2\text{)}$$

so that

$$D_2 = 1.200 \text{ (in.)}$$

NOZZLE DIFFUSER

Assume we have a converging-diverging nozzle as shown in Fig. 18.10. In Example 18.10 the section from the throat to the exit acted as a nozzle since the Mach number at the throat was critical or $M^* = 1$, and $p_2 < p^*$. Let us now assume that the pressure at 2 is slowly raised until $p_2 > p^*$. The section from the throat to the exit will now act as a subsonic diffuser. Leaving conditions at the exit are $M_2 < 1$, $p_2 > p^*$, and $T_2 > T^*$.

Example 18.11. Rework Example 18.10 where $p_1 = 100$ psia, $t_1 = 540$ F, $D_1 = 1.2$ in., $D^* = 1$ in., and $p_2 = 100$ psia. The Mach number at the throat $= 1$. Determine, using air as a perfect gas, m, $\overline{V}_2$, a_2, M_2, and D_2. From the inlet to the throat this problem is identical with Example 18.10, so that $T_0 = 1041$ R, $p_0 = 115.1$ psia, and $m = 1.491$ lbm/sec.

At the nozzle exit

$$T_2 = T_1(p_2/p_1)^{(k-1)/k} = 1000(100/100)^{0.286} = 1000 \text{ (R)}$$

$$\rho_2 = p_2/RT_2 = [100(144)]/[53.3(1000)]$$

$$= 0.2702 \text{ (lbm/ft}^3\text{)}$$

$$\overline{V}_2^{\,2} = 2g_0Jc_p(T_0 - T_2) = 2(32.2)(778)(0.24)(1041 - 1000)$$

$$= 4.93 \times 10^5 \text{ (ft}^2\text{/sec}^2\text{)}$$

$$\overline{V}_2 = 702 \text{ (ft/sec)}$$

$$a_2 = (kg_0RT_2)^{1/2} = [1.4(32.2)(53.3)(1000)]^{1/2} = 1550 \text{ (ft/sec)}$$

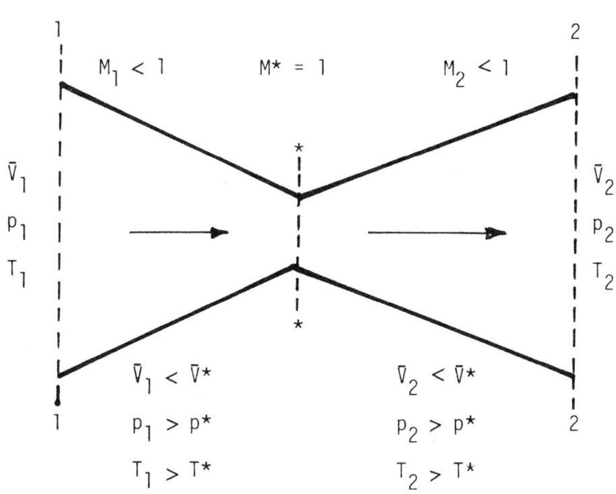

Figure 18.10

$$M_2 = \overline{V}_2/a_2 = 702/1550 = 0.453$$

$$A_2 = m/\rho_2 V_2 = 1.491/[0.2702(702)]$$

$$= 0.00786\ (\text{ft}^2)$$

$$= 1.1319\ (\text{in.}^2)$$

$$D_2^2 = (4/\pi)A_2 = 1.4412\ (\text{in.}^2)$$

$$D_2 = 1.20\ (\text{in.})$$

Note that the thermodynamic properties are the same at the inlet and outlet, which indicates the process is fully reversible. The nozzle used in Example 18.10 is also reversible, but since the back pressure p_2 was not returned to its initial value, the thermodynamic properties at the exit are different.

The same inlet, throat, and exit diameters can produce a subsonic or a supersonic discharge depending on the leaving pressure. Both configurations produce the same mass flow.

CONVERGING-DIVERGING NOZZLE DESIGN

Example 18.12. Determine the diameter of an isentropic nozzle at various pressure locations for a mass flow of 2.0 lbm/sec. The working substance is air as a perfect gas. Given are $T_0 = 600$ R, $p_0 = 100$ psia, and $p_2 = 20$ psia. From Eqs. (18.20) and (18.21)

$$T^* = 0.833T_0 = 0.833(600)$$

$$= 500\ (\text{R})$$

$$p^* = 0.528p_0 = 52.8\ (\text{psia})$$

Since $p_2 < p^*$, the flow from the throat to the exit will be supersonic. From Eq. (18.12)

$$a^* = \overline{V}^* = (kg_0RT^*)^{1/2} = [1.4(32.2)(53.3)(500)]^{1/2}$$

$$= 1096\ (\text{ft/sec})$$

The density at the throat is

$$\rho^* = p^*/RT^* = [52.8(144)]/[53.3(500)]$$

$$= 0.2853\ (\text{lbm/ft}^3)$$

The area and diameter at the throat are

$$A^* = m/\rho^*\overline{V}^* = 2/[0.2853(1096)]$$

$$= 0.00640\ (\text{ft}^2) = 0.921\ (\text{in.}^2)$$

and

$$D^* = 1.083\ (\text{in.})$$

For the section where p = 95 psia

$$T = T_0(p/p_0)^{(k-1)/k} = 600(95/100)^{0.286}$$

$$= 591\ (\text{R})$$

so that

$$\rho = p/RT = [95(144)]/[53.3(591)]$$

$$= 0.4343\ (\text{lbm/ft}^3)$$

$$\overline{V}^2 = 2g_0 Jc_p(T_0 - T)$$

$$= 2(32.2)(778)(0.24)(600 - 591)$$

$$= 108{,}220\ (\text{ft}^2/\text{sec}^2)$$

$$\overline{V} = 329\ (\text{ft/sec})$$

The diameter is

$$D = [(4/\pi)(m/\rho\overline{V})]^{1/2}$$

$$= \{(4/\pi)[2/0.4343(329)]\}^{1/2}$$

$$= 0.1335\ (\text{ft}) = 1.602\ (\text{in.})$$

For other values of the pressure the values of the diameter are

p (psia):	95	90	80	70	60
T (R):	591	582	563	542	518
$\overline{V}$ (ft/sec):	329	465	667	835	993
M:	0.28	0.39	0.57	0.73	0.89
D (in.):	1.602	1.374	1.197	1.122	1.086

p (psia):	52.8	50	40	30	20
T (R):	500	492	462	425	379
$\overline{V}$ (ft/sec):	1097	1140	1288	1451	1630
M:	1.00	1.05	1.22	1.44	1.71
D (in.):	1.083	1.083	1.103	1.151	1.256

A plot of the pressure versus the diameter is shown in Fig. 18.11.

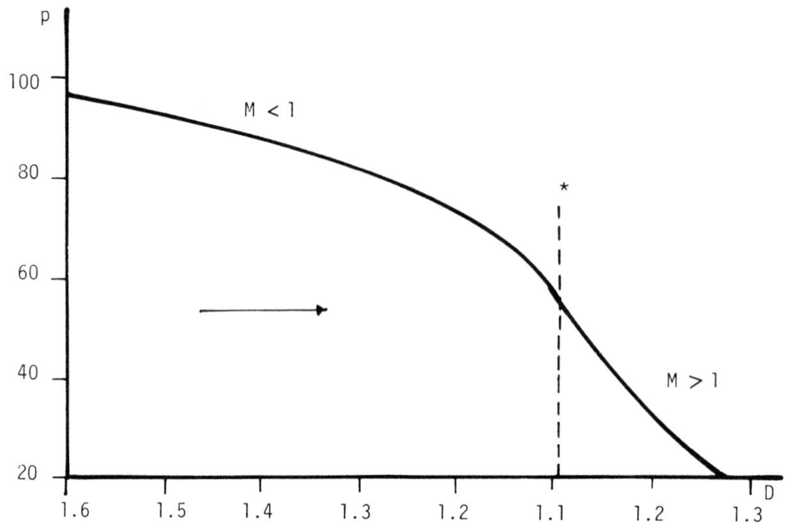

Figure 18.11

If atmospheric air containing water vapor were used in the nozzle, freezing would occur in the exit section since the leaving temperature is -81 F. This also suggests the nozzle could be used as a cooling device.

Example 18.13. Given the data in Example 18.12, determine the diameter in the diverging section when this section acts as a diffuser. For a pressure of 80 psia where $T_0 = 600$ R and $p_0 = 100$ psia the temperature is

$$T = T_0(p/p_0)^{(k-1)/k} = 600(80/100)^{0.286} = 563 \text{ (R)}$$

so that

$$\rho = p/RT = [80(144)]/[53.3(563)]$$

$$= 0.3839 \text{ (lbm/ft}^3\text{)}$$

and

$$\bar{V}^2 = 2g_0 Jc_p(T_0 - T)$$

$$= 2(32.2)(778)(0.24)(600 - 563)$$

$$= 44{,}500 \text{ (ft}^2\text{/sec}^2\text{)}$$

$$\bar{V}_2 = 667 \text{ (ft/sec)}$$

The acoustic velocity is

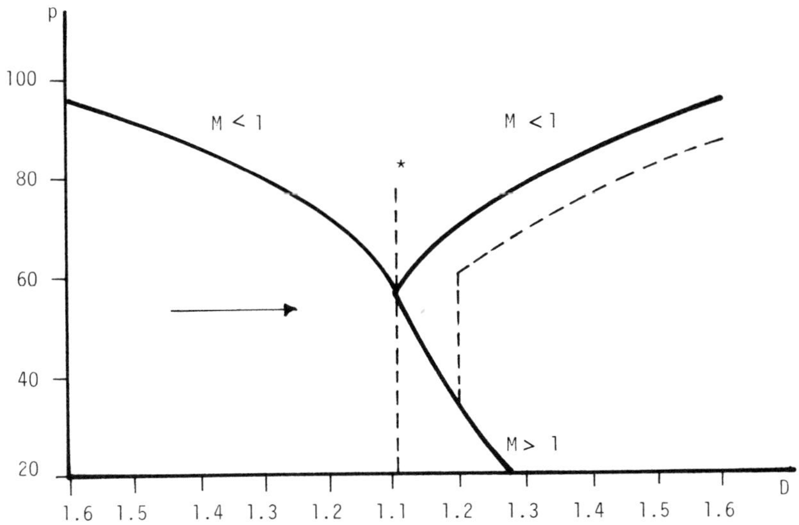

Figure 18.12

$$a = (kg_0RT)^{1/2} = [1.4(32.2)(53.3)(563)]^{1/2}$$

$$= 1163 \text{ (ft/sec)}$$

yielding a Mach number of

$$M = \overline{V}/a = 667/1163 = 0.57$$

The exit diameter is

$$D = [(4/\pi)(m/\rho\overline{V})]^{1/2}$$

$$= \{(4/\pi)[2/0.3839(667)]\}^{1/2}$$

$$= 0.0997 \text{ (ft)} = 1.197 \text{ (in.)}$$

Note that these diameter property values are identical with those of the converging section in Example 18.12 for the same pressure. This must be true since the process has been assumed to be reversible.

Figure 18.12 is a plot of pressure versus diameter for both supersonic and subsonic flow in the diverging section.

NORMAL SHOCK

A shock wave is an irreversible change of state that causes a sudden increase in the pressure. In Fig. 18.12 a normal shock is illustrated by means of dashed lines. The cause of shock may be a too abrupt change in the cross-sectional area of the nozzle or roughness on the nozzle surface. The pressure leaving the nozzle will be between the values found for the two reversible processes. Proper design and machining of the nozzle will prevent the occurrence of shock waves.

NOZZLE DESIGN FOR A SEMIPERFECT GAS

The properties of a converging-diverging nozzle may be determined by use of the gas tables. The results will be more accurate than those assuming a perfect gas, since the variation in the specific heats with temperature will be included in the calculations.

Example 18.14. Given a nozzle with $T_0 = 1000$ R and $p_0 = 100$ psia having a mass flow of 2.0 lbm/sec, determine the diameter at the throat and the exit for a discharge pressure of 20 psia for air as a perfect gas and as a semiperfect gas.

For a perfect gas, at the throat,

$$T^* = 0.833T_0 = 833 \text{ (R)}$$

$$p^* = 0.528p_0 = 52.8 \text{ (psia)}$$

Therefore,

$$\rho^* = p^*/RT^* = [52.8(144)]/[53.3(833)]$$
$$= 0.1712 \text{ (lbm/ft}^3\text{)}$$

and

$$\bar{V}^* = a^* = (kg_0RT^*)^{1/2}$$
$$= [1.4(32.2)(53.3)(833)]^{1/2}$$
$$= 1415 \text{ (ft/sec)}$$

The throat area is

$$A^* = m/\rho^*a^* = 2/[0.1712(1415)]$$
$$= 0.00826 \text{ (ft}^2\text{)} = 1.189 \text{ (in.}^2\text{)}$$

giving a diameter of

$$D^{*2} = 4A^*/\pi = [4(1.189)]/\pi$$
$$= 1.514 \text{ (in.}^2\text{)}$$
$$D^* = 1.230 \text{ (in.)}$$

For a perfect gas with an exit pressure of 20 psia

$$T = T_0(p/p_0)^{(k-1)/k} = 1000(20/100)^{0.286}$$

$$= 631 \text{ (R)}$$

yielding a density of

$$\rho = p/RT = [20(144)]/[53.3(631)]$$

$$= 0.0856 \text{ (lbm/ft}^3\text{)}$$

The exit velocity is

$$\overline{V}^2 = 2g_0 J c_p(T_0 - T) = 2(32.2)(778)(0.24)(1000 - 631)$$

$$= 4{,}437{,}000 \text{ (ft}^2\text{/sec}^2\text{)}$$

or

$$\overline{V} = 2106 \text{ (ft/sec)}$$

The exit area and diameter are

$$A = m/\overline{V}\rho = 2/[2106(0.0856)]$$

$$= 0.0111 \text{ (ft}^2\text{)} = 1.598 \text{ (in.}^2\text{)}$$

and

$$D = 1.426 \text{ (in.)}$$

For a semiperfect gas, at the throat

$$M^* = 1$$

By assuming T = 833 R, as for a perfect gas, from the gas tables by interpolation,

$$h = 199.86 \text{ (B/lbm)}$$

From Eq. (18.1)

$$\overline{V}^2 = 2g_0 J(h_0 - h) = 2(32.2)(778)(240.98 - 199.86)$$

where

$$h_0 = 240.98 \text{ (B/lbm)}$$

at

$$T_0 = 1000 \text{ (R)}$$

The velocity at the throat would be

$$V = 1435 \text{ (ft/sec)}$$

The value of k at T = 833 R may be approximated from the gas tables, where

$$k = \Delta h/\Delta u = (201.56 - 196.69)/(143.98 - 140.47)$$

$$= 1.39$$

Values of h and u are based on temperatures of 820 and 840 R. The acoustic velocity is

$$a = (kg_0RT)^{1/2}$$

$$= [1.39(32.2)(53.3)(833)]^{1/2}$$

$$= 1410 \text{ (ft/sec)}$$

giving a Mach number of

$$M = \overline{V}/a = 1435/1410 = 1.018$$

However, at the throat

$$M = 1.0$$

It will be necessary to assume a different value for T. For T = 840 R, $\overline{V}$ = 1405 ft/sec, and a = 1415 ft/sec, giving M = 0.993. Interpolating between T = 833 and 840 R for M = 1.0 gives T* = 838 R, so that

$$\overline{V}^* = 1414 \text{ (ft/sec)}$$

and

$$a^* = 1413 \text{ (ft/sec)}$$

for

$$M = 1.00$$

Again from the gas tables, $p_r^* = 6.519$ at 838 R, and for the isentropic process

$$p^*/p_0 = p_r^*/p_{r0}$$

and

$$p^* = 100(6.519/12.298)$$

$$= 53.0 \text{ (psia)}$$

where $p_{r0} = 12.298$ at $T_0 = 1000$ R. The density at the throat is

$$\rho^* = p^*/RT^* = [53.0(144)]/[53.3(838)]$$
$$= 0.1708\ (\text{lbm/ft}^3)$$

so that

$$A^* = \dot{m}/\rho^*\overline{V}^* = 2/[0.1708(1414)]$$
$$= 0.00828\ (\text{ft}^2) = 1.192\ (\text{in.}^2)$$

and

$$D^* = 1.232\ (\text{in.})$$

$D^* = 1.230$ in. for a perfect gas. At the exit,

$$p/p_0 = p_r/p_{r0}$$

so that

$$p_r = p_{r0}(p/p_0)$$
$$= 12.298(20/100) = 2.460$$

From the gas tables, T = 636 R and h = 152.11 B/lbm for $p_r = 2.460$. The density and velocity are

$$\rho = p/RT = [20(144)]/[53.3(636)] = 0.08450\ (\text{lbm/ft}^3)$$

and

$$\overline{V} = [2g_0J(h_0 - h)]^{1/2} = [2(32.2)(778)(240.98 - 152.11)]^{1/2}$$
$$= 2110\ (\text{ft/sec})$$

The area and diameter at the exit are

$$A = \dot{m}/\rho\overline{V} = 2/[0.08450(2110)] = 0.01122\ \text{ft}^2$$
$$= 1.615\ (\text{in.}^2)$$

and

$$D = 1.434\ (\text{in.})$$

compared to 1.426 in. for air as a perfect gas.

STEAM NOZZLES (SUPERHEATED VAPOR)

The techniques used in the solution of gas flow nozzle problems may also be applied to the flow of a vapor. If the vapor is in the superheat region, it may be possible to use the perfect gas relationships. However, the most accurate solution to these problems is obtained by use of either the steam tables or the Mollier diagram.

Example 18.15. Given a nozzle for which $t_0 = 1000$ F, $p_0 = 100$ psia, and m = 2 lbm/sec, determine the diameter at the throat and exit for an exit pressure of 20 psia using steam as a perfect gas and a vapor.

For steam as a perfect gas, R = 85.8 ft-lbf/lbm-R, k = 1.329, c_p = 0.445, and c_v = 0.335 B/lbm-R. From Eq. (18.18)

$$T^* = T_0[2/(k + 1)] = 1460[2/(1.329 + 1)] = 1254 \text{ (R)}$$

and from Eq. (18.19)

$$p^* = p_0[2/(k + 1)]^{k/(k-1)} = 100(0.859)^{4.040}$$

$$= 54.1 \text{ (psia)}$$

From the steam tables, the stagnation and throat conditions of the steam are in the superheat region.

The velocity at the throat is

$$\bar{V}^* = [2g_0Jc_p(T_0 - T^*)]^{1/2} = [2(32.2)(778)(0.445)(1460 - 1254)]^{1/2}$$

$$= 2143 \text{ (ft/sec)}$$

or by using the acoustic velocity,

$$a^* = (kg_0RT^*)^{1/2} = [1.329(32.2)(85.8)(1254)]^{1/2}$$

$$= 2146 \text{ (ft/sec)}$$

$\bar{V}$ and a are equal at the throat so that

$$M^* = 1.00$$

The density at the throat is

$$\rho^* = p^*/RT^* = [54.1(144)]/[85.8(1254)]$$

$$= 0.0724 \text{ (lbm/ft}^3\text{)}$$

giving an area and diameter of

$$A^* = m/\rho^*\bar{V}^* = 2/[0.0724(2143)]$$

$$= 0.0129 \text{ (ft}^2\text{)} = 1.856 \text{ (in.}^2\text{)}$$

and

$$D^* = [(4/\pi)A^*]^{1/2} = [(4/\pi)1.856]^{1/2}$$

$$= 1.537 \text{ (in.)}$$

The temperature at the exit is

$$T = T_0(p/p_0)^{(k-1)/k} = 1460(20/100)^{0.248}$$

$$= 980 \text{ (R)}$$

The steam at the exit is also in the superheat region.

The density at the exit is

$$\rho = p/RT = [20(144)]/[85.8(980)]$$

$$= 0.0342 \text{ (lbm/ft}^3\text{)}$$

and the velocity is

$$\overline{V} = [2g_0Jc_p(T_0 - T)]^{1/2} = [2(32.2)(778)(0.445)(1460 - 980)]^{1/2}$$

$$= 3271 \text{ (ft/sec)}$$

giving an area of

$$A = m/\rho\overline{V} = 2/[0.0342(3271)]$$

$$= 0.0178 \text{ (ft}^2\text{)} = 2.574 \text{ (in.}^2\text{)}$$

and

$$D = 1.810 \text{ (in.)}$$

For steam, as a vapor, by use of the steam tables, h_0 = 1530.8 B/lbm, s_0 = 1.9193 B/lbm-R, t_0 = 1000 F, and p_0 = 100 psia. Assume p = 55 psia; by interpolation from the superheat tables for $s = s_0$ = 1.9193 B/lbm-R, h = 1440.4 B/lbm, t = 812 F, and v = 14.108 ft^3/lbm. The velocity at the throat is

$$\overline{V} = [2g_0J(h_0 - h)]^{1/2} = [2(32.2)(778)(1530.8 - 1440.4)]^{1/2}$$

$$= 2128 \text{ (ft/sec)}$$

For the approximate value of k, let

$$k = \Delta h/\Delta u$$

along a line of constant pressure; from the superheat tables where p = 60 psia,

t (F)	h (B/lbm)	u (B/lbm)
800	1430.5	1292.2
900	1480.8	1331.4
	Δh = 50.3	Δu = 39.2

so that

$$k = 50.3/39.2 = 1.28$$

The acoustic velocity is

$$a = (kg_0RT)^{1/2} = [1.28(32.2)(85.8)(812 + 460)]^{1/2}$$

$$= 2121 \text{ (ft/sec)}$$

giving a Mach number of

$$M = \overline{V}/a = 2128/2121 = 1.003$$

The pressure value of the pressure at the throat, p = 55 psia, produces a Mach number of 1.003. This is sufficiently close to the required value of 1.00, so that further calculations are not necessary. If the value of the Mach number had been substantially different from unity, it would have been necessary to assume a different throat pressure and interpolate the properties for the desired sonic velocity at the throat.

The area of the throat is

$$A^* = mv^*/\overline{V}^* = [2(14.108)]/2128$$

$$= 0.01326 \text{ (ft}^2\text{)} = 1.909 \text{ (in.}^2\text{)}$$

and

$$D^* = 1.559 \text{ (in.)}$$

For the exit, after the isentropic expansion from the stagnation state, p = 20 psia and s = 1.9193 B/lbm-R. By interpolation in the steam tables, t = 558 F, h = 1314.3 B/lbm, and v = 30.202 ft^3/lbm. The velocity is

$$\overline{V} = [2g_0J(h_0 - h)]^{1/2} = [2(32.2)(778)(1530.8 - 1314.3)]^{1/2}$$

$$= 3294 \text{ (ft/sec)}$$

and the area is

$$A = mv/\overline{V} = [2(30.202)]/3294$$

$$= 0.01834 \text{ (ft}^2\text{)} = 2.641 \text{ (in.}^2\text{)}$$

and

$$D = 1.833 \text{ (in.)}$$

A comparison of the dimensions and properties for the perfect gas and tabular solution is presented in the following table:

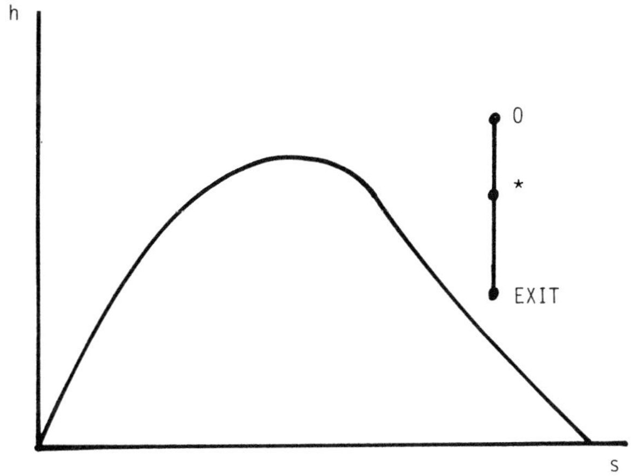

Figure 18.13

	Throat		Exit	
	Perfect Gas	Vapor	Perfect Gas	Vapor
p (psia)	54.1	55.0	20.0	20.0
t (F)	794.0	812.0	520.0	558.0
v (ft^3/lbm)	13.81	14.10	29.24	30.20
$\overline{V}$ (ft/sec)	2143.0	2128.0	3271.0	3294.0
a (ft/sec)	2146.0	2121.0	1897.0	1916.0
M	1.00	1.00	1.724	1.719
D (in.)	1.537	1.559	1.810	1.833
Error in D	-1.4%	–	-1.2%	–

Since the steam remains in the superheat region throughout the nozzle, the results of the two solutions are reasonably close to one another. For superheated steam, the ratio of

$$p^*/p_0 = 54.6/100 = 0.546$$

is a reasonable method of obtaining p*. The process is sketched on the hs diagram in Fig. 18.13.

STEAM NOZZLES (MIXED FLOW)

The following example will illustrate the solution for a nozzle whose stagnation properties are in the superheat region and whose throat and exit properties are in the wet region.

Example 18.16. Given a nozzle for which t_0 = 350 F, p_0 = 100 psia, and m = 2 lbm/sec, determine the diameter at the throat and exit for an exit pressure of 20 psia, using steam as a perfect gas and as a vapor. The solution for the perfect gas assumption is identical to Example 18.15, so that $t^* = 236$ F and $p^* = 54.1$ psia, which is in the wet region. The velocity and acoustic velocity are $\overline{V}^* = 1594$ ft/sec and $a^* = 1599$ ft/sec. The density at the throat is $\rho^* = 0.1305$ lbm/ft^3, giving an area and diameter of $A^* = 1.384$ in.^{2} and $D^* = 1.328$ in. For the exit, $t = 83$ F and $\rho = 0.0618$ lbm/ft^3, so that $\overline{V} = 2440$ ft/sec and $A = 1.910$ in.^{2}, resulting in $D = 1.559$ in. For steam as a vapor, assume $p = 60$ psia and $s = s_0 = 1.6187$ B/lbm-R, where s_0 is interpolated from the superheat tables for $p_0 = 100$ psia and $t_0 = 350$ F. For $p = 60$ psia, $s_g = 1.6440$ B/lbm-R, indicating that the steam is in the wet region. The property values at this condition are $t = 293$ F and $h = 1158.6$ B/lbm for $x = 97.9\%$. The velocity is

$$\overline{V} = [2g_0J(h_0 - h)]^{1/2} = [2(32.2)(778)(1199.9 - 1158.6)]^{1/2}$$

$$= 1438 \text{ (ft/sec)}$$

where $h_0 = 1199.9$ B/lbm at $p_0 = 100$ psia and $t_0 = 350$ F. To approximate the value of k at a constant pressure of 60 psia,

$x = 1.000$	$h = 1177.6$ (B/lbm)	$u = 1097.9$ (B/lbm)
$x = 0.950$	$h = 1131.8$ (B/lbm)	$u = 1055.9$ (B/lbm)
	$\Delta h = 45.8$ (B/lbm)	$\Delta u = 42.0$ (B/lbm)

$$k = \Delta h/\Delta u = 45.8/42.0 = 1.09$$

so that

$$a = (kg_0RT)^{1/2} = [1.09(32.2)(85.8)(293 + 460)]^{1/2}$$

$$= 1505 \text{ (ft/sec)}$$

and

$$M = \overline{V}/a = 1438/1505 = 0.955$$

Assume that $p = 50$ psia gives $h = 1144.5$ B/lbm, $\overline{V} = 1666$ ft/sec, $k = 1.09$, $t = 281$ F, and $a = 1493$ ft/sec, so that $M = 1.116$. By interpolation between $p = 50$ psia and 60 psia for $M = 1$, $p = 57.2$ psia. By interpolation for $p = 57.2$ psia and $s = 1.6187$ B/lbm-R, $h = 1154.6$ B/lbm, $\overline{V} = 1507$ ft/sec, and $a = 1502$ ft/sec for $t = 289$ F, which gives $M = 1.00$. Again, by interpolation, the specific volume for $p^* = 57.2$ psia and $s^* = 1.6187$ B/lbm-R is

$$v^* = 7.364 \text{ (ft}^3\text{/lbm)}$$

and

$$A^* = mv^*/\overline{V}^* = [2(7.364)]/1507$$

$$= 0.00977\ (\text{ft}^2) = 1.407\ (\text{in.}^2)$$

so that

$$D^* = 1.338\ (\text{in.})$$

At the exit, p = 20 psia and s = 1.6187 B/lbm-R, and from the tables x = 0.918, h = 1078.4 B/lbm, and v = 18.459 ft^3/lbm. The velocity at the exit is

$$\overline{V} = [2g_0J(h_0 - h)]^{1/2} = [2(32.2)(778)(1199.9 - 1078.4)]^{1/2}$$

$$= 2467\ (\text{ft/sec})$$

The area at the exit is

$$A = mv/\overline{V} = [2(18.459)]/2467$$

$$= 0.01496\ (\text{ft}^2) = 2.155\ (\text{in.}^2)$$

and

$$D = 1.656\ (\text{in.})$$

A comparison of the dimensions and properties for the perfect gas and vapor solution is presented in the following table:

	Throat		Exit	
	Perfect Gas	Vapor	Perfect Gas	Vapor
p (psia)	54.1	57.2	20.0	20.0
t (F)	236.0	289.0	83.0	228.0
v (ft^3/lbm)	7.67	7.36	16.18	18.46
$\overline{V}$ (ft/sec)	1594.0	1507.0	2440.0	2467.0
D (in.)	1.328	1.338	1.559	1.656
Error in D	-0.7%	–	-5.8%	–

The error, with reference to the diameter, has increased for the mixed flow as compared to the superheat flow. More important, physical properties such as specific volume and temperature show considerable deviation from the values obtained in the perfect gas solution.

In general, tabular values must be used for flow calculations involving vapors. For steam, the ratios of

$$p^*/p_0 = 0.546$$

for superheated steam and

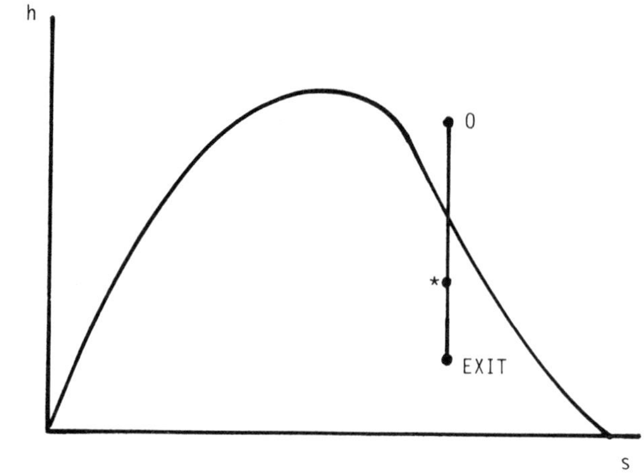

Figure 18.14

$$p^*/p_0 = 0.572$$

for wet steam are good approximations for the critical pressure.

The process for this example is sketched in Fig. 18.14.

SUBSONIC FLOW

Consider the converging-diverging section sketched in Fig. 18.15. The pressure at the exit is greater than the critical pressure and has not been

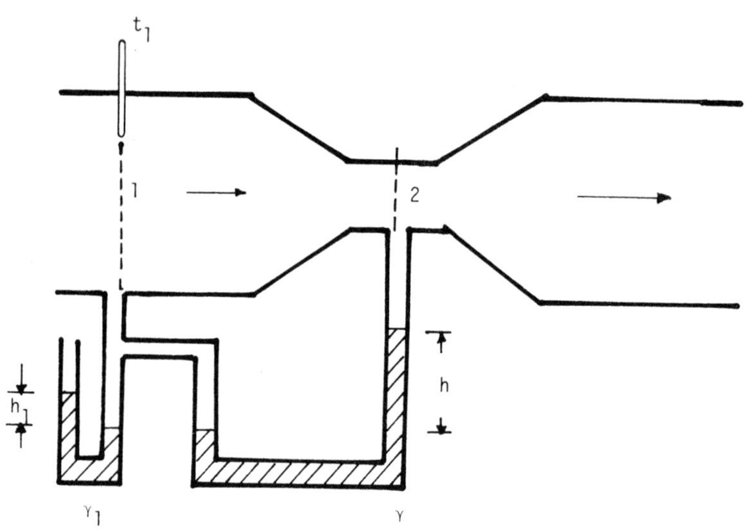

Figure 18.15

previously reduced to a level sufficient to establish critical conditions at the throat.

In this situation the flow will remain subsonic, $M < 1$, in both sections of the system. The converging section will act as a nozzle, and the diverging section will act as a diffuser.

Example 18.17. Given $T_0 = 600$ R, $p_0 = 100$ psia, and that the working substance is air acting as a perfect gas, determine the mass flow for the nozzle in Example 18.12 for a throat pressure of 80 psia (Fig. 18.16).

From Example 18.12 the throat diameter is 1.083 in. The temperature at the throat, for the isentropic expansion, is

$$T = T_0(p/p_0)^{(k-1)/k} = 600(80/100)^{0.286}$$

$$= 563 \text{ (R)}$$

and the density is

$$\rho = p/RT = [80(144)]/[53.3(563)]$$

$$= 0.3839 \text{ (lbm/ft}^3\text{)}$$

The velocity is

$$\overline{V} = [2g_0Jc_p(T_0 - T)]^{1/2} = [2(32.2)(778)(0.24)(600 - 563)]^{1/2}$$

$$= 667 \text{ (ft/sec)}$$

yielding a mass flow of

$$m = \rho A\overline{V} = 0.3839(\pi/4)(1.083/12)^2(667)$$

$$= 1.638 \text{ (lbm/sec)}$$

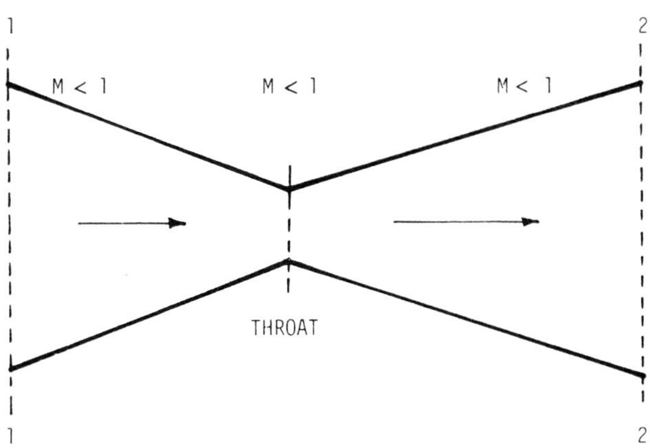

Figure 18.16

If the exit pressure were reduced, the mass flow would increase until it reached 2.0 lbm/sec, resulting in critical conditions at the nozzle throat. Further reduction in the exit pressure for the given stagnation properties would not result in an increase in the mass flow.

The section diameter for p = 90 may be obtained from

$$T = T_0(p/p_0)^{(k-1)/k} = 600(90/100)^{0.286}$$
$$= 582 \text{ (R)}$$

and

$$\rho = p/RT = [90(144)]/[53.3(582)]$$
$$= 0.4178 \text{ (lbm/ft}^3\text{)}$$

so that

$$\overline{V} = [2g_0Jc_p(T_0 - T)]^{1/2} = [2(32.2)(778)(0.24)(600 - 582)]^{1/2}$$
$$= 465 \text{ (ft/sec)}$$

and the area is

$$A = m/\rho\overline{V} = 1.638/[0.4178(465)]$$
$$= 0.00843 \text{ (ft}^2\text{)} = 1.214 \text{ (in.}^2\text{)}$$

or

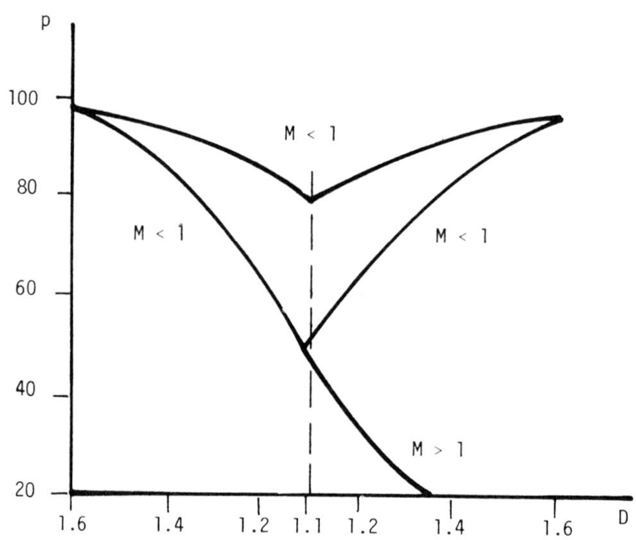

Figure 18.17

$D = 1.243$ (in.)

For other values of the pressure the values of the diameters are

p (psia):	97.5	95	92.5	90
T (R):	596	591	587	582
V (ft/sec):	219	329	395	465
M:	0.18	0.28	0.33	0.39
D (in.):	1.76	1.450	1.336	1.243

p (psia):	87.5	85	82.5	80
T (R):	578	573	568	563
V (ft/sec):	514	570	620	667
M:	0.44	0.49	0.53	0.57
D (in.):	1.195	1.147	1.111	1.083

Since the flow through the nozzle is reversible, the preceding property values apply to either the converging or diverging sections.

In Fig. 18.17, values of pressure and diameter are plotted for this example. The data from Example 18.12 are also included in this figure.

INSTRUMENTATION, INCOMPRESSIBLE FLOW

One important application of the subsonic converging-diverging section is in the measurement of the flow of compressible or incompressible fluids. We shall assume that the process is reversible, which in the case of fluid flow precludes the existence of either fluid friction or turbulence.

Figure 18.15 illustrates the system usually encountered when measuring the flow of either compressible or incompressible liquids.

For the incompressible fluid, Bernoulli's equation may be used, where

$$Z_1 + \bar{V}_1^2/2g + p_1/\gamma_1 = Z_2 + \bar{V}_2^2/2g + p_2/\gamma_2$$

In most practical applications the change in potential energy is negligible so that Bernoulli's equation reduces to

$$(\bar{V}_1^2/2g) + (p_1/\gamma_1) = (\bar{V}_2^2/2g) + (p_2/\gamma_2) \tag{18.22}$$

where the units of each term in the equation are feet. Since the flow is incompressible,

$$\gamma_1 = \gamma_2$$

To eliminate the inlet velocity [Eq. (2.5)] the equation of continuity may be applied, so that

$$m_1 = \rho_1 A_1 \bar{V}_1 = \rho_2 A_2 \bar{V}_2$$

$$\bar{V}_1 = \rho_2 A_2 \bar{V}_2 / \rho_1 A_1 = (A_2/A_1)\bar{V}_2 \tag{18.23}$$

for

$$\rho_1 = \rho_2$$

To determine the pressure differential between the inlet and the throat, the manometer arrangement shown in Fig. 18.15 may be used, where

$$p_1 - p_2 = h(\gamma - \gamma_1)$$

where h = the difference in the level of the manometer fluid (ft), γ = the specific weight of the manometer fluid (lbf/ft^3), and γ_1 = the specific weight of the fluid flowing in the system (lbf/ft^3).

Solving for $\bar{V}_2$ in Eq. (18.22) gives

$$\bar{V}_2^2 = 2g[(p_1 - p_2)/\gamma_1] + \bar{V}_1^2$$

By use of Eq. (18.23)

$$\bar{V}_2^2[1 - (A_2/A_1)^2] = 2g[(p_1 - p_2)/\gamma_1]$$

$$= 2gh(\gamma - \gamma_1)/\gamma_1$$

Defining F, the velocity of approach factor, such that

$$F = [1 - (A_2/A_1)^2]^{1/2} \tag{18.24}$$

which is identical to Eq. (18.17), where $\rho_1 = \rho_2$ for incompressible flow, then

$$\bar{V}_2 = (1/F)[2gh(\gamma - \gamma_1)/\gamma_1]^{1/2} \tag{18.25}$$

The mass flow for the system in Fig. 18.15 is

$$m = \rho_2 A_2 \bar{V}_2$$

Example 18.18. Given are the following data for the system in Fig. 18.15: h_1 = 4 in., h = 6 in., t_1 = 60 F, D_1 = 4 in., fluid in system S.G. = 0.8, and fluid in manometers is mercury. Determine the mass flow for a barometric pressure of 14.50 psia. Throat diameter, D_2 = 1 in.

From Eq. (18.24)

$$F = [1 - (A_2/A_1)^2]^{1/2} = [1 - (D_2/D_1)^4]^{1/2}$$

$$= [1 - (1/4)^4]^{1/2} = 0.998$$

and

$$(\gamma - \gamma_1)/\gamma_1 = [13.6(62.4) - 0.8(62.4)]/[0.8(62.4)]$$

$$= (13.6 - 0.8)/0.8 = 16.0$$

From Eq. (18.25)

$$\bar{V}_2 = (1/F)[2gh(\gamma - \gamma_1)/\gamma_1]^{1/2}$$

$$= (1/0.998)[2(32.2)(6/12)(16)]^{1/2}$$

$$= 22.74 \text{ (ft/sec)}$$

and

$$m = \rho_2 A_2 \bar{V}_2 = 0.8(62.4)(\pi/4)(1/12)^2(22.74)$$

$$= 6.19 \text{ (lbm/sec)}$$

The manometer used to determine the pressure at the inlet was not needed to determine the mass flow. Also, the barometric pressure and temperature of the flowing fluid were not used. Generally, temperature readings are required since the specific weight of the flowing substance is dependent on the temperature.

COMPRESSIBLE FLOW

We shall use the system in Fig. 18.15 to determine the mass flow of a compressible fluid. Equations (18.16) and (18.17) may be used for the converging section or the nozzle, where

$$\bar{V}_2 = [2g_0 J c_p (T_1 - T_2)]^{1/2}/F$$

and

$$F = \{1 - [(\rho_2/\rho_1)(A_2/A_1)]^2\}^{1/2}$$

To eliminate the temperature at the throat

$$(T_1 - T_2) = T_1[1 - (T_2/T_1)] = T_1[1 - (p_2/p_1)^{(k-1)/k}] \qquad (18.26)$$

for an isentropic process.

Example 18.19. Repeat Example 18.18 where the flowing fluid is air and the fluid in the manometers is water. Since there is air in both legs of the manometer used to measure either the inlet or differential pressure,

$$p_1 - p_2 = \gamma h \quad \text{where } \gamma >> \gamma 1$$

For the inlet pressure (Fig. 18.14)

$$p_1 - p_{bar} = \gamma h_1$$

$$p_1 - 14.50 = [62.4(4/12)]/144$$

$$p_1 = 14.50 + 0.144 = 14.644 \text{ (psia)}$$

For the differential manometer, between the inlet and the throat,

$$p_1 - p_2 = \gamma h$$

$$p_2 = p_1 - \gamma h = 14.644 - [62.4(6/12)]/144$$

$$= 14.644 - 0.217 = 14.427 \text{ (psia)}$$

Solving for $T_1 - T_2$ in Eq. (18.26) gives

$$T_1 - T_2 = T_1[1 - (p_2/p_1)^{(k-1)/k}]$$

$$= 520[1 - (14.427/14.644)^{0.286}]$$

$$= 2.216 \text{ (R)}$$

The density ratio, for the isentropic process, is

$$\rho_2/\rho_1 = v_1/v_2 = (p_2/p_1)^{1/k} = (14.43/14.64)^{0.714} = 0.989$$

The velocity of approach factor is

$$F = [1 - (\rho_2 A_2/\rho_1 A_1)^2]^{1/2} = [1 - (\rho_2/\rho_1)^2(D_2/D_1)^4]^{1/2}$$

$$= [1 - (0.989)^2(1/4)^4]^{1/2} = 0.998$$

The velocity at the throat is

$$\overline{V}_2 = [2g_0 J c_p(T_1 - T_2)]^{1/2}/F = (1/0.998)[2(32.2)(778)(0.24)(2.216)]^{1/2}$$

$$= 163.6 \text{ (ft/sec)}$$

The densities at the inlet and throat are

$$\rho_1 = p_1/RT_1 = [14.64(144)]/[53.3(520)] = 0.0761 \text{ (lbm/ft}^3\text{)}$$

and

$$\rho_2 = 0.989\rho_1 = 0.989(0.0761) = 0.0753 \text{ (lbm/ft}^3\text{)}$$

The mass flow is

$$m = \rho_2 A_2 \bar{V}_2 = 0.0753(\pi/4)(1/12)^2(163.6)$$

$$= 0.0671 \text{ (lbm/sec)}$$

In the above solution the velocity of approach factor was 0.998. Depending on the value of the diameters at the inlet and throat, the correction for the inlet velocity may often be omitted. In this problem

$$m = 0.0670 \text{ (lbm/sec)}$$

for

$$F = 1.0$$

APPROXIMATE VELOCITY SOLUTION BY EXPANSION EQUATIONS

The ratio

$$(p_2/p_1)^{(k-1)/k}$$

encountered in Eq. (18.26) is difficult to evaluate accurately because of the small difference between the values of p_1 and p_2 in the ratio p_2/p_1. Rewriting the ratio as

$$\left(\frac{p_2}{p_1}\right)^{(k-1)/k} = \left(\frac{p_2}{p_1} + 1 - 1\right)^{(k-1)/k} = \left(1 + \frac{p_2 - p_1}{p_1}\right)^{(k-1)/k}$$

and by applying the binomial theorem, where $n = (k - 1)/k$ and $x = (p_2 - p_1)/p_1$,

$$(1 + x)^n = 1 + nx + \frac{n(n - 1)}{2} x^2 + \frac{n(n - 1)(n - 2)}{3} x^3$$

and limiting the expansion to second-order terms, one obtains

$$\left(\frac{p_2}{p_1}\right)^{(k-1)/k} = 1 + \frac{k - 1}{k}\left(\frac{p_2 - p_1}{p_1}\right) - \left(\frac{k - 1}{2k}\right)\left(\frac{1}{k}\right)\left(\frac{p_2 - p_1}{p_1}\right)^2$$

The following table lists the results of using the expansion equation for various values of the pressure ratio when $k = 1.4$. The last two columns of the table represent the approximate value of $(p_2/p_1)^{(k-1)/k}$ using the first- and second-order terms in $(p_2 - p_1)/p_1$, respectively.

p_2/p_1	$(p_2/p_1)^{(k-1)/k}$	x	1 + nx	$1 + nx + [n(n-1)/2]x^2$
1.0	1.0	0.0	1.0	1.0
0.95	0.9855	-0.05	0.9857	0.9854
0.90	0.9703	-0.10	0.9714	0.9704
0.85	0.9546	-0.15	0.9571	0.9548
0.80	0.9382	-0.20	0.9429	0.9388
0.75	0.9211	-0.25	0.9286	0.9222
0.70	0.9031	-0.30	0.9143	0.9051

The error in calculating the velocity by use of the approximate equation is

$$\text{Error} = 1 - \frac{[1 - (p_2/p_1)^{(k-1)/k}_{\text{approx}}]^{1/2}}{[1 - (p_2/p_1)^{(k-1)/k}_{\text{exact}}]^{1/2}}$$

The error is tabulated for various pressure ratios in the following table:

p_2/p_1	Error, 1st Order (%)	Error, 2nd Order (%)
1.0	0	0
0.95	0.70	-0.34
0.90	1.87	0.17
0.85	2.79	0.22
0.80	3.88	0.49
0.75	4.87	0.70
0.70	5.96	1.03

By using the second-order term, the velocity equation

$$\bar{V}_2 = (1/F)\{2g_0 Jc_p T_1[1 - (p_2/p_1)^{(k-1)/k}]\}^{1/2}$$

may be written as

$$\bar{V}_2 = \frac{1}{F}\left\{2g_0 Jc_p T_1\left(\frac{k-1}{k}\right)\left[\left(\frac{p_1 - p_2}{p_1}\right) + \left(\frac{1}{2k}\right)\left(\frac{p_1 - p_2}{p_1}\right)\right]^2\right\}^{1/2} \tag{18.27}$$

which has an error of 0.7% or less for $p_2/p_1 > 0.75$.

By omitting the second-order term

$$\bar{V}_2 = \frac{1}{F}\left[2g_0 Jc_p (T_1)\left(\frac{k-1}{k}\right)\left(\frac{p_1 - p_2}{p_1}\right)\right]^{1/2} \tag{18.28}$$

which has an error of 0.7% or less for $p_2/p_1 > 0.95$.

Since, for the perfect gas, from Eq. (5.8),

$$c_p = kR/[J(k - 1)]$$

Eqs. (18.27 and (18.28) may be rewritten as

$$\bar{V}_2 = \frac{1}{F}\left\{ 2g_0RT_1\left[\left(\frac{p_1 - p_2}{p_1}\right) + \left(\frac{1}{2k}\right)\left(\frac{p_1 - p_2}{p_1}\right)^2 \right]\right\}^{1/2} \tag{18.29}$$

and

$$\bar{V}_2 = (1/F)[2g_0RT_1(p_1 - p_2)/p_1]^{1/2} \tag{18.30}$$

Example 18.20. Determine the errors in calculating the mass flow by use of the approximate equations for Example 18.19.

From Example 18.19, m = 0.0671 lbm/sec, F = 0.998, and T_1 = 520 R. Also, p_1 = 14.64 psia, h = 6 in.-H_2O and ρ_2 = 0.0753 lbm/ft^3.

Given the above values,

$$\frac{p_1 - p_2}{p_1} = \frac{\gamma h}{p_1} = \frac{62.4(6/12)}{14.64(144)} = 0.01480$$

and

$$\left(\frac{1}{2k}\right)\left(\frac{p_1 - p_2}{p_1}\right)^2 = 0.00008$$

From Eq. (18.29)

$$\bar{V}_2 = (1/0.998)[2(32.2)(53.3)(520)(0.01480 + 0.00008)]^{1/2}$$

$$= 163.3 \text{ (ft/sec)}$$

and

$$m = \rho_2A_2\bar{V}_2 = 0.0753(\pi/4)(1/12)^2(163.3)$$

$$= 0.0671 \text{ (lbm/sec)}$$

which is the same as the exact value. From Eq. (18.30)

$$\bar{V}_2 = (1/0.998)[2(32.2)(53.3)(520)(0.01480)]^{1/2}$$

$$= 162.9 \text{ (ft/sec)}$$

and

$$m = 0.0669 \text{ (lbm/sec)}$$

which is 0.3% below the exact value. Since

$$p_2/p_1 = 14.43/14.64 = 0.98$$

Eq. (18.30) is both accurate and easier to apply than the exact technique. Omitting the velocity of approach factor gives

$$m = 0.0668 \text{ (lbm/sec)}$$

for a -0.4% error in the mass flow.

DISCHARGE COEFFICIENT

In the preceding discussions we have assumed that the flow through the system is reversible and adiabatic; that is, the process is isentropic. In actual systems used for the measurement of fluid flow, friction and turbulence reduce the mass of material passing through the system. The discharge coefficient is defined by the equation

$$c_d = \text{(actual mass flow)}/\text{(theoretical mass flow)} \tag{18.31}$$

Representative values for the coefficient of discharge for some common configurations of flow measuring instruments (Fig. 18.18) are as follows: venturi meter, 0.97-0.99; flow nozzle, 0.96-0.99; and orifice, 0.57-0.63.

The above values serve only to indicate the usual range of values that occur in these instruments. For accurate measurements, the instruments must be carefully installed, which includes precise location and fabrication of the instrument itself. The ASME Performance Test Codes describe in detail the installation requirements for the various instruments.

In the case of liquid flow, where the material can be safely handled, calibration of the instrument to obtain the discharge coefficient is possible by either mass or volume measurements. For gases or vapors, especially for high rates of flow, careful following of the test code is mandatory.

The converging-diverging section, sketched in Fig. 18.18, is commonly called the venturi meter. The advantages of the venturi meter include greater accuracy in fluid measurement as compared to the other two configurations and higher pressure recovery between the inlet and outlet.

Loss of the pressure head, due to pressure drops in the instrument, must be recovered by installation of a larger pump or blower. In the venturi the recovery is high because the turbulence in the exit part of the instrument is carefully controlled.

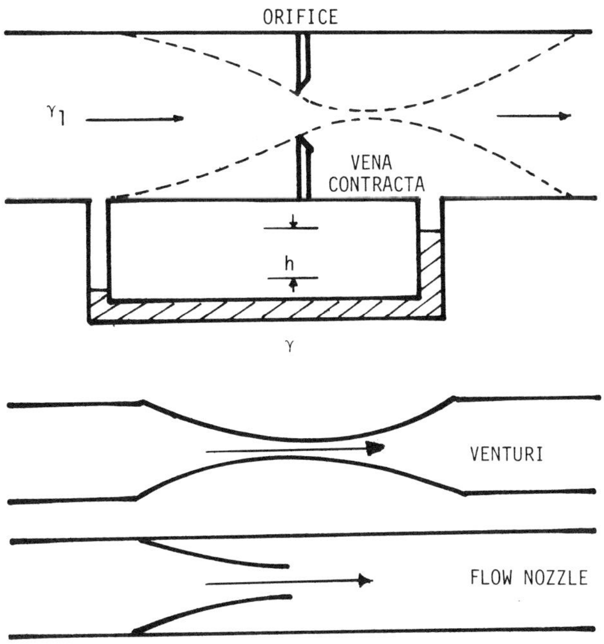

Figure 18.18

The disadvantage of the venturi lies in the cost of accurately machining the converging-diverging section. Another possible problem is the space required for its installation since it is the largest of the three systems illustrated in Fig. 18.18.

The flow nozzle directs the flow from the inlet to the throat, insuring that the pressure measurements are accurate. Pressure recovery is less, due to turbulence, and the other advantages and disadvantages are intermediate to those of the venturi and orifice.

The orifice is the least expensive of the systems and requires the least space in the flow system. The narrowest section of flow, the vena contracta, occurs downstream from the orifice opening. It is also a variable and tends to change location with varying flow rates. This means the location of the downstream pressure tap is arbitrary and is not a measurement of the pressure at the narrowest section of flow. The orifice is the most difficult of the three instruments to use accurately, and extreme care must be exercised in its use.

Example 18.21. An orifice (Fig. 18.18) is used to measure the flow of water. The differential height for the manometer is 4 in. and the temperature at the orifice is 100 F. For calibration purposes the water is collected in a tank having an internal diameter of 3 ft. The water temperature in the tank averages 80 F. During a 15-min test the water rises 16 in. in the tank. Determine the discharge coefficient of the orifice. D pipe = 3 in., D orifice = 1 in., and S.G. manometer fluid = 1.4. From Eq. (18.24)

$$F = [1 - (D_2/D_1)^4]^{1/2} = [1 - (1/3)^4]^{1/2} = 0.994$$

From the steam tables for water at 100 F

$$v_f = 0.01613 \text{ (ft}^3\text{/lbm)}$$

and

$$\rho_f = 1/v_f = 1/0.01613 = 62.0 \text{ (lbm/ft}^3\text{)}$$

For g = 32.2 ft/sec

$$\gamma_1 = 62.0 \text{ (lbf/ft}^3\text{)}$$

Solving for the specific weight ratio gives

$$(\gamma - \gamma_1)/\gamma_1 = [1.4(62.4) - 62.0]/62.0$$

$$= 0.409$$

From Eq. (18.25)

$$\overline{V}_2 = (1/F)[2gh(\gamma - \gamma_1)/\gamma_1]^{1/2}$$

$$= (1/0.994)[2(32.2)(4/12)(0.409)]^{1/2}$$

$$= 2.98 \text{ (ft/sec)}$$

and

$$m = \rho_2 A_2 \overline{V}_2 = 62.0(\pi/4)(1/12)^2(2.98)$$

$$= 1.008 \text{ (lbm/sec)}$$

$$= 3630 \text{ (lbm/hr)}$$

For the water in the collection tank

$$\rho_f = 1/v_f = 1/0.01608 = 62.2 \text{ (lbm/ft}^3\text{)}$$

at 80 F. The volume of water collected is

$$V = (\pi/4)D^2h = (\pi/4)(3)^2(16/12)$$

$$= 9.42\ (\text{ft}^3/15\ \text{min}) = 37.70\ (\text{ft}^3/\text{hr})$$

The mass of water actually flowing through the orifice is

$$m = \rho V = 62.2(37.70) = 2340\ (\text{lbm/hr})$$

From Eq. (18.31)

$$c_d = (\text{actual mass flow})/\text{theoretical mass flow}) = 2340/3620 = 0.64$$

The accuracy of the test would be improved by placing the collecting tank on scales so that the mass of water could be calculated directly.

TWO-DIMENSIONAL FLOW

Throughout our previous discussions, not only have we assumed reversible flow but we have also assumed one-dimensional flow (Fig. 18.19). One-dimensional flow assumes that the velocity of the fluid has a constant value across the cross-sectional area through which the flow is occurring. For low flow rates or for viscous materials, such as heavy oils, molasses, and others, the velocity varies with the distance from the tube walls.

Theoretically, the shape of the velocity curve is that of a parabola, and the flow is said to be laminar and two-dimensional. The other type of flow is turbulent flow, which more nearly approaches the one-dimensional assumption and occurs at higher flow rates.

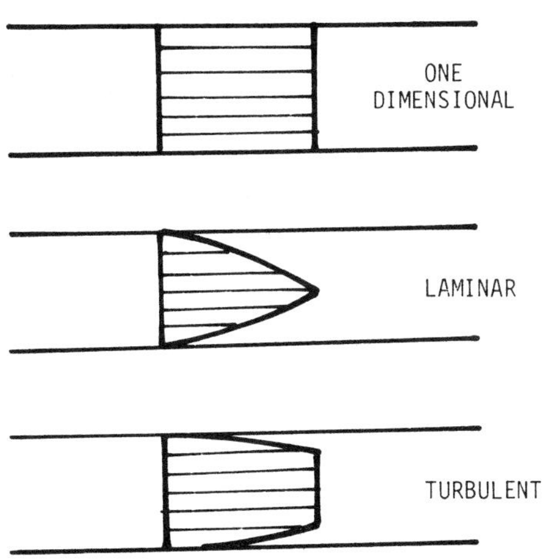

Figure 18.19

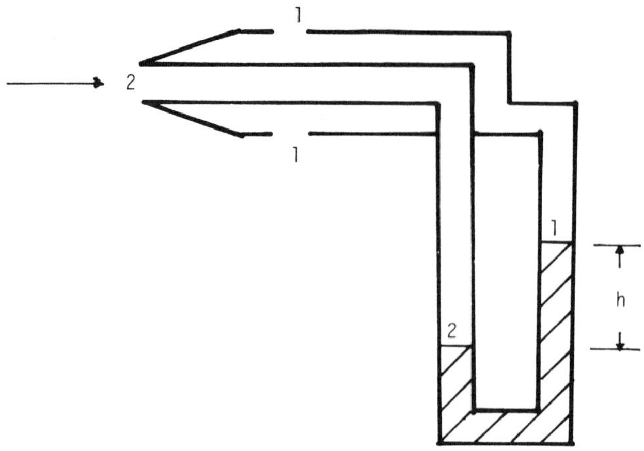

Figure 18.20

To measure two-dimensional flow the venturi, flow nozzle, or orifice may be used, especially for liquid flow where direct calibration is possible. It is also possible to determine the velocity at a particular location or point in the system. If the velocity at a sufficient number of points is determined by means of a traverse, the flow rate can be established. The determination of the velocity at a particular point in a system is important in many problems, such as flow over bodies, flow in irregular sections, and air speed measurements.

For this purpose, the pitot tube is often used (Fig. 18.20).

PITOT TUBE

The pitot tube consists of an inner tube that, in use, is placed parallel to the flow direction. If the flow direction is not known, the tube may be rotated until the maximum height differential is indicated on the manometer; it will then be parallel to, and facing into, the flow stream.

Since the fluid is brought to rest at the tube opening, its state corresponds to the stagnation state discussed earlier in this chapter. The pressure at this point is often called the impact or total pressure.

The outer tube of the pitot instrument contains a number of openings by means of which the static pressure may be measured. The difference between these two readings is the velocity pressure. When these pressure readings are expressed in terms of feet of the fluid flowing, the preceding three terms are described as the total head, static head, and velocity head.

From Bernoulli's equation, for $\Delta PE = 0$,

$$(p_1/\gamma_1) + (\overline{V}_1{}^2/2g) = (p_2/\gamma_2) + (\overline{V}_2{}^2/2g)$$

Let state 1 designate the static condition and state 2 the stagnation condition, so that

$$\overline{V}_2 = 0$$

and

$$\overline{V}_1 = \{2g[(p_2/\gamma_2) - (p_1/\gamma_1)]\}^{1/2} \qquad (18.32)$$

For moderate velocities, the difference between static and total pressure is small, so that

$$\gamma_1 = \gamma_2$$

which assumes that the flow is incompressible; therefore,

$$\begin{aligned}\overline{V}_1 &= [2g(p_2 - p_1)/\gamma_1]^{1/2} \\ &= (2gH)^{1/2}\end{aligned} \qquad (18.33)$$

where H = the velocity head for the flowing fluid.

Example 18.22. The pitot tube in Fig. 18.20 is used to measure the speed of an aircraft. The outside air temperature is 40 F, and the differential pressure is 3 in.-H_2O. What is the air speed of the aircraft? The barometric pressure in and outside of the aircraft is 13.5 psia.

The density of the air through which the plane is flying is

$$\begin{aligned}\rho_1 &= p_1/RT_1 = [13.5(144)]/[53.3(500)] \\ &= 0.0729 \text{ (lbm/ft}^3)\end{aligned}$$

For g = 32.2 ft/sec

$$\gamma_1 = 0.0729 \text{ (lbf/ft}^3)$$

For the manometer

$$p_2 - p_1 = \gamma h = 62.4(3/12) = 15.6 \text{ (lbf/ft}^2)$$

and

$$H = (p_2 - p_1)/\gamma_1 = 15.6/0.0729 = 214 \text{ (ft of air)}$$

From Eq. (18.33)

$$\overline{V}_1 = (2gH)^{1/2} = [2(32.2)(214)]^{1/2} = 117.4 \text{ (ft/sec)}$$

$$= 80.0 \text{ (mph)}$$

Example 18.23. Determine the air speed in Example 18.22 assuming the air at the pitot tube undergoes an isentropic compression process to the stagnation condition. Solving for the total pressure gives

$$p_1 = 13.5 \text{ (psia)} = 1944 \text{ (psfa)}$$

$$p_2 = 1944 + 15.6 = 1959.6 \text{ (psfa)}$$

$$v_1/v_2 = \rho_2/\rho_1 = (p_2/p_1)^{1/k} = (1959.6/1944)^{0.714} = 1.006$$

$$\rho_2 = 1.006(0.0729) = 0.0733 \text{ (lbm/ft}^3\text{)}$$

so that

$$\gamma_2 = 0.0733 \text{ (lbf/ft}^3\text{)}$$

By use of Eq. (18.32)

$$\overline{V}_1 = \{2g[(p_2/\gamma_2) - (p_1/\gamma_1)]\}^{1/2} = \{2(32.2)[(1959.6/0.0733) - (1944/0.0729)]\}^{1/2}$$

$$= 65.8 \text{ (ft/sec)} = 44.9 \text{ (mph)}$$

In Example 18.22, $\overline{V}_1$ = 80 mph, so that the above answer for $\overline{V}_1$ is incorrect. Bernoulli's equation cannot be applied since $\Delta u \neq 0$ for the isentropic compression. We have attempted to use the incompressible flow equation for a compressible process. Including the internal energy in the steady flow equation gives

$$(p_1/\gamma_1) + (V_1^2/2g) + u_1 = (p_2/\gamma_2) + u_2$$

where $\overline{V}_2 = 0$. Therefore,

$$\overline{V}_1^2/2g = (p_2/\gamma_2) - (p_1/\gamma_1) + u_2 - u_1$$

and

$$T_2 = T_1(p_2/p_1)^{(k-1)/k} = 500(1959.6/1944)^{0.286}$$

$$= 501.1 \text{ (R)}$$

Therefore

$$\Delta u = c_v\ \Delta T = 0.1714(501.1 - 500) = 0.1885\ (B/lbm)$$

$$= 146\ (ft\text{-}lbf/lbm)$$

which for $g = 32.2$ ft/sec

$$\Delta u = 146\ (ft)$$

Then

$$\bar{V}_1^{\,2}/2g = (1959.6/0.0733) - (1944/0.0729) + 146$$

$$= 67 + 146 = 214\ (ft)$$

and

$$\bar{V}_1 = 117.2\ (ft/sec) = 79.9\ (mph)$$

This is the same value as obtained assuming incompressible flow in Example 18.22.

APPROXIMATE VELOCITY EQUATIONS

For the pitot tube the steady-flow energy equation is

$$h_1 + \bar{V}_1^{\,2}/2g_0J = h_2$$

so that

$$\bar{V}_1^{\,2} = 2g_0J(h_2 - h_1)$$

and

$$\bar{V}_1^{\,2} = 2g_0Jc_p(T_2 - T_1)$$

For a perfect gas

$$\bar{V}_1^{\,2} = 2g_0Jc_pT_1[(T_2/T_1) - 1]$$

For an isentropic process

$$\bar{V}_1^{\,2} = 2g_0Jc_pT_1[(p_2/p_1)^{(k-1)/k} - 1] = -2g_0Jc_pT_1[1 - (p_2/p_1)^{(k-1)/k}]$$

This is similar to the equation used to derive Eqs. (18.29) and (18.30), where $F \doteq 1$ since $\bar{V}_2 = 0$. For the pitot tube

$$\bar{V}_1 = \left\{2g_0RT_1\left[\left(\frac{p_2 - p_1}{p_1}\right) - \left(\frac{1}{2k}\right)\left(\frac{p_2 - p_1}{p_1}\right)^2\right]\right\}^{1/2} \tag{18.34}$$

and

$$\bar{V}_1 = [2g_0RT_1(p_2 - p_1)/p_1]^{1/2} \tag{18.35}$$

Example 18.24. Repeat Example 18.23 using the approximate velocity equations. From Example 18.23, p_1 = 1944 psfa, p_2 = 1959.6 psfa, and T_1 = 500 R. For the given pressures

$$(p_2 - p_1)/p_1 = 0.008025$$

and

$$(1/2k)[(p_2 - p_1)/p_1]^2 = 0.000023$$

From Eq. (18.34)

$$\bar{V}_1 = [2(32.2)(53.3)(500)(0.008025 - 0.000023)]^{1/2}$$
$$= 117.2 \text{ (ft/sec)} = 79.9 \text{ (mph)}$$

and from Eq. (18.35)

$$\bar{V}_1 = [2(32.2)(53.3)(500)(0.008025)]^{1/2}$$
$$= 117.4 \text{ (ft/sec)} = 80.0 \text{ (mph)}$$

Compared to Example 18.23, the errors due to use of the approximate equations are negligible.

PROBLEMS

18.1 Air, at a temperature and pressure of 540 F and 100 psia, has a velocity of 300 mph. Determine T_0, p_0, v_0, ρ_0, a, and M. The air acts as a perfect gas.

18.2 Repeat Prob. 18.1 for helium and carbon dioxide.

18.3 Repeat Prob. 18.1 for air as a semiperfect gas.

18.4 Air, as a perfect gas, enters a subsonic nozzle (Fig. 18.3) at a pressure of 50 psia and a temperature of 240 F. The throat diameter is 1.5 in., and the Mach number at the throat is 1.0. Determine p_0, T_0, $\bar{V}_1$, M_1, $\bar{V}_2$, p_2, T_2, and m. The inlet diameter is 2.0 in.

18.5 Air, as a perfect gas, enters a supersonic nozzle (Fig. 18.6) at a pressure of 50 psia and a temperature of 240 F. The inlet diameter is 1.5 in., and the entering and leaving Mach numbers are 1.0 and 1.5. Determine m, T_0, p_0, T_2, p_2, $\overline{V}_2$, and D_2.

18.6 A subsonic diffuser, using air as a perfect gas (Fig. 18.7), has an initial pressure and temperature of 50 psia and 240 F. The Mach number at the inlet is 0.8, and the exit pressure is 70 psia. For an inlet diameter of 1.5 in., determine a_1, $\overline{V}_1$, T_0, p_0, m, T_2, $\overline{V}_2$, a_2, M_2, and D_2.

18.7 A supersonic diffuser (Fig. 18.8) uses hydrogen as a perfect gas. The pressure and temperature at the inlet are 50 psia and 240 F. The Mach numbers at the inlet and exit are 1.5 and 1.2. For an inlet diameter of 1.5 in., determine m, T_0, p_0, $\overline{V}_2$, and D_2.

18.8 The nozzle in Prob. 18.4 is used as the converging section of a converging-diverging nozzle (Fig. 18.9). The exit pressure is 15 psia. Determine $\overline{V}_2$, a_2, M_2, and D_2.

18.9 The nozzle in Prob. 18.4 is used as the converging section of a nozzle diffuser (Fig. 18.10). The exit pressure is 52 psia. Determine $\overline{V}_2$, a_2, M_2, and D_2.

18.10 Steam with a total pressure and temperature of 100 psia and 600 F flows through a converging nozzle having a throat diameter of 1.5 in. Determine the mass flow and the throat velocity from (a) tabular values and (b) the approximate equation for p^*.

18.11 The nozzle in Prob. 18.4 is used as the converging section of a nozzle diffuser. The Mach number at the throat is 0.5. The exit pressure is 48 psia. Determine m, the exit velocity, the inlet and exit Mach numbers, and the exit diameter.

18.12 Air flows in the system shown in Fig. 18.15. h_1 = 15 in., h = 10 in., the fluid in the upstream manometer is mercury, and in the throat manometer the fluid is water. The inlet diameter is 2 in., and the throat diameter is 0.5 in. The barometric pressure is 29 in.-Hg. The inlet temperature is 100 F. Determine the mass flow by (a) the exact method, (b) first-order approximation, and (c) second-order approximation.

18.13 A pitot tube is used to measure the speed of an aircraft. The outside air temperature is -60 F, and the barometric pressure in the aircraft is 20 in.-Hg. The differential pressure (Fig. 18.20) is 30 in.-H_2O. Determine the velocity of the aircraft by (a) the exact solution, (b) first-order approximation, and (c) second-order approximation.

APPENDIXES

APPENDIX A

Conversion to International System of Units (SI)

To Convert from	To	Multiply by
British thermal unit (B)	Joule (J)	1.054E + 03
B/pound-mass (B/lbm)	Joule/kilogram (J/kg)	2.326E + 03
B/lbm-R	Joule/kilogram-Kelvin (J/kg-K)	4.184E + 03
B/hour	Watt (W)	2.931E - 01
B/minute	Watt (W)	1.757E + 01
B/second	Watt (W)	1.054E + 03
Degree Fahrenheit	Degree Celsius (C)	$t_c = (t_f - 32)/1.8$
	Kelvin (K)	$T_K = (t_f + 460)/1.8$
Degree Rankine	Kelvin (K)	$T_K = t_R/1.8$
Foot	Meter (m)	3.048E - 01
Cubic foot/minute	Cubic meter/second (m^3/s)	4.719E - 04
Cubic foot/second	Cubic meter/second (m^3/s)	2.832E - 02
Cubic foot	Cubic meter (m^3)	2.832E - 02
Square foot	Square meter (m^2)	9.290E - 02
Foot/hour	Meter/second (m/s)	8.467E - 05
Foot/minute	Meter/second (m/s)	5.080E - 03
Foot/second	Meter/second (m/s)	3.048E - 01
Square foot/second	Square meter/second (m^2/s)	9.290E - 02
Foot of water (39.2 F)	Pascal (Pa)	2.989E + 03
Foot-pound-force	Joule (J)	1.356E + 00
Foot-pound-force/hour	Watt (W)	3.766E - 04
Foot-pound-force/minute	Watt (W)	2.260E - 02
Foot-pound-force/second	Watt (W)	1.356E + 00
Foot/square second	Meter/square second (m/s^2)	3.048E - 01
Gravity standard, 32.2 feet/square second	9.806 second (m/s^2)	
Gallon (U.S. liquid)	Cubic meter (m^3)	3.785E - 03
Gallon (U.S. liquid) /minute	Cubic meter/second (m^3/s)	6.309E - 05
Grain (1/7000 lbm avoirdupois)	Kilogram (kg)	6.480E - 05

Conversion to International System of Units (continued)

To Convert from	To	Multiply by
Horsepower (550 ft-lbf/sec)	Watt (W)	7.457E + 02
Inch	Meter (m)	2.540E - 02
Square inch	Square meter (m^2)	6.451E - 04
Cubic inch	Cubic meter (m^3)	1.639E - 05
Inch of mercury (32 F)	Pascal (Pa)	3.386E + 03
Inch of mercury (60 F)	Pascal (Pa)	3.377E + 03
Inch of water (39.2 F)	Pascal (Pa)	2.491E + 02
Inch of water (60 F)	Pascal (Pa)	2.488E + 02
Inch/square second	Meter/square second (m/s^2)	2.540E - 02
Kilowatt-hour	Joule (J)	3.600E + 06
Mile (U.S. statute)	Meter (m)	1.609E + 03
Mile/hour (U.S. statute) (mph)	Meter/second (m/s)	4.470E - 01
Pound-force (lbf avoirdupois)	Newton (N)	4.448E + 00
Pound-force-foot	Newton-meter (N-m)	1.356E + 00
Pound-force/square foot	Pascal (Pa)	4.788E + 01
Pound-force/square inch (psi)	Pascal (Pa)	6.895E + 03
Pound-mass (lbm avoirdupois)	Kilogram (kg)	4.536E - 01
Pound-mass/square foot	Kilogram/square meter (kg/m^2)	4.882E + 00
Pound-mass/second	Kilogram/second (kg/s)	4.536E - 01
Pound-mass/minute	Kilogram/second (kg/s)	7.560E - 03
Pound-mass/cubic foot	Kilogram/cubic meter (kg/m^3)	1.602E + 01
Pound-mass/foot-second	Pascal-second (Pa-s)	1.488E + 00
Ton (2000 lbm)	Kilogram (kg)	9.072E + 02

Examples of the Conversion of Thermodynamic Terms from the English to the SI System

The following basic conversions will be employed in all examples:

Length: foot to meter	1 ft = 3.048E - 01 m
Heat: British thermal unit to Joule	1 B = 1.054E + 03 J
Mass: pound-mass to kilogram	1 lbm = 4.536E - 01 kg
Force: pound-force to Newton	1 lbf = 4.448 N
Temperature: degree Rankine to degree Kelvin	1 R = 5.556E - 01 K

A, area, ft^2 to m^2:

$$1\ ft^2 = (3.048E - 01)^2 = 9.290E - 02\ m^2$$

a, acceleration, ft/sec^2 to m/s^2:

$$1\ ft/sec^2 = 3.0481E - 01\ m/s^2$$

a, acoustic velocity, ft/sec to m/s:

$$1\ ft/sec = 3.0481E - 01\ m/s$$

BHP, brake horsepower, HP to W:

$$\begin{aligned} 1\ HP &= 33{,}000\ ft\text{-}lbf/min = 550\ ft\text{-}lbf/sec \\ &= 550(3.048E - 01)(4.448) = 745.7\ N\text{-}m/s \\ &= 7.457E + 02\ W \end{aligned}$$

c, specific heat, B/lbm-R to J/kg-K:

$$\begin{aligned} 1\ B/lbm\text{-}R &= (1.054E + 03)/[(4.536E - 01)(5.556E - 01)] \\ &= 4.182E + 03\ J/kg\text{-}K \end{aligned}$$

There is a slight difference from the tabular values due to round off.

D, diameter, ft to m:

$$1\ ft = 3.048E - 01\ m$$

E, total energy, B to J:

$$1\ B = 1.054E + 03\ J$$

E, total energy, ft-lbf to J:

$$\begin{aligned} 1\ ft\text{-}lbf &= (3.048E - 01)4.448 = 1.356\ N\text{-}m \\ &= 1.356E + 00\ J \end{aligned}$$

F, force, lbf to N:

$$1\ lbf = 4.448E + 00\ N$$

FE, flow energy: See total energy, E.

FHP, friction horsepower: See brake horsepower, BHP.

g, acceleration due to gravity: See acceleration, a.

H, total enthalpy: See total energy, E.

HHV, higher heating value: See specific enthalpy, h.

HP, horsepower: See brake horsepower, BHP.

h, specific enthalpy, B/lbm to J/kg:

1 B/lbm = (1.054E + 03)/(4.536E - 01) = 2.324E + 03 J/kg

h, height: See diameter, D.

IHP, indicated horsepower: See brake horsepower, BHP.

L, length: See diameter, D.

LHV, lower heating value: See specific enthalpy, h.

m, mass, lbm to kg:

1 lbm = 4.536E - 01 kg

P, power, B/sec to W:

1 B/sec = 1.054E + 03 J/s
= 1.054E + 03 W

P, power, ft-lb/sec to W:

1 ft-lbf/sec = (3.048E - 01)4.448 = 1.356E + 00 N-m/s
= 1.356E + 00 W

PE, potential energy: See total energy, E.

p, pressure, psi to Pa:

1 lbf/in.^{2} = 144(4.448)/(3.048E - 01)^2
= 6.89E + 03 N/m^2
= 6.894E + 03 Pa

p, pressure, psf to Pa:

1 lbf/ft^2 = 4.448/(3.048E - 01)^2
= 4.788E + 01 N/m^2
= 4.788E + 01 Pa

p, pressure, in.-Hg to Pa:

1 in.-Hg = 0.491 psi = 0.491(6.894E + 03)
= 3.385E + 03 N/m^2
= 3.385E + 03 Pa

p_m, mean effective pressure: See pressure, p.

Q, heat, B to J:

1 B = 1.054E + 03 J

Q_A, available heat: See heat, Q.

Q_U, unavailable heat: See heat, Q.

R, specific gas constant, ft-lbm-R to J/kg-R:

$$1 \text{ ft-lbf/lbm-R} = [(3.048E - 01)4.448]/[(4.536E - 01)(5.556E - 01)]$$
$$= 5.380E + 00 \text{ J/kg-K}$$

S, total entropy, B/R to J/K:

$$1 \text{ B/R} = (1.054E + 03)/(5.556E - 01) = 1.897E + 03 \text{ J/K}$$

s, specific entropy: See specific heat, c.

T, temperature, R to K:

$$1 \text{ R} = 5.556E - 01 \text{ K}$$

T, torque, ft-lbf to N-m:

$$1 \text{ ft-lbf} = (3.048E - 01)4.448 = 1.356E + 00 \text{ N-m}$$

U, total internal energy: See total energy, E.

u, specific internal energy: See specific enthalpy, h.

V, total volume, ft^3 to m^3:

$$1 \text{ ft}^3 = (3.048E - 01)^3 = 2.832E - 02 \text{ m}^3$$

$\overline{V}$, velocity: See acoustic velocity, a.

V′, actual air supplied: See volume, V.

V_c, clearance volume: See volume, V.

v, specific volume, ft^3/lbm to m^3/kg:

$$1 \text{ ft}^3/\text{lbm} = (3.048E - 01)^3/(4.536E - 01)$$
$$= 6.243E - 02 \text{ m}^3/\text{kg}$$

W, work, B to J:

$$1B = 1.054E + 03 \text{ J}$$

W, work, ft-lbf to J:

$$1 \text{ ft-lbf} = (3.048E - 01)4.448 = 1.356 \text{ N-m}$$
$$= 1.356E + 00 \text{ J}$$

W_c, compressor work: See specific enthalpy, h.

W_t, turbine work: See specific enthalpy, h.

Z, elevation: See diameter, D.

γ, specific weight, lbf/ft^3 to N/m^3:

$$1 \text{ lbf/ft}^3 = 4.448/(3.048E - 01)^3$$
$$= 1.571E + 02 \text{ N/m}^3$$

ΔH_R, enthalpy of reaction: See specific enthalpy, h.

ΔS_p, entropy production: See total entropy, S.

ρ, density, lbm/ft^3 to kg/m^3:

$$1\ lbm/ft^3 = (4.536E - 01)/(3.048E - 01)^3$$
$$= 1.602E + 01\ kg/m^3$$

APPENDIX B

Gas Constants*

Gas	Chemical Equation	Molecular Weight, M	Specific Gas Constant, R	Specific Heat at Constant Pressure, c_p	Specific Heat at Constant Volume, c_v	$k = c_p/c_v$
Acetylene	C_2H_2	26.0	59.4	0.405	0.329	1.232
Air	—	29.0	53.3	0.240	0.1714	1.400
Argon	A	40.0	38.7	0.124	0.075	1.667
n-Butane	C_4H_{10}	58.1	26.6	0.395	0.361	1.093
Carbon dioxide	CO_2	44.0	35.1	0.202	0.157	1.288
Carbon monoxide	CO	28.0	55.2	0.249	0.178	1.399
Chlorine	Cl_2	70.9	21.8	0.115	0.086	1.324
Ethylene	C_2H_4	28.1	55.1	0.366	0.295	1.240
Helium	He	4	386.0	1.234	0.741	1.667
Hydrogen	H_2	2	767.0	3.450	2.465	1.400
Hydrogen sulfide	H_2S	34.1	45.3	0.248	0.188	1.321
Methane	CH_4	16.0	96.3	0.509	0.386	1.321
Methanol	CH_4O	32.0	48.2	0.336	0.274	1.226
Nitrogen	N_2	28.0	55.2	0.249	0.178	1.399
Oxygen	O_2	32.0	48.3	0.219	0.157	1.395
Water	H_2O	18.0	85.8	0.445	0.335	1.329

*Units: M, lbm/mole; R, ft-lbf/lbm-R; c_p, c_v, B/lbm-R.

APPENDIX C

Properties of Air at Low Pressure (for 1 lb)*

T (R)	h (B/lbm)	u (B/lbm)	p_r	v_r	φ (B/lbm-R)
300	71.61	51.04	0.17795	624.5	0.46007
380	90.75	64.70	0.4061	346.6	0.51663
400	95.53	68.11	0.4858	305.0	0.52890
420	100.32	71.52	0.5760	270.1	0.54058
440	105.11	74.93	0.6776	240.6	0.55172
460	109.90	78.36	0.7913	215.33	0.56235
480	114.69	81.77	0.9182	193.65	0.57255
500	119.48	85.20	1.0590	174.90	0.58233
520	124.27	88.62	1.2147	158.58	0.59173
537	128.34	91.53	1.3593	146.34	0.59945
540	129.06	92.04	1.3860	144.32	0.60078
560	133.86	95.47	1.5742	131.78	0.60950
580	138.66	98.90	1.7800	120.70	0.61793
600	143.47	102.34	2.005	110.88	0.62607
620	148.28	105.78	2.249	102.12	0.63395
640	153.09	109.21	2.514	94.30	0.64159
660	157.92	112.67	2.801	87.27	0.64902
680	162.73	116.12	3.111	80.96	0.65621
700	167.56	119.58	3.446	75.25	0.66321
720	172.39	123.04	3.806	70.07	0.67002
740	177.23	126.51	4.193	65.38	0.67665
760	182.08	129.99	4.607	61.10	0.68312
780	186.94	133.47	5.051	57.20	0.68942
800	191.81	136.97	5.526	53.63	0.69558
820	196.69	140.47	6.033	50.35	0.70160
840	201.56	143.98	6.573	47.34	0.70747
860	206.46	147.50	7.149	44.57	0.71323
880	211.35	151.02	7.761	42.01	0.71886
900	216.26	154.57	8.411	39.64	0.72438
920	221.18	158.12	9.102	37.44	0.72979
940	226.11	161.68	9.834	35.41	0.73509
960	231.06	165.26	10.610	33.52	0.74030
980	236.02	168.83	11.430	31.76	0.74540
1000	240.98	172.43	12.298	30.12	0.75042
1020	245.97	176.04	13.215	28.59	0.75536
1040	250.95	179.66	14.182	27.17	0.76019
1060	255.96	183.29	15.203	25.82	0.76496
1080	260.97	186.93	16.278	24.58	0.76964
1100	265.99	190.58	17.413	23.40	0.77426
1120	271.03	194.25	18.604	22.30	0.77880
1140	276.08	197.94	19.858	21.27	0.78326
1160	281.14	201.63	21.18	20.293	0.78767
1180	286.21	205.33	22.56	19.377	0.79201
1200	291.30	209.05	24.01	18.514	0.79628

Properties of Air at Low Pressure (for 1 lb)(continued)

T (R)	h (B/lbm)	u (B/lbm)	p_r	v_r	φ (B/lbm-R)
1220	296.41	212.78	25.23	17.700	0.80050
1240	301.52	216.53	27.13	16.932	0.80466
1260	306.65	220.28	28.80	16.205	0.80876
1280	311.79	224.05	30.55	15.518	0.81280
1300	316.94	227.83	32.39	14.868	0.81680
1320	322.11	231.68	34.31	14.253	0.82075
1340	327.29	235.43	36.31	13.670	0.82464
1360	332.48	239.25	38.41	13.118	0.82848
1380	337.68	243.08	40.59	12.593	0.83229
1400	342.90	246.93	42.88	12.095	0.83604
1420	348.14	250.79	45.26	11.622	0.83975
1440	353.37	254.66	47.75	11.172	0.84341
1460	358.63	258.54	50.34	10.743	0.84704
1480	363.89	262.44	53.04	10.336	0.85062
1500	369.17	266.34	55.86	9.948	0.85416
1520	374.47	270.26	58.78	9.578	0.85767
1540	379.77	274.20	61.83	9.226	0.86113
1560	385.08	278.13	65.00	8.890	0.86456
1580	390.40	282.09	68.30	8.569	0.86794
1600	395.74	286.06	71.73	8.263	0.87130
1620	401.09	290.04	75.29	7.971	0.87462
1640	406.45	294.03	78.99	7.691	0.87791
1660	411.82	298.02	82.83	7.424	0.88116
1680	417.20	302.04	86.82	7.168	0.88439
1700	422.59	306.06	90.95	6.924	0.88758
1720	428.00	310.09	95.24	6.690	0.89074
1740	433.41	314.13	99.69	6.465	0.89387
1760	438.83	318.18	104.30	6.251	0.89697
1780	444.26	322.24	109.08	6.045	0.90003
1800	449.71	326.32	114.03	5.847	0.90308
1820	455.17	330.40	119.16	5.658	0.90609
1840	460.63	334.50	124.47	5.476	0.90908
1860	466.12	338.61	129.95	5.302	0.91203
1880	471.60	342.73	135.64	5.134	0.91497
1900	477.09	346.85	141.51	4.974	0.91788
1920	482.60	350.98	147.59	4.819	0.92076
1940	488.12	355.12	153.87	4.670	0.92362
1960	493.64	359.28	160.37	4.527	0.92645
1980	499.17	363.43	167.07	4.390	0.92926
2000	504.71	367.61	174.00	4.258	0.93205
2020	510.26	371.79	181.16	4.130	0.93481
2040	515.82	375.98	188.54	4.008	0.93756
2060	521.39	380.18	196.16	3.890	0.94026
2080	526.97	384.39	204.02	3.777	0.94296
2100	532.55	388.60	212.1	3.667	0.94564

Properties of Air at Low Pressure (for 1 lb)(continued)

T (R)	h (B/lbm)	u (B/lbm)	p_r	v_r	φ (B/lbm-R)
2150	546.54	399.17	233.5	3.410	0.95222
2200	560.59	409.78	256.6	3.176	0.95868
2250	574.69	420.46	281.4	2.961	0.96501
2300	588.82	431.16	308.1	2.765	0.97123
2350	603.00	441.91	336.8	2.585	0.97732
2400	617.22	452.70	367.6	2.419	0.98331
2450	631.48	463.54	400.5	2.266	0.98919
2500	645.78	474.40	435.7	2.125	0.99497
2550	660.12	485.31	473.3	1.9956	1.00064
2600	674.49	496.26	513.5	1.8756	1.00623
2650	688.90	507.25	556.3	1.7646	1.01172
2700	703.35	518.26	601.9	1.6617	1.01712
2750	717.83	529.31	650.4	1.5662	1.02244
2800	732.33	540.40	702.0	1.4775	1.02767
2850	746.88	551.52	756.7	1.3951	1.03282
2900	761.45	562.66	814.8	1.3184	1.03788
2950	776.05	573.84	876.4	1.2469	1.04288
3000	790.68	585.04	941.4	1.1803	1.04779
3500	938.40	698.48	1,829.3	0.7087	1.09332
4000	1088.26	814.06	3,280	0.4518	1.13334
4500	1239.86	931.39	5,521	0.3019	1.16905
5000	1392.87	1050.12	8,837	0.20959	1.20129
6000	1702.29	1291.00	20,120	0.11047	1.25769
6500	1858.44	1412.87	28,974	0.08310	1.28268

*Extracted from Gas Tables by J. H. Keenan and J. Kaye, with permission of the authors and the publisher, John Wiley & Sons, Inc., New York, 1945.

APPENDIX D

TABLE 1. Steam Tables - Temperatures*

Saturation: Temperatures†

Temperature (t)	Pressure (p)	Specific volume			Enthalpy			Entropy		
		v_f	v_{fg}	v_g	h_f	h_{fg}	h_g	s_f	s_{fg}	s_g
32	0.0885	0.01602	3306	3306	0.00	1075.8	1075.8	0.0000	2.1877	2.1877
35	0.1000	0.01602	2947	2947	3.02	1074.1	1077.1	0.0061	2.1709	2.1770
40	0.1217	0.01602	2444	2444	8.05	1071.3	1079.3	0.0162	2.1435	2.1597
45	0.1475	0.01602	2036	2036	13.06	1068.4	1081.5	0.0262	2.1167	2.1429
50	0.1781	0.01603	1703	1703	18.07	1065.6	1083.7	0.0361	2.0903	2.1264
60	0.2563	0.01604	1207	1207	28.06	1059.9	1088.0	0.0555	2.0393	2.0948
70	0.3631	0.01606	867.8	867.9	38.04	1054.3	1092.3	0.0745	1.9902	2.0647
80	0.5069	0.01608	633.1	633.1	48.02	1048.6	1096.6	0.0932	1.9428	2.0360
90	0.6982	0.01610	468.0	468.0	57.99	1042.9	1100.9	0.1115	1.8972	2.0087
100	0.9492	0.01613	350.3	350.4	67.97	1037.2	1105.2	0.1295	1.8531	1.9826
110	1.2748	0.01617	265.3	265.4	77.94	1031.6	1109.5	0.1471	1.8106	1.9577
120	1.6924	0.01620	203.2	203.3	87.92	1025.8	1113.7	0.1645	1.7694	1.9339
130	2.2225	0.01625	157.3	157.3	97.90	1020.0	1117.9	0.1816	1.7296	1.9112
140	2.8886	0.01629	123.0	123.0	107.9	1014.1	1122.0	0.1984	1.6910	1.8894
150	3.718	0.01634	97.06	97.07	117.9	1008.2	1126.1	0.2149	1.6537	1.8685
160	4.741	0.01639	77.27	77.29	127.9	1002.3	1130.2	0.2311	1.6174	1.8485
170	5.992	0.01645	62.04	62.06	137.9	996.3	1134.2	0.2472	1.5822	1.8293
180	7.510	0.01651	50.21	50.23	147.9	990.2	1138.1	0.2630	1.5480	1.8109
190	9.339	0.01657	40.94	40.96	158.0	984.1	1142.0	0.2785	1.5147	1.7932
200	11.53	0.01663	33.62	33.64	168.0	977.9	1145.9	0.2938	1.4824	1.7762
210	14.12	0.01670	27.80	27.82	178.0	971.6	1149.7	0.3090	1.4508	1.7598
212	14.70	0.01672	26.78	26.80	180.1	970.3	1150.4	0.3120	1.4446	1.7566
220	17.19	0.01677	23.13	23.15	188.1	965.2	1153.4	0.3239	1.4201	1.7440
230	20.78	0.01684	19.36	19.38	198.2	958.8	1157.0	0.3387	1.3901	1.7288
240	24.97	0.01692	16.31	16.32	208.3	952.2	1160.5	0.3531	1.3609	1.7140
250	29.82	0.01700	13.80	13.82	218.5	945.5	1164.0	0.3675	1.3323	1.6998

TABLE 1 (Continued)

Saturation: Temperatures†

Temperature (t)	Pressure (p)	Specific volume			Enthalpy			Entropy		
		v_f	v_{fg}	v_g	h_f	h_{fg}	h_g	s_f	s_{fg}	s_g
260	35.43	0.01709	11.75	11.76	228.6	938.7	1167.3	0.3817	1.3043	1.6860
270	41.86	0.01717	10.04	10.06	238.8	931.8	1170.6	0.3958	1.2769	1.6727
280	49.20	0.01726	8.628	8.645	249.1	924.7	1173.8	0.4096	1.2501	1.6597
290	57.56	0.01735	7.444	7.461	259.3	917.5	1176.8	0.4234	1.2238	1.6472
300	67.01	0.01745	6.449	6.466	269.6	910.1	1179.7	0.4369	1.1980	1.6350
310	77.68	0.01755	5.609	5.626	279.9	902.6	1182.5	0.4504	1.1727	1.6231
320	89.66	0.01765	4.896	4.914	290.3	894.9	1185.2	0.4637	1.1478	1.6115
330	103.1	0.01776	4.289	4.307	300.7	887.0	1187.7	0.4769	1.1233	1.6002
340	118.0	0.01787	3.770	3.778	311.1	879.0	1190.1	0.4900	1.0992	1.5891
350	134.6	0.01799	3.324	3.342	321.6	870.7	1192.3	0.5029	1.0754	1.5783
360	153.0	0.01811	2.939	2.957	322.2	862.2	1194.4	0.5158	1.0519	1.5677
370	173.4	0.01823	2.606	2.625	342.8	853.5	1196.3	0.5286	1.0287	1.5573
380	195.8	0.01836	2.317	2.335	353.4	844.6	1198.1	0.5413	1.0059	1.5471
390	220.4	0.01850	2.065	2.084	364.2	835.4	1199.6	0.5539	0.9832	1.5371
400	247.3	0.01864	1.845	1.863	375.0	826.0	1201.0	0.5664	0.9608	1.5272
410	276.8	0.01878	1.651	1.670	385.8	816.3	1202.1	0.5788	0.9386	1.5174
420	308.8	0.01894	1.481	1.500	396.8	806.3	1203.1	0.5912	0.9166	1.5078
430	343.7	0.01910	1.331	1.350	407.8	796.0	1203.8	0.6035	0.8947	1.4982
440	381.6	0.01926	1.198	1.217	418.9	785.4	1204.3	0.6158	0.8730	1.4887
450	422.6	0.0194	1.080	1.099	430.1	774.5	1204.6	0.6280	0.8513	1.4793
460	466.9	0.0196	0.975	0.994	441.4	763.2	1204.6	0.6402	0.8298	1.4700
470	514.7	0.0198	0.881	0.901	452.8	751.5	1204.3	0.6523	0.8083	1.4606
480	566.1	0.0200	0.797	0.817	464.4	739.4	1203.7	0.6645	0.7868	1.4513
490	621.4	0.0202	0.722	0.742	476.0	726.8	1202.8	0.6766	0.7653	1.4419
500	680.8	0.0204	0.654	0.675	487.8	713.9	1201.7	0.6887	0.7438	1.4325
550	1045	0.0218	0.402	0.424	549.3	640.8	1190.0	0.7497	0.6346	1.3843
600	1543	0.0236	0.243	0.267	617.0	548.5	1165.5	0.8131	0.5176	1.3307

650	2208	0.027	0.135	0.162	695.7	422.8	1118.5	0.8828	0.3809	1.2637
700	3094	0.037	0.039	0.076	823.3	172.1	995.4	0.9905	0.1484	1.1389
705.4	3206	0.050	0.000	0.050	902.7	0.0	902.7	1.058	0.000	1.058

*Extracted from Thermodynamic Properties of Steam by J. H. Keenan and F. G. Keyes, with permission of the estate of Mr. Keenan; published by John Wiley & Sons, Inc., New York, 1936.

†Units: t, F; p, psia; v, ft^3/lbm; h, B/lbm; s, B/lbm-R.

TABLE 2. Steam Tables — Pressure

Saturation: Pressures*

Pressure (p)	Temperature (t)	Specific volume		Enthalpy			Entropy			Internal energy	
		v_f	v_g	h_f	h_{fg}	h_g	s_f	s_{fg}	s_g	u_f	u_g
1.0	101.7	0.01614	333.6	69.70	1036.3	1106.0	0.1326	1.8456	1.9782	69.70	1044.3
2.0	126.1	0.01623	173.7	93.99	1022.2	1116.2	0.1749	1.7451	1.9200	93.98	1051.9
3.0	141.5	0.01630	118.7	109.4	1013.2	1122.6	0.2008	1.6855	1.8863	109.4	1056.7
4.0	153.0	0.01636	90.63	120.9	1006.4	1127.3	0.2198	1.6427	1.8625	120.8	1060.2
5.0	162.2	0.01640	73.52	130.1	1001.0	1131.1	0.2347	1.6094	1.8441	130.1	1063.1
6.0	170.1	0.01645	61.98	138.0	996.2	1134.2	0.2472	1.5820	1.8292	137.9	1065.4
7.0	176.8	0.01649	53.64	144.8	992.1	1136.9	0.2581	1.5586	1.8167	144.7	1067.4
8.0	182.9	0.01653	47.34	150.8	988.5	1139.3	0.2674	1.5383	1.8057	150.8	1069.2
9.0	188.3	0.01656	42.40	156.2	985.2	1141.4	0.2759	1.5203	1.7962	156.2	1070.8
10.0	193.2	0.01659	38.42	161.2	982.1	1143.3	0.2835	1.5041	1.7876	161.1	1072.2
14.7	212.0	0.01672	26.80	180.1	970.3	1150.4	0.3120	1.4446	1.7566	180.0	1077.5
15.0	213.0	0.01672	26.29	181.1	969.7	1150.8	0.3135	1.4415	1.7549	181.1	1077.8
20.0	228.0	0.01683	20.09	196.2	960.1	1156.3	0.3356	1.3962	1.7319	196.1	1081.9
25.0	240.1	0.01692	16.30	208.4	952.1	1160.6	0.3533	1.3606	1.7139	208.4	1085.1
30.0	250.3	0.01701	13.75	218.8	945.3	1164.1	0.3680	1.3313	1.6993	218.7	1087.8
35.0	259.3	0.01708	11.90	227.9	939.2	1167.1	0.3807	1.3063	1.6870	227.8	1090.1
40.0	267.2	0.01715	10.50	236.0	933.7	1169.7	0.3919	1.2844	1.6763	235.9	1092.0
45.0	274.4	0.01721	9.401	243.4	928.6	1172.0	0.4019	1.2650	1.6669	243.2	1093.7
50.0	281.0	0.01727	8.515	250.1	924.0	1174.1	0.4110	1.2474	1.6585	249.9	1095.3

TABLE 2 (Continued)

Saturation: Pressures*

Pressure (p)	Temperature (t)	Specific volume		Enthalpy			Entropy			Internal energy	
		v_f	v_g	h_f	h_{fg}	h_g	s_f	s_{fg}	s_g	u_f	u_g
55.0	287.1	0.01732	7.787	256.3	919.6	1175.9	0.4193	1.2316	1.6509	256.1	1096.7
60.0	292.7	0.01738	7.175	262.1	915.5	1177.6	0.4270	1.2168	1.6438	261.9	1097.9
65.0	298.0	0.01743	6.655	267.5	911.6	1179.1	0.4342	1.2032	1.6374	267.3	1099.1
70.0	302.9	0.01748	6.206	272.6	907.9	1180.6	0.4409	1.1906	1.6315	272.4	1100.2
75.0	307.6	0.01753	5.816	277.4	904.5	1181.9	0.4472	1.1787	1.6259	277.2	1101.2
80.0	312.0	0.01757	5.472	282.0	901.1	1183.1	0.4531	1.1676	1.6207	281.8	1102.1
85.0	316.2	0.01761	5.168	286.4	897.8	1184.2	0.4587	1.1571	1.6158	286.1	1102.9
90.0	320.3	0.01766	4.896	290.6	894.7	1185.3	0.4641	1.1471	1.6112	290.3	1103.7
95.0	324.1	0.01770	4.652	294.6	891.7	1186.2	0.4692	1.1376	1.6068	294.2	1104.5
100.0	327.8	0.01774	4.432	298.4	888.8	1187.2	0.4740	1.1286	1.6026	298.1	1105.2
110.0	334.8	0.01782	4.049	305.7	883.2	1188.9	0.4832	1.1117	1.5948	305.3	1106.5
120.0	341.2	0.01789	3.728	312.4	877.9	1190.4	0.4916	1.0962	1.5878	312.0	1107.6
130.0	347.3	0.01796	3.455	318.8	872.9	1191.7	0.4995	1.0817	1.5812	318.4	1108.6
140.0	353.0	0.01802	3.220	324.8	868.2	1193.0	0.5069	1.0682	1.5751	324.4	1109.6
150.0	358.4	0.01809	3.015	330.5	863.6	1194.1	0.5138	1.0556	1.5694	330.0	1110.5
160.0	363.5	0.01815	2.834	335.9	859.2	1195.1	0.5204	1.0436	1.5640	335.4	1111.2
170.0	368.4	0.01822	2.675	341.1	854.9	1196.0	0.5266	1.0324	1.5590	340.5	1111.9
180.0	373.1	0.01827	2.532	346.0	850.8	1196.9	0.5325	1.0217	1.5542	345.4	1112.5
190.0	377.5	0.01833	2.404	350.8	846.8	1197.6	0.5381	1.0116	1.5497	350.2	1113.1
200.0	381.8	0.01839	2.288	355.4	843.0	1198.4	0.5435	1.0018	1.5453	354.7	1113.7
250.0	401.0	0.01865	1.844	376.0	825.1	1201.1	0.5675	0.9588	1.5263	375.1	1115.8
300.0	417.3	0.01890	1.543	393.8	809.0	1202.8	0.5879	0.9225	1.5104	392.8	1117.1
350.0	431.7	0.01913	1.326	409.7	794.2	1203.9	0.6056	0.8910	1.4966	408.4	1118.0
400.0	444.6	0.0193	1.161	424.0	780.5	1204.5	0.6214	0.8630	1.4844	422.6	1118.5
450.0	456.3	0.0195	1.032	437.2	767.4	1204.6	0.6356	0.8378	1.4734	435.5	1118.7
500.0	467.0	0.0197	0.928	449.4	755.0	1204.0	0.6487	0.8147	1.4634	447.6	1118.6
600.0	486.2	0.0201	0.770	471.6	731.6	1203.2	0.6720	0.7734	1.4454	469.4	1117.7

700.0	503.1	0.0205	0.655	491.5	709.7	1201.2	0.6925	0.7371	1.4296	488.8	1116.3
800.0	518.2	0.0209	0.569	509.7	688.9	1198.6	0.7108	0.7045	1.4153	506.6	1114.4
900.0	532.0	0.0212	0.501	526.6	668.8	1195.4	0.7275	0.6744	1.4020	523.1	1112.1
1000.0	544.6	0.0216	0.456	542.4	649.4	1191.8	0.7430	0.6467	1.3897	538.4	1109.4
1500.0	596.2	0.0235	0.276	611.6	556.3	1167.9	0.8082	0.5269	1.3351	605.1	1091.2
2000.0	635.8	0.0257	0.188	671.7	463.4	1135.1	0.8619	0.4230	1.2849	662.2	1065.6
2500.0	668.1	0.0287	0.131	730.6	360.5	1091.1	0.9126	0.3197	1.2322	717.3	1030.6
3000.0	695.4	0.0346	0.086	802.5	217.8	1020.3	0.9731	0.1885	1.1615	783.4	972.7
3206.2	705.4	0.0503	0.050	902.7	0.0	902.7	1.0580	0.0000	1.0580	872.9	872.9

*Units: p, psia; t, F, v, ft^3/lbm; h, B/lbm; u, B/lbm; s, B/lbm-R.

TABLE 3. Superheated Vapor*

Pressure (p)		Temperature (F) 200	300	400	500	600	700	800	900	1000	1200
	v	392.6	452.3	572.0	571.6	631.2	690.8	750.4	809.9	869.5	988.7
1	h	1150.4	1195.8	1241.7	1288.3	1335.7	1383.8	1432.8	1482.7	1533.5	1637.7
	s	2.0512	2.1153	2.1720	2.2233	2.2702	2.3137	2.3542	2.3923	2.4283	2.4952
	v	78.16	90.25	102.26	114.22	126.16	138.10	150.03	161.95	173.87	197.71
5	h	1148.8	1195.0	1241.2	1288.0	1335.4	1383.6	1432.7	1482.6	1533.4	1637.7
	s	1.8718	1.9370	1.9942	2.0456	2.0927	2.1361	2.1767	2.2148	2.2509	2.3178
	v	38.85	45.00	51.04	57.05	63.03	69.01	74.98	80.95	86.92	98.84
10	h	1146.6	1193.9	1240.6	1287.5	1335.1	1383.4	1432.5	1482.4	1533.2	1637.6
	s	1.7927	1.8595	1.9172	1.9689	2.0160	2.0596	2.1002	2.1383	2.1744	2.2413
	v		30.53	34.68	38.78	42.86	46.94	51.00	55.07	59.13	67.25
14.7	h		1192.8	1239.9	1287.1	1334.8	1383.2	1432.3	1482.3	1533.1	1637.5
	s		1.8160	1.8743	1.9261	1.9734	2.0170	2.0576	2.0958	2.1319	2.1989
	v		22.36	25.43	28.46	31.47	34.47	37.46	40.45	43.44	49.41
20	h		1191.6	1239.2	1286.6	1334.4	1382.9	1432.1	1482.1	1533.0	1637.4
	s		1.7808	1.8396	1.8918	1.9392	1.9829	2.0235	2.0618	2.0978	2.1648
	v		11.040	12.628	14.168	15.688	17.198	18.702	20.20	21.70	24.69
40	h		1186.8	1236.5	1284.8	1333.1	1381.9	1431.3	1481.4	1532.4	1637.0
	s		1.6994	1.7608	1.8140	1.8619	1.9058	1.9467	1.9850	2.0212	2.0833

TABLE 3 (Continued)

Pressure (p)		Temperature (F) 200	300	400	500	600	700	800	900	1000	1200
60	v		7.259	8.357	9.403	10.427	11.441	12.449	13.452	14.454	16.451
	h		1181.6	1233.6	1283.0	1331.8	1380.9	1430.5	1480.8	1531.9	1636.6
	s		1.6492	1.7135	1.7678	1.8162	1.8605	1.9015	1.9400	1.9762	2.0434
80	v			6.220	7.020	7.797	8.562	9.322	10.077	10.830	12.332
	h			1230.7	1281.1	1330.5	1379.9	1429.7	1480.1	1531.3	1636.2
	s			1.6791	1.7346	1.7836	1.8281	1.8694	1.9079	1.9442	2.0115
100	v			4.937	5.589	6.218	6.835	7.446	8.052	8.656	9.860
	h			1227.6	1279.1	1329.1	1378.9	1428.9	1479.5	1530.8	1635.7
	s			1.6518	1.7085	1.7581	1.8029	1.8443	1.8829	1.9193	1.9867
120	v			4.081	4.636	5.165	5.683	6.195	6.702	7.207	8.212
	h			1224.4	1277.2	1327.7	1377.8	1428.1	1478.8	1530.2	1635.3
	s			1.6287	1.6869	1.7370	1.7822	1.8237	1.8625	1.8990	1.9664
140	v			3.468	3.954	4.413	4.861	5.301	5.738	6.172	7.035
	h			1221.1	1275.2	1326.4	1376.8	1427.3	1478.2	1529.7	1634.9
	s			1.6087	1.6683	1.7190	1.7645	1.8063	1.8451	1.8817	1.9493
160	v			3.008	3.443	3.849	4.244	4.631	5.015	5.396	6.152
	h			1217.6	1273.1	1325.0	1375.7	1426.4	1477.5	1529.1	1634.5
	s			1.5908	1.6519	1.7033	1.7491	1.7911	1.8301	1.8667	1.9344
180	v			2.649	3.044	3.411	3.764	4.110	4.452	4.792	5.466
	h			1214.0	1271.0	1323.5	1374.7	1425.6	1476.8	1528.6	1634.1
	s			1.5745	1.6373	1.6894	1.7355	1.7776	1.8167	1.8534	1.9212
200	v			2.361	2.726	3.060	3.380	3.693	4.002	4.309	4.917
	h			1210.3	1268.9	1322.1	1373.6	1424.8	1476.2	1528.0	1633.7
	s			1.5594	1.6240	1.6767	1.7232	1.7655	1.8048	1.8415	1.9094
220	v			2.125	2.465	2.772	3.066	3.352	3.634	3.913	4.467
	h			1206.5	1266.7	1320.7	1372.6	1424.0	1475.5	1527.5	1633.3
	s			1.5453	1.6117	1.6652	1.7120	1.7545	1.7939	1.8308	1.8987
240	v			1.9276	2.247	2.533	2.804	3.068	3.327	3.584	4.093
	h			1202.5	1264.5	1319.2	1371.5	1423.2	1474.8	1526.9	1632.9
	s			1.5319	1.6003	1.6546	1.7017	1.7444	1.7839	1.8209	1.8889

260	v	2.063	2.330	2.582	2.827	3.067	3.305	3.776
	h	1262.3	1317.7	1370.4	1422.3	1474.2	1526.3	1632.5
	s	1.5897	1.6447	1.6922	1.7352	1.7748	1.8118	1.8799
300	v	1.7675	2.005	2.227	2.442	2.652	2.859	3.269
	h	1257.6	1314.7	1368.3	1420.6	1472.8	1525.2	1631.7
	s	1.5701	1.6268	1.6751	1.7184	1.7582	1.7954	1.8638
350	v	1.4923	1.7036	1.8980	2.084	2.266	2.445	2.798
	h	1251.5	1310.9	1365.5	1418.5	1471.1	1523.8	1630.7
	s	1.5481	1.6070	1.6563	1.7002	1.7403	1.7777	1.8463
400	v	1.2851	1.4770	1.6508	1.8161	1.9767	2.134	2.445
	h	1245.1	1306.9	1362.7	1416.4	1469.4	1522.4	1629.6
	s	1.5281	1.5894	1.6398	1.6842	1.7247	1.7623	1.8311

Pressure (p)		Temperature (F)										
		500	540	600	620	640	660	680	700	800	900	1000
450	v	1.1231	1.1977	1.3005	1.3332	1.3652	1.3967	1.4278	1.4584	1.6074	1.7516	1.8928
	h	1238.4	1265.6	1302.8	1314.6	1326.2	1337.5	1348.8	1359.9	1414.3	1467.7	1521.0
	s	1.5095	1.5373	1.5735	1.5845	1.5951	1.6054	1.6153	1.6250	1.6699	1.7108	1.7486
500	v	0.9927	1.0633	1.1591	1.1893	1.2188	1.2478	1.2763	1.3044	1.4405	1.5715	1.6996
	h	1231.3	1260.0	1298.6	1310.7	1322.6	1334.2	1345.7	1357.0	1412.1	1466.0	1519.6
	s	1.4919	1.5213	1.5588	1.5701	1.5810	1.5915	1.6016	1.6115	1.6571	1.6982	1.7363
550	v	0.8852	0.9528	1.0431	1.0714	1.0989	1.1259	1.1523	1.1783	1.3038	1.4241	1.5414
	h	1223.7	1254.2	1294.3	1306.8	1318.9	1330.8	1342.5	1354.0	1409.9	1464.3	1518.2
	s	1.4571	1.5061	1.5451	1.5568	1.5680	1.5787	1.5890	1.5991	1.6452	1.6868	1.7250
600	v	0.7947	0.8602	0.9463	0.9729	0.9988	1.0241	1.0489	1.0732	1.1899	1.3013	1.4096
	h	1215.7	1248.1	1289.9	1302.7	1315.2	1327.4	1339.3	1351.1	1407.7	1462.5	1516.7
	s	1.4586	1.4916	1.5323	1.5443	1.5558	1.5667	1.5773	1.5875	1.6343	1.6762	1.7147
700	v		0.7134	0.7934	0.8177	0.8411	0.8639	0.8860	0.9077	1.0108	1.1082	1.2024
	h		1235.0	1280.6	1294.3	1307.5	1320.3	1332.8	1345.0	1403.2	1459.0	1513.9
	s		1.4640	1.5084	1.5212	1.5333	1.5449	1.5559	1.5665	1.6147	1.6573	1.6963
800	v		0.6015	0.6779	0.7006	0.7223	0.7433	0.7635	0.7833	0.8763	0.9633	1.0470
	h		1220.5	1270.7	1285.4	1299.4	1312.9	1325.9	1338.6	1398.6	1455.4	1511.0
	s		1.4375	1.4863	1.5000	1.5129	1.5250	1.5366	1.5476	1.5972	1.6407	1.6801
900	v		0.5124	0.5873	0.6089	0.6294	0.6491	0.6680	0.6863	0.7716	0.8506	0.9262
	h		1204.4	1260.1	1275.9	1290.9	1305.1	1318.8	1332.1	1393.9	1451.8	1508.1
	s		1.4111	1.4653	1.4800	1.4938	1.5066	1.5187	1.5303	1.5814	1.6257	1.6656

TABLE 3 (Continued)

Pressure (p)		Temperature (F) 500	540	600	620	640	660	680	700	800	900	1000
1000	v			0.5140	0.5350	0.5546	0.5733	0.5912	0.6084	0.6878	0.7604	0.8294
	h			1248.8	1265.9	1281.9	1297.0	1311.4	1325.3	1389.2	1448.2	1505.1
	s			1.4450	1.4610	1.4757	1.4893	1.5021	1.5141	1.5670	1.6121	1.6525
1100	v			0.4532	0.4738	0.4929	0.5110	0.5281	0.5445	0.6191	0.6866	0.7503
	h			1236.7	1255.3	1272.4	1288.5	1303.7	1318.3	1384.3	1444.5	1502.2
	s			1.4251	1.4425	1.4583	1.4728	1.4862	1.4989	1.5535	1.5995	1.6405
1200	v			0.4016	0.4222	0.4410	0.4586	0.4752	0.4909	0.5617	0.6250	0.6843
	h			1223.5	1243.9	1262.4	1279.6	1295.7	1311.0	1379.3	1440.7	1499.2
	s			1.4052	1.4243	1.4413	1.4568	1.4710	1.4843	1.5409	1.5879	1.6293
1400	v			0.3174	0.3390	0.3580	0.3753	0.3912	0.4062	0.4714	0.5281	0.5805
	h			1193.0	1218.4	1240.4	1260.3	1278.5	1295.5	1369.1	1433.1	1493.2
	s			1.3639	1.3877	1.4079	1.4258	1.4419	1.4567	1.5177	1.5666	1.6093
1600	v				0.2733	0.2936	0.3112	0.3271	0.3417	0.4034	0.4553	0.5027
	h				1187.8	1215.2	1238.7	1259.6	1278.7	1358.4	1425.3	1487.0
	s				1.3489	1.3741	1.3952	1.4137	1.4303	1.4964	1.5476	1.5914
1800	v					0.2407	0.2597	0.2760	0.2907	0.3502	0.3986	0.4421
	h					1185.1	1214.0	1238.5	1260.3	1347.2	1417.4	1480.8
	s					1.3377	1.3638	1.3855	1.4044	1.4765	1.5301	1.5752
2000	v					0.1936	0.2161	0.2337	0.2489	0.3074	0.3532	0.3935
	h					1145.6	1184.9	1214.8	1240.0	1335.5	1409.2	1474.5
	s					1.2945	1.3300	1.3564	1.3783	1.4576	1.5139	1.5603
2500	v							0.1484	0.1686	0.2294	0.2710	0.3061
	h							1132.3	1176.8	1303.6	1387.8	1458.4
	s							1.2687	1.3073	1.4127	1.4772	1.5273
3000	v								0.0984	0.1760	0.2159	0.2476
	h								1060.7	1267.2	1365.0	1441.8
	s								1.1966	1.3690	1.4439	1.4984
3206.2	v									0.1583	0.1981	0.2288
	h									1250.5	1355.2	1434.7
	s									1.3508	1.4309	1.4874

*Units: p, psia; t, F; v, ft^3/lbm; h, B/lbm; s, B/lbm-R.

TABLE 4. Compressed Liquid*

Pressure (p)		Temperature (F)								
		32	100	200	300	400	500	600	700	705.4
200	Δv	-1.1	-1.1	-1.1	-1.1					
	Δh	+0.61	+0.54	+0.41	+0.23					
	Δs	+0.03	-0.05	-0.21	-0.21					
1000	Δv	-5.7	-5.1	-5.4	-6.9	-8.7	-6.4			
	Δh	2.99	+2.70	+2.21	+1.75	+0.84	-0.14			
	Δs	+0.15	-0.53	-1.20	-1.64	-2.00	-1.41			
3000	Δv	-16.3	-14.7	-16.0	-20.7	-30.0	-47.1	-87.9		
	Δh	+9.00	+7.88	+6.76	+5.49	+3.33	-0.41	-6.9		
	Δs	+0.28	-1.79	-3.56	-5.12	-7.03	-9.42	-12.4		
3206.2	Δv	-17.5	-15.7	-17.1	-22.2	-32.1	-51.0	-98.0	-354	0
	Δh	+9.61	+8.45	+7.25	+5.90	+3.62	-0.40	-7.6	-21.6	0
	Δs	+0.29	-1.93	-3.80	-5.50	-7.54	-10.19	-14.0	-19.2	0
5000	Δv	-26.7	-23.6	-26.0	-34.0	-49.6	-80.5	-169.3	-1017	-2309
	Δh	+14.75	+13.08	+11.30	+9.36	+6.08	+0.25	-12.1	-76.9	-146.7
	Δs	+0.22	-3.07	-5.92	-8.40	-11.74	-16.47	-25.3	-75.3	-134.6

$v - v_f = \Delta v/10^5$ $\quad h - h_f = \Delta h$ $\quad s - s_f = \Delta s/10^3$

*Units: p, psia; t, F; Δv, ft^3/lbm; Δh, B/lbm; Δs, B/lbm-R.

APPENDIX E

TABLE 1. Refrigerant Ammonia
Properties of the Saturated Liquid and Saturated Vapor

T	P	v_f	v_g	h_f	h_g	s_f	s_g
-105	0.996	0.0219	223.2	-68.5	570.3	-0.1774	1.6243
-100	1.24	0.0220	182.4	-63.3	572.5	-0.1626	1.6055
-95	1.52	0.0221	150.0	-58.0	574.7	-0.1481	1.5874
-90	1.86	0.0222	124.1	-52.8	576.9	-0.1338	1.5699
-85	2.27	0.0223	103.3	-47.5	579.1	-0.1197	1.5531
-80	2.74	0.0224	86.50	-42.2	581.2	-0.1057	1.5368
-75	3.29	0.0225	72.81	-37.0	583.3	-0.0920	1.5211
-70	3.94	0.0226	61.60	-31.7	585.5	-0.0784	1.5059
-65	4.69	0.0227	52.37	-26.5	587.5	-0.0650	1.4911
-60	5.55	0.0228	44.73	-21.2	589.6	-0.0517	1.4769
-55	6.54	0.0229	38.38	-15.9	591.6	-0.0386	1.4631
-50	7.67	0.0230	33.08	-10.6	593.7	-0.0256	1.4497
-45	8.95	0.0231	28.62	-5.3	595.6	-0.0127	1.4368
-40	10.41	0.0232	24.86	0.0	597.6	0.0000	1.4242
-35	12.05	0.0233	21.68	5.3	599.5	0.0126	1.4120
-30	13.90	0.0234	18.97	10.7	601.4	0.0250	1.4001
-25	15.98	0.0236	16.66	16.0	603.2	0.0374	1.3886
-20	18.30	0.0237	14.68	21.4	605.0	0.0497	1.3774
-15	20.88	0.0238	12.97	26.7	606.7	0.0618	1.3664
-10	23.74	0.0239	11.50	32.1	608.5	0.0738	1.3558
-5	26.92	0.0241	10.23	37.5	610.1	0.0857	1.3454
0	30.42	0.0242	9.116	42.9	611.8	0.0975	1.3352
5	34.27	0.0243	8.150	48.3	613.3	0.1092	1.3253
10	38.51	0.0245	7.304	53.8	614.9	0.1208	1.3157
15	43.14	0.0246	6.562	59.2	616.3	0.1323	1.3062
20	48.21	0.0247	5.910	64.7	617.8	0.1437	1.2969
25	53.73	0.0249	5.334	70.2	619.1	0.1551	1.2879
30	59.74	0.0250	4.825	75.7	620.5	0.1663	1.2790
32	62.29	0.0251	4.637	77.9	621.0	0.1708	1.2755
35	66.26	0.0252	4.373	81.2	621.7	0.1775	1.2704
40	73.72	0.0253	3.971	86.8	623.0	0.1885	1.2618
45	80.96	0.0255	3.614	92.3	624.1	0.1996	1.2535
50	89.19	0.0256	3.294	97.9	625.2	0.2105	1.2453
55	98.06	0.0258	3.008	103.5	626.3	0.2214	1.2373
60	107.6	0.0260	2.751	109.2	627.3	0.2322	1.2294
65	117.8	0.0261	2.520	114.8	628.2	0.2430	1.2216
70	128.8	0.0263	2.312	120.5	629.1	0.2537	1.2140
75	140.5	0.0265	2.125	126.2	629.9	0.2643	1.2065
80	153.0	0.0267	1.955	132.0	630.7	0.2749	1.1991
85	166.4	0.0269	1.801	137.8	631.4	0.2854	1.1918

TABLE 1 (Continued)

T	P	v_f	v_g	h_f	h_g	s_f	s_g
90	180.6	0.0271	1.661	143.5	632.0	0.2958	1.1846
95	195.8	0.0273	1.534	149.4	632.6	0.3062	1.1775
100	211.9	0.0275	1.419	155.2	633.0	0.3166	1.1705
105	228.9	0.0277	1.313	161.1	633.4	0.3269	1.1635
110	247.0	0.0279	1.217	167.0	633.7	0.3372	1.1566
115	266.2	0.0281	1.128	173.0	633.9	0.3474	1.1497
120	286.4	0.0284	1.047	179.0	634.0	0.3576	1.1427
125	307.8	0.0286	0.973	185.1	634.0	0.3679	1.1358

TABLE 2. Refrigerant Ammonia
Properties of Superheated Vapor

Pressure (Sat. Temp.)		Sat. Vapor	Temperature (F) -40	-20	0	20	40	60	80
5	v	49.31	52.36	54.97	57.55	60.12	62.69	65.24	67.79
(-63.11)	h	588.3	600.3	610.4	620.4	630.4	640.4	650.5	660.6
	s	1.4857	1.5149	1.5385	1.5608	1.5821	1.6026	1.6223	1.6413
10	v	25.81		27.26	28.58	29.90	31.20	32.49	33.78
(-41.34)	h	597.1		608.5	618.9	629.1	639.3	649.5	659.7
	s	1.4276		1.4542	1.4773	1.4992	1.5200	1.5400	1.5593
15	v	17.67		18.01	18.92	19.82	20.70	21.58	22.44
(-27.29)	h	602.4		606.4	617.2	627.8	638.2	648.5	658.9
	s	1.3938		1.4031	1.4272	1.4497	1.4709	1.4912	1.5108
20	v	13.50			14.09	14.78	15.45	16.12	16.78
(-16.64)	h	606.2			615.5	626.4	637.0	647.5	658.0
	s	1.3700			1.3907	1.4138	1.4356	1.4562	1.4760
25	v	10.96			11.19	11.75	12.30	12.84	13.37
(-7.96)	h	609.1			613.8	625.0	635.8	646.5	657.1
	s	1.3515			1.3616	1.3855	1.4077	1.4287	1.4487
30	v	9.236			9.250	9.731	10.20	10.65	11.10
(-0.57)	h	611.6			611.9	623.5	634.6	645.5	656.2
	s	1.3364			1.3371	1.3618	1.3845	1.4059	1.4261
35	v	7.991				8.287	8.695	9.093	9.484
(5.89)	h	613.6				622.0	633.4	644.4	655.3
	s	1.3236				1.3413	1.3646	1.3863	1.4069
40	v	7.047				7.203	7.568	7.922	8.268
(11.66)	h	615.4				620.4	632.1	643.4	654.4
	s	1.3125				1.3231	1.3470	1.3692	1.3900
50	v	5.710					5.988	6.280	6.564
(21.67)	h	618.2					629.5	641.2	652.6
	s	1.2939					1.3169	1.3399	1.3613
60	v	4.805					4.933	5.184	5.428
(30.31)	h	620.5					626.8	639.0	650.7
	s	1.2787					1.2913	1.3152	1.3373
70	v	4.151					4.177	4.401	4.615
(37.70)	h	622.4					623.9	636.6	648.7
	s	1.2658					1.2688	1.2937	1.3166
80	v	3.655						3.812	4.005
(44.40)	h	624.0						634.3	646.7
	s	1.2545						1.2745	1.2981
90	v	3.266						3.353	3.529
(50.47)	h	625.3						631.8	644.7
	s	1.2445						1.2571	1.2814
100	v	2.952						2.985	3.149
(56.10)	h	626.5						629.3	642.6
	s	1.2356						1.2409	1.2661
110	v	2.693							2.837
(61.21)	h	627.5							640.5
	s	1.2275							1.2519
120	v	2.476							2.576
(66.02)	h	628.4							638.3
	s	1.2201							1.2386

100	120	140	160	180	200	220	240	260
70.33	72.87	75.41	77.95					
670.7	680.9	691.2	701.6					
1.6598	1.6778	1.6952	1.7122					
35.07	36.35	37.62	38.90					
670.0	680.3	690.6	701.1					
1.5779	1.5960	1.6136	1.6307					
23.31	24.17	25.03	25.88	26.74	27.59	28.44		
669.2	679.6	690.0	700.5	711.1	721.7	732.4		
1.5296	1.5478	1.5655	1.5827	1.5995	1.6158	1.6318		
17.43	18.08	18.73	19.37	20.02	20.66	21.30	21.94	
668.5	678.9	689.4	700.0	710.6	721.2	732.0	742.8	
1.4950	1.5133	1.5312	1.5485	1.5653	1.5817	1.5978	1.6135	
13.90	14.43	14.95	15.47	15.99	16.50	17.02	17.53	18.04
667.7	678.2	688.8	699.4	710.1	720.8	731.6	742.5	753.4
1.4679	1.4864	1.5043	1.5217	1.5387	1.5552	1.5713	1.5870	1.6025
11.55	11.99	12.43	12.87	13.30	13.73	14.16	14.59	15.02
666.9	677.5	688.2	698.8	709.6	720.3	731.1	742.0	753.0
1.4456	1.4642	1.4823	1.4998	1.5168	1.5334	1.5495	1.5653	1.5808
9.869	10.25	10.63	11.00	11.38	11.75	12.12	12.49	12.86
666.1	676.8	687.6	698.3	709.1	719.9	730.7	741.7	752.7
1.4265	1.4453	1.4635	1.4811	1.4982	1.5148	1.5311	1.5469	1.5642
8.609	8.945	9.278	9.609	9.938	10.27	10.59	10.92	11.24
665.3	676.1	686.9	697.7	708.5	719.4	730.3	741.3	752.3
1.4098	1.4288	1.4471	1.4648	1.4820	1.4987	1.5150	1.5309	1.5465
6.843	7.117	7.387	7.655	7.921	8.185	8.448	8.710	8.970
663.7	674.7	685.7	696.6	707.5	718.5	729.4	740.5	751.6
1.3816	1.4009	1.4195	1.4374	1.4548	1.4716	1.4880	1.5040	1.5197
5.665	5.897	6.126	6.352	6.576	6.798	7.019	7.238	7.457
662.1	673.3	684.4	695.5	706.5	717.5	728.6	739.7	750.9
1.3581	1.3778	1.3966	1.4148	1.4323	1.4493	1.4658	1.4819	1.4976
4.822	5.025	5.224	5.420	5.615	5.807	5.998	6.187	6.376
660.4	671.8	683.1	694.3	705.5	716.6	727.7	738.9	750.1
1.3378	1.3579	1.3770	1.3954	1.4131	1.4302	1.4469	1.4631	1.4789
4.190	4.371	4.548	4.722	4.893	5.063	5.231	5.398	5.565
658.7	670.4	681.8	693.2	704.4	715.6	726.9	738.1	749.4
1.3199	1.3404	1.3598	1.3784	1.3963	1.4136	1.4304	1.4467	1.4626
3.698	3.862	4.021	4.178	4.332	4.484	4.635	4.785	4.933
657.0	668.9	680.5	692.0	703.4	714.7	726.0	737.3	748.7
1.3038	1.3247	1.3444	1.3633	1.3813	1.3988	1.4157	1.4321	1.4481
3.304	3.454	3.600	3.743	3.883	4.021	4.158	4.294	4.428
655.2	667.3	679.2	690.8	702.3	713.7	725.1	736.5	747.9
1.2891	1.3104	1.3305	1.3495	1.3678	1.3854	1.4024	1.4190	1.4350
2.981	3.120	3.255	3.386	3.515	3.642	3.768	3.892	4.015
653.4	665.8	677.8	689.6	701.2	712.8	724.3	735.7	747.2
1.2755	1.2972	1.3176	1.3370	1.3555	1.3732	1.3904	1.4070	1.4232
2.712	2.842	2.967	3.089	3.209	3.326	3.442	3.557	3.671
651.6	664.2	676.5	688.4	700.2	711.8	723.4	734.9	746.5
1.2628	1.2850	1.3058	1.3254	1.3441	1.3620	1.3793	1.3960	1.4123

TABLE 2 (Continued)

Pressure (Sat. Temp.)		Sat. Vapor	-40	-20	0	20	40	60	80
			Temperature (F)						
130	v	2.291							2.355
(70.53)	h	629.2							636.0
	s	1.2132							1.2260
140	v	2.132							2.166
(74.79)	h	629.9							633.8
	s	1.2068							1.2140
150	v	1.994							2.001
(78.81)	h	630.5							631.4
	s	1.2009							1.2025
160	v	1.872							
(82.64)	h	631.1							
	s	1.1952							
170	v	1.764							
(86.29)	h	631.6							
	s	1.1900							
180	v	1.667							
(89.78)	h	632.0							
	s	1.1850							
190	v	1.581							
(93.13)	h	632.4							
	s	1.1802							
200	v	1.502							
(96.34)	h	632.7							
	s	1.1756							
220	v	1.367							
(102.42)	h	633.2							
	s	1.1671							
240	v	1.253							
(108.09)	h	633.6							
	s	1.1592							
260	v	1.155							
(113.42)	h	633.9							
	s	1.1518							
280	v	1.072							
(118.45)	h	634.0							
	s	1.1449							
300	v	0.999							
(123.21)	h	634.0							
	s	1.1383							

*Saturated and superheated property values for ammonia were extracted by permission from National Bureau of Standards Circular 142, 1945.

100	120	140	160	180	200	220	240	260
2.484	2.606	2.724	2.838	2.949	3.059	3.167	3.273	3.379
649.7	662.7	675.1	687.2	699.1	710.9	722.5	734.1	745.7
1.2509	1.2736	1.2947	1.3146	1.3335	1.3516	1.3690	1.3858	1.4022
2.288	2.404	2.515	2.622	2.727	2.830	2.931	3.030	3.129
647.8	661.1	673.7	686.0	698.0	709.9	721.6	733.3	745.0
1.2396	1.2628	1.2843	1.3045	1.3236	1.3418	1.3594	1.3763	1.3928
2.118	2.228	2.334	2.435	2.534	2.631	2.726	2.820	2.912
645.9	659.4	672.3	684.8	696.9	708.9	720.7	732.5	744.3
1.2289	1.2526	1.2745	1.2949	1.3142	1.3327	1.3504	1.3675	1.3840
1.969	2.075	2.175	2.272	2.365	2.457	2.547	2.635	2.723
643.9	657.8	670.9	683.5	695.8	707.9	719.9	731.7	743.5
1.2186	1.2429	1.2652	1.2859	1.3054	1.3240	1.3419	1.3591	1.3757
1.837	1.939	2.035	2.127	2.216	2.303	2.389	2.473	2.555
641.9	656.1	669.4	682.3	694.7	706.9	719.0	730.9	742.8
1.2087	1.2336	1.2563	1.2773	1.2971	1.3159	1.3338	1.3512	1.3679
1.720	1.818	1.910	1.999	2.084	2.167	2.248	2.328	2.407
639.9	654.4	668.0	681.0	693.6	705.9	718.1	730.1	742.0
1.1992	1.2247	1.2477	1.2691	1.2891	1.3081	1.3262	1.3436	1.3605
1.615	1.710	1.799	1.884	1.966	2.045	2.123	2.199	2.274
637.8	652.6	666.5	679.7	692.5	704.9	717.2	729.3	741.3
1.1899	1.2160	1.2396	1.2612	1.2815	1.3007	1.3189	1.3365	1.3534
1.520	1.612	1.698	1.780	1.859	1.935	2.009	2.082	2.154
635.6	650.9	665.0	678.4	691.3	703.9	716.3	728.4	740.5
1.1809	1.2077	1.2317	1.2537	1.2742	1.2935	1.3120	1.3296	1.3467
	1.433	1.525	1.601	1.675	1.745	1.814	1.881	1.947
	647.3	662.0	675.8	689.1	701.9	714.4	726.8	739.0
	1.1917	1.2167	1.2394	1.2604	1.2801	1.2989	1.3168	1.3340
	1.302	1.380	1.452	1.521	1.587	1.651	1.714	1.775
	643.5	658.8	673.1	686.7	699.8	712.6	725.1	737.5
	1.1764	1.2025	1.2259	1.2475	1.2677	1.2867	1.3049	1.3224
	1.182	1.257	1.326	1.391	1.453	1.514	1.572	1.630
	639.5	655.6	670.4	684.4	697.7	710.7	723.4	736.0
	1.1617	1.1889	1.2132	1.2354	1.2560	1.2754	1.2938	1.3115
	1.078	1.151	1.217	1.279	1.339	1.396	1.451	1.505
	635.4	652.2	667.6	681.6	695.6	708.8	721.8	734.4
	1.1473	1.1759	1.2011	1.2239	1.2449	1.2647	1.2834	1.3013
		1.058	1.123	1.183	1.239	1.294	1.346	1.397
		648.7	664.7	679.5	693.5	706.9	720.0	732.9
		1.1632	1.1894	1.2129	1.2344	1.2546	1.2736	1.2917

APPENDIX F

TABLE 1. Refrigerant 12 (Dichlorodifluoromethane)
Properties of Saturated Liquid and Vapor

T	p	v_f	v_g	h_f	h_g	s_f	s_g
-120	0.64190	0.00982	46.741	-16.565	64.052	-0.043723	0.19359
-110	0.97034	0.00990	31.777	-14.518	65.145	-0.037786	0.19002
-100	1.4280	0.01000	22.164	-12.466	66.248	-0.032005	0.18683
-95	1.7163	0.01003	18.674	-11.438	66.801	-0.029169	0.18536
-90	2.0509	0.01007	15.821	-10.409	67.355	-0.026367	0.18398
-85.	2.4371	0.01012	13.474	-9.3782	67.911	-0.023599	0.18267
-80	2.8807	0.01016	11.533	-8.3451	68.467	-0.020862	0.18143
-75	3.3879	0.01021	9.9184	-7.3101	69.023	-0.018156	0.18027
-70	3.9651	0.01026	8.5687	-6.2730	69.580	-0.015481	0.17916
-65	4.6139	0.01031	7.4347	-5.2336	70.137	-0.012834	0.17812
-60	5.3575	0.01036	6.4774	-4.1919	70.693	-0.010214	0.17714
-55	6.1874	0.01041	5.6656	-3.1477	71.249	-0.007622	0.17621
-50	7.1168	0.01046	4.9742	-2.1011	71.805	-0.005056	0.17533
-45	8.1540	0.01051	4.3828	-1.0519	72.359	-0.002516	0.17451
-40	9.3076	0.01056	3.8750	0.0000	72.913	0.000000	0.17373
-35	10.586	0.01062	3.4373	1.0546	73.464	0.002492	0.17299
-30	11.999	0.01067	3.0585	2.1120	74.015	0.004961	0.17229
-25	13.556	0.01073	2.7295	3.1724	74.563	0.007407	0.17164
-20	15.267	0.01079	2.4429	4.2357	75.110	0.009831	0.17102
-15	17.141	0.01085	2.1924	5.3020	75.654	0.012234	0.17043
-10	19.189	0.01091	1.9727	6.3716	76.196	0.014617	0.16989
-5	21.422	0.01097	1.7794	7.4444	76.842	0.016979	0.16937
0	23.849	0.01103	1.6089	8.5207	77.271	0.019323	0.16888
5	26.483	0.01109	1.4580	9.6005	77.805	0.021647	0.16842
10	29.335	0.01116	1.3241	10.684	78.335	0.023954	0.16798
15	32.415	0.01123	1.2050	11.771	78.861	0.026243	0.16758
20	35.736	0.01130	1.0988	12.863	79.385	0.028515	0.16719
25	39.310	0.01138	1.0039	13.958	79.904	0.030772	0.16683
30	43.148	0.01144	0.91880	15.058	80.419	0.033013	0.16648
35	47.263	0.01152	0.84237	16.163	80.930	0.035240	0.16616
40	51.667	0.01158	0.77357	17.273	81.436	0.037453	0.16586
45	56.373	0.01167	0.71149	18.387	81.937	0.039652	0.16557
50	61.394	0.01175	0.65537	19.507	82.433	0.041839	0.16530
55	66.743	0.01183	0.60453	20.634	82.924	0.044015	0.16504
60	72.433	0.01191	0.55839	21.766	83.409	0.046180	0.16479
65	78.477	0.01200	0.51642	22.905	83.887	0.048336	0.16456
70	84.888	0.01209	0.47818	24.050	84.359	0.050482	0.16434
75	91.682	0.01218	0.44327	25.204	84.825	0.052620	0.16412
80	98.870	0.01228	0.41135	26.365	85.282	0.054751	0.16392
85	106.47	0.01238	0.38212	27.534	85.732	0.056877	0.16372
90	114.49	0.01248	0.35529	28.713	86.174	0.058997	0.16353
95	122.95	0.01258	0.33063	29.901	86.606	0.061113	0.16334
100	131.86	0.01269	0.30794	31.100	87.029	0.063227	0.16315
105	141.25	0.01281	0.28701	32.310	87.442	0.065339	0.16297

TABLE 1 (Continued)

T	p	v_f	v_g	h_f	h_g	s_f	s_g
110	151.11	0.01292	0.26769	33.531	87.844	0.067451	0.16279
115	161.47	0.01305	0.24982	34.765	88.233	0.069564	0.16260
120	172.35	0.01317	0.23326	36.013	88.610	0.071680	0.16241
125	183.76	0.01331	0.21791	37.275	88.973	0.073800	0.16222
130	195.71	0.01345	0.20364	38.553	89.321	0.075927	0.16202
140	221.32	0.01375	0.17799	41.162	89.967	0.080205	0.16159
150	249.31	0.01408	0.15564	43.850	90.534	0.084531	0.16110
160	279.82	0.01445	0.13604	46.633	91.006	0.088927	0.16053
170	313.00	0.01487	0.11873	49.529	91.359	0.093418	0.15985
180	349.00	0.01536	0.10330	52.562	91.561	0.098039	0.15900
190	387.98	0.01594	0.089418	55.769	91.561	0.10284	0.15793
200	430.09	0.01689	0.076728	59.203	91.278	0.10789	0.15651
210	475.52	0.01760	0.064843	62.959	90.558	0.11332	0.15453
220	524.43	0.01899	0.053140	67.246	89.036	0.11943	0.15149

TABLE 2. Refrigerant 12 (Dichlorodifluoromethane)*

A. Properties of Superheated Vapor

Pressure (Sat. Temp.)		Sat. Vapor	Temperature (F) -80	-40	0	40	80	120	160	200	240	280
1	v	30.896	33.531	37.119	40.695	44.263	47.826	51.386				
(-109.24)	h	65.229	68.679	73.633	78.845	84.297	89.971	95.853				
	s	0.18977	0.19922	0.21162	0.22347	0.23484	0.24576	0.25628				
5	v	6.9069		7.3239	8.0611	8.7903	9.5142	10.234				
(-62.35)	h	70.432		73.291	78.582	84.090	89.806	95.717				
	s	0.17759		0.18459	0.19663	0.20812	0.21912	0.22968				
10	v	3.6246			3.9809	4.3556	4.7248	5.0903	5.4533			
(-37.23)	h	73.219			78.246	83.828	89.596	95.546	101.669			
	s	0.17331			0.18471	0.19635	0.20746	0.21809	0.22830			
14.696	v	2.5315			2.6764	2.9382	3.1942	3.4464	3.6960			
(-21.62)	h	74.933			77.924	83.577	89.396	95.383	101.534			
	s	0.17121			0.17788	0.18967	0.20087	0.21157	0.22183			
20	v	1.8977			1.9390	2.1373	2.3295	2.5179	2.7036	2.8874		
(-8.13)	h	76.397			77.550	83.289	89.168	95.198	101.380	107.712		
	s	0.16969			0.17222	0.18419	0.19550	0.20628	0.21659	0.22649		
30	v	1.2964				1.3969	1.5306	1.6600	1.7868	1.9116		
(11.11)	h	78.452				82.730	88.729	94.843	101.086	107.464		
	s	0.16789				0.17671	0.18826	0.19918	0.20960	0.21957		
40	v	0.98743				1.0258	1.1306	1.2309	1.3282	1.4236	1.5176	
(25.93)	h	80.000				82.148	88.277	94.480	100.788	107.212	113.757	
	s	0.16676				0.17112	0.18292	0.19401	0.20453	0.21457	0.22420	
50	v	0.79824				0.80248	0.89025	0.97313	1.0529	1.1307	1.2070	
(38.15)	h	81.249				81.540	87.811	94.110	100.485	106.958	113.539	
	s	0.16597				0.16655	0.17862	0.18988	0.20051	0.21064	0.22032	
60	v	0.67005					0.72964	0.80110	0.86928	0.93531	0.99988	
(48.64)	h	82.299					87.330	93.731	100.176	106.700	113.319	
	s	0.16537					0.17497	0.18641	0.19716	0.20736	0.21710	
80	v	0.50680					0.52795	0.58556	0.63943	0.69095	0.74090	0.78975
(66.21)	h	84.003					86.316	92.945	99.542	106.174	112.872	119.652
	s	0.16450					0.16885	0.18070	0.19170	0.20207	0.21193	0.22135

B. Properties of Liquid and Saturated Vapor

Pressure (Sat. Temp.)		Sat. Vapor	Temperature (F) 100	120	140	160	180	200	250	300	350	400
100	v	0.40674	0.43138	0.45562	0.47881	0.50118	0.52291	0.54413	0.59549	0.64518		
(80.76)	h	85.351	88.694	92.116	95.507	98.884	102.257	105.633	114.119	122.707		
	s	0.16389	0.16996	0.17597	0.18172	0.18726	0.19262	0.19782	0.21022	0.22191		
120	v	0.33886	0.34655	0.36841	0.38901	0.40870	0.42766	0.44606	0.49025	0.53267		
(93.29)	h	86.459	87.675	91.237	94.736	98.199	101.642	105.076	113.670	122.333		
	s	0.16340	0.16559	0.17184	0.17778	0.18346	0.18892	0.19421	0.20677	0.21856		
140	v	0.28964		0.30549	0.32445	0.34232	0.35939	0.37584	0.41499	0.45226		
(104.35)	h	87.389		90.297	93.923	97.483	101.003	104.501	113.212	121.953		
	s	0.16299		0.16808	0.17423	0.18007	0.18566	0.19104	0.20377	0.21567		
160	v	0.25222		0.25764	0.27558	0.29224	0.30797	0.32301	0.35847	0.39190		
(114.30)	h	88.180		89.283	93.059	96.732	100.340	103.907	112.743	121.567		
	s	0.16263		0.16454	0.17094	0.17697	0.18270	0.18819	0.20110	0.21312		
180	v	0.22276			0.23710	0.25297	0.26775	0.28176	0.31442	0.34492	0.37409	
(123.38)	h	88.857			92.136	95.940	99.647	103.291	112.263	121.174	130.113	
	s	0.16228			0.16783	0.17407	0.17995	0.18556	0.19868	0.21081	0.22220	
200	v	0.19891			0.20579	0.22121	0.23535	0.24860	0.27911	0.30730	0.33408	
(131.74)	h	89.439			91.137	95.100	98.921	102.652	111.771	120.775	129.778	
	s	0.16195			0.16480	0.17130	0.17737	0.18311	0.19644	0.20870	0.22017	
300	v	0.12510					0.13482	0.14697	0.17230	0.19402	0.21388	
(166.18)	h	91.240					94.556	98.975	109.102	118.670	128.036	
	s	0.16012					0.16537	0.17217	0.18697	0.20000	0.21195	
400	v	0.085587						0.091005	0.11743	0.13680	0.15353	0.16885
(192.93)	h	91.513						93.718	105.950	116.343	126.177	135.792
	s	0.15755						0.16092	0.17882	0.19298	0.20552	0.21704
500	v	0.058917							0.082606	0.10190	0.11712	0.13054
(215.10)	h	89.924							101.970	113.729	124.183	134.166
	s	0.15317							0.17063	0.18666	0.19999	0.21196
600	v								0.056059	0.078059	0.092697	0.10498
	h								96.105	110.729	122.035	132.466
	s								0.16057	0.18053	0.19496	0.20746

*Property values for saturated and superheated Freon vapor extracted, by permission, from tables published by "Freon" Products Division of E. I. duPont and Company, Inc., Wilmington, Del.

APPENDIX G

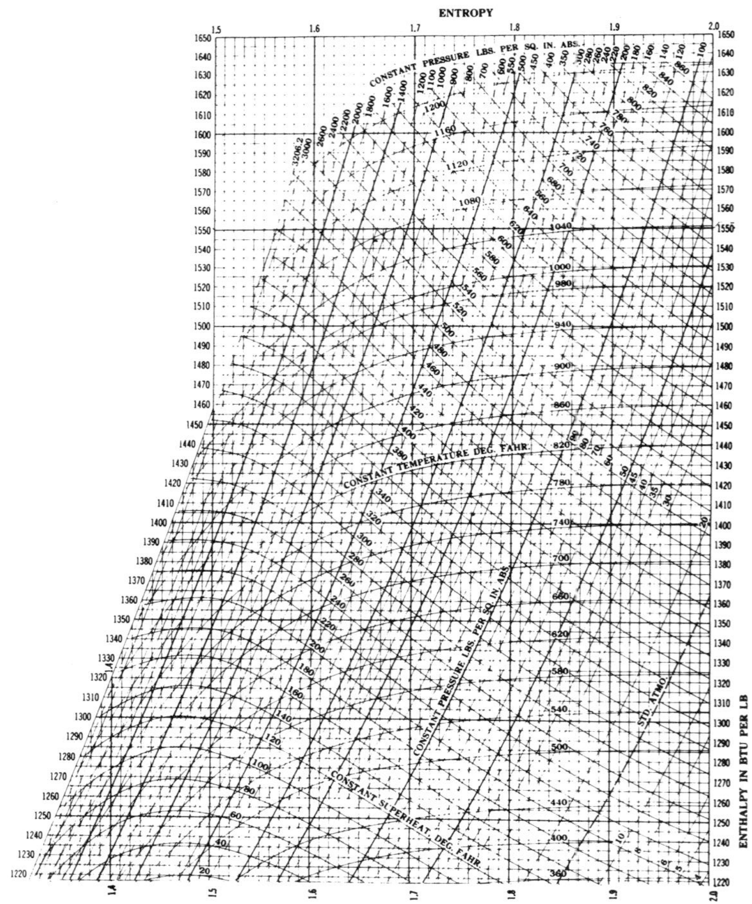

ENTROPY
ENTHALPY IN BTU PER LB
CONSTANT PRESSURE LBS. PER SQ. IN. ABS.
CONSTANT TEMPERATURE DEG. FAHR.
CONSTANT SUPERHEAT DEG. FAHR.
STD. ATMO.

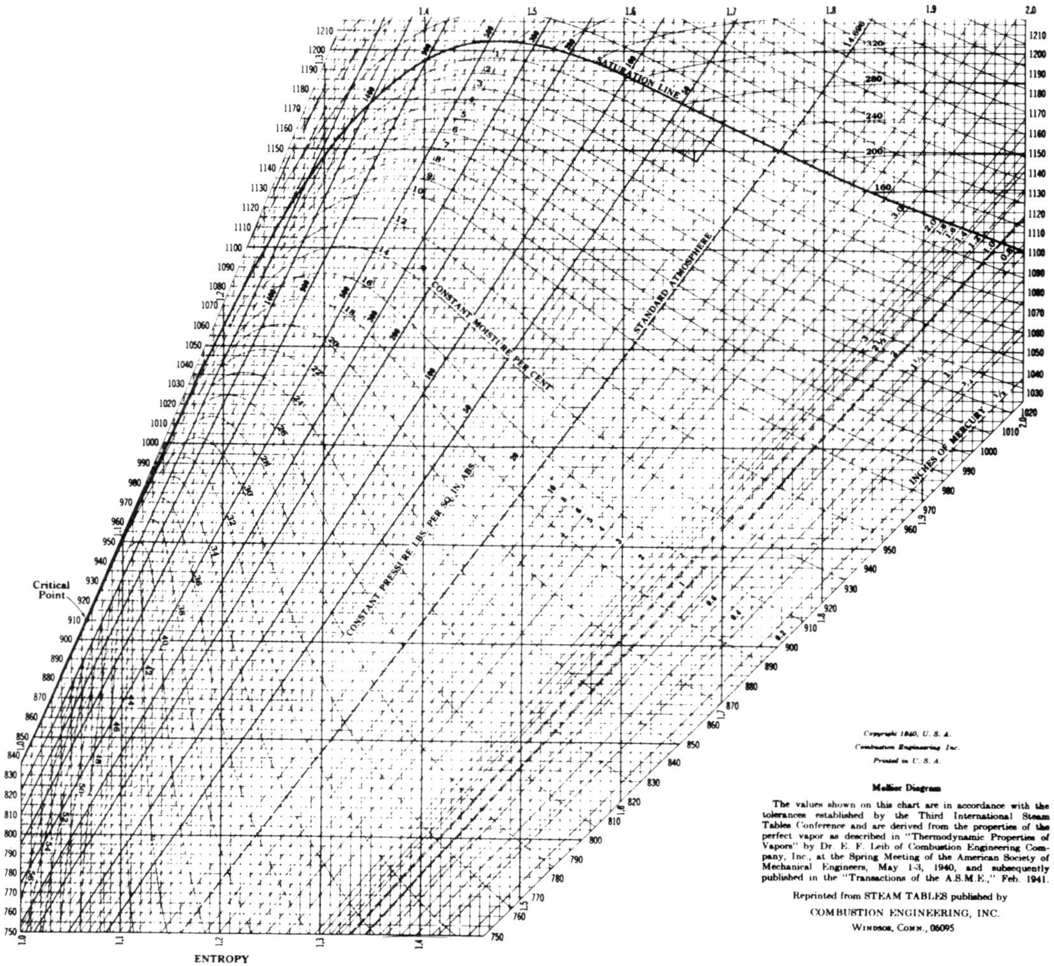

Copyright 1940, U. S. A.
Combustion Engineering Inc.
Printed in U. S. A.

Mollier Diagram

The values shown on this chart are in accordance with the tolerances established by the Third International Steam Tables Conference and are derived from the properties of the perfect vapor as described in "Thermodynamic Properties of Vapors" by Dr. E. F. Leib of Combustion Engineering Company, Inc., at the Spring Meeting of the American Society of Mechanical Engineers, May 1-3, 1940, and subsequently published in the "Transactions of the A.S.M.E.," Feb. 1941.

Reprinted from STEAM TABLES published by COMBUSTION ENGINEERING, INC. Windsor, Conn., 06095

Mollier Diagram

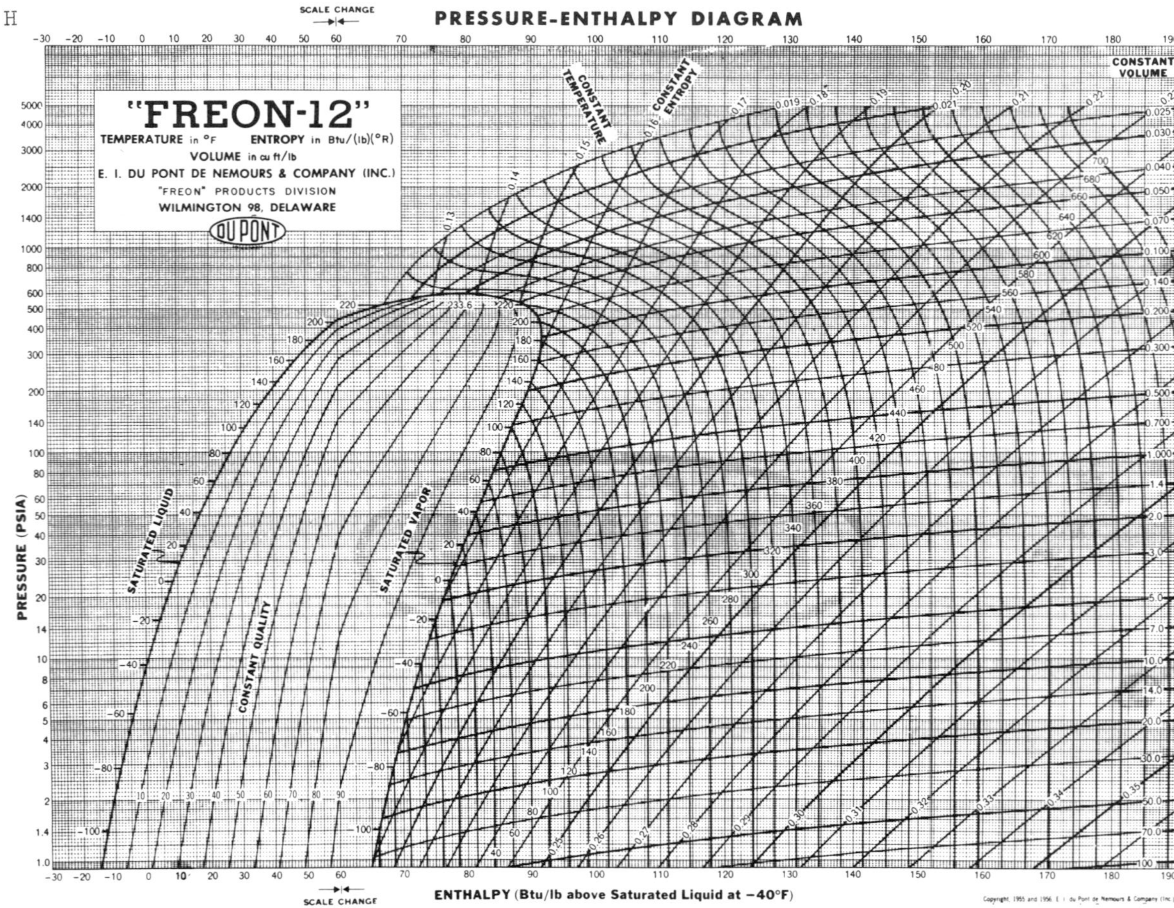

Published by permission of Freon Products Division of E.I. duPont and Co., Inc.

APPENDIX I

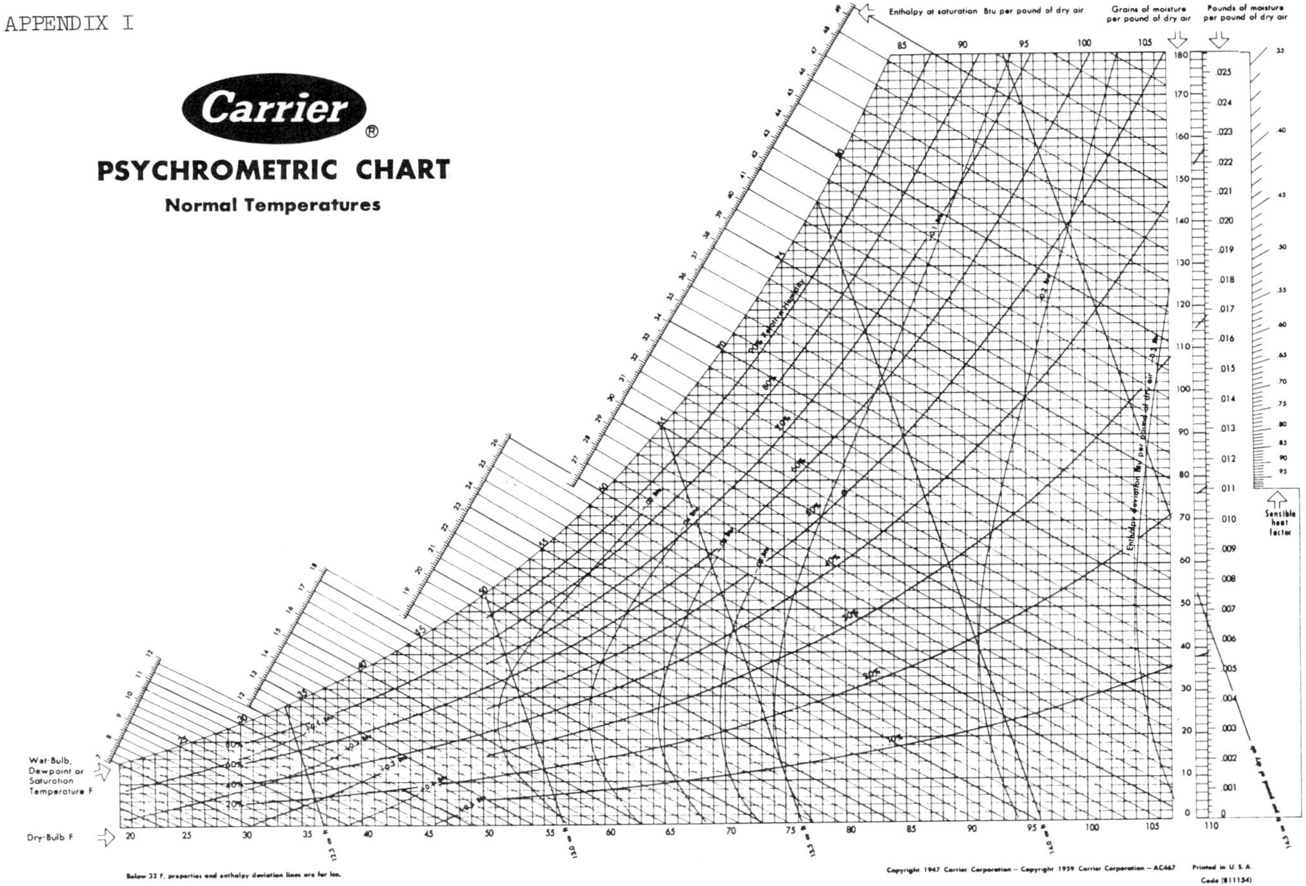

Below 32 F, properties and enthalpy deviation lines are for ice. Published by permission of Carrier Corporation.

APPENDIX J

Enthalpy of Reaction at 77 F (B/lbm)

Type	Formula*	$\Delta h_R[H_2O(l)]$	$\Delta h_R[H_2O(g)]$
n-Pentane	$C_5H_{12}(l)$	-20,910	-19,340
Benzene	$C_6H_6(g)$	-18,180	-17,450
	$C_6H_6(l)$	-17,990	-17,260
n-Hexane	$C_6H_{14}(g)$	-20,970	-19,420
Tuolene	$C_7H_8(g)$	-18,500	-17,670
n-Heptane	$C_7H_{16}(g)$	-20,820	-19,310
	$C_7H_{16}(l)$	-20,670	-19,160
Xylene	$C_8H_{10}(g)$	-18,650	-17,760
Octcnc	$C_8H_{16}(l)$	-20,350	-19,000
n-Octane	$C_8H_{18}(g)$	-20,750	-19,200
	$C_8H_{18}(l)$	-20,590	-19,100
Methyl alcohol	$CH_3OH(g)$	-10,260	-9,070
Ethyl alcohol	$C_2H_5OH(g)$	-13,160	-11,920
Hydrogen	$H_2(g)$	-61,090	-51,620
Ammonia	$NH_3(g)$	-9,670	-7,980
Hydrogen sulfide	$H_2S(g)$	-7,100	-6,540
Sulfur	$S(s)$	-3,980	-3,980

*g, gas; l, liquid; s, solid.

APPENDIX K

Heats of Formation at 77 F

Type	Formula	Δh_f (B/mole)
Carbon monoxide	$CO(g)$	-47,560
Carbon dioxide	$CO_2(g)$	-169,290
Methane	$CH_4(g)$	-32,200
Acetylene	$C_2H_2(g)$	97,540
Ethylene	$C_2H_4(g)$	22,570
Ethane	$C_2H_6(g)$	-36,410
Propane	$C_3H_8(g)$	-44,660
n-Butane	$C_4H_{10}(g)$	-54,270
n-Pentane	$C_5H_{12}(g)$	-62,960
Benzene	$C_6H_6(g)$	35,660
n-Hexane	$C_6H_{14}(g)$	-71,880
n-Heptane	$C_7H_{16}(g)$	-80,760
n-Octane	$C_8H_{18}(g)$	-89,640
	$C_8H_{18}(l)$	-107,530
Methyl alcohol	$CH_3OH(g)$	-86,320
Ethyl alcohol	$C_2H_5OH(g)$	-100,840
Water	$H_2O(g)$	-104,040
	$H_2O(l)$	-122,930
Ammonia	$NH_3(g)$	-19,750
Hydrogen sulfide	$H_2S(g)$	-14,390
Sulfur dioxide	$SO_2(g)$	-127,700

APPENDIX L

Internal Energy and Enthalpy of Various Gases at Low Pressure (B/mole)*

	Carbon dioxide		Carbon monoxide		Nitrogen		Oxygen		Water		200% Theor. air	
R	$\bar{h}$	$\bar{u}$	$\bar{h}$	$\bar{u}$	$\bar{h}$	$\bar{u}$	$\bar{h}$	$\bar{u}$	$\bar{h}$	$\bar{u}$	$\bar{h}$	$\bar{u}$
300	2108.2	1512.4	2081.9	1486.1	2082.0	1486.2	2073.5	1477.8	2367.6	1771.8	2096.7	1500.9
380	2716.4	1961.8	2637.9	1883.3	2638.0	1883.4	2630.0	1875.3	3004.4	2249.8	2660.3	1905.7
400	2874.7	2080.4	2776.9	1982.6	2777.0	1982.6	2769.1	1974.8	3163.8	2369.4	2801.4	2007.0
420	3035.7	2201.7	2916.0	2081.9	2916.1	2082.0	2908.3	2074.3	3323.2	2489.1	2943.0	2108.9
440	3199.4	2325.6	3055.0	2181.2	3055.1	2181.3	3047.5	2173.8	3482.7	2608.9	3084.8	2211.0
460	3365.7	2452.2	3194.0	2280.5	3194.1	2280.6	3186.9	2273.4	3642.3	2728.8	3226.8	2313.3
480	3534.7	2581.5	3333.0	2379.8	3333.1	2379.9	3326.5	2373.3	3802.0	2848.8	3368.9	2415.7
500	3706.2	2713.3	3472.1	2479.2	3472.2	2479.3	3466.2	2473.2	3962.0	2969.1	3511.2	2518.3
520	3880.3	2847.7	3611.2	2578.6	3611.3	2578.6	3606.1	2573.4	4122.0	3089.4	3653.7	2621.0
537	4030.2	2963.8	3729.5	2663.1	3729.5	2663.1	3725.1	2658.7	4258.3	3191.9	3774.9	2708.5
540	4056.8	2984.4	3750.3	2677.9	3750.3	2678.0	3746.2	2673.8	4282.4	3210.0	3796.3	2723.9
560	4235.8	3123.7	3889.5	2777.4	3889.5	2777.4	3886.6	2774.5	4442.8	3330.7	3939.4	2827.3
580	4417.2	3265.4	4028.7	2876.9	4028.7	2876.9	4027.3	2875.5	4603.7	3451.9	4082.7	2930.9
600	4600.9	3409.4	3168.0	2976.5	4167.9	2976.4	4168.3	2976.8	4764.7	3573.2	4226.3	3034.8
620	4786.8	3555.6	4307.4	3076.2	4307.1	3075.9	4309.7	3078.4	4926.1	3664.4	4370.1	3138.9
640	4974.9	3704.0	4446.9	3175.9	4446.4	3175.5	4451.4	3180.4	5087.8	3816.8	4514.0	3243.0
660	5165.2	3854.6	4586.5	3275.8	4585.8	3275.2	4593.5	3282.9	5250.0	3939.3	4658.6	3347.9
680	5357.6	4007.2	4726.2	3375.8	4725.3	3374.9	4736.2	3385.8	5412.5	4062.1	4802.8	3452.4
700	5552.0	4161.9	4866.0	3475.9	4864.9	3474.8	4879.3	3489.2	5575.4	4185.3	4947.7	3557.5
720	5748.4	4318.6	5006.1	3576.3	5004.5	3574.7	5022.9	3593.1	5738.8	4309.0	5092.6	3662.8
740	5946.8	4477.3	5146.4	3676.9	5144.3	3674.7	5167.0	3697.4	5902.6	4433.1	5237.9	3768.4
760	6147.0	4637.9	5286.8	3777.5	5284.1	3774.9	5311.4	3802.2	6066.9	4557.6	5383.7	3874.4
780	6349.1	4800.1	5427.4	3878.4	5424.2	3875.2	5456.4	3907.5	6231.7	4682.7	5529.9	3980.9
800	6552.9	4964.2	5568.2	3979.5	5564.4	3975.7	5602.0	4013.3	6396.9	4808.2	5676.3	4087.6
820	6758.3	5129.9	5709.4	4081.0	5704.7	4076.3	5748.1	4119.7	6562.6	4934.2	5823.1	4194.7
840	6965.7	5297.6	5850.7	4182.6	5845.3	4177.1	5894.8	4226.6	6728.9	5060.8	5969.9	4301.8
860	7174.7	5466.9	5992.3	4284.5	5985.9	4278.1	6041.9	4334.1	6895.6	5187.8	6117.5	4409.6

880	7385.3	5637.7	6134.2	4386.6	6126.9	4379.4	6189.6	4442.0	7062.9	5315.3	6264.9	4517.3
900	7597.6	5810.3	6276.4	4489.1	6268.1	4480.8	6367.9	4550.6	7230.9	5443.6	6413.0	4625.7
920	7811.4	5984.4	6419.0	4592.0	6409.6	4582.6	6486.7	4659.7	7399.4	5572.4	6561.5	4734.5
940	8026.8	6160.1	6561.7	4695.0	6551.2	4684.5	6636.1	4769.4	7568.4	5701.7	6710.4	4843.7
960	8243.8	6337.4	6704.9	4798.5	6693.1	4786.7	6786.0	4879.5	7738.0	5831.6	6859.8	4953.4
980	8462.2	6516.1	6848.4	4902.3	6835.4	4889.3	6936.4	4990.3	7908.2	5962.0	7009.7	5063.6
1000	8682.1	6696.2	6992.2	5006.3	6977.9	4992.0	7087.5	5101.6	8078.9	6093.0	7159.8	5173.9
1020	8903.4	6877.8	7136.4	5110.8	7120.7	5095.1	7238.9	5213.3	8250.4	6224.8	7310.5	5284.9
1040	9126.2	7060.9	7281.0	5215.7	7263.8	5198.5	7391.0	5325.7	8422.4	6357.1	7461.1	5395.8
1060	9350.3	7245.3	7425.9	5320.9	7407.2	5302.2	7543.6	5438.6	8595.0	6490.0	7612.7	5507.7
1080	9575.8	7431.1	7571.1	5426.4	7551.0	5406.2	7696.8	5552.1	8768.2	6623.5	7764.3	5619.6
1100	9802.6	7618.1	7716.8	5532.3	7695.0	5510.5	7850.4	5665.9	8942.0	6757.5	7916.4	5731.9
1120	10030.6	7806.4	7862.9	5638.7	7839.3	5615.2	8004.5	5780.3	9116.4	6892.2	8069.0	5844.8
1140	10260.1	7996.2	8009.2	5745.4	7984.0	5720.1	8159.1	5895.2	9291.4	7027.5	8222.1	5958.2
1160	10490.6	8187.0	8156.1	5852.5	8129.0	5825.4	8314.2	6010.6	9467.1	7163.5	8375.5	6071.9
1180	10722.3	8379.0	8303.3	5960.0	8274.4	5931.0	8469.8	6126.5	9643.4	7300.1	8529.2	6185.9
1200	10955.3	8572.3	8450.8	6067.8	8420.0	6037.0	8625.8	6242.8	9820.4	7437.4	8683.6	6300.6
1220	11189.4	8766.6	8598.8	6176.0	8566.1	6143.4	8782.4	6359.6	9998.0	7575.2	8838.6	6415.8
1240	11424.6	8962.1	8747.2	6284.7	8712.6	6250.1	8939.4	6476.9	10176.1	7713.6	8993.7	6531.2
1260	11661.0	9158.8	8896.0	6393.8	8859.3	6357.2	9096.7	6594.5	10354.9	7852.7	9149.3	6647.1
1280	11898.4	9356.5	9045.0	6503.1	9006.4	6464.5	9254.6	6712.7	10534.4	7992.5	9305.3	6763.4
1300	12136.9	9555.3	9194.6	6613.0	9153.9	6572.3	9412.9	6831.3	10714.5	8132.9	9461.7	6880.1
1320	12376.4	9755.0	9344.6	6723.2	9301.8	6680.4	9571.6	6950.2	10895.3	8274.0	9618.8	6997.4
1340	12617.0	9955.9	9494.8	6833.7	9450.0	6788.9	9730.7	7069.6	11076.6	8415.5	9776.2	7115.1
1360	12858.5	10157.7	9645.5	6944.7	9598.6	6897.8	9890.2	7189.4	11258.7	8557.9	9933.8	7233.0
1380	13101.0	10360.5	9796.6	7056.1	9747.5	7007.0	10050.1	7309.6	11441.4	8700.9	10091.9	7351.4
1400	13344.7	10564.5	9948.1	7167.9	9896.9	7116.7	10210.4	7430.1	11624.8	8844.6	10250.7	7470.5
1420	13589.1	10769.2	10100.0	7280.1	10046.6	7226.7	10371.0	7551.1	11808.8	8988.9	10410.0	7590.1
1440	13834.5	10974.8	10252.2	7392.6	10196.6	7337.0	10532.0	7672.4	11993.4	9133.8	10569.3	7709.7
1460	14080.8	11181.4	10404.8	7505.4	10347.0	7447.6	10693.3	7793.9	12178.8	9279.4	10729.3	7829.9
1480	14328.0	11388.9	10557.8	7618.7	10497.8	7558.7	10855.1	7916.0	12364.8	9425.7	10889.5	7950.4
1500	14576.0	11597.2	10711.1	7732.3	10648.9	7670.1	11017.1	8038.3	12551.4	9572.7	11050.2	8071.4
1520	14824.9	11806.4	10864.9	7846.4	10800.4	7781.9	11179.6	8161.1	12738.8	9720.3	11211.4	8192.9
1540	15074.7	12016.5	11019.0	7960.8	10952.2	7893.9	11342.4	8284.2	12926.8	9868.6	11372.8	8314.6

Internal Energy and Enthalpy of Various Gases at Low Pressure (continued)

R	Carbon dioxide $\bar{h}$	Carbon dioxide $\bar{u}$	Carbon monoxide $\bar{h}$	Carbon monoxide $\bar{u}$	Nitrogen $\bar{h}$	Nitrogen $\bar{u}$	Oxygen $\bar{h}$	Oxygen $\bar{u}$	Water $\bar{h}$	Water $\bar{u}$	200% Theor. air $\bar{h}$	200% Theor. air $\bar{u}$
1560	15325.3	12227.3	11173.4	8075.4	11104.3	8006.4	11505.4	8407.4	13115.6	10017.6	11534.7	8436.7
1580	15576.7	12439.0	11328.2	8190.5	11256.9	8119.2	11668.8	8531.1	13305.0	10167.3	11696.8	8559.1
1600	15829.0	12651.6	11483.4	8306.0	11409.7	8232.3	11832.5	8655.1	13494.9	10317.6	11859.6	8682.2
1620	16081.9	12864.8	11638.9	8421.8	11562.8	8345.7	11996.6	8779.5	13685.7	10468.6	12022.7	8805.6
1640	16335.7	13078.9	11794.7	8537.9	11716.4	8459.6	12160.9	8904.1	13877.0	10620.2	12186.2	8929.4
1660	16590.2	13293.7	11950.9	8654.4	11870.2	8573.6	12325.5	9029.0	14069.2	10772.7	12350.0	9053.5
1680	16845.5	13509.2	12107.5	8771.2	12024.3	8688.1	12490.4	9154.1	14261.9	10925.6	12514.2	9177.9
1700	17101.4	13725.4	12264.3	8888.3	12178.9	8802.9	12655.6	9279.6	14455.4	11079.4	12678.6	9302.6
1720	17358.1	13942.4	12421.4	9005.7	12333.7	8918.0	12821.1	9405.4	14649.5	11233.8	12843.8	9428.1
1740	17615.5	14160.1	12579.0	9123.6	12488.8	9033.4	12986.9	9531.5	14844.3	11388.9	13009.0	9553.6
1760	17873.5	14378.4	12736.7	9241.6	12644.3	9149.2	13153.0	9657.9	15039.8	11544.7	13174.6	9679.5
1780	18132.2	14597.4	12894.9	9360.0	12800.2	9265.3	13319.2	9784.4	15236.1	11701.2	13340.3	9805.5
1800	18391.5	14816.9	13053.2	9478.6	12956.2	9381.7	13485.8	9911.2	15433.0	11858.4	13507.0	9932.4
1820	18651.5	15037.2	13212.0	9597.7	13112.7	9498.4	13652.5	10038.2	15630.6	12016.3	13673.8	10059.5
1840	18912.2	15258.2	13371.0	9717.0	13269.5	9615.5	13819.6	10165.6	15828.7	12174.7	13840.5	10186.5
1860	19173.4	15479.7	13530.2	9836.5	13426.5	9732.8	13986.8	10293.1	16027.6	12333.9	14008.4	10314.7
1880	19435.3	15701.9	13689.8	9956.4	13583.9	9850.5	14154.4	10421.0	16227.2	12493.8	14176.1	10442.7
1900	19697.8	15924.7	13849.8	10076.6	13741.6	9968.4	14322.1	10549.0	16427.5	12654.4	14344.1	10571.0
1920	19960.8	16147.9	14009.9	10197.0	13899.5	10086.7	14490.1	10677.2	16628.5	12815.6	14512.5	10699.6
1940	20224.4	16371.8	14170.3	10317.7	14057.8	10205.2	14658.2	10805.6	16830.0	12977.4	14681.4	10828.8
1960	20488.6	16596.3	14331.0	10438.7	14216.4	10324.0	14826.6	10934.3	17032.4	13140.1	14850.4	10958.1
1980	20753.3	16821.3	14492.0	10560.0	14375.2	10443.2	14995.2	11063.2	17235.3	13303.3	15019.8	11087.8
2000	21018.7	17046.9	14653.2	10681.5	14534.4	10562.6	15164.0	11192.3	17439.0	13467.3	15189.3	11217.6
2020	21284.4	17272.9	14814.8	10803.3	14693.8	10682.3	15332.9	11321.4	17643.2	13631.8	15359.3	11347.9
2040	21550.7	17499.5	14976.5	10925.3	14853.4	10802.2	15502.1	11450.9	17848.1	13796.9	15529.7	11478.5
2060	21817.6	17726.7	15138.5	11047.6	15013.3	10922.5	15671.5	11580.6	18053.6	13962.7	15700.3	11609.4
2080	22084.9	17954.3	15300.8	11170.2	15173.5	11042.9	15841.1	11710.5	18259.9	14129.3	15871.3	11740.7
2100	22352.7	18182.4	15463.3	11293.0	15334.0	11163.7	16010.9	11840.6	18466.9	14296.6	16042.4	11872.1

2140	22889.9	18640.1	15789.0	11539.3	15655.6	11405.9	16350.9	12101.2	18882.5	14632.7	16385.6	12135.8
2200	23699.0	19330.1	16279.4	11910.5	16139.8	11770.9	16862.6	12493.7	19510.8	15141.9	16902.5	12533.6
2240	24240.6	19792.2	16607.4	12159.0	16463.7	12015.4	17204.5	12756.2	19932.8	15484.4	17248.6	12800.3
2300	25056.3	20488.8	17101.0	12533.5	16951.2	12383.7	17718.8	13151.3	20570.6	16003.1	17769.3	13201.8
2340	25602.2	20955.3	17431.0	12784.1	17277.2	12630.3	18062.4	13415.5	20998.7	16351.8	18117.8	13470.9
2400	26424.0	21657.9	17927.4	13161.3	17767.9	13001.8	18579.2	13813.1	21645.7	16879.6	18642.1	13876.0
2440	26973.7	22128.2	18259.4	13413.9	18095.9	13250.5	18924.4	14078.9	22079.8	17234.3	18992.8	14147.3
2500	27801.2	22836.5	18758.8	13794.1	18589.5	13624.9	19443.4	14478.7	22735.4	17770.7	19520.7	14556.0
2540	28354.5	23310.4	19092.5	14048.4	18919.4	13875.3	19790.2	14746.1	23175.3	18131.2	19873.7	14829.6
2600	29187.1	24023.8	19594.3	14431.0	19415.8	14252.5	20311.4	15148.1	23839.5	18676.2	20404.6	15241.3
2640	29743.8	24501.1	19929.7	14687.0	19747.5	14504.8	20659.5	15416.8	24284.8	19042.1	20759.5	15516.8
2700	30581.2	25219.4	20434.0	15072.2	20246.4	14884.5	21182.9	15821.0	24957.2	19595.4	21293.8	15932.0
2740	31141.0	25699.7	20770.8	15329.5	20579.8	15138.5	21532.5	16091.2	25407.9	19966.6	21659.9	16216.6
2800	31982.8	26422.4	21277.2	15716.8	21081.1	15520.6	22057.8	16497.4	26088.0	20527.6	22187.5	16627.1
2840	32545.5	26905.6	21615.4	15975.5	21416.0	15776.1	22408.7	16768.8	26543.8	20903.9	22546.4	16906.5
2900	33391.5	27632.5	22123.8	16364.8	21919.5	16160.5	22936.1	17177.1	27231.2	21472.2	23086.0	17327.0
2940	33956.8	28118.3	22463.3	16624.8	22255.8	16417.4	23288.3	17449.9	27691.9	21853.4	23446.7	17608.2
3000	34806.6	28849.0	22973.4	17015.8	22761.5	16803.9	23817.7	17860.1	28386.3	22428.7	23988.5	18030.9
3500	41965.2	35014.7	27261.8	20311.2	27015.9	20065.3	28273.3	21322.8	34323.5	27373.0	28556.8	21606.2
4000	49231.4	41287.9	31605.2	23661.7	31329.4	23386.0	32806.1	24862.6	40489.1	32545.6		
4500	56581.0	47644.6	35988.6	27052.2	35687.8	26751.4	37411.8	28475.4	46835.9	37899.5		
5000	64000.0	54070.6	40402.7	30473.4	40079.8	30150.5	42086.3	32157.0	53327.4	43398.0		
5380	69676.5	58992.5	43774.1	33090.1	43436.0	32752.1	45682.1	34998.1	58338.7	47654.7		

*Extracted from Gas Tables by J. H. Keenan and J. Kaye with permission of the authors and the publisher, John Wiley & Sons, Inc., New York, 1945.

INDEX